Informatik-Fachberichte 154

Herausgegeben von W. Brauer
im Auftrag der Gesellschaft für Informatik (GI)

U. Herzog M. Paterok (Hrsg.)

Messung, Modellierung und Bewertung von Rechensystemen

4. GI/ITG-Fachtagung

Erlangen, 29. September – 1. Oktober 1987

Proceedings

Springer-Verlag
Berlin Heidelberg New York
London Paris Tokyo

Herausgeber

U. Herzog
M. Paterok
Institut für Mathematische Maschinen und Datenverarbeitung
(Informatik VII), Universität Erlangen-Nürnberg
Martensstraße 3, 8520 Erlangen

CR Subject Classifications (1987): C.4, I.6

ISBN-13: 978-3-540-18406-5 e-ISBN-13: 978-3-642-73016-0
DOI: 10.1007/ 978-3-642-73016-0

Repro– u Druckarbeiten: Weihert-Druck GmbH, Darmstadt
2145/3140–543210

Vorwort

Die 4. Fachtagung "Messung, Modellierung und Bewertung von Rechensystemen" findet vom 29. September bis zum 1. Oktober 1987 an der Universität Erlangen statt. Sie wird alle zwei Jahre an wechselnden Orten von der gleichnamigen Interessengruppe des Fachbereichs 3 der Gesellschaft für Informatik und des Fachbereichs 4 der Informationstechnischen Gesellschaft veranstaltet.

Ziel der Tagungsreihe ist der Austausch neuer Ideen und Erfahrungen bei der quantitativen Untersuchung von Rechensystemen in Theorie und Praxis. Dies beinhaltet zum einen die reine Leistungsuntersuchung mit ihren Bereichen Messung, Modellierung, Modellanalyse sowie Bewertung und Synthese, zum anderen ihre Verbindung mit Zuverlässigkeitsaspekten, inzwischen unter dem Namen Performability etabliert.

Der Schwerpunkt der Tagung liegt wie gewohnt auf den Gebieten Modellanalyse und Modellierung, wohingegen der Bereich Messung zahlenmäßig nur schwach vertreten ist. Zur Auswahl von Alternativen bzw. der Synthese von (optimalen) Strukturen sind keine Beiträge eingereicht worden, was sicherlich an der heutzutage vorrangigen Behandlung der Analyse von Alternativen liegt. Liegt hier erst einmal ein Grundstock an Techniken nebst einer Methodik zu ihrer Anwendung vor, so wird der reiche Fundus an Auswahl- und Optimierungstechniken aus dem Operations Research und der Mathematik verstärkt in die Leistungsbewertung einfließen. Etwas überraschend war hingegen das geringe Interesse an Zuverlässigkeitsaspekten.

Auf dem Gebiet der Modellierung werden deutliche Fortschritte sichtbar. Wurden auf der vorhergehenden Tagung noch die Konzepte einiger Modellierungswerkzeuge geschildert, so sind diese Konzepte inzwischen zum Teil nebst Modellanalyse in Programmpakete umgesetzt. Über die Erfahrungen mit der Anwendung dieser Pakete wird nun zu berichten sein. Währenddessen drängen die Werkzeuge der nächsten Generation bereits nach. Deshalb war die Zeit reif, einige Programmpakete aus Hochschule und Industrie im Rahmenprogramm der diesjährigen Tagung vorzustellen.

Anzahl und insbesondere Qualität der eingereichten Beiträge waren recht gut : von 30 anonym begutachteten wurden 19 angenommen. Das Programm wird abgerundet durch 5 eingeladene Vorträge, die durchweg einen weitreichenden Überblick über aktuelle Forschungsbereiche geben.

Die Durchführung einer solchen Tagung erfordert den Einsatz und die Hilfsbereitschaft zahlreicher Personen und Stellen. An erster Stelle sei hier Herrn Dr. G. Bolch und Herrn R. Rimane gedankt, die uns im lokalen Organisationskomitee tatkräftig unterstützten; ferner den Mitgliedern des Programmausschusses, den Gutachtern der Beiträge, den unterstützenden Institutionen, dem Springer-Verlag sowie zahlreichen Stellen innerhalb und außerhalb der Universität. Undenkbar ist die Organisation auch ohne ein funktionierendes Sekretariat : hierfür gilt unserer Dank den Damen G. Lindhoff, S. Wilfer, G. Morgenstern und G. Pastore.

Erlangen, im Juli 1987

U. Herzog
M. Paterok

Dem Programmausschuß der Tagung gehören an:

W. Ameling, Uni Aachen
H. Beilner, Uni Dortmund
R. Bordewisch, Nixdorf Paderborn
G. Bolch, Uni Erlangen
U. Herzog, Uni Erlangen
L. Hieber, Datenzentrale Württemberg
W. Hoffmann, Siemens Erlangen
R. Klar, Uni Erlangen
P. Kühn, Uni Stuttgart
F. Lehmann, Uni BuWe München
R. Lehnert, Philips Nürnberg
B. Mertens, KFA Jülich
M. Paterok, Uni Erlangen
B. Schmidt, Uni Erlangen
H. Schmutz, IBM Heidelberg
O. Spaniol, Uni Aachen
P. Spies, Uni Bonn
B. Walke, Uni Hagen
S. Zorn, Siemens München
W. Zorn, Uni Karlsruhe

Neben den Programmausschußmitgliedern wirkten als Gutachter mit:

H. Decker
W. Dulz
G. Erpenbeck
G. Fleischmann
K. Geihs
W. Gürich
H. Jüchter
H. Jung
W. Krämer
W. Kowalk
D. Litzba
H. de Meer
F. Noack
G. Rosentreter
H. Schickle
H. Scholten
B. Schwärmer
R. Simon
Spieker
M. Vering
E. Walther-Klaus
G. Werner

INHALTSVERZEICHNIS / TABLE OF CONTENTS

Simulation 1

Wartenetze 1 / Queueing Networks 1

Lokale Netze / Local Area Networks

Simulation 2

Datenbanksysteme / Data Base Systems

Wartenetze 2 / Queueing Networks 2

FROM PERFORMANCE TO PERFORMABILITY.

Raymond Marie
IRISA
Campus Universitaire de Beaulieu
35042 RENNES Cédex, France.

1. Evolution of Performance Evaluation.

In order to follow the evolution of the computer systems, the specialists in performance evaluation have been forced to develop more and more complex models. In general, the exhibited models are discrete models and use the notions of the queueing theory. It is indeed very natural to associate the resources of a computer system to the servers of the theory, and the software processes to the customers.

If a unique queue was a robust model for a computer of the first generation, models of the present generation are much more sophisticated. In fact, needs for new models in order to evaluate the computers have been a real factor in developing researches which have given new results for queueing network theory. The results on the so called multiclass product form queueing networks are such an example.

The features of interest for the evaluator are in general the throughputs and response times of the customers and the utilization factor of the servers.

Among these models, the steady state (the equilibrium behaviour) has been the more studied; by looking only at the steady state, it is possible to deal with larger models with more stations, more customers or more classes.

At the level of the entire computer system, the stations of the model represent resources such as the central units,the central memory, the disk units, the input/output units. The customers represent the tasks such as the user commands. If the system is an interactive one, each user terminal is represented by a server, the service time of which corresponds to the thinking time of the user.

If the obtained results concern mostly the steady state behaviour, it is undoubtedly because it seems very difficult to obtain tractable results on the transient behaviour for large models. But it is worthwhile to note that, from a practical point of view, results in steady state behaviour are in general sufficient when we are dealing with the performance of a computing system. In fact, because of the speed of the execution of the operations, the model reaches rapidly a behaviour close to its steady state.

Let us remark that in general the performance studies have been done under the assumption that neither the hardware nor the software will breakdown; this other aspect of a computing system beeing left to an other specialist: the reliability man.

2. The way of the reliability man.

On his side, the reliability man concerned with the potential breakdowns of the computer system constructs a reliability diagram where the elements are in serie if the breakdown of an element involves the crash of the system (fig. 1).

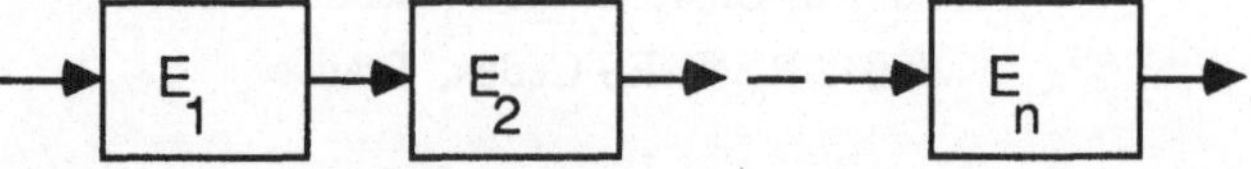

Fig. 1: Reliability diagram.

Some of the elements of this diagram correspond to the resources mentioned above, i.e., the central unit, the memory, etc,... But in addition to the elements which have been taken into account for the performance evaluation, some other elements are introduced on this diagram; it is the case for example of the power supply or of the bus connecting the processor to the memory.

A basic hypothesis consists here in considering that the breakdowns are independent from one element to another. Thanks to this hypothesis, it is relatively easy to get the reliability of such a system. If R(t) denotes the reliability of the system, i.e., the probability that the system has not failed yet at time t, and if $R_i(t)$ denotes the reliability of element i, then the reliability of the system consisting of n elements in serie can be written as follows:

$$R(t) = \prod_{i=1}^{n} R_i(t)$$

It is well known that adding elements in serie induces a decrease of the reliability of the system. In order to avoid this handicap, redondancies of elements are used. The reliability diagram of the figure 2 illustrates the use of r/ k redondancies where a subsystem made of k identical elements is considered operating as long as at least r elements have not failed yet.

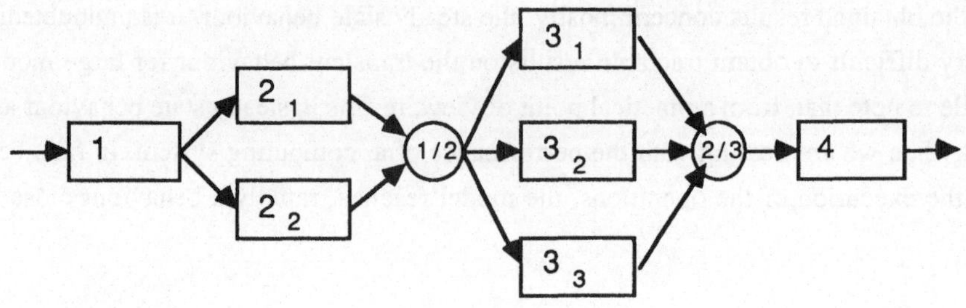

Fig. 2. Reliability diagram with r/k redondancies.

Under the assumption on the independence of the breakdowns between the differents elements, the reliability of a subsystem consisting of a r/k redondancy with identical elements is given by the following expression:

$$R(t) = \sum_{i=r}^{k} \binom{k}{i} R_e^i(t)(1-R_e(t))^{k-i}$$

where $R_e(t)$ is the reliability at time t of one element.

With this kind of redondancy, we get a reliability diagram with a "serie-parallel" type structure where an element in parallel may in fact be made of a serie of smaller modules. By successive reductions, it is possible to compute the reliability of any system with a serie-parallel structured diagram under the hypothesis on the independence of the breakdowns of the different elements.

Many of the already launched satellites have computer systems with such a redondant structure. So far, these systems are considered as not reparable and we are only interested by their reliability function (only the switches are teleoperated from the ground).

When the system is reparable, then the specialist prefers to calculate A(t), the availability of the system; the availability is the probability that the system is up at time t. If the repair processes are also independent, then the expression of the availability will be quite similar to the one of the reliability. For example, if two modules of type i are put in parallel, i.e., the subsystem is good as long as at least one element is good, and if $A_i(t)$ denotes the availability of a module of type i, then the availability of the corresponding subsystem is written as follows:

$$A(t) = 1-(1-A_i(t))^2 = A_i(t)(2-A_i(t)).$$

For reparable systems, the availability tends to a limit as t tends to the infinity. For that, let us use the following notations:

$$A_i = \lim_{t\to\infty} A_i(t), \qquad A = \lim_{t\to\infty} A(t)$$

When dealing with very large systems, the specialist is generally satisfied with these asymptotic values because of the fact that the computation of the transient solutions $A_i(t)$'s are difficult to deal with. Consequently, we first determine the asymptotic values A_i's and then obtain relatively easily the value of A through the successive reductions if the structure is of a serie-parallel type.

Very often, the user of a large computing system wants to obtain another set of values from the asymptotic behaviour: the MUT (mean up time), the MDT (mean down time) and the MTBF (mean time between failures). Indeed, the life of a reparable system is a succession of up and down epochs. In steady state behaviour, the expectations of these epochs are respectively MUT and MDT; and MTBF is equal to the sommation of these two values.

For any serie-parallel structure the MUT, MDT and MTBF can be can determined from the elementary MUT_i and MDT_i of the modules from the Buzacott's formulae:

a/ for n modules in serie:

$$\frac{1}{MUT} = \sum_{i=1}^{n} \frac{1}{MUT_i}$$

and MTBF = MUT/A, MDT = MTBF(1-A),

b/ for n modules in parallel:

$$\frac{1}{MDT} = \sum_{i=1}^{n} \frac{1}{MDT_i}$$

and MTBF = MDT/(1-A), MUT = MTBF.A,

c/ for n identical modules in partial redondancy r/n:

$$\frac{1}{MTBF} = \binom{n}{r} A_i^r . A_i^{n-r} . \frac{r}{MUT_i}$$

and MUT = MTBF.A, MDT = MTBF(1-A).

These formulae are valid for very general failure and repair time distributions; it is sufficient that the failure and repair processes are recurrent and aperiodic renewal processes.

When the reliability diagram is no longer of a serie-parallel type, it is possible to extend these kinds of results but we will have to pay the price through more computation. In order to present the new expressions, let us introduce a few notations:

S: a system which is defined by is reliability diagram (its structure and the parameters of the modules),

S^i: a system which is deduced from the system S by replacing the module i by a short circuit,

S^{-i}: a system which is deduced from the system S by replacing the module i by an open circuit,

A(S,t): availability of system S at time t,

A(S): asymptotic availability of system S,

G(S): inverse of the MTBF of system S,

G_i: for a module i, $G_i = 1/(MUT_i + MDT_i)$.

Using these notations, we have for a general structure the following recursive relations:

$$A(S,t) = A_i(t).A(S^i,t) + (1-A_i(t)).A(S^{-i},t),$$

$$A(S) = A_i.A(S^i) + (1-A_i).A(S^{-i}),$$

$$G(S) = A_i.G(S^i) + (1-A_i).G(S^{-i}) + (A(S^i)-A(S^{-i})).G_i.$$

$$MUT(S) = A(S)/G(S), \qquad MDT(S) = (1-A(S))/G(S).$$

Of course, as soon as in this recursive procedure S has a serie-parallel type structure, the former results should be used.

3. First rendez-vous.

A naturel reaction of the user who receives the results of a performance evaluation in terms of throughput and response time is to wonder what would be the effect of a failure on these values.

A first step of the specialists in order to take into account the failures of the computing system was to introduce fictitious customers into their model in order to "simulate" the failures. By using multiclass queueing networks, it is possible to add new classes of fictitious customers without increasing too much the complexity of the network. The beginning of a service of a fictitious customer represents the arrival of a breakdown and the service time corresponds to the repair time. The interval of time between the instant the fictitious customer frees the server and the next time he gets him back represents the up time of the server. The idea is artful but, however, we have to remark that the only way to master the up time distribution is to introduce a preemptive scheduling discipline for the fictitious customer and by doing so, the model has no longer a product form solution.

Another step consists of including the eventual breakdowns of the server within the service process of the real customers. For that purpose, a completion time is defined [GAV-62]. It is the time which elapses between the the first time the server takes a customer into account and the end of the service. Different models corresponding to different hypotheses have been studied. Generally, breakdowns are rare events; so, if we assume that no more than one breakdown can occur during a service process, then we obtain simpler models and larger networks can be handled [ALT-83].

4. Generalization of fault tolerant systems.

At the beginning, a few systems from the aerospace or from the military domain had redondant modules; the error detections where compendious and in case of failure a human intervention was needed to operate the switches in order to deconnect the bad element and to set on the good one. Thanks to periodic check points and to recovery processes, the information was not lost; but it could take fifteen minutes to get the situation back to normal.

A fault tolerant system (FTS) is a system where all the actions are taken automatically, by the system itself. Since the level of redondancy is not infinite, because of the cost of it, there is still a chance for the system to crash; but this chance is very small, compared to the original one. The figure 3 illustrates the main features of a fault tolerant architecture with two processors.

For a system to be fault tolerant, in addition to the hardware and/or the software redondancy, the system must be able to:

-detect a failure because of the occurence of an error,

-isolate the failing element,

-reconfigure itself and recover quickly from a recovery point.

Nowadays, use of computers have diffused over all activity branches and need for fault tolerant systems have been generalized (banks, airlines,etc,...).

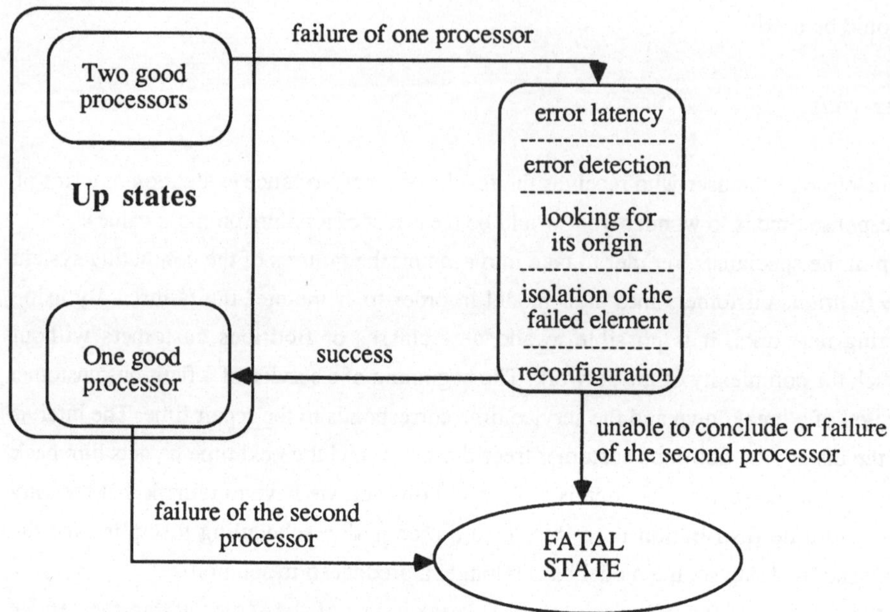

Fig. 3. Schema of the behaviour of a fault tolerant architecture with two processors.

Fault tolerant systems can be partitionned in two classes:

a/ the non reparable FTS: these systems are built in order to achieve a mission over an finite interval of time [0,T]. T can be expressed in a number of days (e.g., for a space shuttle) or in a number of years (e.g., for a satellite).

b/ the reparable FTS: with this class, the failing element can be repaired or exchanged on line almost immediately or after periodic inspection.

When we are concerned with the performance evaluation of fault tolerant systems, the main difference with classic performance evaluation is that it is in general necessary to study the model of the system in **transient behaviour**. For the non reparable FTS, this is the only possibility since:

$$\lim_{t \to \infty} \text{Prob}\{\text{"the system is down at time t"}\} = 1$$

With the reparable FTS, even if the steady state analysis gives interesting results on the characteristics of complex systems, the transient analyse is also desired; because of the numerical values of the parameters of the system, the system characteristics on the interval [0,T] are quite different from the equilibrium characteristics; they still depends on the initial state of the sytem. The more the system is reliable, the more the difference is high. For example if we are interested in building a system with an availability greater than $(1\text{-}10^{-4})$, then it may happen that the characteristics depends on the initial conditions after several years of use.

We may also think of the example of a ditributed system where a resource is shared thanks to a token through a logical ring. For a lot of reasons, the token may disappear. In order to get this subsystem fault tolerant, we can implement k other tokens with the charge of supervising the former token and restoring it when it is lost. For such a subsystem, we may construct a model where the MUT of the subsystem is of the order of 10^{10} hours...

With the determination of building such reliable FTS, needs for sharper characteristics araise.

5. Definition of new characteristics.

Let us assume that if X(t) denotes the state of a FTS at time t, $\{X(s), s\geq 0\}$ is a stochastic process which takes its values from a set $\mathbb{X}$. The set $\mathbb{X}$ is partitioned into two classes: $\mathbb{X}_u$ (the subset of states for which the system is up) and $\mathbb{X}_d$ (the subset of states for which the system is down). Let us consider the indicator process $\{I(t), t\geq 0\}$ such that:

$$I(t) = \begin{cases} 1 & \text{if } X(t) \text{ belongs to } \mathbb{X}_u, \\ 0 & \text{else} \end{cases}$$

We also assume the system is a coherent system, i.e., if the system is down, a new failure (from another element) does not put the system up. Note that, because of the redondancies, some elements can have failed while the FTS is still in a up state.

If the system is not reparable (during a mission), then once the system reaches a state of $\mathbb{X}_d$, the system stays down. While, if the system is reparable, it has a chance to go back to a state of $\mathbb{X}_u$ after visiting a state of $\mathbb{X}_d$.

For a reparable system, the expectation E[I(t)] is nothing but the classic availability A(t) we already introduce in section 2; to be more accurate, we may call it here the "instantaneous availability". If the system is not reparable, then the expectation E[I(t)] is equal to the reliability R(t). Note that even for the reparable case, the reliability measure R(t) is still meaningful: it is the probability that X(.) never reaches a state of $\mathbb{X}_d$ on the interval [0,t].

Another characteristic of interest takes the cumulative operational time into account:

$$O(t) = \int_0^t I(s)ds$$

In fact, a few companies have already offered a guaranteed availability to their customers. By accepting to pay a penalty if the cumulative operational time over the interval [0,T] is lower than $\alpha.T$, they are taking a risk, quantified by the probability $\Pr\{O(T) < \alpha.T\}$, to pay the penalty.

From this characteristic, we may get the percentage of this cumulative operational time: O(t)/t. Note that the expectation of O(t)/t is related to the instantaneous availability by the following relation:

$$E[O(t)/t] = \frac{1}{t}\int_0^t E[I(ts)]ds = \frac{1}{t}\int_0^t A(s)ds.$$

This expectation can be called, in a natural way, the mean availability over the interval [0,t].

Sometimes, we want to make a distinction among the up states of the set $\mathbb{X}_u$ and introduce a reward process. For that, let us assume that $\mathbb{X}_u$ have been partitioned in J classes denoted $\mathbb{X}_{u1}$,..., $\mathbb{X}_{uJ}$, and let us introduce the process $\{R(s), s \geq 0\}$ such that, if X(t) belongs to $\mathbb{X}_{uj}$, then R(t) takes an positive real value r_j, j=1,..., J (if X(t) belongs to $\mathbb{X}_d$, then R(t) takes the value zero).

For example, the FTS has J processor units and the set $\mathbb{X}_u$ is partitioned according to the number of processor units which are not down; the value of r_j is set equal to this integer. In that case, the measure

$$CR(t) = \int_0^t R(s)ds$$

gives the cumulative potential of processing power. Then, we may be interested in the cumulative distribution function of the random variable CR(t).

No need to say that a lot of these measures have no already known closed form solution, even in the case of a small model!

6. Basic model for the transient solutions.

In order to get a beginning of an answer to their problem, that is to build a probabilistic model, the specialists have been looking once again towards Markovian models. The memoryless property of the Markov process is really a key point which allows sometimes the user to get tractable results. On one hand, with such a model, the holding time of every state has to be exponentially distributed; but, on the other hand, it is generally possible to deal with non exponentially distributed random variables by increasing the state space. Even if very often, when the number of states becomes large, no more tractable results are available.

But the construction of a Markovian model is generally an easy thing to do when we get used to it; moreover, it makes sense for the specialist. For the type of problems we are concerned with, the state space is discrete and the Markov process is in fact a continuous time Markov chain (CTMC). Quite often, a state is a n-tuple of boolean variables, each variable corresponding to the two possible conditions of each module of the system (good or failed).

As we know, the infinitesimal generator of the CTMC is a matrix which is, except for the diagonal elements, made of the transition rates from one state to another. These transition rates are arithmetic expressions of failure and repair rates, therefore, they are obtained easily from examination of the problem. Even if the construction of the matrix may become quite tedious when dealing with large models!

From a mathematical point of view, the solution is well known. If A denotes the infinitesimal generator and $\pi(0)$ denotes the initial state probability vector, then π (t), the state probability vector at time t, is given by the following expression:

$$\pi(t) = \pi(0)e^{At}$$

where

$$e^{At} = \sum_{k=0}^{\infty} \frac{(At)^k}{k!}$$

With the help of this expression, it is worthwhile to remark that the Markov process is characterized by its infinitesimal generator and its initial state probability vector.

Some characteristics may be obtained directly from the vector $\pi(t)$; for example:

$$E[I(t)] = \sum_{i \in X_u} \pi_i(t)$$

So, we could think of getting some measures by the computation of e^{At}. The first idea in order to do that is to compute a truncation of the serie expansion, i.e., to calculate

$$\sum_{k=0}^{N} \frac{(At)^k}{k!},$$

but it can be shown that, as t increases, this approximation can be made arbitrarily poor [GRA-77]. This is because of the fact that the absolute values $|(At)^k|/k!$ of the terms of the remainder, i.e., of

$$\sum_{k=N+1}^{\infty} \frac{(At)^k}{k!},$$

can be made greater than any positive value by increasing t.

As it is known, there are other possibilities for computing e^{At}, but Moler et al [MOL-78] concluded, after selecting nineteen ways to compute the exponential of a matrix, that none of them were completely satisfactory. In our case, the main reason for getting into troubles comes from the fact that the matrix A contains both positive and negative coefficients (the negative ones are those of the diagonal). This involves roundoff errors and loss of accuracy.

But it seems that for the case of the Markov process, i.e., when A is an infinitesimal generator, we may succeed thanks to the artifice of the so called Uniformization.

From an algebraic point of view, the progress is the following: we define the matrix P as

$$P = A/v + I,$$

where v is any positive real such that $v \geq \max_i |a_{ii}|$.

Then e^{At} can be rewritten as

$$e^{At} = e^{vPt - vIt} = e^{-vt} e^{vPt},$$

or

$$e^{At} = e^{-vt} \sum_{k=0}^{\infty} P^k \frac{(vt)^k}{k!},$$

From this we get

$$\pi(t) = \sum_{k=0}^{\infty} \pi(0)\, P^k \frac{(vt)^k}{k!},$$

Note that from the complexity point of view, $\pi(0)\, P^k$, for k>0, can be computed in the following recurrent way:

$\pi^k = \pi(0)\, P^k = \pi^{k-1} P$, with $\pi^1 = \pi(o)\, P$.

The nice fact in this expression is that every term is non-negative; also, the matrix P is a stochastic matrix and so are its powers. Consequently, the roundoff errors will be smaller than those of the former case. Of course, it is still necessary to realize a truncation of the summation, i.e., we compute

$$\sum_{k=0}^{N} \pi(0)\, P^k \frac{(vt)^k}{k!} e^{-vt},$$

but this time,we are able to exhibit an upper bound of the error which is introduced by the truncation. First we have (by convention, an inequality between two vectors means that the inequality stands for each component):

$$\sum_{k=0}^{N} \pi(0)\, P^k \frac{(vt)^k}{k!} e^{-vt} \leq \sum_{k=0}^{\infty} \pi(0)\, P^k \frac{(vt)^k}{k!} e^{-vt},$$

because every term is non negative; secondly, if 1 denotes the unity vector,

$$\sum_{k=N+1}^{\infty} \pi(0)\, P^k \frac{(vt)^k}{k!} e^{-vt} \leq \sum_{k=N+1}^{\infty} 1 \frac{(vt)^k}{k!} e^{-vt},$$

because $\pi(0)\, P^k = \pi^k$ is a probability vector.

Therefore, it follows that

$$\sum_{k=0}^{N} \pi(0)\, P^k \frac{(vt)^k}{k!} e^{-vt} \leq \sum_{k=0}^{\infty} \pi(0)\, P^k \frac{(vt)^k}{k!} e^{-vt} \leq \sum_{k=0}^{N} \pi(0)\, P^k \frac{(vt)^k}{k!} e^{-vt} + \sum_{k=N+1}^{\infty} 1 \frac{(vt)^k}{k!} e^{-vt},$$

So we get an error bound which is the tail of the Poisson distribution.

As we are going to see, this procedure of the Uniformization can be also presented through a stochastic approach. Let us consider the transition graph of the CTMC. To each node i, let us add a loop with a rate $(v+a_{ii})$ as illustrated by the figure 4 (remember a_{ii} is non-positive).

With the new CTMC, there are "transitions" which do not involve changing a state (as it is the case with discret time Markov chains). We call this new CTMC the uniformized CTMC canonically associated with the original one. In the uniformized CTMC, the "transitions" occur with rate v, whatever the state of the process is; but, given a transition occurs while the process is in state i, the process will stay in state i with probability $1 - a_{ii}/v$.

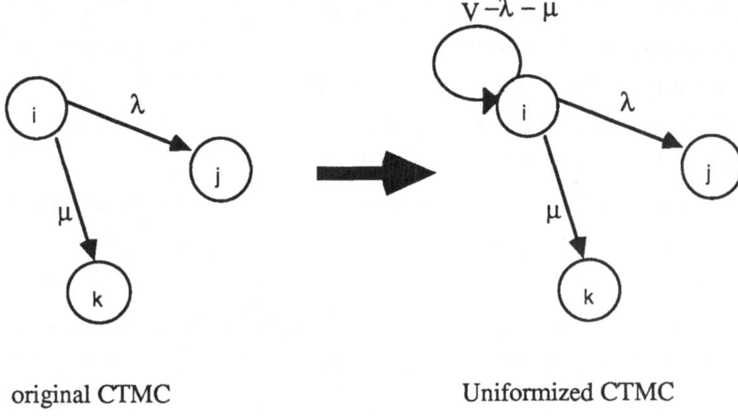

original CTMC Uniformized CTMC

Fig. 4. Construction of the uniformized CTMC.

Note that if v has been taken as $v = \max_i |a_{ii}|$, then there is at least one node of the transition graph with no loop. Also note that it is possible to show by standard probability calculus that the time the uniformized CTMC will stay in state i is also exponentially distributed with rate $-a_{ii}$.

The two CTMC are the same stochastic process because the joint probability distribution of $\{X(t), t \geq 0\}$ remains unchanged. However, with the new CTMC, we get a new way to compute the state probabilities [ROS-83]. If we look at the probability that the process visits state j after the next transition, given the process is in state i, we obtain the probability a_{ij}/v if $j \neq i$, and $1 + a_{ii}/v$ if $j = i$. So these probabilities are the coefficients of the matrix P. Consequently, P was also the transition matrix of the discret time Markov chain (DTMC) embedded in the uniformized CTMC.

Let N(t) denotes the number of transitions of the uniformized CTMC on [0,t]. Since the inter-transition times are independent and identically exponentially distributed, then $\{N(t), t \geq 0\}$ is a Poisson process with rate v. In order to get the transient state probability vector, we first determine the transition probability functions $P_{ij}(t)$ by conditioning on N(t):

$$P_{ij}(t) = P\{\, X(t) = j \,/\, X(0) = i \,\}$$

$$= \sum_{n=0}^{\infty} P\{\, X(t) = j \,/\, X(0) = i, N(t) = n \,\}\, P\{\, N(t) = n \,/\, X(0) = i \,\}$$

$$= \sum_{n=0}^{\infty} P\{\, X(t) = j \,/\, X(0) = i, N(t) = n \,\}\, \frac{(vt)^n}{n!} e^{-vt}$$

But the conditional probability $P\{\, X(t) = j \,/\, X(0) = i, N(t) = n \,\}$ is nothing but $P_{ij}{}^n$, where P^n is the matrix of the n stage transition probabilities of the embedded DTMC.

Using the fact that $\pi(t) = \pi(0)P(t)$, we get the result we already deduced by help of the algebra.

In fact, it could be also possible to formally construct the CTMC $\{X(t), t \geq 0\}$ directly from the DTMC with transition matrix P, and from an independent Poisson process N(t) with rate v.

So, this technique, which consists basically in the artful of using the matrix P instead of the matrix A, gives the way to very good results. We still have noticed some numerical difficulties but only for very stiff Markov chains (with models of FTS, we may sometimes have a scale ratio between different repair and failure rates of the order of 10^{10}...).

As we already mentioned it, some measures of interest are directly obtained from the vector $\pi(t)$. In [SOU-86], there is an interesting way of getting other measures such as the cumulative operational time distribution. This is done numerically through recurrent formulae and thanks to the Uniformization approach. A key point in this paper is to start to look for the distribution of O(t) given N(t) and $N_u(t)$, where $N_u(t)$ denotes the number of transitions occuring while the process is in the subset $\mathbb{X}_u$ of the up states.

Very often, for FTS, the Markov chain has an absorbing state (like the fatal state on figure 3) and it makes sense to look at the cumulative operational time over $[0,\infty[$, i.e., $O(\infty)$. The expression of it distribution can be found in [MAR-86].

The computation of the distribution of some cumulative reward is difficult to obtain and, to the best of our knowledge, this is not done for a general Markov chain.

7. Conclusions.

We have presented in this paper how reliability preoccupation has been introduced in the field of the performance evaluation of computer systems. Because of the development of fault tolerant systems, it has been necessary to investigate the transient behaviour of the Markovian models. We emphasized on the basics models which give tractable results. However, for several case studies (cf. J. Meyer or J.C. Laprie's works), approximations had to be made in order to reduce the state space.

References

[ALT-83] T. Altiok and S. Stidham, "The allocation of interstage buffer capacities in production lines", IIE Transactions, Vol. 15, n°4, pp 292-299, 1983.

[SOU-86] E. de Souza e Silva and H. R. Gail, "Calculating cumulative operational time Distributions of repairable computer systems", IEEE T/C, Vol. C-35, n° 4, 1986.

[GAV-62] D.P. Gaver, "A waiting line with interrupted service, including priorities", J. Roy. Statist. Soc., Ser. B24, pp 73-90, 1962.

[GRA-77] W. K. Grassmann, "Transient solutions in Markovian queueing systems", Compt. Oper. Res., Vol. 4, pp 47-53, 1977.

[MAR-86] R.A. Marie and B. Sericola, "Distribution du temps total de sejour dans un sous ensemble d'états transitoires d'un processus markovien homogène à espace d'état fini", INRIA research report n° 585, 1986.

[MOL-78] C. Moler and C. Van Loan, "Nineteen dubious ways to compute the exponential of a matrix", Siam Review, Vol. 20, n° 4, pp 801-836, 1978.

[ROS-83] S. M. Ross, Stochastic Processes, John Wiley & sons, 1983.

Wahrscheinlichkeitsverteilungen maximaler Entropie für die Wartezeiten in G/G/1-Systemen

Johann Christoph Strelen
Institut für Informatik, Universität Bonn
Wegelerstr. 6, D-53 Bonn

Zusammenfassung

Es wird ein Verfahren vorgestellt, mit dem die Wartezeitverteilungen in G/G/1-Systemen bei mittlerer bis hoher Verkehrsintensität näherungsweise berechnet werden können, der Reihe nach für jeden Kunden, ausgehend von einer Verteilung für die Wartezeit des ersten Kunden, also nicht nur im statistischen Gleichgewicht. Angewendet werden die Prinzipien der maximalen Entropie und der minimalen Kreuzentropie. Die dafür benötigten Erwartungswerte ermittelt das Verfahren selbst, sie brauchen nicht mit anderen Näherungsverfahren bestimmt zu werden. Vergleiche mit bekannten Werten zeigen, daß für die berechneten Wartezeitverteilungen drei bis fünf Momente gut mit den exakten Zahlen übereinstimmen.

1. Einleitung

In einer Reihe von Arbeiten wird das Prinzip der maximalen Entropie für Wahrscheinlichkeitsverteilungen ([SJ]) angewendet, um stochastische Modelle im statistischen Gleichgewicht näherungsweise zu analysieren und so Verteilungen für interessierende Zufallsvariablen zu berechnen. Dazu müssen Erwartungswerte vorgegeben werden, die aus anderen (Näherungs-) Verfahren stammen. Meistens werden diskrete Verteilungen untersucht. Einfache Wartesysteme werden in [AK] (M/G/1, G/M/1), [K2], [TZ], [T], [S] (G/G/1, M/G/1) betrachtet, Wartenetze in [K1], [W], [TZ], [T].

In der vorliegenden Arbeit werden Dichten für die Wartezeiten in G/G/1-Systemen mit mittlerer bis hoher Verkehrsintensität näherungsweise berechnet, indem Verteilungen mit maximaler Entropie (VmE) oder minimaler Kreuzentropie (VmC) ermittelt werden. Dazu werden für den ersten Kunden im System die Verteilung der Wartezeit, oder zumindest einige Momente davon, vorgegeben. Daraus werden dann sukzessive für die nachfolgenden Kunden die Wartezeitverteilungen berechnet, indem die Chapman-Kolmogoroff-Gleichungen ausgewertet werden. Dabei werden drei bzw. fünf Momente der Zwischenankunftszeit und der Bedienzeit berücksichtigt. Bei einer Verkehrsintensität kleiner als eins stellt sich nach einiger Zeit statistisches Gleichgewicht ein, und man hat dann die Wartezeitverteilung auch dafür näherungsweise berechnet; sonst läuft das System voll. Folgendes sind wesentliche Merkmale der Vorgehensweise: Es werden VmE und VmC für Dichten berechnet; das Verfahren ermittelt die erforderlichen Momente selbst, sie brauchen nicht vorgegeben zu werden; es wird auch das transiente Verhalten des Systems analysiert.

In Abschnitt 2 werden die Prinzipien der maximalen Entropie und der minimalen Kreuzentropie auf die Berechnung von Dichten angewendet, in Abschnitt 3 werden solche Dichten für die Wartezeiten in G/G/1-Systemen angegeben, und in Abschnitt 4 werden Beispiele betrachtet.

2. Verteilungen maximaler Entropie und minimaler Kreuzentropie

In jüngerer Zeit wurden die Prinzipien der minimalen Kreuzentropie und der maximalen Entropie angewendet, um Wahrscheinlichkeitsverteilungen zu definieren ([SJ]). Dabei wird zunächst eine Menge von Verteilungen festgelegt, indem einige Erwartungswerte vorgegeben werden. Dann wird aus dieser Menge eine Verteilung P ausgewählt durch die Forderung, daß ihre Kreuzentropie minimal bzw. ihre Entropie maximal ist. Wir nennen P dann "Verteilung mit minimaler Kreuzentropie" (VmC) bzw. "Verteilung mit maximaler Entropie" (VmE). Im ersten Fall wird eine weitere Annahme über die Verteilung berücksichtigt.

Man kann diese Vorgehensweise auch als Näherungsverfahren zur Bestimmung von Verteilungen interpretieren: Gegeben seien eine Verteilung P_X und einige Erwartungswerte dafür. Mit deren Hilfe wird nach den genannten Prinzipien die Verteilung P berechnet und als Näherung für P_X angesehen.

Zunächst befassen wir uns mit dem Prinzip der maximalen Entropie. Sei X eine reelle Zufallsvariable mit der unbekannten Wahrscheinlichkeitsverteilung P_X. Bekannt seien die Erwartungswerte

$$E(g_k(X)) = G_k, \quad k \in [0:K], \quad K \in \mathbb{N}_0. \tag{1}$$

Dabei seien g_k bekannte Funktionen, insbesondere $g_0(x) = 1$, und G_k reelle Zahlen, insbesondere $G_0 = 1$.

Aus der Menge der durch (1) bestimmten Verteilungen wird eine ausgewählt durch die Forderung, daß ihre Entropie H maximal ist. Bei abzählbarem Ergebnisraum $\Omega \subset \mathbb{Z}$ ist die Entropie einer Verteilung $(p_n, n \in \Omega)$

$$H \stackrel{\text{def}}{=} -\sum_{n\in\Omega} p_n \ln p_n, \tag{2}$$

bei $\Omega \subset \bar{\mathbb{R}}$ ($\mathbb{R} \cup \{-\infty\} \cup \{\infty\}$) ist die Entropie einer Dichte $f(x)$

$$H \stackrel{\text{def}}{=} -\int_\Omega f(x) \ln f(x) dx. \tag{2'}$$

Für die Verteilungen muß

$$p_n \geq 0 \text{ für alle } n \in \Omega \text{ und } \sum_{n\in\Omega} p_n = 1$$

bzw.

$$f(x) \geq 0 \text{ für } x \in \Omega \text{ und } \int_\Omega f(x)dx = 1$$

gelten. Wir nehmen im folgenden an, daß die Verteilungen mit maximaler Entropie im Innern dieser Bereiche liegen. In [HA] wird für den diskreten Fall gezeigt, wie man das durch lineare Programmierung feststellen kann, und es wird betrachtet, wie die Verteilungen auf den Rändern gleichmäßig approximiert werden können.

Wir interessieren uns hier für Dichten. Der Ergebnisraum Ω sei ein Intervall aus $\bar{\mathbb{R}}$. Gegeben seien die Momente

$$M_j = \int_\Omega x^j f_X(x)dx, \quad j \in [0:K], \tag{3}$$

einer unbekannten Dichte $f_X(x)$ mit $M_0 = 1$. Eine Näherung für $f_X(x)$ mit denselben Momenten soll durch Maximierung der Entropie

$$H = -\int_\Omega f(x) \ln f(x) dx \tag{4}$$

berechnet werden.

Wir bezeichnen diese Näherung als "von der Ordnung K".

Es liegt ein isoperimetrisches Variationsproblem vor, siehe [SM]. Eine Lösungsfunktion $y = f(x)$ dafür genügt der partiellen Eulerschen Differentialgleichung

$$h_y - y'' h_{y'y'} - y' h_{y'y} - h_{y'x} = 0 \tag{5}$$

($'$ bezeichnet eine totale Ableitung, z. B. $y' = \frac{d}{dx} f(x)$, und ein Index eine partielle Ableitung, z. B. $h_y = \frac{\partial}{\partial y} h$). $h(x, y, y')$ ist die Summe des Integranden des zu maximierenden Integrals und der Integranden der Nebenbedingungen, jeweils mit einem Lagrange-Faktor versehen. In unserem Fall haben wir also

$$h(x, y, y') = -y \ln y + \sum_{j=0}^{K} \lambda_j^* x^j y; \tag{6}$$

hier sind die λ_j^* die Lagrange-Faktoren. Die Eulersche Differentialgleichung entartet damit zu

$$-\ln y - 1 + \sum_{j=0}^{K} \lambda_j^* x^j = 0,$$

und wir haben als Lösung

$$f(x) = e^{-\lambda(x)} \tag{7}$$

mit den Bezeichnungen

$$\lambda_0 \stackrel{\text{def}}{=} 1 - \lambda_0^*, \quad \lambda_j \stackrel{\text{def}}{=} -\lambda_j^*, \quad j \in [1:K],$$

$$\lambda(x) \stackrel{\text{def}}{=} \sum_{j=0}^{K} \lambda_j x^j. \tag{8}$$

Das eigentliche Problem liegt nun darin, die λ_j aus den nichtlinearen Gleichungen

$$M_k = \mu_k \stackrel{\text{def}}{=} \int_\Omega x^k e^{-\lambda(x)} dx, \quad k \in [0:K], \tag{9}$$

die sich aus (3) ergeben, zu berechnen. Dabei betrachten wir nur solche λ_j, mit denen die Integrale existieren.

Wir betrachten drei Beispiele.

Beispiel 1 Sei $\Omega = I\!R_+, K = 1$ und M_1 gegeben. Dann ist

$$\mu_0 = \frac{1}{\lambda_1} e^{-\lambda_0}$$

und

$$\mu_1 = \frac{1}{\lambda_1^2} e^{-\lambda_0},$$

falls $\lambda_1 \neq 0$. Wir bekommen als Lösung

$$f(x) = \lambda_1 e^{-\lambda_1 x}$$

mit

$$\lambda_1 = 1/M_1,$$

also die Exponentialverteilung.

Beispiel 2 Sei $\Omega = I\!R$, $K = 2$ und $M_2 \neq 0$. Man erhält nach längerer Rechnung

$$\mu_0 = e^{-\lambda_0} e^{\lambda_1^2/(4\lambda_2)} \frac{1}{\sqrt{\lambda_2}} \sqrt{\pi},$$

$$\mu_1 = -\mu_0 \frac{\lambda_1}{2\lambda_2},$$

$$\mu_2 = \mu_0 \left(\frac{1}{2\lambda_2} + \frac{\lambda_1^2}{4\lambda_2^2}\right).$$

Es gelingt leicht, die Gleichungen (9) aufzulösen:

$$\begin{aligned} e^{-\lambda_0} &= e^{-M_1^2/(2\sigma^2)} / \sqrt{2\pi\sigma^2}, \\ \lambda_1 &= -M_1/\sigma^2, \\ \lambda_2 &= \frac{1}{2\sigma^2} \end{aligned} \tag{10}$$

mit

$$\sigma^2 = M_2 - M_1^2.$$

Damit ist

$$f(x) = \frac{1}{\sqrt{2\pi\sigma^2}} e^{-(x-M_1)^2/(2\sigma^2)},$$

also die Dichte der Normalverteilung $N(M_1, \sigma)$.

Beispiel 3 Sei $\Omega = [0,1], K = 0$ oder $1, M_1 = 1/2$. Dann ist

$$\mu_0 = e^{-\lambda_0}$$

bzw.

$$\mu_0 = e^{-\lambda_0} \begin{cases} 1 & \text{falls } \lambda_1 = 0, \\ (1 - e^{-\lambda_1})/\lambda_1 & \text{sonst,} \end{cases}$$

$$\mu_1 = e^{-\lambda_0} \begin{cases} 1/2 & \text{falls } \lambda_1 = 0, \\ (1 - (1 + \lambda_1)e^{-\lambda_1})/\lambda_1^2 & \text{sonst.} \end{cases}$$

Damit haben wir als Lösung $e^{-\lambda_0} = 1, (\lambda_1 = 0)$, also in beiden Fällen

$$f(x) = 1,$$

die Gleichverteilung, die im Falle der Ordnung 0 ohne Verwendung des ersten Momentes berechnet wurde.

Beim Prinzip der minimalen Kreuzentropie wird eine Vorausinformation über die Form der Verteilung eingearbeitet durch eine Verteilung $(v_n, n \in \Omega)$ im diskreten Fall bzw. eine Dichte $v(x), x \in \Omega$, im stetigen Fall. Damit wird die Kreuzentropie definiert:

$$H^+ \stackrel{\text{def}}{=} \sum_{n \in \Omega} p_n(\ln p_n - \ln v_n) \tag{11}$$

bzw.

$$H^+ \stackrel{\text{def}}{=} \int_\Omega f(x)[\ln f(x) - \ln v(x)]dx. \tag{11'}$$

Eine VmC wird dann aus den durch (1), speziell (3) definierten Mengen von Verteilungen ausgewählt, indem H^+ minimiert wird.

Für Dichten liegt wieder ein isoperimetrisches Variationsproblem vor. Die Eulersche Differentialgleichung muß für

$$h = y \ln y - y \ln v + \sum_{j=0}^{K} \lambda_j^* x^j y \tag{12}$$

erfüllt sein. Sie entartet zu

$$\ln y + 1 - \ln v + \sum_{j=0}^{K} \lambda_j^* x^j = 0,$$

und hat demnach die Lösung

$$f(x) = v(x)e^{-1-\lambda_0^*-\lambda_1^* x - \ldots - \lambda_K^* x^K}. \tag{13}$$

Wir geben aus praktischen Gründen für unsere Anwendung

$$v(x) = x^{p-1} e^{c_0 + c_1 x + \ldots + c_K x^K}, p > 0,$$

vor, ohne p festzulegen. Damit hat die Dichte die Form

$$f(x) = x^{p-1} e^{a_0 + a_1 x + \ldots + a_K x^K} \tag{14}$$

mit den Koeffizienten p und $a_0, a_1, \ldots, a_K$.

Beispiel 4 Sei $\Omega = \bar{I\!R}_+$ und $K = 1$. Mit den Momenten M_1 und M_2 ist

$$\lambda_1 = \frac{M_1}{M_2 - M_1^2}$$

und

$$p = \frac{M_1^2}{M_2 - M_1^2} - 1;$$

$f(x)$ ist die Gammaverteilung mit den Momenten M_1 und M_2.

Es ist ungeklärt, wie genau eine VmE eine vorgegebene Verteilung approximiert. Wir wollen jedoch auf zwei Gegebenheiten hinweisen, bei deren Vorliegen diese Approximation nicht gut sein kann, ohne näher darauf einzugehen, was "gut" hier bedeutet.

VmE der Form (7) haben offenbar höchstens $K - 1$ Extrema im Endlichen, und sie sind stetig und glatt. Somit können Dichten mit einer höheren Zahl von Extrema oder mit Sprüngen oder Knicken nicht gut angenähert werden.

Betrachtet man zu gegebenen Verteilungen numerischer Zufallsvariablen über Ω VmE, so läßt sich beobachten, daß i. allg. nicht die relativen Fehler von $P\{A\}$ für verschiedene $A \subset \Omega$ die gleiche Größenordnung haben, sondern eher die absoluten. Somit sind bei kleineren $P\{A\}$ die relativen Fehler i. allg. größer. Es ist dann also ungünstig, bedingte Wahrscheinlichkeiten $P\{B|A\}$, $B \subset \Omega$, zu benutzen.

Wie man weiter unten bemerken kann, liegt dieser Sachverhalt bei unseren Verfahren vor, wobei $P\{A\} = \rho$ ist; ρ ist die Auslastung. Daher sind die Verfahren für kleine ρ ungeeignet.

Wir benutzen im folgenden VmE der Ordnungen 1 und 2, also Exponential- und Normalverteilungen, und VmC der Form (14), Ordnung 1, also Γ-Verteilungen.

3. Wartezeiten in G/G/1-Systemen

Wir betrachten G/G/1-Systeme, die die üblichen Voraussetzungen erfüllen, siehe z. B. [KL]. Die Zwischenankunftszeiten seien wie eine Zufallsvariable T verteilt, die Bedienzeiten wie eine Zufallsvariable S, beide über $\bar{I\!R}_+$. Ferner seien $C \stackrel{\text{def}}{=} S - T$, und $W^{(n)}, n \in I\!N$, die Wartezeiten des n-ten Kunden. Mit den "Pseudowartezeiten"

$$\widetilde{W}^{(n)} \stackrel{\text{def}}{=} W^{(n-1)} + C \tag{1}$$

ist

$$W^{(n)} = \max(0, \widetilde{W}^{(n)}), n = 2, 3, \ldots . \tag{2}$$

Wir nehmen an, daß S und T Dichten besitzen, und daß hinreichend viele Momente dafür existieren.

In diesem Abschnitt zeigen wir, wie man Näherungen für die Verteilungen der Wartezeiten $W^{(n)}$ berechnen kann, wobei man mit einer bekannten Verteilung für den ersten Kunden beginnt.

Seien $T_j, S_j, C_j, \widetilde{W}_j^{(n)}$ und $W_j^{(n)}, j \in I\!N_0$, die j-ten Momente der o.g. Zufallsvariablen, und $\widetilde{w}^{(n)}(t), t \in \bar{I\!R}$, die Dichten der Pseudowartezeiten. Die Verteilungen der Wartezeiten $W^{(n)}$ werden durch die Wahrscheinlichkeiten $p_0^{(n)} \stackrel{\text{def}}{=} P\{W^{(n)} = 0\}$ und die Dichten $w^{(n)}(t), t > 0$, beschrieben. Dafür ist

$$w^{(n)}(t) = \widetilde{w}^{(n)}(t), t > 0,$$

$$p_0^{(n)} = \int_{-\infty}^{0} \widetilde{w}^{(n)}(t)dt. \tag{3}$$

Wir geben nun an, wie man für die $w^{(n)}(t)$ Näherungen berechnen kann. Dazu definieren wir Zufallsvariablen U über $\bar{I\!R}_-$ und V über $\bar{I\!R}_+$,

$$U \stackrel{\text{def}}{=} \min(0, \widetilde{W}^{(n)}), \ V \stackrel{\text{def}}{=} \max(0, \widetilde{W}^{(n)}),$$

und

$$\alpha \stackrel{\text{def}}{=} P\{\widetilde{W}^{(n)} \leq 0\}, \ \beta \stackrel{\text{def}}{=} 1 - \alpha.$$

Die Dichten dieser Zufallsvariablen unter der Bedingung, daß letztere verschieden von null sind, seien $u(t)$ bzw. $v(t)$, und ihre Momente U_j bzw. $V_j, j \in I\!N_0$. Dann gilt

$$\widetilde{W}_j = \alpha U_j + \beta V_j,$$

$$\widetilde{w}(t) = \alpha u(t) + \beta v(t) \text{ für } t \neq 0, \widetilde{w}(0) = \lim_{\tau \uparrow 0} u(\tau). \tag{4}$$

(Die oberen Indizes $^{(n)}$ bei U,V,..., zur Kennzeichnung des Kunden lassen wir i.allg. weg)

Für die Dichten $u(t)$ und $v(t)$ berechnen wir VmE und VmC der Ordnung 1 als Näherung, und sprechen dann von M-Näherungen und Γ-Näherungen. Wir nehmen diese Aufteilung der Pseudowartezeiten vor, weil deren Dichten im Nullpunkt i.allg. nicht stetig differenzierbar sind, und daher durch eine glatte VmE schlecht angenähert werden.

Als VmE der Ordnung 1 nehmen wir

$$u(t) = ae^{at}, v(t) = be^{-bt}, \tag{4'}$$

wobei $a > 0$ und $b > 0$ sein muß, und als VmC Gammaverteilungen,

$$u(t) = \frac{a^p}{\Gamma(p)}(-t)^{p-1}e^{at},$$

$$v(t) = \frac{b^q}{\Gamma(q)} t^{q-1} e^{-bt} \tag{4''}$$

mit $p, a, q, b > 0$. Dafür ist dann

$$U_j = (-1)^j \frac{j!}{a^j}, V_j = \frac{j!}{b^j}$$

bzw.

$$U_j = (-1)^j \frac{p(p+1)\dots(p+j-1)}{a^j},$$

$$V_j = \frac{q(q+1)\dots(q+j-1)}{b^j}, j \in \mathbb{N},$$

$$U_0 = 1, V_0 = 1.$$

Wir geben nun an, wie die Koeffizienten a, b, α bzw. p, a, q, b, α für jeden Kunden berechnet werden können. Mit (4) hat man dann Näherungen für die Pseudowartezeiten, und mit (3) auch für die Wartezeiten.

Dazu definieren wir Funktionen φ_j dieser Koeffizienten,

$$\varphi_j \stackrel{\text{def}}{=} \alpha U_j + \beta V_j - \widetilde{W}_j,$$

also im Fall der VmE

$$\varphi_j \stackrel{\text{def}}{=} \alpha(-1)^j \frac{j!}{a^j} + (1-\alpha)\frac{j!}{b^j} - \widetilde{W}_j.$$

Dann ist

$$\varphi_j = 0, \quad j = 1, 2, 3, \tag{5'}$$

ein nichtlineares Gleichungssystem für die Koeffizienten a, b, α der VmE für U und V. Um sie zu berechnen, werden wir es für jeden Kunden lösen. Dabei sind die Beschränkungen

$$a > 0, b > 0, 0 < \alpha < 1 \tag{6'}$$

einzuhalten.

Die Momente $\widetilde{W}_j^{(n)}$ bekommt man wegen der Unabhängigkeit der Wartezeit des $(n-1)$-ten Kunden, seiner Bedienzeit und der Zwischenankunftszeit bis zum Eintreffen des n-ten Kunden nach (1) aus

$$\widetilde{W}_j^{(n)} = \sum_{i=0}^{j} \binom{j}{i} W_{j-i}^{(n-1)} C_i \tag{7}$$

mit

$$C_j = \sum_{i=0}^{j} \binom{j}{i} (-1)^i S_{j-i} T_i$$

und

$$W_j^{(n-1)} = (1-\alpha^{(n-1)}) V_j^{(n-1)}, j = 1, 2, \dots.$$

Im Falle der VmC für U und V ist mit

$$\varphi_j \stackrel{\text{def}}{=} \alpha(-1)^j \frac{p(p+1)\dots(p+j-1)}{a^j} + (1-\alpha)\frac{q(q+1)\dots(q+j-1)}{b^j} - \widetilde{W}_j$$

$$\varphi_j = 0, \quad j = 1, \dots, 5, \tag{5''}$$

ein nichtlineares Gleichungssystem für die Koeffizienten p, a, q, b, α. Die Beschränkungen sind hier

$$p, a, q, b > 0, \ 0 < \alpha < 1. \tag{6''}$$

Eine besonders einfache Möglichkeit, Näherungen für die Wartezeitverteilungen zu berechnen, besteht darin, die Dichten der Pseudowartezeiten überall durch VmE der Ordnung 2 zu approximieren:

$$\widetilde{w}^{(n)}(t) = e^{-\lambda_0 - \lambda_1 t - \lambda_2 t^2},$$

wobei die Koeffizienten nach (10), Abschnitt 1, berechnet werden. Die Momente der Wartezeit sind dann

$$W_j^{(n)} = \int_0^\infty t^j \widetilde{w}^{(n)}(t) dt.$$

Diese Näherungen nennen wir N-Näherungen.

Für die Lösung der Gleichungen (5′) und (5″) benutzen wir ein Newtonverfahren. Um es zu beschreiben, fassen wir die Koeffizienten in einem Tupel zusammen,

$$x = (x_1, \dots, x_K) = \begin{cases} (a, b, \alpha) & \text{für die M-Näherung,} \\ (p, a, q, b, \alpha) & \text{für die } \Gamma\text{-Näherung,} \end{cases}$$

und schreiben die nichtlinearen Gleichungssysteme in der Form

$$\varphi_j(x) = 0, j \in \mathcal{K}, \tag{5}$$

wobei $\mathcal{K} = [1 : K]$ mit $K = 3$ bzw. 5 ist. Die Beschränkungen sind dann

$$x_i > 0, i \in \mathcal{K}, \ x_K < 1. \tag{6}$$

Ausgehend von einer Anfangsnäherung, die die Beschränkungen (6) erfüllt, wird in einem Schritt des Newtonverfahrens ein neuer Wert x_{neu} aus einem alten, x_{alt}, gemäß

$$x_{neu} := x_{alt} - c\delta \tag{8}$$

berechnet.

$\delta = (\delta_1, \dots, \delta_K)$ ist Lösung des lineares Gleichungssystems

$$J\delta = \varphi(x_{alt}), \tag{9}$$

wobei J die Jacobimatrix des Systems an der Stelle x_{alt} ist, und $\varphi = (\varphi_1, \ldots, \varphi_K)$. Erstere berechnen wir näherungsweise mit Differenzenquotienten, wobei die Beschränkungen (6) beachtet werden:

$$J_{j,i} = \frac{\varphi_j(y) - \varphi_j(x)}{y_i - x_i}$$

mit

$$\begin{aligned} x &= x_{alt}, \\ y_k &= x_k, k \in \mathcal{K} - \{i\}, \\ y_i &= (1-\epsilon)x_i, j, i \in \mathcal{K}, \\ \epsilon &> 0, \epsilon << 1. \end{aligned}$$

Die Zahl $c \in (0,1]$ wird so festgelegt, daß x_{neu} die Beschränkungen (6) erfüllt, und daß die Euklidische Norm des Defektes abnimmt,

$$||\varphi(x_{neu})||_E < ||\varphi(x_{alt})||_E. \tag{10}$$

Um ersteres zu erreichen, wird für alle $j \in \mathcal{K}$, für die $x_j - \delta_j \leq 0$ ist,

$$c_j = \frac{x_j - \epsilon}{\delta_j}$$

gesetzt, und falls $x_K - \delta_K \geq 1$,

$$c_{K+1} = \frac{x_K - 1 + \epsilon}{\delta_K},$$

$\epsilon < 0$, $\epsilon << 1$. c' ist dann das Minimum dieser Zahlen.

Die Abnahme des Defektes wird mit

$$c = 2^{-k} c'$$

erreicht, wobei $k \in I\!N_0$ die kleinste Zahl ist, mit der (10) gilt (Bisektion, siehe [ST]).

In den Beispielen haben wir $\epsilon = 10^{-4}$ gewählt und die Bisektion nach 13 Schritten abgebrochen. Den Newtonschritt (8) haben wir solange durchgeführt, bis $||\varphi||_E < 10^{-4}$ war, höchstens 20 mal.

Wichtig für Newtonverfahren ist die Wahl guter Ausgangsnäherungen. Ab dem dritten Kunden haben wir dafür die Koeffizienten des vorausgehenden Kunden genommen. Beim zweiten Kunden wurde für eine Γ-Näherung zunächst eine M-Näherung a, b, α oder eine N-Näherung berechnet, und daraus die Koeffizienten der Ausgangsnäherung ($x_{alt} = (1, a, 1, b, \alpha)$ bei Initialisierung mit einer M-Näherung).

Für die Wartezeit des ersten Kunden haben wir entweder $W^{(1)} = 0$ gewählt oder die Wartezeit $Y^{(s)}$ des "zugeordneten" M/M/1-Systems im statistischen Gleichgewicht; damit meinen wir das M/M/1-System mit denselben Erwartungswerten für die Zwischenankunftszeit und die Bedienzeit.

Für die Berechnung einer M-Näherung des zweiten Kunden haben wir die Ausgangsnäherung aus der Pseudowartezeit des zugeordneten M/M/1-Systems bestimmt: Bei $W^{(1)} = 0$ aus der Pseudowartezeit des zweiten Kunden dieses Systems, ebenfalls leer gestartet, bei $W^{(1)} = Y^{(s)}$ aus der Pseudowartezeit dieses Systems im statistischen Gleichgewicht. Beide Pseudowartezeiten haben exakt die Form (4), (4'), und man kann die Koeffizienten angeben. Für den zweiten Kunden sind diese bei Start mit leerem System

$$a = 1/T_1,\ b = 1/S_1,\ \alpha = \frac{T_1}{T_1 + S_1}, \tag{11}$$

und im stationären Fall

$$a = 1/T_1,\ b = 1/S_1 - 1/T_1,\ \alpha = \frac{T_1 - S_1}{T_1}. \tag{12}$$

Die Auswahl der VmE und VmC zur Approximation hat auch praktische Aspekte. Sicher kommt es der Genauigkeit zugute, wenn viele freie Parameter bestimmt werden. So erhält man bessere Ergebnisse, wenn man statt N-Näherungen (zwei Parameter) M-Näherungen verwendet, und noch bessere mit den Γ-Näherungen (fünf Parameter). Deswegen haben wir überhaupt die VmC herangezogen.

Andererseits bringen größere Ordnungen K, als hier vorgesehen, Probleme mit sich: Die Integrationen zur Berechnung der Momente sind nicht mehr geschlossen möglich, und der Koeffizient bei x_K in (8) und (12), Abschnitt 1, ist oft sehr klein, muß aber genau berechnet werden, allein schon, weil bei falschem Vorzeichen keine Dichte vorliegt. Wir hatten deshalb keinen Erfolg mit größeren K.

Bei den M- und den Γ-Näherungen approximieren wir $\tilde{w}(t)$ über $\bar{\mathbb{R}}_+$ und $\bar{\mathbb{R}}_-$ mit verschiedenen VmE bzw. VmC, um wie gesagt einen Knick von $\tilde{w}(t)$ im Nullpunkt zu berücksichtigen, aber auch um mehr Parameter zu haben; letzteres ist vermutlich wichtiger. Da $\tilde{w}(t)$ stetig ist, liegt es eigentlich nahe, statt (4)

$$\tilde{w}(t) = \frac{ab}{a+b} \begin{cases} e^{at}, & x \leq 0, \\ e^{-bt}, & x > 0, \end{cases}$$

zu nehmen. Damit sind die berechneten Ergebnisse jedoch wesentlich ungenauer, sicherlich weil nur zwei Parameter berücksichtigt werden.

4. Beispiele

Wir haben die beschriebenen Verfahren an Beispielen auch für $\rho = S_1/T_1 > 1$ erprobt und die Resultate mit solchen aus anderen Näherungsverfahren und, wenn möglich, mit den exakten Werten verglichen. Es zeigt sich, daß die ersten K Momente der Wartezeit mit mittleren bis geringen Fehlern berechnet wurden.

Nicht so gut waren die berechneten Werte für die Wahrscheinlichkeiten $P\{W = 0\}$.

Die Tabelle gibt eine Übersicht über eine Reihe von Beispielen. Für vier Werte von ρ wurden je 13 G/G/1-Systeme betrachtet. Für jeden dieser Werte und für verschiedene errechnete Näherungen wird der größte bei den 13 Systemen beobachtete Fehler angegeben.

Größte beobachtete Fehler in %

bei	$\rho = 0.5$	$\rho = 0.6$	$\rho = 0.8$	$\rho = 0.9$
M-Näherungen:				
W_1	33	15.4	6	2.9
W_2	50	21.7	2.2	1.6
W_3	88.9	47.1	5.7	0.7
Γ-Näherungen:				
W_1	3.7	3.4	1.8	0.7
W_2	10	2.5	1.2	0.7
W_3	4.2	0.9	1.1	0.7
W_4	25.3	4.7	1.3	0.7
W_5	73.1	13.2	1.7	0.7
W_{AC}	44.5	42.6	14.9	6.5
W_{KLB}	31.6	25.5	30	35.6
W_{MG}	17.3	13.2	5.9	2.7

Die Zwischenankunftszeiten und die Bedienzeiten hatten Exponentialverteilung (M), Gammaverteilung bzw. Erlangverteilung (Γ_p bzw. E_p) mit Parameter b und p,

$$\frac{b^p}{\Gamma(p)} t^{p-1} e^{-bt},$$

hyperexponentielle Verteilung (H) mit den Parametern λ_1, λ_2, q,

$$q\lambda_1 e^{-\lambda_1 t} + (1-q)\lambda_2 e^{-\lambda_2 t},$$

oder Gleichverteilung (U) über $[0, c]$, $c \in \mathbb{R}_+$. Damit sind die genannten 13 Systeme:
M/G/1 für G ∈ {M, H, U, $\Gamma_{0.5}$, E_4 } und
G/G'/1 für G ∈ {H, U, $\Gamma_{0.5}$, E_4 } und G' ∈ {M, E_2}.

Bei der Betrachtung des statistischen Gleichgewichts wurden so viele Kunden untersucht, bis die Momente der Wartezeit zweier aufeinanderfolgender Kunden sich nur noch wenig unterschieden, genauer: bis $\Delta < 10^{-5}$ mit

$$\Delta \stackrel{\text{def}}{=} \sum_{j \in \mathcal{K}} \left| \frac{W_j^{(n)} - W_j^{(n-1)}}{W_j^{(n)}} \right|.$$

Dabei wurde für die Folge der errechneten Werte jedes Koeffizienten x_j mit der Δ^2-Methode von Aitken (siehe z. B. [ST]) eine transformierte Folge berechnet, die immer deutlich schneller konvergierte.

Die Vergleichsverfahren waren die Näherungsformeln für den Erwartungswert der Wartezeit von Allen und Cunneen, siehe [AC], von Krämer und Langenbach-Belz, siehe [GK], und von Marchall und Gross:

$$W_{AC} = \frac{\rho S_1}{2(1-\rho)}(C_T^2 + C_S^2),$$

$$W_{KLB} = \max\{0, \frac{\rho(2-\rho)\sigma_T^2 + 2\sigma_S^2}{4T_1(1-\rho)} - S_1/4\}$$

bzw.

$$W_{MG} = \frac{1 + C_S^2}{1/\rho^2 + C_S^2} \frac{\sigma_T^2 + \sigma_S^2}{2T_1(1-\rho)}$$

mit

$$\sigma_T^2 = T_2 - T_1^2,\ \sigma_S^2 = S_2 - S_1^2,$$
$$C_T^2 = \sigma_T^2/T_1^2,\ C_S^2 = \sigma_S^2/S_1^2.$$

Wir wollen noch Ergebnisse für ein $\Gamma_{0.5}$ / E_2 / 1 - System im einzelnen angeben:
$T_1 = 1,\ S_1 = 0.6,\ \rho = 0.6$
Momente der Wartezeit im statistischen Gleichgewicht:

$W_1 = 1.29$ $W_2 = 4.39$ $W_3 = 22.32$ $W_4 = 151.2$ $W_5 = 1281$

$P\{W = 0\} = 0.274$

M-Näherungen:

$x_1^{(2)} = 0.597$	$x_2^{(2)} = 2.546$	$x_3^{(2)} = 0.384$
$W_1^{(2)} = 0.242$	$W_2^{(2)} = 0.190$	$W_3^{(2)} = 0.224$
$x_1^{(3)} = 0.584$	$x_2^{(3)} = 1.729$	$x_3^{(3)} = 0.322$
$W_1^{(3)} = 0.392$	$W_2^{(3)} = 0.454$	$W_3^{(3)} = 0.787$
$x_1^{(77)} = 0.597$	$x_2^{(77)} = 0.610$	$x_3^{(77)} = 0.239$
$W_1^{(AIT)} = 1.25$	$W_2^{(AIT)} = 4.11$	$W_3^{(AIT)} = 20.3$

Die Momente $W_j^{(AIT)}$ wurden aus den o.g. Aitken-Folgen berechnet. Ihre relativen Fehler gegenüber den exakten Werten sind:

3% 6.8% 10.1%

$|P\{W = 0\} - x_3^{(77)}| = 0.035$

Γ-Näherungen:

$x_1^{(2)} = 0.609$	$x_2^{(2)} = 0.509$	$x_3^{(2)} = 2.450$	$x_4^{(2)} = 4.040$	$x_5^{(2)} = 0.558$
$W_1^{(2)} = 0.268$	$W_2^{(2)} = 0.229$	$W_3^{(2)} = 0.252$	$W_4^{(2)} = 0.340$	$W_5^{(2)} = 0.542$
$x_1^{(3)} = 0.630$	$x_2^{(3)} = 0.510$	$x_3^{(3)} = 2.488$	$x_4^{(3)} = 3.000$	$x_5^{(3)} = 0.465$
$W_1^{(3)} = 0.443$	$W_2^{(3)} = 0.515$	$W_3^{(3)} = 0.771$	$W_4^{(3)} = 1.410$	$W_5^{(3)} = 3.048$
$x_1^{(109)} = 0.716$	$x_2^{(109)} = 0.521$	$x_3^{(109)} = 1.101$	$x_4^{(109)} = 0.611$	$x_5^{(109)} = 0.291$
$W_1^{(AIT)} = 1.28$	$W_2^{(AIT)} = 4.39$	$W_3^{(AIT)} = 22.3$	$W_4^{(AIT)} = 149.8$	$W_5^{(AIT)} = 1252$

Die relativen Fehler der Momente $W_j^{(AIT)}$ sind:

0.9% 0.1% 0.1% 1% 2.4%

$|P\{W=0\} - x_5^{(109)}| = 0.017$

Relativer Fehler bei W_{AC}: 12.7%, W_{KLB}: 12.7%, W_{MG}: 3.2%.

Bei $\rho \leq 0.6$ hatte in seltenen Fällen für einen einzelnen Kunden das Iterationsverfahren zur Lösung des nichtlinearen Gleichungssystems (5), Abschnitt 3, keinen Erfolg; wir haben dann die Werte dieses Kunden aus den N-Näherungen berechnet, die immer angegeben werden können. Numerische Schwierigkeiten traten sonst gelegentlich auf, wenn die Momente sehr verschieden groß waren. Diese konnten durch eine Skalierung der Zeiteinheit überwunden werden. Außerdem dürfen die Koeffizienten nicht kleiner als ϵ sein. Das kann vorkommen, wenn die Verkehrsintensität größer als eins ist und das System volläuft: Die Wahrscheinlichkeit $P\{W^{(n)} = 0\}$ und damit $\alpha = x_K$ wird beliebig klein, und die Gleichungen (5) des Abschnitt 3 hängen nicht mehr von a und p ab. In diesem Fall können jedoch die Momente der Wartezeiten leicht aus $W_j^{(n)} \approx W_j^{(n-1)} + C$ berechnet werden, mit (7), Abschnitt 3, oder als N-Näherungen.

Schlußbetrachtung

Wir haben in der vorliegenden Arbeit aufgezeigt, wie ein spezieller Markoff-Prozeß, nämlich der der Wartezeiten in einem G/G/1-System, für mittlere und hohe Auslastung analysiert werden kann, indem die Dichten der "Pseudowartezeit" näherungsweise durch VmE oder VmC dargestellt werden. Sie werden aus einigen Momenten berechnet. Somit werden die Dichten immer durch wenige Parameter beschrieben, wechselweise durch die Parameter der VmE bzw. VmC und durch Momente, je nachdem, was damit gemacht werden soll: Für Faltungen sind Momente geeignet, für die Berechnung bedingter Momente die Verteilungen.

Es wurde gezeigt, daß man auf diese Weise die Systeme genau und detailliert analysieren kann.

Es bleibt eine Reihe von Fragen, die unbefriedigend oder gar nicht geklärt sind und somit weiter bearbeitet werden könnten: Wie groß sind die Fehler, die mit den VmE gemacht werden? Wie kann man die große Ungenauigkeit bedingter Wahrscheinlichkeiten bei VmE vermeiden? Sind die Iterationen konvergent? Kann durch Approximation mit anderen Funktionssystemen das Lösen nichtlinearer Gleichungen vermieden werden, wie bei den N-Näherungen? Wenn nein, wie steht es mit der Existenz und der Eindeutigkeit der Lösungen?

Literatur

[AC] A.O. Allen: Probability, Statistics, and Queueing Theory. Academic Press, New York, 1978

[AK] M.A. El-Affendi and D.D. Kouvatsos: A Maximum Entropy Analysis of the M/G/1 and G/M/1 Queueing Systems at Equilibrium. Acta Informatica 19 (1983), S.339

[GK] Gnedenko, König: Handbuch der Bedienungstheorie, Band 2. Akademie-Verlag, Berlin, 1984

[HA] G. Haßlinger: Das Näherungsverfahren nach dem Prinzip der maximalen Entropie für Warteschlangen-Verteilungen in geschlossenen Netzwerken. Dissertation, Technische Hochschule Darmstadt, 1986

[KL] L. Kleinrock: Queueing Systems, Vol. 1, J. Wiley, New York, 1975

[K1] D.D. Kouvatsos: A Universal Maximum Entropy Algorithm for the Analysis of General Closed Networks. In: T.Hasegawa, H.Takagi, Y.Takahashi (Hrsg.), Computer Networking and Performance Evaluation, Elsvier Science Publishers, Amsterdam, 1986, S. 113

[K2] D.D. Kouvatsos: A Maximum Entropy Queue Length Distribution for a G/G/1 Finite Capacity Queue. ACM Performance Evaluation Review, Vol.14, No.1 (1986), S. 224

[S] J. E. Shore: Information Theoretic Approximations for M/G/1 and G/G/1 Queueing Systems. Acta Informatica 17, 1982, S. 43

[SJ] J. E. Shore and R. W. Johnson: Axiomatic Derivation of the Principle of Maximum Entropy and the Principle of Minimum Cross-Entropy. IEEE Trans. Inf. Theory, Vol. IT-26, No. 1, Jan 1980, S. 26-37

[SM] W. J. Smirnow: Lehrgang der höheren Mathematik, Teil IV, Berlin 1961.

[ST] J. Stoer: Einführung in die Numerische Mathematik. Springer, Berlin, Heidelberg, New York, 1976

[TZ] H. Tzschach: The Principle of Maximum Entropy applied to Queueing Systems. Bericht TI-2, Fachbereich Informatik, Technische Hochschule Darmstadt, 1983.

[T] H. Tzschach: Information Theoretic Approximations of Queueing Systems and Queueing Networks. Bericht des Fachbereichs Informatik, Technische Hochschule Darmstadt, 1986

[W] R.J. Walstra: Nonexponential Networks of Queues: A Maximum Entropy Analysis. ACM Sigmetrics, 1985, S.27

Performance Analysis and Optimal Control of an M/M/1/k Queueing System with Impatient Customers

Peter R. de Waal

Centre for Mathematics and Computer Science
P O Box 4079, 1009 AB Amsterdam, The Netherlands

A simple M/M/1/k queue with impatient customers is presented as a model for communications systems operating under overload conditions The performance analysis and optimal control problem for this model are discussed An efficient algorithm for computing the optimal control is presented along with numerical results

Key Words & Phrases Communication systems, queues, impatient customers, stochastic control
1980 Mathematics Subject Classification 60K25, 68M20, 90B22, 93E20

1. Introduction

In this paper we discuss the performance analysis and optimal control of an M/M/1/k queueing system with impatient customers. The motivation of this investigation is the problem of overload for communication systems.

The most familiar communication system is the telephone network. The operational units in this network are the so-called telephone-switches or -exchanges. During the last few years sophisticated exchanges of the *Stored Program Controlled (SPC)* type have been developed and installed. Most SPC-exchanges are composed of several modules: the *Central Module (CM)*, the *Switching Module (SM)* and one or more *Peripheral Modules (PM)* (for example see figure 1.1). The Central Module takes care of the overall control functions of the system, the Peripheral Module is the interface to the various types of digital and analog user equipment and the Switching Module connects the modules to each other.

The operations in the Central Module are carried out by one or more processors according to a stored program. Besides typical system operations, such as error handling, maintenance and I/O functions, the main part of the workload of the processor is initiated through subscribers' call requests. Examples of such operations are generating a dial tone, receiving the requested phone number digits, checking the validity of the received digits and allocating the available hardware resources.

An SPC-exchange in operation is a typical example of a queueing system where customers compete for a number of limited resources. The limitations stem from both the finite processor capacity and the limited number of hardware resources in the exchange. The performance of an SPC-exchange may therefore degrade significantly during periods in which the demands for service exceed the design capacity (cf. [6]). The response time of tasks scheduled for the processor may become relatively long and this may cause impatient customers to abandon their call request prematurely. Another type of impatient behaviour arises from the use of time-out mechanisms, for instance in the search for free allocatable hardware resources such as senders and receivers. When the time limit that was set for such an operation, expires, a call request may be abandoned before the connection is established. In both cases processor capacity and memory space are wasted on tasks that do not lead to a successful connection. The aim of this investigation is to develop a control mechanism that regulates admission to the exchange to maximize the successful throughput under conditions of overload.

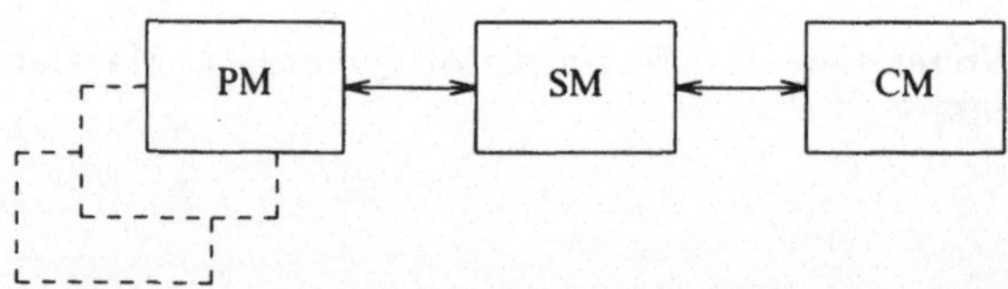

FIGURE 1.1. A Stored Program Controlled Exchange.

In [12] an M/G/1 queue with batch arrivals and service time discretization is presented as a model for switching systems. The successful throughput is expressed in the steady-state probabilities and a suggestion for an overload control is given. In [11] an approximate analysis is presented of an M/G/c queue where customers' call requests become successful according to a probability distribution that is dependent on the waiting time of a customer. A GI/G/1 queue is introduced in [1] with limitations on the waiting and sojourn times of customers. Functional equations for the distribution functions of waiting times and stability conditions are established. The model we present differs from the models in [1], since we allow impatient customers to offer some workload to the server even if they abandon the queue. In [2, 3] models for call request processing and the stochastic control problem for these models are presented.

In this paper we present a simple queueing system model of an SPC-exchange with impatient customers. In Section 2 we give a description of the model. It consists of an M/M/1/k queueing system where customers are served according to the First-Come-First-Served discipline. A newly arriving customer is admitted only if the waiting space of the queue is not fully occupied upon arrival, otherwise he is lost. Once a customer has been accepted, his sojourn time limit is generated according to a general probability distribution, and the customer joins the queue if the server is busy or is served immediately otherwise. Subsequently the customer stays in the queueing system until his service is completed. The service completion of a customer is defined to be *successful* if the actual sojourn time of the customer has not exceeded his sojourn time limit. Our *objective* is to maximize the call completion rate which is defined as the mean number of successful service completions in equilibrium per time unit. In section 3 this performance measure is expressed as a function of the steady-state probabilities of an M/M/1/k queueing system and its dependence of the parameters is discussed. In section 4 we introduce the optimal control problem for an approximating queueing system and derive the Hamilton-Jacobi equations for this problem. It is shown that the optimal control is *bang-bang*, i.e. arriving customers are either accepted or rejected without randomization. Furthermore sufficient conditions are given to ensure that the optimal control is of the form where newly arriving customers are accepted if and only if the current number of customers does not exceed a certain level. In section 5 an efficient algorithm for computing this optimal queue size is given and some numerical results are presented. We conclude this paper with some remarks and suggestions for further study.

ACKNOWLEDGEMENTS
I would like to thank J.H. van Schuppen for his help and support in the preparation of this paper.
These investigations are supported by the Technology Foundation (STW).

2. Description of the Model.

In this section we propose a model for analysing the influence of impatient behaviour of customers in a simple queueing system. The service center consists of one server and *k-1* waiting places. Customers are served in order of arrival and their service demands are exponentially distributed with mean $1/\mu$. Arrivals are generated by a Poisson process with rate λ. An arriving customer is admitted to the queueing system if the waiting space is not fully occupied, otherwise he is rejected. Rejected customers are assumed not to return.

We now extend this ordinary M/M/1/k queueing system with a concept of impatience. When a customer is admitted to the queueing system, he generates a random sojourn time limit with distribution function F_S (to be discussed later in this section) on the set of positive real numbers with mean $1/\sigma$. All random variables are assumed to be independent. Subsequently he joins the queue, if the server is busy, or is served immediately otherwise. At the moment of his service completion his actual *sojourn time* - i.e. waiting time plus service time - is compared with the sojourn time limit that was set at the time of his arrival. If the actual sojourn time has not exceeded this limit, then the service completion is called *successful.*

We can view this M/M/1/k queueing system with impatient customers alternatively as a queueing system consisting of two parallel queues (see figure 2.1). Queue 1 is an M/G/∞ queue used to model the sojourn of customers and queue 2 is an M/M/1/k queue for service demands of customers. Both queues have identical arrival processes. Upon completion of a task queue 1 is examined to check if the customer who generated the task is still present. If so then the service completion is successful.

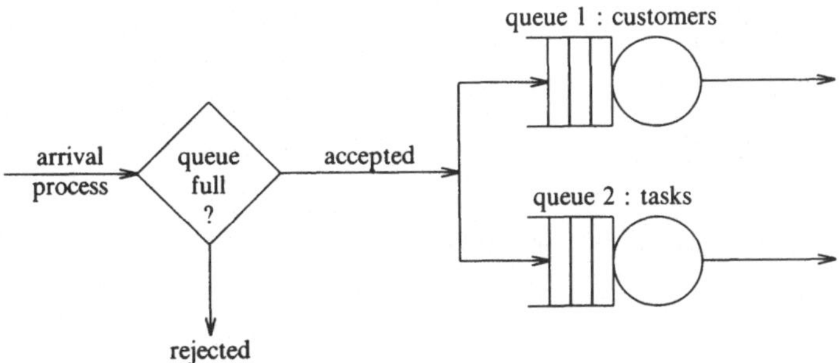

FIGURE 2.1. An M/M/1/k queueing system with impatient customers.

Note that, although the sojourn time limit may already have been exceeded while waiting in the queue, the customer still remains in the queueing system to have his service demand fullfilled. This approach has been chosen, since in telephone exchanges even abandoned call requests offer some load to the processor. Furthermore, note that the time limit may be exceeded while the customer is being served. This is to account for the fact that a calling subscriber is not aware of the moment his service begins (e.g. when he is waiting for a dial tone, the exchange is executing tasks, although the subscriber is not aware of this).

We conclude this description with a remark about the probability distribution function F_S of the sojourn time limit. We consider only *Erlang* and *deterministic* distributions, mainly because the Erlang-3 distribution seems a reasonable choice for describing customer impatience in telephone exchanges [1] and the deterministic distribution because this is convenient for describing time-out mechanisms.

3. Performance Analysis.

In this section we give a performance analysis of the queueing system we introduced in the preceding section. The performance measures we are interested in are the call completion rate and the rejection probability.

The *rejection probability* R is simply the probability that an arbitrary arriving customer will be rejected, or equivalently finds the queue fully occupied. This measure will become of interest when considering so-called reattempts, i.e. attempts by customers to reenter the communication system after having been refused access. In this report we will however not consider reattempts.

The *call completion rate* Λ is defined as the throughput of successful service completions, i.e. the mean number of successful service completions in equilibrium per time unit. We will express both measures in the equilibrium probabilities of an M/M/1/k queue.

In this section we will take the viewpoint of the queueing system as in figure 2, i.e. customers have a *sojourn* time probability distribution function F_S and tasks have a *service* time with an exponential distribution with mean $1/\mu$. Let $p(n)$, $n=0,...,k$, denote the probability that n tasks are present in the queue in equilibrium. It is well known that $p(n)$ is given by

$$p(n)=\frac{1-\rho}{1-\rho^{k+1}}\rho^n, \qquad n=0,\cdots,k, \tag{3.1}$$

where $\rho=\lambda/\mu$.

Due to *PASTA (Poisson Arrivals See Time Averages)* [13] the rejection probability R is equal to $p(k)$, i.e. the probability that k tasks are present.

When considering the call completion rate it is more convenient to look at arriving customers (or tasks) rather than departing tasks. Suppose a customer arrives and is accepted in the queue, where he finds $n, n=0,...,k-1$ tasks in front of him. The sojourn time of his task is the sum of $n+1$ independent exponentially distributed random variables, each with mean $1/\mu$, or equivalently an Erlang-(n+1) distributed random variable with mean $(n+1)/\mu$. Denote this variable by S_{n+1}. Let S denote the random variable corresponding to the sojourn time of the customer, i.e. a random variable with probability distribution function F_S and mean $1/\sigma$. Furthermore let $q(n)$ denote the probability that the sojourn time of the customer exceeds the sojourn time of his task. It is clear that $q(n)=\mathrm{P}\{S>S_{n+1}\}$. Since in equilibrium the mean number of successful service completions per time unit equals the mean number of arriving customers per time unit whose sojourn times exceed the sojourn times of their corresponding tasks, we have

$$\Lambda=\lambda\sum_{n=0}^{k-1}p(n)q(n). \tag{3.2}$$

The following lemma gives expressions for the probabilities $q(n)$.

Lemma 2.1. *If S has an Erlang-m distribution then*

$$q(n)=1-\left[\frac{m\sigma}{m\sigma+\mu}\right]^m\sum_{j=0}^{n}\binom{m+j-1}{j}\left[\frac{\mu}{m\sigma+\mu}\right]^j, \qquad n=0,\cdots,k-1. \tag{3.3}$$

If S has a deterministic distribution then

$$q(n)=1-e^{-\mu/\sigma}\sum_{j=0}^{n}\frac{\mu^j}{\sigma^j j!}, \qquad n=0,\cdots,k-1. \tag{3.4}$$

Proof

Let F_S be an Erlang-m distribution function with mean $1/\sigma$, i.e.

$$F_S(t)=\frac{1}{(m-1)!}\int_0^t(m\sigma)^m x^{m-1}e^{-m\sigma x}\,dx, \qquad t\geqslant 0 \tag{3.5}$$

and $F_{S_{n+1}}$ an Erlang-(n+1) distribution function with mean $(n+1)/\mu$, i.e.

$$F_{S_{n+1}}=\frac{1}{n!}\int_0^t\mu^{n+1}x^n e^{-\mu x}\,dx, \qquad t\geqslant 0 \tag{3.6}$$

We will prove by induction that

$$q(n)=\begin{cases} 1-\left[\dfrac{m\sigma}{m\sigma+\mu}\right]^m, & n=0 \\ q(n-1)-\binom{n+m-1}{n}\left[\dfrac{m\sigma}{m\sigma+\mu}\right]^m\left[\dfrac{\mu}{m\sigma+\mu}\right]^n, & n>0 \end{cases} \tag{3.7}$$

For $n=0$ we have

$$\begin{aligned} q(0) &= \int_0^\infty \int_0^t dF_{S_1}(x)\,dF_S(t) \\ &= \frac{1}{(m-1)!}\int_0^\infty \int_0^t \mu e^{-\mu x}(m\sigma)^m t^{m-1} e^{-m\sigma t}\,dx\,dt \\ &= 1-\frac{1}{(m-1)!}\int_0^\infty (m\sigma)^m t^{m-1} e^{-(m\sigma+\mu)t}\,dt \\ &= 1-\left[\frac{m\sigma}{m\sigma+\mu}\right]^m \end{aligned}$$

Let $n>0$, then

$$\begin{aligned} q(n) &= \int_0^\infty \int_0^t dF_{S_{n+1}}(x)\,dF_S(t) \\ &= \frac{1}{n!(m-1)!}\int_0^\infty \int_0^t \mu^{n+1} x^n e^{-\mu x}(m\sigma)^m t^{m-1} e^{-m\sigma t}\,dx\,dt \\ &= q(n-1)-\frac{(m\sigma)^m \mu^n}{n!(m-1)!}\int_0^\infty t^{n+m-1} e^{-(m\sigma+\mu)t}\,dt \\ &= q(n-1)-\frac{(n+m-1)!}{n!(m-1)!}\left[\frac{m\sigma}{m\sigma+\mu}\right]^m\left[\frac{\mu}{m\sigma+\mu}\right]^n. \end{aligned}$$

From (7) one can easily derive (3).
Let S be equal to $1/\sigma$ (deterministic), then for $n=0$ we have

$$\begin{aligned} q(0) &= \int_0^{1/\sigma} \mu e^{-\mu x}\,dx \\ &= 1-e^{-\mu/\sigma}. \end{aligned}$$

For $n>0$ we have

$$\begin{aligned} q(n) &= \frac{1}{n!}\int_0^{1/\sigma} \mu^{n+1} x^n e^{-\mu x}\,dx \\ &= q(n-1)-e^{-\mu/\sigma}\frac{\mu^n}{n!\,\sigma^n}. \end{aligned}$$

□

With the expressions (1)-(4) we can easily evaluate the influence of the parameters of the model on Λ and R. Some numerical examples are presented in section 5.

4. Optimal Control

In this section we discuss the optimal control of an M/M/1 queueing system with impatient customers. We first introduce some definitions and notations of counting processes and subsequently discuss the optimal control of an M/M/1 queue with a general reward structure. A queueing system with impatient customers is then treated as a special example.

4.1. Counting Processes.

We let N denote the set of natural numbers $\{0,1,2,...\}$, $(\Omega,\mathcal{F},P)$ be some probability space and $(\mathcal{F}_t, t\geq 0)$ a family of increasing σ-algebra's on $\mathcal{F}$, i.e. for all $0\leq s\leq t$ $\mathcal{F}_s\subset\mathcal{F}_t\subset\mathcal{F}$. A stochastic process $(X_t, t\geq 0)$ is called *adapted* to $\mathcal{F}_t$ if for all $t>0$ $\mathcal{F}_t^X=\sigma(X_s, s\in[0,t])\subset\mathcal{F}_t$. In the sequel all stochastic processes are assumed to be adapted to $\mathcal{F}_t$.

A *counting process* $(n_t, t \geq 0)$ is a process taking values in N that has unit jumps. An example of a counting process is a *Poisson process*, where the intervals between successive jumps are independent and exponentially distributed with a constant parameter.

In most applications counting processes can be represented as

$$n_t = \int_0^t \lambda_s \, ds + m_t, \qquad t \geq 0$$

or

$$dn_t = \lambda_t \, dt + dm_t, \qquad t \geq 0$$

where $(\lambda_t, t \geq 0)$ is a nonnegative process adapted to $\mathcal{F}_t$ and $(m_t, t \geq 0)$ a $(P, \mathcal{F}_t)$-*martingale*. The process $(\lambda_t, t \geq 0)$ is called the *rate* or *intensity* process of $(n_t, t \geq 0)$. For example a $(P, \mathcal{F}_t)$-Poisson process is a counting process with a deterministic rate process $(\lambda(t), t \geq 0)$. Finally a counting process $(n_t, t \geq 0)$ is called *non-explosive* if

$$n_t < \infty, \qquad t \geq 0, \ \mathrm{P}-a.s.$$

A thorough treatment of counting processes can be found in [5].

4.2. The optimal control of an M/M/1-queueing system.
Consider the M/M/1-queueing system as shown in figure 4.1.

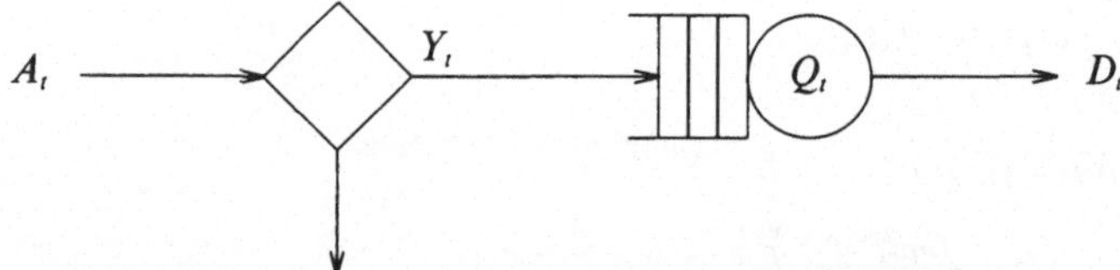

FIGURE 4.1. An M/M/1-queueing system with regulated arrivals.

The number of customers present in the queue at time t is denoted as Q_t. This queueing process $(Q_t, t \geq 0)$ is an N-valued process defined on $(\Omega, \mathcal{F}, P)$ and is of the form

$$Q_t = Q_0 + Y_t - D_t, \qquad t \geq 0 \tag{4.1}$$

where Y_t and D_t are non-explosive counting processes *without common jumps*. The above definition implies that $\mathrm{P}-a.s.$, $D_t \leq Q_0 + Y_t$, $t \geq 0$ since Q_t is nonnegative. Q_t is called the *state process* and Q_0 is the *initial state*. For each $t \geq 0$ Y_t is the number of admitted arrivals in $(0,t]$ and D_t the number of departures in $(0,t]$. These processes are constructed as follows.

Customers arrive according to a point process $(A_t, t \geq 0)$ which admits a constant deterministic intensity $\lambda > 0$. Let $\{t_n\}_{n \in N}$ denote the sequence of stopping times that correspond to arrival times, then

$$0 < t_1 < t_2 < \cdots < \infty$$

Upon arrival customers may be admitted to the queue or refused access. Let $\{X_n\}_{n \in N}$ be a sequence of $\{0,1\}$-valued random variables, where if $X_n = 1$, then the customer arriving at time t_n is admitted to the queueing system; otherwise he is refused access. The number of admitted customers over the interval $(0,t]$ is represented by the counting process $(Y_t, t \geq 0)$ where

$$Y_t = \sum_{n \geq 1} X_n I_{(t_n \leq t)}. \tag{4.2}$$

The service times $\{s_n\}_{n \in N}$ of customers at the queue are assumed to be a sequence of i.i.d. random variables which are independent of A_t (and hence of Y_t), but which may depend on Q_t. If we let the service time of a customer, when k customers are present, be exponentially distributed with mean $1/\mu_k > 0$, then the departure process D_t is a counting process that admits the intensity $\mu_{Q_{t-}}$, where $\mu_0 = 0$.

For the information patterns of these counting processes let $\mathcal{F}_s^A = \sigma(A_s, 0 \leq s \leq t)$, $\mathcal{F}_s^Q = \sigma(Q_s, 0 \leq s \leq t)$ and $\mathcal{F}_t = \mathcal{F}_t^A \vee \mathcal{F}_t^Q$. Denote $G_n = \mathcal{F}_{t_n-}^Q \vee \mathcal{F}_{t_n}^A$, $n \geq 1$, where

$$\mathcal{F}_{t_n-}^Q = \sigma(A_s \cap \{s < t_n\} ; A_s \in \mathcal{F}_s^Q, s \geq 0) \tag{4.3}$$

and

$$\mathcal{F}_{t_n}^A = \{A \in \mathcal{F} | A \cap \{t_n \leq t\} \in \mathcal{F}_t^A\}. \tag{4.4}$$

We can interpret G_n as the cumulative information of $(A_t, t \geq 0)$ up to the instant t_n as well as that of $(Q_t, t \geq 0)$

up to the instant right before t_n. Based on the information pattern G_n, $n \geq 1$, the admissible admission policies can be defined as follows.

DEFINITION 4.1. *An $\mathcal{F}_t$-predictable process $(u_t, t \geq 0)$ with $u_t \in [0,1]$ is said to be an admissible admission policy if there exists a probability measure P^u on $(\Omega, \mathcal{F})$ such that $(A_t, t \geq 0)$, $(D_t, t \geq 0)$ and $(Y_t, t \geq 0)$ are counting processes with $(P^u, \mathcal{F}_t)$-intensities λ, $\mu_{Q_{t-}}$ and λu_t respectively.*

The class of all admissible (random) admission policies will be denoted by U. With this definition and t_n an arrival time u_{t_n} can be interpreted as the probability of admitting the nth arriving customer given the information G_n. Moreover definition 4.1 ensures that any admissible policy u induces a probability measure P^u under which the original distributions of $(A_t, t \geq 0)$ and $(D_t, t \geq 0)$ are not altered. For details on this approach see [4, 5 VII].

Consider now the following optimal admission problem.

PROBLEM P. *Given $Q_0 = x \in N$, $\alpha > 0$, and a function $g: N \to [0, \infty)$, find a $u^* \in U$ such that*

$$J_x^\alpha(u^*) = \sup_{u \in U} J_x^\alpha(u) \tag{4.5}$$

where

$$J_x^\alpha(u) = \lim_{t \to \infty} E^u\left[\int_0^t e^{-\alpha s} g(Q_{s-})\, dD_s\right] \tag{4.6}$$

and E^u denotes the expectation with respect to P^u.

Note that problem P is formulated as an optimization problem for a discounted reward, although the call completion rate is formulated as a performance measure of the queueing system in equilibrium, i.e. a time-average reward. We study discounted rewards, however, mainly because the optimal control for this problem can easily be solved and because the average reward can be treated as the limit of the discounted reward as $\alpha \downarrow 0$.

The function g corresponds with a reward associated with the departure of customers. One can view $g(k)$ as the probability that a service completion (or departure) is successful, given there are k customers in the service center immediately before the departure. In principle this probability can be determined exactly, but it will be a function of the past of the state process. Therefore it is approximated by

$$g(k) = I_{(k>0)} \left(\frac{\mu_{k-1}}{\mu_{k-1} + \sigma}\right)^{k-1}, \qquad k \in N \tag{4.7}$$

where $1/\sigma$ is the mean sojourn time of a customer as defined in section 3. Equation (4.7) may be a reasonable approximation, because in equilibrium a departing customer on the average leaves behind as many customers as he faced on arrival. Considering this we can approximate $g(k)$ by the probability that a customer's service will become successful, given he finds $k-1$ customers in the queue at his arrival epoch, i.e. $q(k-1)$ as defined in section 3. Since $g(Q_{t-})$ is a non-negative $\mathcal{F}_t$-predictable process, (4.6) can also be written as

$$J_x^\alpha(u) = \lim_{t \to \infty} E^u\left[\int_0^t e^{-\alpha s} g(Q_{s-})\mu_{Q_{s-}}\, ds\right]$$

$$= \lim_{t \to \infty} E^u\left[\int_0^t e^{-\alpha s} g(Q_s)\mu_{Q_s}\, ds\right]$$

since D_t admits the $(P^u, \mathcal{F}_t)$-intensity $\mu_{Q_{t-}}$.

Note that in the original description of the problem, the admission policy was of the impulsive control type, meaning that a decision X_n had to be made at time t_n for each $n \geq 1$. With an appropriate transformation (see e.g. [5] VII.3) the optimal admission problem is equivalent to the intensity control problem formulated above.

4.3. The dynamic programming equation.

In this subsection sufficient conditions for an optimal admission policy for problem P are given. These conditions are expressed in terms of a dynamic programming equation. With this equation we show that the optimal control is of *bang-bang* type, i.e. new customers are either accepted or rejected without randomisation. Furthermore we give sufficient conditions for the optimal control being of the type where new customers are admitted if and only if the present number of customers does not exceed a certain number (like window flow-control in communication networks, cf [8]).

First we define the following two transition maps A, $D: N \to N$ by $Ak = k+1$ and $Dk = \max(0, k-1)$ respectively. Admission of a customer at time t then corresponds to a transition $Q_{t-} \to A\, Q_{t-}$ and a departure corresponds to a transition $Q_{t-} \to D\, Q_{t-}$.

The next theorem, which is presented without proof, gives sufficient conditions for the optimal admission policy.

THEOREM 4.2. *(Dynamic programming equation) If the function $V: N \to [0, \infty)$ solves the following equation*

$$0 = -\alpha V(k) + g(k) + \lambda \max_{u \in [0,1]} \{ u [V(Ak) - V(k)] \} + \mu_k [V(Dk) - V(k)], \tag{4.8}$$

then $J_x^\alpha(u^*) = V(x)$ and the optimal control u^* is given by

$$u_t^* = \begin{cases} 1 & \text{if } V(AQ_{t-}) - V(Q_{t-}) > 0 \\ 0 & \text{if } V(AQ_{t-}) - V(Q_{t-}) \leq 0 \end{cases} \tag{4.9}$$

The solution V is called the *value function* of problem P. The proof of the theorem is analogous to that of ([4] Lemma 3), and can be found by standard dynamic programming techniques. It can be shown that the solution to equations (4.8) and (4.9) exists and is unique (cf. [10]). Furthermore one can see that the optimal control is bang-bang, since the term in (4.8) that has to be maximized is linear in u. From (4.9) one can also see that the optimal control value depends only on the number of customers present. We can therefore also represent the optimal control as a control law $u^*: N \to \{0, 1\}$, which uses only the state of the queueing process. From now on we won't distinguish between the control process and the control law.

Although theorem 4.2 gives us a sufficient condition for the optimal control, solving the equations is a formidable computational task. Firstly one has to impose a sufficiently large maximum queue size, say l, to avoid dealing with an infinite set of equations. Secondly, finding the optimal policy amounts to proposing a possible control $\tilde{u}$ (i.e. a function $\tilde{u}: N_l \to \{0, 1\}$, a total of 2^{l+1} possibilities), solving equation (4.8) with $\max_{u \in [0,1]} u [V(Ak) - V(k)]$ replaced by $\tilde{u}_k [V(Ak) - V(k)]$ and checking the solution $\tilde{V}(k)$ whether it is consistent with the control $\tilde{u}$, i.e. $\tilde{u}_k = 1$ if and only if $\tilde{V}(Ak) - \tilde{V}(k) > 0$. Without any knowledge about the optimal control, finding the optimal control in the worst case amounts to solving 2^{l+1} sets of $l+1$ linear equations.

From these considerations we may deduce that any preliminary knowledge about the structure or form of the optimal control might be very useful in finding the optimal control. For instance, if we know that the optimal control is of the type $u_t^* = I_{(Q_{t-} < l^*)}$ for some l^*, $0 \leq l^* \leq l$, then we only have to solve $l+1$ sets of $l+1$ equations, a significant reduction in computational effort.

The following lemma gives sufficient conditions for the optimal control being of this type.

LEMMA 4.3. *Let $u^*: N \to \{0, 1\}$ be the optimal control for problem P.*

(i) *If there exists an $l_1 \in N$ such that for all k, $1 \leq k \leq l_1$, $g(k) \geq g(k-1)$, then for all k, $0 \leq k < l_1$, we have $u_k^* = 1$.*

(ii) *If there exists an $l_2 \in N$ such that $u_{l_2}^* = 0$ and for all k, $k \geq l_2$, $g(k) \geq g(k+1)$, then for all k, $k \geq l_2$, we have $u_k^* = 0$.*

PROOF. Both proofs are given by complete induction.

Let $V^*: N \to R$ be the solution of (4.8) and (4.9) corresponding to the optimal control u^*. Define $x^*: N \to R$ by

$$x^*(k) = \begin{cases} V^*(0) & , k = 0 \\ V^*(k) - V^*(k-1) & , k > 0. \end{cases} \tag{4.10}$$

Equations (4.8) and (4.9) can now be written as

$$0 = -\alpha \sum_{i=0}^{k} x^*(i) + g(k) + \lambda u_k^* x^*(k+1) - \mu_k x^*(k), \quad k \in N \tag{4.11}$$

and

$$u_k^* = \begin{cases} 1 & \text{if } x^*(k+1) > 0 \\ 0 & \text{if } x^*(k+1) \leq 0. \end{cases} \tag{4.12}$$

(i) Let $l_1 \in N$ be such that for all k, $1 \leq k \leq l_1$, $g(k) > g(k-1)$. We have to prove that $x^*(k) > 0$ for all k, $1 \leq k \leq l_1$. Equation (4.11) for $k = 0$ and $k = 1$ reads

$$0 = -\alpha x^*(0) + g(0) + \lambda u_0^* x^*(1) \tag{4.13}$$

$$0 = -\alpha (x^*(0) + x^*(1)) + g(1) + \lambda u_1^* x^*(2) - \mu_1 x^*(1) \tag{4.14}$$

respectively. Subtracting (4.13) from (4.14) gives

$$\lambda (u_0^* x^*(1) - u_1^* x^*(2)) = -\alpha x^*(1) + (g(1) - g(0)) - \mu_1 x^*(1). \tag{4.15}$$

Suppose now that $u_0^* = 0$, or, equivalently, $x^*(1) \leq 0$. Since $u_1^* x^*(2) \geq 0$, we then have the left-hand side of (4.15) smaller than or equal to zero and the right-hand side strictly greater than zero. From this contradiction we may deduce that $u_0^* = 1$. The proof of $u_k^* = 1$ for k, $1 \leq k < l_1$ proceeds in a similar way.

(ii) Let $l_2 \in N$ be such that $u_{l_2}^* = 0$, or, equivalently, $x^*(l_2 + 1) \leq 0$, and $g(k) \geq g(k+1)$ for all k, $k \geq l_2$.

Define $x : N \to R$ by

$$x(k)=\begin{cases} x^*(k) & 0 \leqslant k \leqslant l_2 \\ [g(k)-\alpha \sum_{i=0}^{k-1} x(i)][\alpha+\mu_k]^{-1} & k > l_2 \end{cases} \tag{4.16}$$

We first prove that $x(k) \leqslant 0$ for $k > l_2$.

(ii.a) Let $k = l_2 + 1$. From the definition of x and x^* we have

$$\begin{aligned} & x(l_2+1)(\alpha+\mu_{l_2+1}) \\ & = g(l_2+1) - \alpha \sum_{i=0}^{l_2} x(i) \\ & = g(l_2+1) - \alpha \sum_{i=0}^{l_2} x^*(i) \\ & = \alpha x^*(l_2+1) - \lambda u^*_{l_2+1} x^*(l_2+2) + \mu_{l_2+1} x^*(l_2+1) \\ & \leqslant 0 \end{aligned}$$

by the induction assumption $x^*(l_2+1) \leqslant 0$ and by $u^*_k x^*(k+1) \geqslant 0$ for all k.

(ii.b) Suppose that for $k \in N$, $k > l_2$, $x(k) \leqslant 0$. It will be shown that then $x(k+1) \leqslant 0$.

$$\begin{aligned} & x(k+1)(\alpha+\mu_{k+1}) \\ & = g(k+1) - g(k) - \alpha x(k) + g(k) - \alpha \sum_{i=0}^{k-1} x(i) \\ & = g(k+1) - g(k) - \alpha x(k) + (\alpha+\mu_k) x(k) \\ & = g(k+1) - g(k) + \mu_k x(k) \\ & \leqslant 0 \end{aligned}$$

by the induction assumption and $g(k+1) - g(k) \leqslant 0$, hence $x(k+1) \leqslant 0$. Now the definition of x and the fact that $x(k) \leqslant 0$ for $k > l_2$ imply that x is the solution of the system of equations

$$0 = -\alpha \sum_{i=0}^{k} x(i) + g(k) + \lambda \max_{u_k \in [0,1]} u_k x(k+1) - \mu_k x(k)$$

Because this system of equations has a unique solution, it follows that $x(k) = x^*(k)$ for all $k \in N$ so $u^*_t = 0$ if $Q_{t-} \geqslant l_2$. ▯

Corollary 4.4. *If there exists an $l^* \in N$ such that*

$$\begin{aligned} &\text{(i)} \quad g(k-1) < g(k) \quad \text{for all } k,\ 1 \leqslant k \leqslant l^* \\ &\text{(ii)} \quad g(k-1) \geqslant g(k) \quad \text{for all } k,\ k > l^* \end{aligned} \tag{4.17}$$

then there exists a $k^* \geqslant l^*$ such that the optimal control u^* of problem P is of the form

$$u^*_t = \begin{cases} 1 & \text{if } Q_{t-} < k^* \\ 0 & \text{if } Q_{t-} \geqslant k^* \end{cases}$$

Proof. Part (i) of lemma 4.6 ensures that $u^*_t = 1$ if $Q_{t-} < l^*$ and part (ii) ensures that $u^*_t = 0$ if $Q_{t-} \geqslant k^*$ with $k^* = \inf_{k \in N}\{k \mid u^*(k) = 0\}$.

For instance the example of g given in equation (4.7) satisfies (4.17) if $\mu_k \leqslant \mu_{k-1}$, for all k, $k \geqslant 1$. This can easily be seen, since $g(0) = 0$ and for $k \geqslant 1$ we have

$$g(k+1) - \left[\frac{\mu_k}{\mu_k + \sigma}\right]^k$$

$$= \left[\frac{1}{1+\sigma/\mu_k}\right]^k$$
$$\leqslant \left[\frac{1}{1+\sigma/\mu_{k-1}}\right]^k$$
$$\leqslant \left[\frac{1}{1+\sigma/\mu_{k-1}}\right]^{k-1}$$
$$= g(k)$$

The condition $\mu_k \leqslant \mu_{k-1}$ is not an unrealistic one for one-server queues with queue-dependent service rates, since the service rate is likely to decrease with the number of customers in the queue, mainly because of the increasing amount of overhead. In the next subsection we present an algorithm for computing k^* if g satisfies (4.17).

As we stated in subsection 4.2, the average reward can be treated as the limit of the discounted reward as $\alpha \downarrow 0$. To do this we first have to extend the notation of the value function to $V_\alpha : N \to R$ to denote its dependency on the discount factor α. The result is stated in the next theorem.

THEOREM. 4.5. *If g is bounded and if there exists an $M < \infty$ such that*

$$|V_\alpha(i) - V_\alpha(0)| < M \tag{4.18}$$

for all $\alpha > 0$ and $i \in N$, then the optimal average reward is equal to $\lim_{\alpha \downarrow 0} \alpha V_\alpha(0)$.

The proof of the theorem can be found in [9], Theorem 7.7. We have not been able to prove whether (4.18) holds, but we have observed this bound in numerical examples. Furthermore, in most cases, as presented in section 5, the optimal policies for all $\alpha \leqslant 10^{-10}$ are equal.

4.4. An algorithm for determining the optimal policy.
Suppose we have a reward rate $g : N \to [0, \infty)$ that satisfies $g(0) = 0$ and $0 < g(k+1) \leqslant g(k)$ for $k \geqslant 1$. These conditions are satisfied for example if $g(k) = I_{\{k>0\}}\, q(k-1)$ with $q(k)$ defined as in (3.3) or (3.4). From corollary 4.4 we know that the optimal control is of the form $u_t^* = I_{\{Q_{t-} < k^*\}}$ for some $k^* \geqslant 1$.

Let us suppose that we know that $k^* \geqslant k$ for some k. In order to check whether $k^* = k$ we have to solve (4.8) with the corresponding u. This can be rewritten as the matrix equation

$$g_k = A_k V_k$$

with $g_k,\ V_k : N_{k+1} \to [0, \infty)$ defined by

$$g_k = [g(0), \ldots, g(k+1)]^T$$

and

$$V_k = [V_k(0), \ldots, V_k(k+1)]^T$$

and $A_k : N_{k+1} \times N_{k+1} \to R$ given by

$$\begin{bmatrix} \alpha+\lambda & -\lambda & & & & & \\ -\mu_1 & \alpha+\mu_1+\lambda & -\lambda & & & & \\ & -\mu_2 & \alpha+\mu_2+\lambda & -\lambda & & & \\ & & \ddots & \ddots & \ddots & & \\ & & & -\mu_{k-1} & \alpha+\mu_{k-1}+\lambda & -\lambda & \\ & & & & -\mu_k & \alpha+\mu_k & \\ & & & & & -\mu_{k+1} & \alpha+\mu_{k+1} \end{bmatrix} \tag{4.19}$$

and check the solution whether

$$V_k(l) > V_k(l-1), \qquad 1 \leqslant l \leqslant k$$

and

$$V_k(k) \geqslant V_k(k+1).$$

Solving (4.18) can be done by standard *LU-decomposition* [7], using the diagonal entries of A_k as pivots and starting with column zero.

Let $A_k^{(n)} = M_n \cdots M_0 A_k$, $0 \leqslant n \leqslant k+1$, denote the matrix that is obtained after the nth step of the LU-decomposition, i.e. the matrix where all the entries under the diagonal in the columns 0 to n are eliminated. $A_k^{(k-1)}$ has the following form

$$\begin{bmatrix} A_k^{(k-1)}[0,0] & -\lambda & & & & \\ & A_k^{(k-1)}[1,1] & -\lambda & & & \\ & & \ddots & \ddots & & \\ & & & A_k^{(k-1)}[k-1,k-1] & -\lambda & \\ & & & & A_k^{(k-1)}[k,k] & \\ & & & & -\mu_{k+1} & \alpha+\mu_{k+1} \end{bmatrix} \tag{4.20}$$

Let $g_k^{(n)} = M_n \cdots M_0 g_k$. Then from (4.20) one can easily check whether the solution V_k satisfies $V_k(k+1) \leqslant V_k(k)$ since V_k must also satisfy $A_k^{(k-1)} V_k = g_k^{(k-1)}$, and consequently

$$V_k[k] = g_k^{(k-1)}[k] \,/\, A_k^{(k-1)}[k,k]$$

and

$$V_k[k+1] = [g(k+1) + \mu_{k+1} V_k(k)] \,/\, \{\alpha + \mu_{k+1}\}$$

If this solution would not satisfy the inequality, then our next guess for k^* would be $k+1$, in which case we have to solve the matrix equation $A_{k+1} V_{k+1} = g_{k+1}$. One can easily check that $A_{k+1}^{(k-1)}$ is almost equal to $A_k^{(k-1)}$, namely

$$\begin{bmatrix} A_k^{(k-1)}[0,0] & -\lambda & & & & & \\ & A_k^{(k-1)}[1,1] & -\lambda & & & & \\ & & \ddots & \ddots & & & \\ & & & A_k^{(k-1)}[k-1,k-1] & -\lambda & & \\ & & & & A_k^{(k-1)}[k,k]+\lambda & -\lambda & \\ & & & & -\mu_{k+1} & \alpha+\mu_{k+1} & \\ & & & & & -\mu_{k+2} & \alpha+\mu_{k+2} \end{bmatrix}$$

$A_{k+1}^{(k)}$ can thus easily be computed from $A_k^{(k-1)}$, enabling us to solve the dynamic programming equation through one set of linear equations by incorporating the inequality check into the LU-decomposition of the solution. From the above discussion one can easily see that we can compute the optimal value k^* by the following algorithm.

ALGORITHM 4.6.

```
{declarations}
HUGE      : {arbitray large integer}
i,k       : integer;
g,diag,V  : array[0..HUGE] of real;
found     : boolean;

{initialisation}
found: = false;
k: = 0;
for i: = 0 to HUGE
do
  g[i]: = g_i;
  diag[i]: = α + μ_i
od;
```

```
{LU decomposition}
while ((k<HUGE) and not found)
do
  if (g[k+1]+μ_{k+1} g[k]/diag[k])/diag[k+1]≤g[k]/diag[k]
  then
    found: = true;
    g[k+1]: = g[k+1]+μ_{k+1} g[k]/diag[k]
  else
    diag[k]: = diag[k]+λ;
    diag[k+1]: = diag[k+1]-λ μ_{k+1}/diag[k];
    g[k+1]: = g[k+1]+μ_{k+1} g[k]/diag[k]
  fi;
  if not found then k: = k+1 fi;
od;

{computation of V(i)}
V[k+1]: = g[k+1]/diag[k+1]
V[k]: = g[k]/diag[k]
for i: = k-1 downto 0
do
  V[i]: = (g[i]+λV[i+1])/diag[i];
od
```

5. Numerical Results.

In this section we present some numerical examples of the performance analysis and computation of the optimal control law.

The first system is an M/M/1/k queue with mean service time equal to 1.0 and the sojourn times of customers Erlang-3 distributed with mean 20.0. The optimal value for k as derived from the dynamic programming equation is equal to 8, where we have used in the equations an arrival rate $\lambda = 0.8$, discount factor $\alpha = 10^{-10}$ and reward rate $g(k)$ as defined in equation (4.7).

The call completion rates and blocking probabilities of this queue for various values of k and ρ (or equivalently λ since $\mu = 1.0$) are shown in figures 5.1 and 5.2 respectively. From figure 5.1 we may conclude that, although $k = 8$ is optimal for $\rho = 0.8$, the M/M/1/8 queue behaves well also under overload, i.e. for values of ρ ranging from 1.0 to 2.0.

The second system we present is equal to the first one, with the exception that the mean sojourn time is now equal to 50.0. The optimal value for k, using the same λ, α and $g(k)$, in this case is equal to 15. The call completion rate is given in figure 5.3. The graph of the blocking probability is the same as of the first system, since the equilibrium distribution of the number of tasks in both queues does not depend on the sojourn time distribution of the customers.

References

1. F. Baccelli (1983). *Modèles probabilistes de systèmes informatiques distribués,* Thèse d'état, INRIA, Paris.
2. R.K. Boel and J.H. van Schuppen (1985). Overload control for SPC telephone exchanges - refined models and stochastic control, in *Proceedings of the 3rd Bad Honnef Conference*, 100-110, ed. N. Christopeit, K. Helmes, M. Kohlmann, Springer-Verlag, Berlin.
3. R.K. Boel and J.H. van Schuppen (1986). Overload control for switches of communication systems - A two-phase model for call request processing, in *Proceedings of the International Seminar on Teletraffic Analysis and Computer Performance Evaluation*, 209-224, ed. O.J. Boxma, J.W. Cohen, H.C. Tijms, Elsevier, Amsterdam.
4. P. Brémaud (1979). Optimal thinning of a point process, *SIAM J.Control and Optimization*, 17, 222-230.
5. P. Brémaud (1981). *Point processes and queues - martingale dynamics*, Springer-Verlag, Berlin.
6. L.J. Forys (1982). Performance analysis of a new overload strategy, in *10th International Teletraffic Congres.*
7. G.H. Golub and C.F. van Loan (1983). *Matrix computations*, The Johns Hopkins University Press, Baltimore.
8. T.G. Robertazzi and A.A. Lazar (1985). On the modeling and optimal flow control of the Jacksonian network, *Perf.Eval.*, 5, 29-43.
9. S.M. Ross (1970). *Applied probability models with optimization applications*, Holden-Day, San Francisco.
10. C. Striebel (1975). *Optimal control of discrete time stochastic systems*, Springer-Verlag, Berlin.
11. D.Y. Sze (1986). A queueing model for overload analysis, in *Computer Networking and Performance Evaluation*, 413-422, ed. T. Hasegawa, H. Takagi, Y. Takahashi, Elsevier (North-Holland).
12. Phuoc Tran-Gia and M.H. van Hoorn (1986). Dependency of service time on waiting time in switching systems - A queueing analysis with aspects of overload control, *IEEE Trans. Comm.*, 34, 357-364.
13. R. Wolff (1982). Poisson arrivals see time averages, *Oper.Res.*, 30, 223-231.

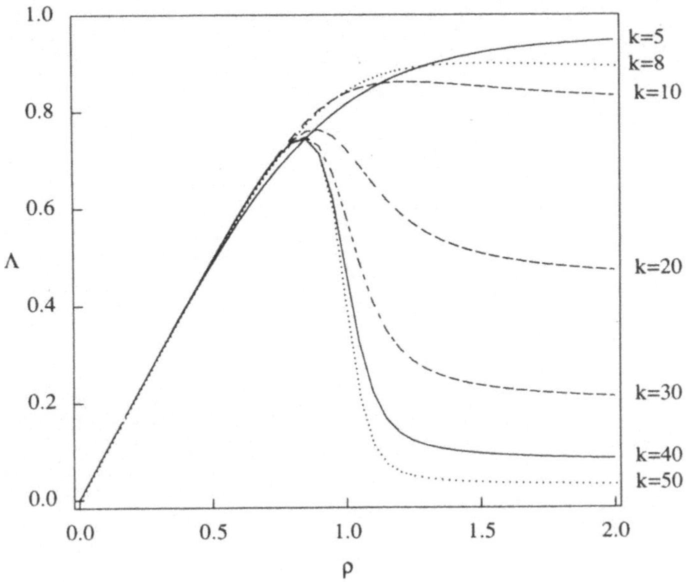

FIGURE 5.1. Call completion rate for an M/M/1/k queue with Erlang-3 distributed sojourn times with mean 20.0.

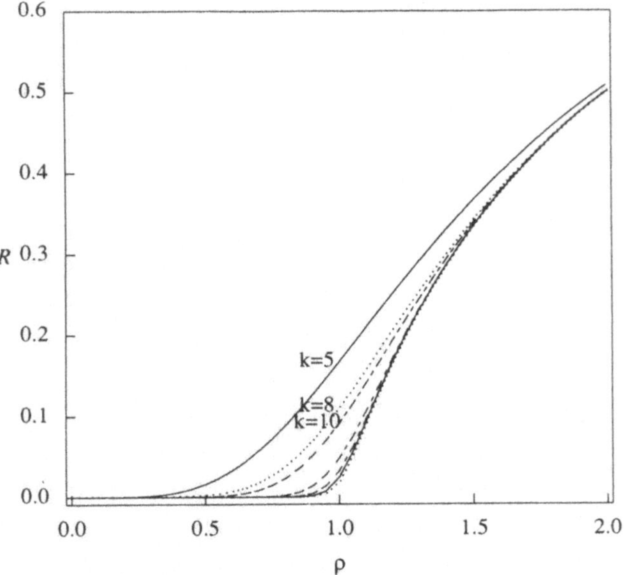

FIGURE 5.2. Blocking probabilities for an M/M/1/k queue.

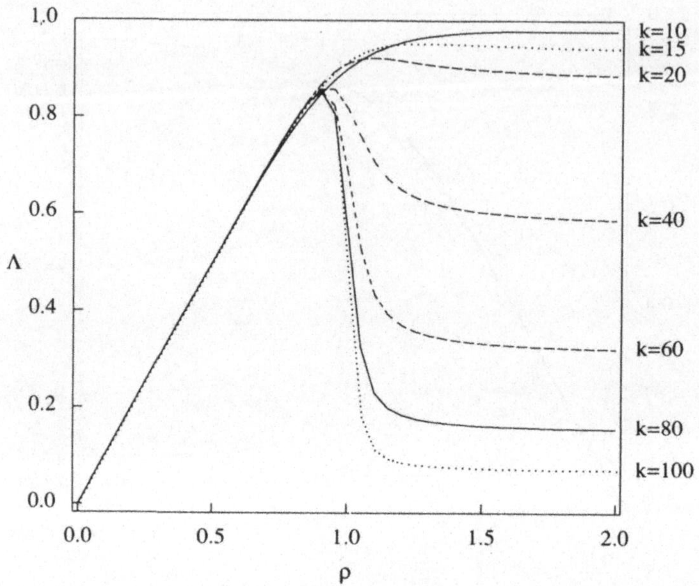

FIGURE 5.3. Call completion rate for an M/M/1/k queue with Erlang-3 distributed sojourn times with mean 50.0.

INTEGRATED SERVICES DIGITAL NETWORKS - BASIC PERFORMANCE MODELLING AND TRAFFIC ENGINEERING

Paul J. Kuehn
Institute of Communication Switching and Data Technics
University of Stuttgart

Kurzfassung: Diensteintegrierende Nachrichtennetze - Grundlagenprobleme der Modellierung und des Traffic Engineering

Die Einführung diensteintegrierender Nachrichtennetze wie das diensteintegrierende digitale Fernmeldenetz ISDN wirft eine Reihe neuer Fragestellungen auf im Hinblick auf die Modellierung und die verkehrsgerechte Auslegung. Nach einer kurzen Einführung in das ISDN und seine Dienste werden verschiedene Klassen von Verkehrsmodellen eingeführt. Da reale Verkehrsmodelle in der Regel implementierungsabhängig sind, werden hier zunächst nur sog. "generic models" entwickelt, die sich auf die wesentlichsten Wirkzusammenhänge beschränken. Die Anlyse derartiger Modelle erlaubt die Einsicht in das prinzipielle Verhalten und bildet die Grundlage realistischer Methoden des Traffic Engineering. Es zeigt sich, daß mit den bekannten Methoden nur ein Teil der Probleme beherrscht werden kann und daß noch viele Probleme offen sind.

1. DEVELOPMENT OF COMMUNICATION NETWORKS AND SERVICES

Communication networks of the past have developed individually and were characterized by one dominating service as, e.g., voice telephony, telex or data transmission. Through the advances of digital transmission, microcomputer control and software technology future communication networks will be able to support a wide spectrum of services. Bases for these developments are cheap VLSI hardware components for call processing, signal processing, switching and storage, high-speed digital transmission, efficient high-level languages, and the international standardization of services and protocols. For the latter, the ISO Basic Reference Model for Open Systems Interconnection has greatly enhanced the development of ISDN.

1.1 Physical Structure and ISDN-Capabilities

Fig. 1 shows the basic structure of the public ISDN and its major capabilities. The main aspects are:

a) Multiple terminal configuration on the customer premises comprizing
   - Terminal Equipment (TE 1) with ISDN compatibility for voice, text, etc.
   - Terminal Equipment (TE 2) not compatible to ISDN with Terminal Adpatation (TA) for protocol conversion
   - Connection of Local Area Networks (LAN) through Gateways (GY)
   - Free connectivity through a local bus system and an ISDN-socket

b) Unified Network Access (Basic Access) through
   - Usage of the existing subscriber loop
   - Network Termination (NT) as the end point of the public network
   - Exchange Termination (ET) as the interface between the subscriber access and the switching exchange

- Provisioning of 2 B-channels (information channels) with 64 kbps FDX each for circuit switched connections or (optionally) packet access
- Provisioning of 1 D-channel with 16 kbps FDX for signalling (s), user packet (p) and teleaction (t) data; s-, p-, and t-connection management is transparent to the NT and subject to the D-channel protocol between TE(TA,GY) and ET (levels 2 and 3).
- Extended Network Access through the introduction of multiplexors (Basic Access Multiplexor) and concentrators in the subscriber area; both are connected with the switching exchange through PCM 30/32 transmission facilities (not shown in Fig. 1).

c) Multiple ISDN Capabilities within the ISDN through
   - Circuit Switched (CS) B-channels
   - CS H-channels of higher bandwidths (H_0: 384 kbps, H_{11} : 1536 kbps, H_{12} : 1920 kbps,...)
   - Non-Switched B- and H-channels
   - A uniform signalling network based on the common signalling channel concept of CCITT No.7
   - Packed Switched (PS) facilities either by providing access to a separate PS-network or by integration of PS-services into the ISDN.

d) Provisioning of facilities for information storage, information processing and interworking with other networks through
   - Data Bases for information (voice, data, text,...) storage and retrieval
   - Hosts for Information Processing (server functions as, e.g., protocol conversion)
   - Gateways (GY) for interconnection of ISDN and non-ISDN networks.

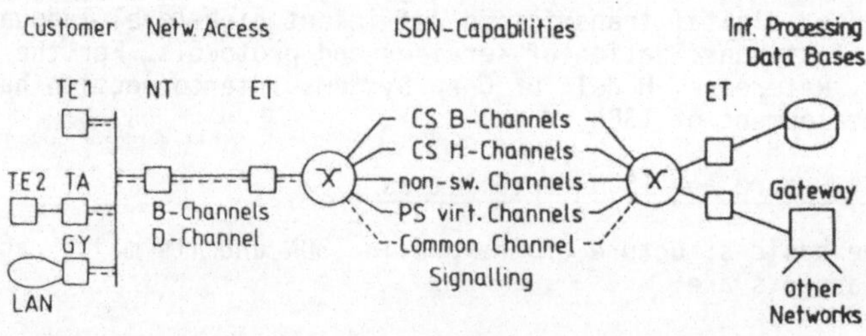

FIGURE 1
The ISDN Network Concept

1.2 ISDN-SERVICE CONCEPT

A "Service" comprises all technical, operational and legal aspects for a particular type of communication between users or between users and the provider of a public network. Within the ISDN various types of services are distinguished:

a) Bearer Services
Bearer Services provide the circuit or packet switched transport of information between two terminal-network interfaces irrespective of the compatibility of the terminals. A typical example of these services are switched or non-switched 64 kbps (B-) channels for text, data and graphic applications.

b) Standard Services

Standard Services provide the transport of information between two terminals with assurance of compatibility. The functionality comprises all 7 layers. Typical representatives are:

- ISDN Telephone
- ISDN Teletex
- ISDN Telefax
- ISDN Textfax.

Again, these services are based on switched B-channels with 64 kbps.

c) Higher Services

Higher Services generally use centralized storage and processing capabilities of the ISDN. Typical examples of such services are

- ISDN Videotex
- ISDN Voice, Text and Fax Mail
- Protocol conversion.

d) Services on the D-channel

The D-channel basically carries signalling information (s-data). These data require only a small part of the available 16 kbps capacity so that low rate user packet data (p-data) may be transferred additionally over virtual connections or telemetric data (t-data) over permanent virtual connections.

e) Supplementary Service Attributes

Additionally to the basic service attributes of the standard services the users may optionally subscribe for further service attributes as, e.g.,

- abbreviated dialling
- automatic repetition of calls in case of blocking
- inward dialling into PBXes
- automatic call back
- call redirection
- conferencing
- reverse charging, etc.

1.3 Protocol Architecture

The exchange of information between users, application programs or any particular level entities is controlled by a set of rules subjected to a protocol definition. Based on the layered protocol architecture of the ISO/CCITT OSI Basic Reference Model CCITT has developed a generalized model for ISDN protocols. Whereas for packetized communication control and user information are combined in the respective PDU's, circuit switched communication with separate signalling channels and networks needs a multidimensional approach where different Protocol Planes are distinguished for User, Control and System Management information.

The multiple plane protocol architecture is illustrated in Fig. 2 for a CS-connection through an ISDN with D-channel signalling for the network access and No.7 common channel signalling network for network-internal signalling.

The relevant standards of CCITT are defined in the I-series for ISDN, particularly I 431, I 441 and I 451 for levels 1, 2 and 3 of the D-channel protocol, and in the No.7 Signalling System. Both, the D-channel and No.7 signalling protocols have been developed for their particular purposes, the control of the network access and network-internal control. Therefore, they differ considerably and an interworking is necessary at the origination and destination exchanges.

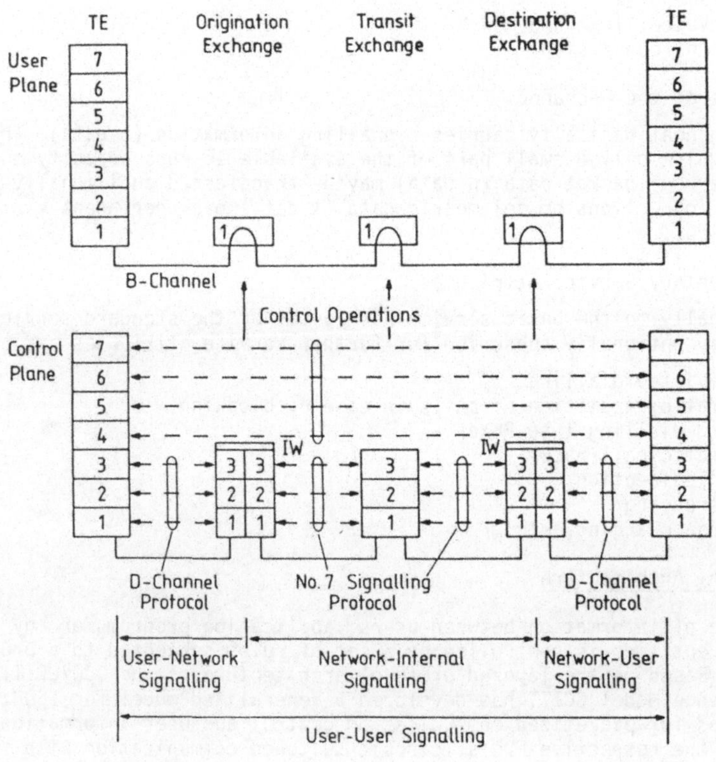

FIGURE 2
ISDN Protocol Architecture. Example: CS-Connection

2. NETWORK PERFORMANCE EVALUATION

2.1 Modelling

Modelling networking performance is essential to evaluate design alternatives and to dimension networking resources to meet throughput and grade of service requirements. Network models can be rather complex consisting of quite heterogeneous components, as

- SERVERS
- QUEUES
- RESOURCE MANAGEMENT
- PROTOCOLS
- TRAFFIC SOURCES.

These components are abstract representations of physical, logical and statistical entities of real systems. SERVER components represent the occupation of a processor or transmission channel for a particular task or call. The component QUEUE may refer to a well organized list of service requests or to a physical buffer for the intermediate storage of data units. RESOURCE MANAGEMENT consists of all strategical properties of computer or networking operating systems through which physical resources are temporarily assigned to requests or users. PROTOCOLS refer to the definition of the functional behavior of real systems with respect to basic services as connection establishment, exchange of data units, error recovery, routing, etc. TRAFFIC SOURCES, finally, characterize the dynamic load to the network including the statistical properties, user or subscriber behavior of human or nonhuman users of particular services.

The phenomenon of traffic generally refers to the system dynamics resulting from the interworking of all of these entities. Since most of the network resources are shared by a large number of independently acting users, bottlenecks may occur resulting in grade of service degradation as losses, blocking, delays or even malfunction.Performance evaluation aims at the experimental or theoretical analysis of the system model with respect to throughput and grade of service figures. These procedures are the basis of network planning and network management procedures to design and to operate networks according to a prescribed grade of service (GOS) under given load figures.

2.2 History of Performance Evaluation

Performance Evaluation of Communication Networks and Systems has a long tradition and reaches more than 80 years back. Its cornerstones are:

- Birth and Death analysis of loss and delay systems
  (Erlang 1917, Engset 1918)
- Analysis of systems with general service times
  (Pollaczek 1930, Crommelin 1932)
- Imbedded Markov Chain analysis
  (Kendall 1953)
- Queues with general input and service processes
  (Lindley 1952)
- Output of queuing systems
  (Burke, Reich, Cohen 1956-1957)
- Queuing networks
  (Jackson 1954, Kleinrock 1964, BCMP 1975)
- Queuing network algorithms
  (Buzen 1972, Reiser/Kobayashi 1976)

- Cyclic queuing systems
  (Cooper, Cohen, Takagi, Boxma 1970-1986)
- Equivalent Random Traffic Theory
  (Bretschneider, Wilkinson, Riordan 1955-1956)
- Analysis of multi-stage switching networks
  (Jacobaeus 1952, Clos 1953, Kharkevich 1960, Lotze 1970)
- Discrete Time simulation techniques
  (Kosten 1948, Neovius 1955).

The methods of analysis cover a wide spectrum, as

- finite state birth and death equations
- integral equation relationships
- phase type representations
- LS and GF transforms and function theory representations
- product form solutions
- convolutional and mean value computational algorithms
- functional and/or statistical independence
- effective accessibility representations
- equivalent random traffic techniques
- decomposition and aggregation techniques
- diffusion and fluid approximations.

The rich field of solutions cannot adequately or exhaustively referred to in this context. Methods of this kind and their results have been successfully applied for network dimensioning and planning.

Performance Analysis has now reached a status with a well-based theoretical framework, a huge number of solutions of specific problems (in particular in Operations Research and Computer Science), a wide spectrum of approximate approaches, and application support by traffic theory tables, numerical procedures and analytical as well as simulation program tools.

2.3 New Challenges for Performance Evaluation

Through the introduction of loosely coupled computer systems interconnected by Local Area or Wide Area Networks (LAN, WAN), distributed end systems and ISDN, a large set of new performance issues has arisen, as

- multiple access to broad band media
- multiplexing of connections and data streams
- data flow control and error recovery
- overload control and adaptive routing
- traffic characteristics of new services
- integration of voice and data
- integration of circuit and packet switching.

The classical solutions of queuing and traffic theory are in general not directly applicable and have to be complemented accordingly. Typically, new problems are characterized by a rather high degree of complexity. Methods for splitting global models into smaller ones with less complexity must be developed to take care of the inter-dependence due to wide-spanning protocol control mechanisms.

2.4 Generic Models for Network Performance Evaluation

To characterize the network performance evaluation issues, elementary structures and mechanisms will be reviewed in the following. This review addresses the most popular models of the past, and especially a number of new basic problems of networks for service integration. The developed models cannot be described rigorously in this short review and should be considered as 'generic' in the sense of their basic purpose.

2.4.1 Switching Networks

Switching networks allow a temporarily switched CS-connection between one input terminal (inlet) and one output terminal (outlet). They can be constructed as single stage connecting arrays with full accessibility (Fig. 3a), limited accessibility (Fig. 3b) or multi-stage connecting arrays with conjugated switching or step-by-step switching (Fig. 3c). Further details refer to the structure of interconnections, the number of stages, the path hunting strategy and the traffic source submodel.

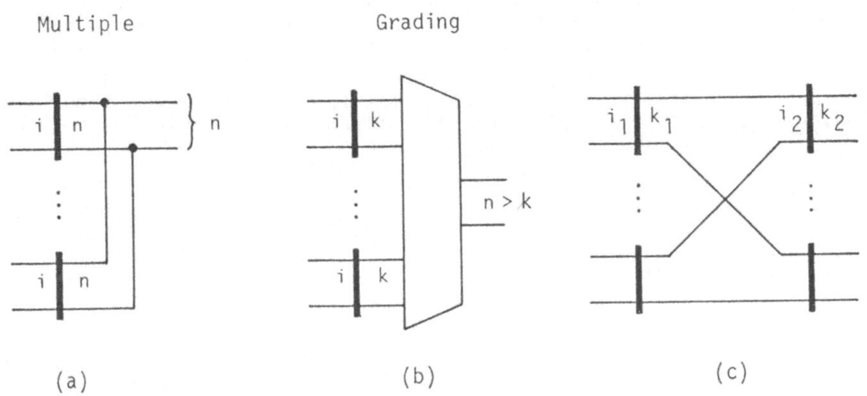

FIGURE 3
Basic Switching Network Structures

2.4.2 Network Models for CS-Traffic Routing

Alternate Routing aims at a simple and robust scheme to select routes within a partially meshed network dependent on the actual state of outgoing trunk groups. Fig. 4a shows the basic model of two primary trunk groups (PTG) and one secondary trunk group (STG). Traffics A_{11} and A_{12} are first offered to their PTG; if these PTG are fully occupied or if there is no access to an idle trunk, the traffic may overflow to the STG. Additionally, the STG may also be offered direct traffic A_2. The overflow changes the traffic characteristic into more "peakedness" which has to be taken into account for GOS calculations.

The signalling network capabilities of the future ISDN allows for more flexible choice of traffic paths according to global load conditions (adaptive routing). These problems are even more important as new networks are structured less hierarchically. Generic models may be defined allowing for a finite number of routes within a prescribed origination-destination graph dependent on the availability of an idle origination-destination path.

2.4.3 Network Models for PS-Traffic Routing

In PS-networks the communication may be either connectionless (datagram,CL) or connectionoriented (virtual circuit,CO). In both cases, adaptive routing is applied aiming at criteria of the shortest expected origination-destination transfer time. Usually, each network node has some information on the actual network state which is periodically or aperiodically updated. There are no simple generic models; the complexity of the problem requires usually simulation techniques to evaluate the performance.

2.4.4 Network Models for CS Service Protection

Within hierarchical networks different traffic types share certain trunk groups. As a classical example, the international inbound traffic shares the downward trunk groups with the national traffic. In future services integrated networks, traffics of various services and eventually different bandwidth or importance may share the same trunk groups. To protect one traffic type against the other, e.g., the inbound international traffic against national traffic, various protection mechanisms have to be applied to meet specific GOS-requirements.

In Fig. 4b two classical approaches are shown. The n trunks are divided into two subgroups of n_1 and n_2 trunks, respectively, where the n_1 trunks are exclusively reserved for traffic A_1 and the n_2 trunks are shared by both traffics A_1 and A_2. Fig. 4c refers to a generalized solution where the acceptance of either traffic A_1 or A_2 depends on the actual occupation state (x_1, x_2) of both traffic types. Methods of this category allow a bandwidth allocation such that the individual GOS-criteria are met under given load.

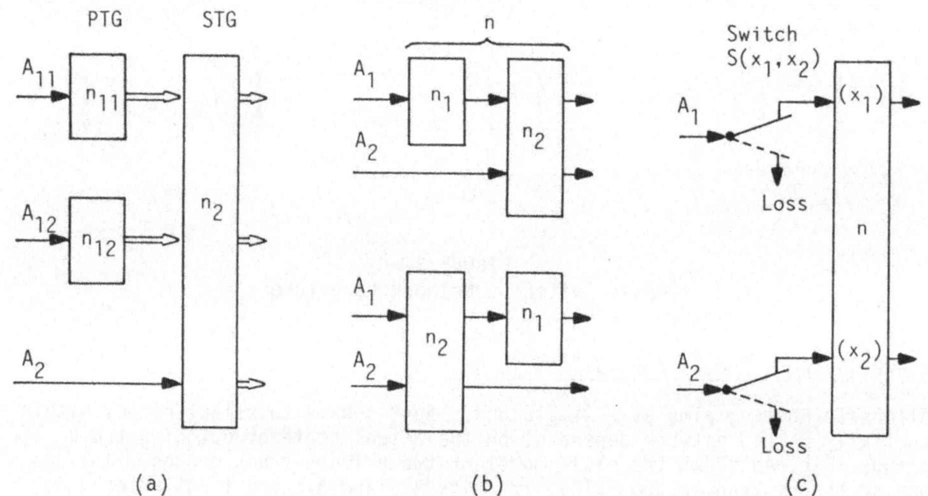

FIGURE 4
a) Overflow Model for Alternate Routing of CS-Connections
b) Service Protection Model by Subdivision of Trunk Groups
c) Service Protection Model by State-Dependent Acceptance Strategy

2.4.5 Variable Bandwidth Allocation

The required bandwidth of various services may vary considerably as, e.g., for

- low speed data (program development, printing)	9.6 - 19.2	kbps
- voice/telematic services	64	kbps
- high resolution graphics	256	kbps
- file transfer	1,000 - 10,000	kbps
- video	64 - 140,000	kbps

Several of these services may share an individual trunk group of a CS-Network. One solution of this problem is to assign single or multiple B-channels with 64 kbps each to these services. Fig. 5 shows the simplest model of a Time Slotted Frame with several arrival streams for multi-slot connections.

For a proper management of multi-slot connections, extended models refer to

- switching networks with multi-slot connections
- subdivided frame structures with classes of connections, e.g., segments for 64 kbps, 384 kbps and 2.048 kbps
- path searching/channel allocation for multi-slot connections under the restriction of bit sequence integrity
- rearrangement of existing connections for purposes of concise packing of remaining multi-slot connections

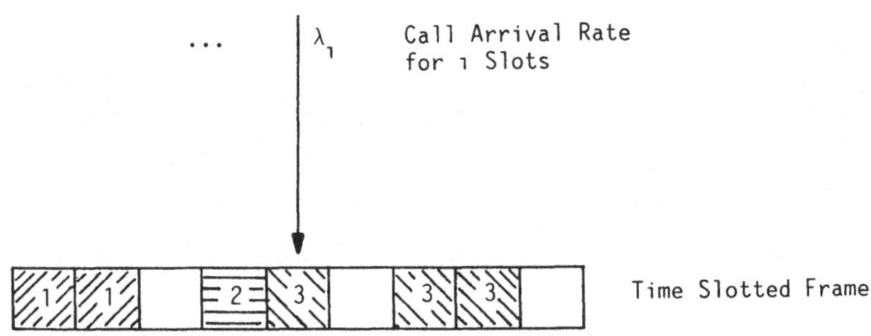

FIGURE 5
Basic Model for Multi-Slot CS-Connections

2.4.6 Advance Reservation

Within the broadband-type ISDN individual videophone connections, video conference connections, switched video program distribution and high speed point-to-point data transmission share the same broadband (optical fiber) channels. Teleconferences differ from individual communications with respect to

- predefined and reserved broadband connections between several teleconferencing studios
- much longer holding times.

Connection reservation schedules depend on various parameters, as

- arrival instant and advance time of reservation requests
- instant and duration of the requested connection
- bandwidth requirements
- sacrificing costs for advance channel blocking or delaying of reservation requests
- time discretization for scheduling algorithm.

Fig. 6 shows the principal model where call reservations and direct calls share a common time slotted transmission channel. The various reservation strategies can only be evaluated by simulation due to the model's complexity.

2.4.7 Single/Multi-Queue Delay Systems

Single/Multi-Queue, Single/Multi-Server delay systems are the most important basic models for applications, as

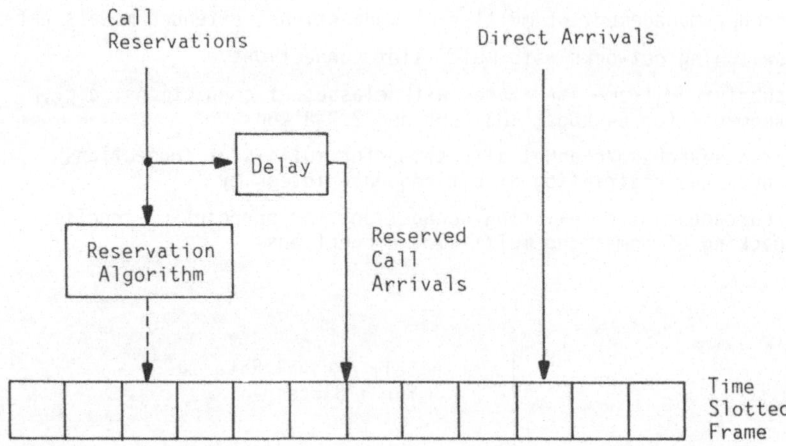

FIGURE 6
Basic Model for Bandwidth Management Including Advance Reservation

- call processing
- packet switching
- multiplexing of control/packet data streams.

Fig. 7 shows various basic models. In particular, requests of various types or originations may share one server; the server allocation may depend on predefined priorities (Fig. 7c) or on a cyclic polling mechanism (Fig. 7d). Polling models have a rather broad application reaching from data collection, packet multiplexing, processor scheduling to distributed multi-access of broadband media as in the case of Token-Passing for Local Area Networks (LAN). More sophisticated Token allocation schemes may be based on Token Priorities and/or Timer devices to guarantee a timely Token issue for real-time or even synchronous communications applications (FDDI).

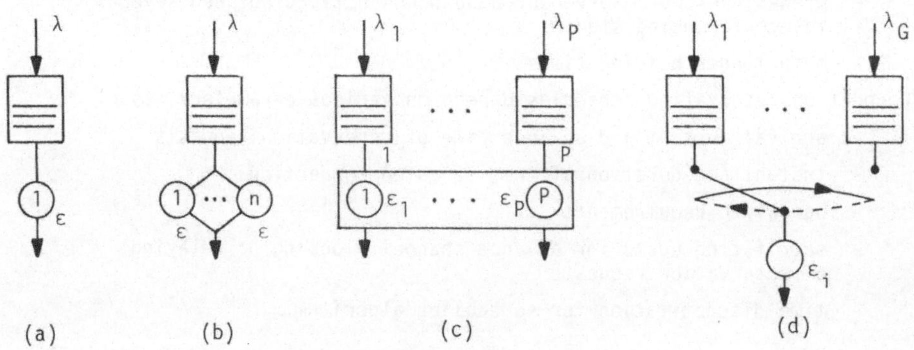

FIGURE 7
Single/Multi-Queue, Single/Multi-Server Delay Models

2.4.8 Packet Switching Nodes

New developments of high-capacity PS nodes direct to highly modularized structures. The modularization follows principles of partial function division and load sharing for purposes of a real-time efficient and protocol layer related implementation as well as load dependent extension, respectively. Fig. 8 shows a typical example of the new generation of high-capacity packet switches.

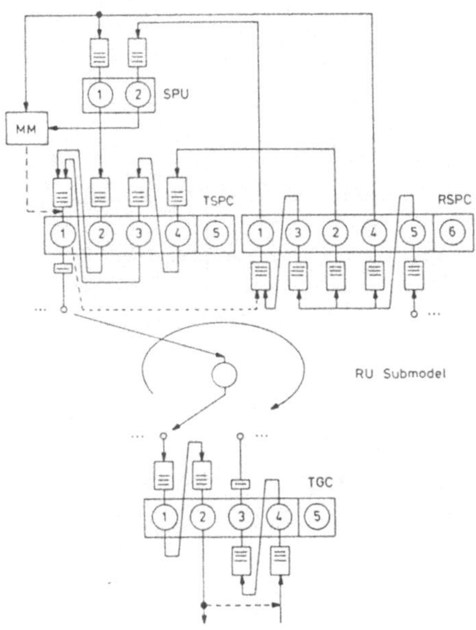

FIGURE 8
High Capacity Packet Switch Node Model

The model consists of peripheral processors (Terminator Group Controller TGC) handling a group of line/trunk terminating units and protocol functions up to layer 2. The number of TGC's may be extended according to the number of connected lines/trunks. At the more central side a variable number of Switching Processor Units SPU are responsible for routing of datagrams and virtual circuits as well as switching of packets (layer 3-functions). The number of SPU's depends on the PS capacity. SPU's and TGC's are interconnected through a high-speed Ring Unit bus system where the inbound and outbound traffic is additionally supported by specialized controllers RSPC and TSPC. All processor devices are of the multi-task types where the waiting requests are scheduled according to the specific task priorities.

The analysis of such models with up to several hundreds of processing modules, internal protocols and LAN-type interconnection schemes exceeds todays performance evaluation tools. The analysis is based on advanced decomposition and aggregation techniques.

2.4.9 CS/PS-Integration

Hybrid Switching systems are designed to support CS as well as PS. This leads to the problem how to assign channels/bandwidth to the various service requests.

Fig. 9 shows a very popular model for a broadband channel within a Local Area Network. The synchronous frame is partitioned into a CS- and a PS-part with a fixed or movable boundary in between. The CS-channels may be assigned by a central device at connection establishment, whereas the PS-subchannel is assigned by a Token Passing protocol. If a packet transmission has not finished at the frame boundary, it is interrupted at this time for the synchronous CS-part. The CS-part is subdivided into basic channels; CS-connections are categorized into classes of bandwidth (multi-slot assignment).

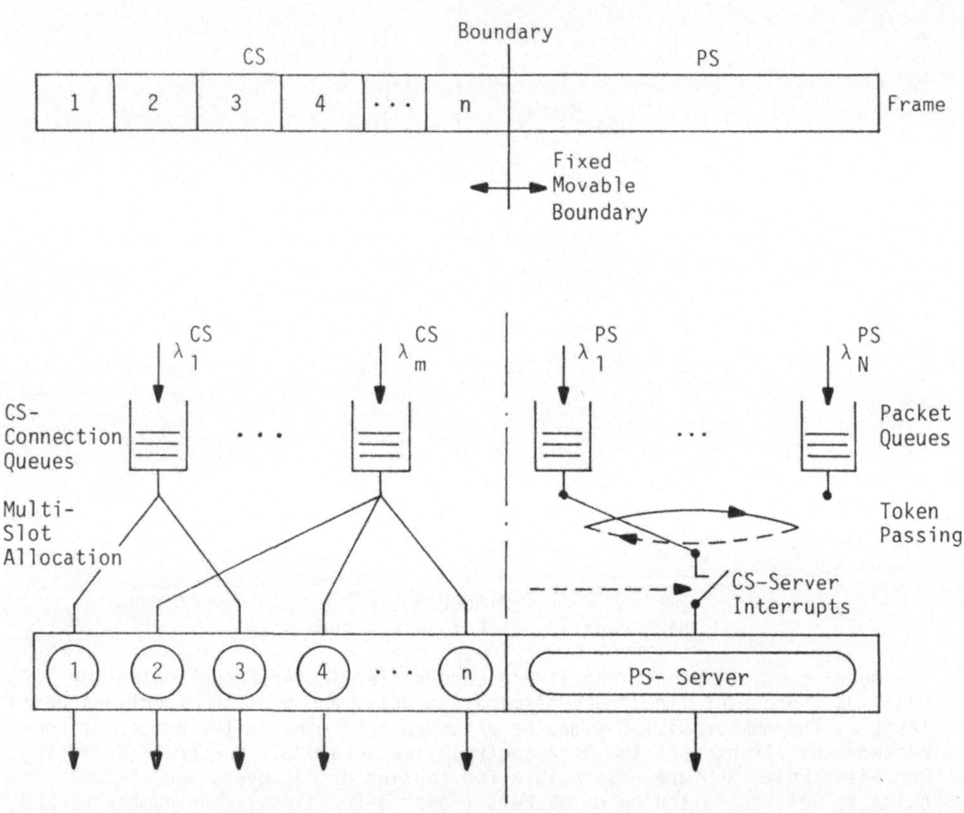

FIGURE 9
CS/PS-Integration by the Partitioned Frame Principle

Another class of hybrid switching systems bases on the Slotted Ring principle, see Fig. 10. The ring system itself owns an inherent distributed ring buffer which is subdivided into fixed size slots. Each slot may carry a CS-connection of fixed bitrate or an individual data unit (packet). In case of a CS-connection, the slot is reserved for this connection at connection establishment, whereas in case of PS, each slot is used individually depending on an idle slot indication. The analysis of the PS-capability depends heavily on the origination-destination relation of individual PS-data units.

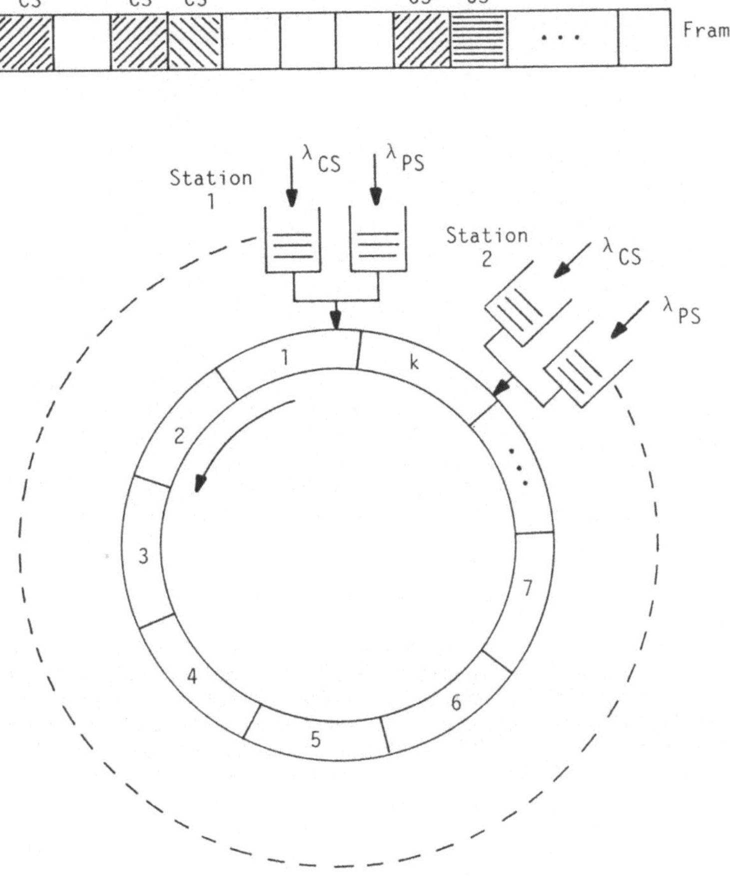

FIGURE 10
CS/PS-Integration by the Slotted Ring Principle

2.4.10 Fast Packet Switching (FPS)

New high speed VLSI circuitry, such as GaAs-Devices allow extremely fast operations which may be used for an all-packetized solution of all forms of communications. Real time, stream-like data flows as for voice or video communications do not require flow control and error recovery. Thus, such packets are just switched through. A packet may even be forwarded as soon as its destination link is known to save buffer space and delays. Such switching techniques are known as Fast Packet Switching (FPS) or Asynchronous Time Division (ATD).

Fig. 11 shows a generic model for FPS. Voice and data packets are multiplexed onto one channel. The packets are switched between different input and output channels by a parallel self-routing Banyan network consisting of elementary 2x2 switch fabrics. Delays may occur at the multiplexing devices and in front of the Binary Switching Network.

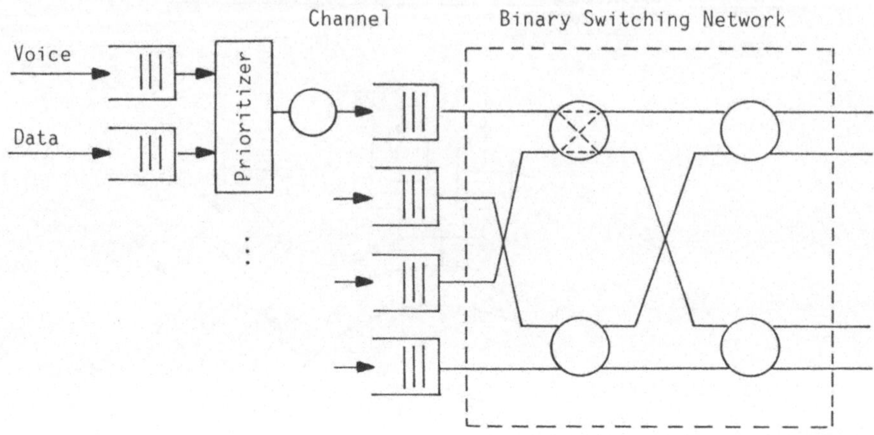

FIGURE 11
Basic Model for Fast Packet Switching

There are two principal types of delays: short delays to phase in simultaneously arriving packets and longer delays due to a temporary overload. In the latter case, dynamic bit dropping for voice packets may be applied to reduce these delays. The traffic performance depends also on the packet arrival stream characteristics. To analyze whole FPS networks two more basic operations on packet data streams have to be treated: merging and splitting, see Fig. 12.

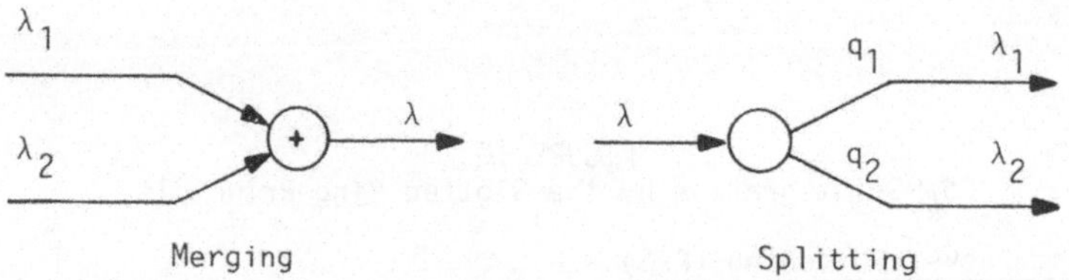

FIGURE 12
Basic Operations for Packet Data Traffic Streams

2.4.11 Queuing Networks for Switching Control

Open or closed queuing networks, such as shown in Fig. 13a, form a powerful means to model computer systems or packet networks and to estimate network delays. Such models, however, are only rough approximations for real system models. Basic models for switching system control fall outside the usual class of queuing networks because of

- clock-type I/O for exchange of commands/responses between peripheral and central control
- prioritized processor scheduling
- branching processes, where, e.g., a command message is echoed

by the CPU through several response messages with different destination, see Fig. 13b.

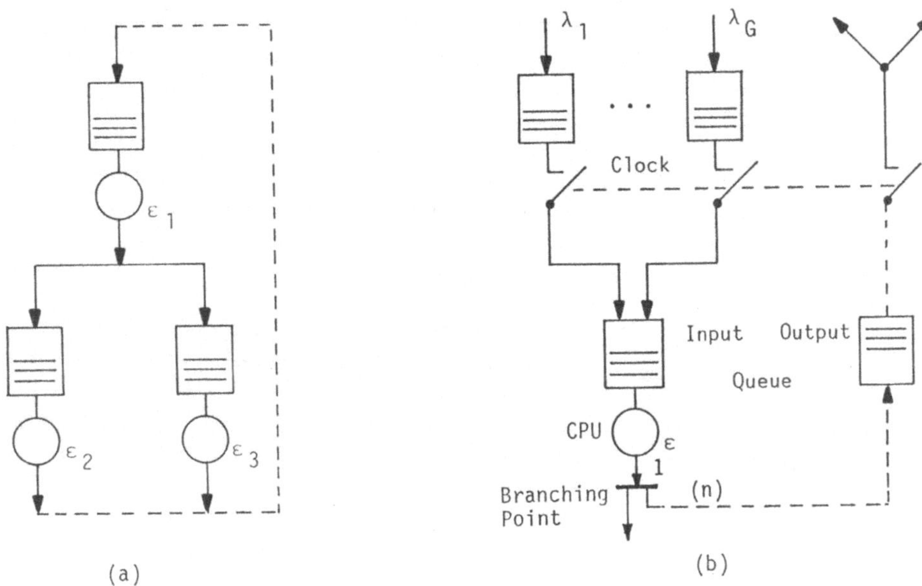

FIGURE 13
a) Open/Closed Model for Computer System or Packet Network
b) Switching System Control Model

2.4.12 Flow-and Congestion Control

Flow Control is used to adapt a slower data sink to a faster data source. Flow control uses various control schemes, such as source on/off-switching, throttling or window mechanisms. Such schemes are part of data transmission protocols. A simple example of a throttling flow control mechanism between remote source and sink models is shown in Fig. 14a. Dependent on the upcrossing (downcrossing) of level L1 (L2) of the sink arrival storage, a throttle (resume) command is sent upon which the source switches to the lower (higher) arrival rate λ_2 (λ_1). The total network delay has been modelled by a simple infinite server station.

Congestion Control aims at the overload protection of a centralized device or a whole switch. This is especially important for systems where human or non-human users repeat their calls in case of blocking: repeated call attempts lead to a high processing load with a low call completion rate which may finally result in a life-lock collapse. Ample control mechanisms have to be used which make sure that a call is accepted only when it can be completed almost sure; otherwise, calls should be blocked before they swamp the control with load of uncompleted calls, see Fig. 14b.

Both, flow and congestion control models have to be analyzed not only with respect to their static behavior but for their transient behavior in response on load variations.

2.4.13 Flow Controlled Virtual Connections

Handshake- or On/Off-protocols for flow control and error recovery lead to much overhead when the propagation delay increases. Therefore, sliding window protocols are used to maintain the throughput of such a connection. Window

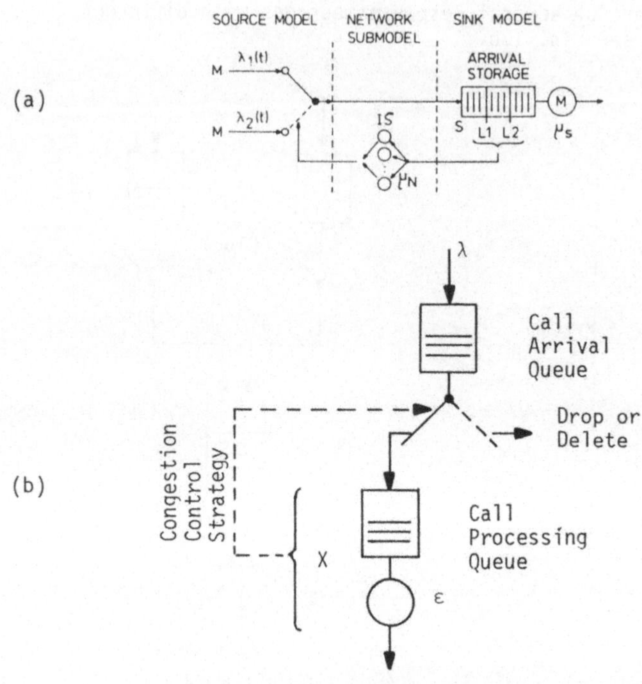

FIGURE 14
Models for Flow and Congestion Control
a) Flow Control Model between Remote Source and Sink
b) Congestion Control Model

flow control mechanisms can be found in almost all protocol layers for layer-specific connections between peer entities or even between entities of adjacent layers across the layered architecture.

Fig. 15 shows the generic model of a window flow controlled simplex connection between two layer (N)-entities. The Petri net symbol stands for the access mechanism: An (N)-Service Data Unit (SDU) from the upper layer is only accepted if the credit queue is nonempty (Y>0); each (N)-SDU is turned into an (N)-Protocol Data Unit (PDU) taking a credit with it from the credit queue. After reception, the (N)-SDU is extracted and passed to the upper layer; simultaneously an acknowledgement (N)-PDU is returned to the origin. Thus, there are always exactly W (window size) PDU's cycling around. This model allows the proper modelling of window flow control including acknowledging as well as transmission and propagation delays.

The simplex virtual connection of Fig. 15 can be extended to a full duplex (FDX) connection by using two separate flow control mechanisms for either direction. Note that both flow control mechanisms may use arbitrary window sizes, but share the same transmission channels, see Fig. 16.

Both models of Figs. 15 and 16 describe the already established virtual connection (VC). To establish the VC the connectionless service is used, i.e. individual datagram type packets are exchanged between both entities.

Both models are simplified; the simplification consists first of an aggregated behavior of the lower layers represented by simple channels. The parameters of these channel service times have to be found from a separate analysis, such that they are equivalent with respect to total throughput and transfer delay. There are several options to replace the lower layers: single server queues

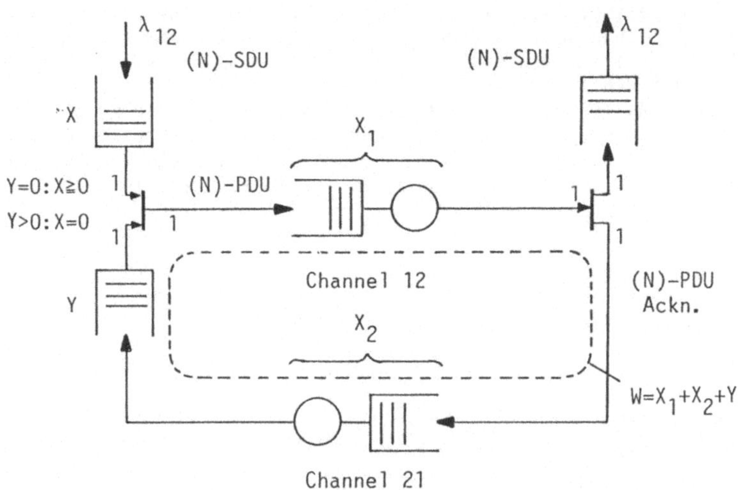

FIGURE 15
Model of a Window Flow Controlled Simplex Virtual Connection

with fixed or state-dependent service rates or infinite server models. Through this procedure, an iterative analysis of the multi-layered protocol architecture can be performed by successive aggregation, bottom-up.

Another simplification has been assumed by neglecting the (N)-entity processing times. There is no problem to extend the model to include these effects. Note, however, that these are implementation-dependent.

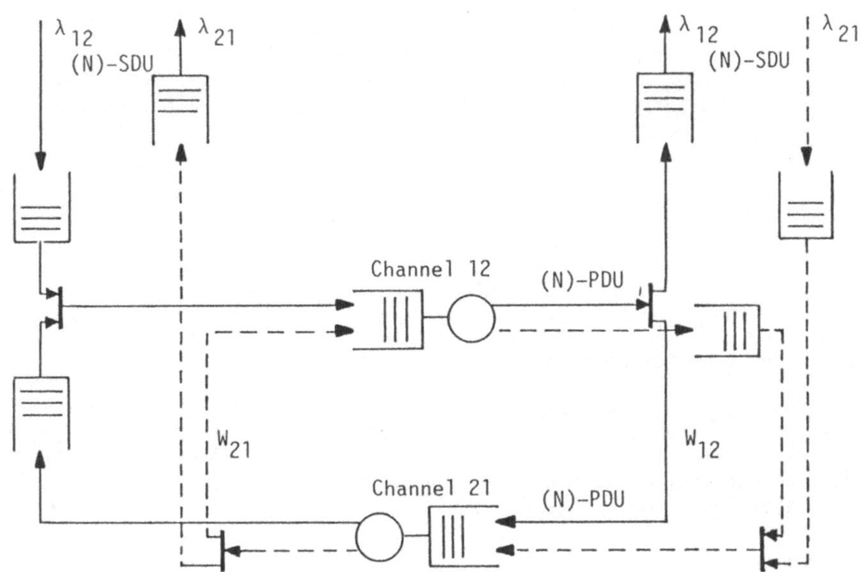

FIGURE 16
Model of a Window Flow Controlled FDX Virtual Connection

Finally, the models of Figs. 15 and 16 do not include the error process and error recovery mechanisms. In some cases as LANS's these may be dropped since their error probabilities are so small that performance is not measurably degraded. Within WAN's and particular on satellite links these processes have to be included. As in the case of HDLC link analyses, an erroneous data unit is rejected and error recovery is invoked upon a sequence error or a time-out. Analytically, these effects can be described by an inflated channel occupation time accounting for enforced idle and retransmission time in the case of an error.

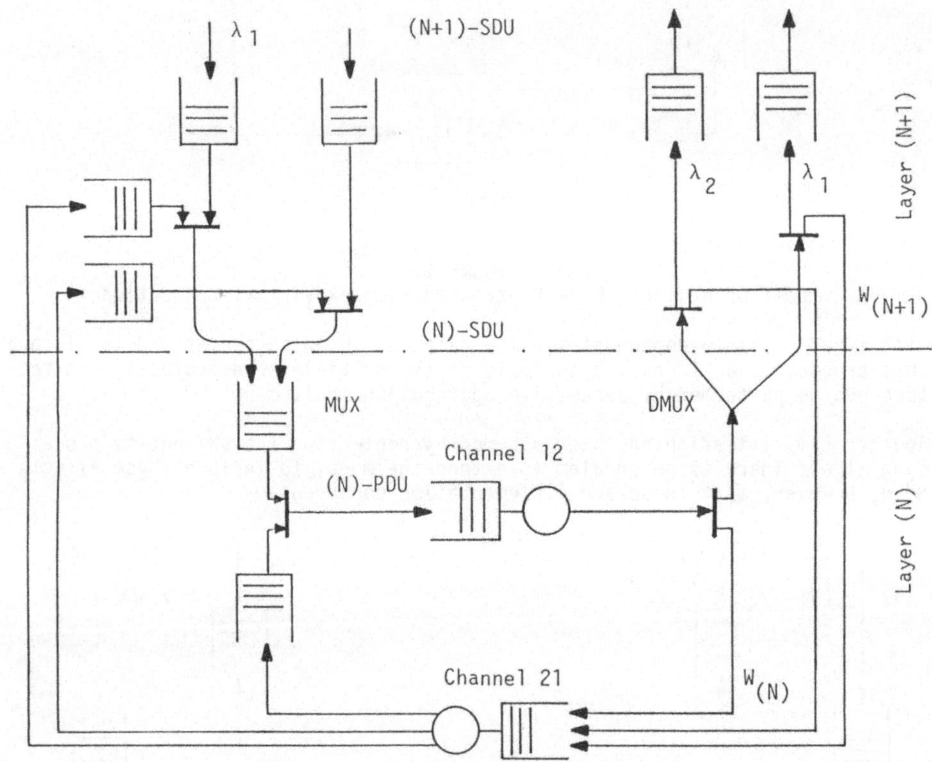

FIGURE 17
Connection Multiplexing (Example: Simplex Connections)

2.4.14 Connection Multiplexing

Several (N+1)-connections may be multiplexed onto one (N)-connection at the origination side and demultiplexed at the destination side. Fig. 17 shows the generic model where the (N+1)- as well as the (N)-connections are window flow controlled. Connection Multiplexing is a basic mechanism and can be found in various layers or network applications:

- Multiplexing of several LAP D-connections onto one Physical Link D-channel for ISDN subscriber access (here, the physical connection is not subjected to flow control)
- Multiplexing of several VC of X.25 layer 3 onto one Dıta Link LAP B-connection of X.25 layer 2

- Multiplexing of several Transport connections for different endsystem applications onto one Network Connection
- Multiplexing of several LLC-connections of layer 2b onto one MAC layer within Local Area Networks.

The analysis can be performed by replacing the whole layer (N)-submodel by an aggregated delay station by which the individual (N+1)-connections fall apart.

Note that in Fig. 17 the acknowledgements of layer (N+1)-connections are transmitted apart from the layer (N)-connection mechanism. There is no principal problem to transfer these acknowledgements within a FDX layer (N)-connection, too.

2.4.15 Connection Splitting

The reverse operation to connection multiplexing is connection splitting by which one (N+1)-connection is divided up into several (N)-connections.

2.4.16 Data Unit Management

In the ISO Basic Reference Model, three principal Data Unit Transformations are defined, see Fig. 18:

- Segmenting/Reassembling
- Blocking/Deblocking
- Concatenation/Separation.

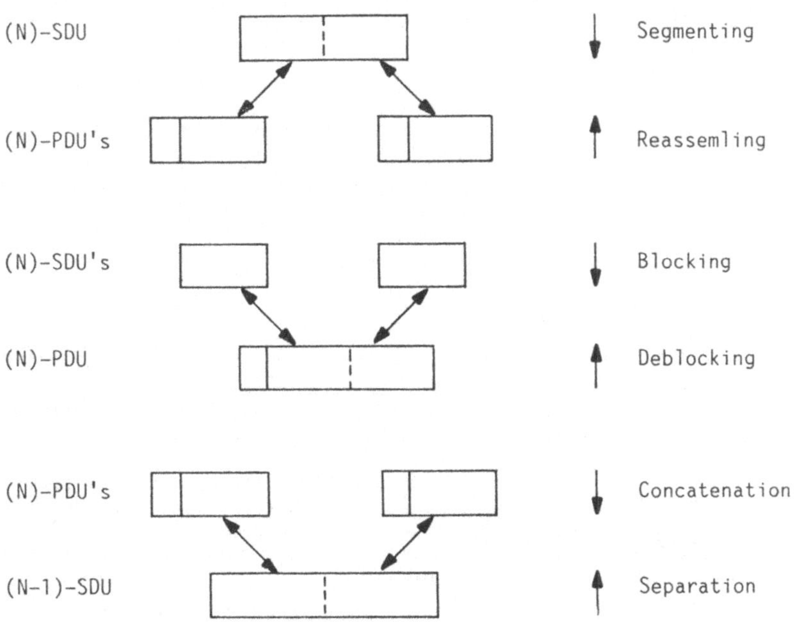

FIGURE 18
Data Unit Management

Data Unit Transformations are necessary when default block sizes of adjacent layers differ as, e.g., in LAN's and PS public data networks. Fig. 19 shows a generic processor model of a layer (N)-entity for Segmenting/Reassembling. Note that another new modelling element enters the stage: Assembly/Reassembly Queues. These queues buffer smaller data units until they can be combined for handover to the next process. Similar models are obtained for Blocking/Deblocking and Concatenation/Separation.

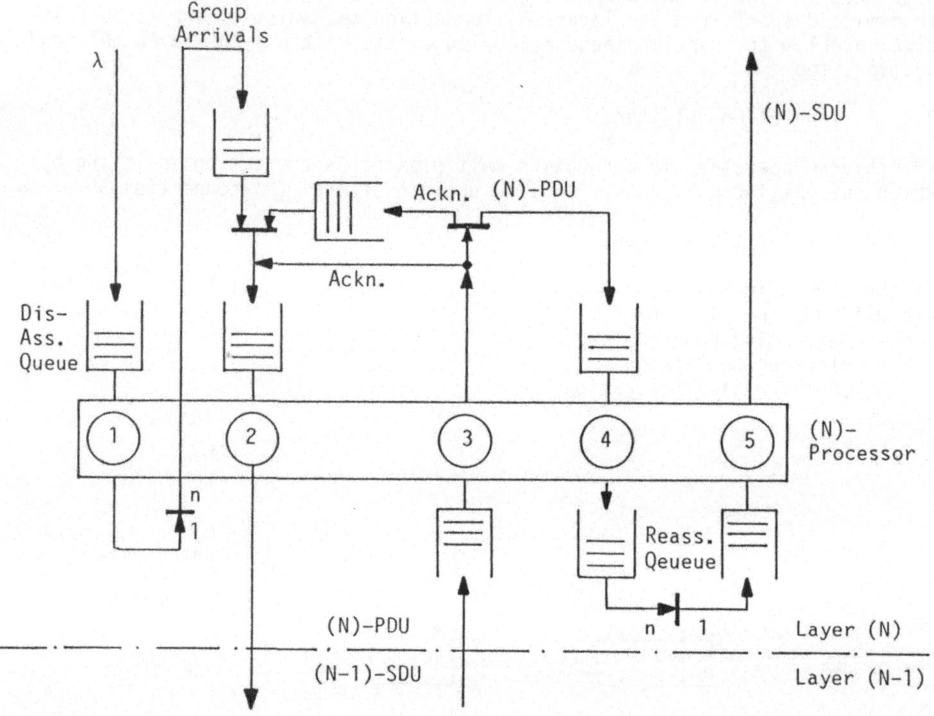

FIGURE 19
Data Unit Management: Segmenting/Reassembling

2.4.17 Network Call Control

To set up a network-wide CS-connection, ISDN uses a separate signalling network for interoffice signalling based on the CCITT Signalling System No.7. The analysis of such networks which are operated thoroughly in the PS-mode, leads to identical or similar basic models as introduced before.

The fast and reliable signalling unloads the information network from inband signalling overhead by approximately 10% of the total load. Besides its basic purpose as signalling vehicle for connection management, the signalling network forms also the basis for intelligent network management and feature support. Future ISDN's will be characterized by

- comparatively simple transport functions as the Bearer and Standard Services
- comparatively complex functions of Higher Services and Supplementary Services resting on network databases and network hosts allowing for feature support and even customer-defined network features.

To give just one example of such an application, Fig. 20 refers to a feature call handling: After initial call processing (phase 1), a Database Access Request is generated to look up or to provide the feature support. Call processing may proceed (phase 2) in parallel to the database access and can only be completed after the arrival of the Database Response (phase 3). This is another example for a model with a branching process and true parallel processing within communication networks.

2.4.18 Internetworking

Internetworking occurs between ISDN and LAN, PSTN, PSDN or between LAN and PSDN. Usually, such networks differ in various points as

- protocol hierarchy
- connection types (CO, CL)
- switching principles (PS, CS)
- protocol data unit sizes
- signalling methods (inband PS, outband CS)

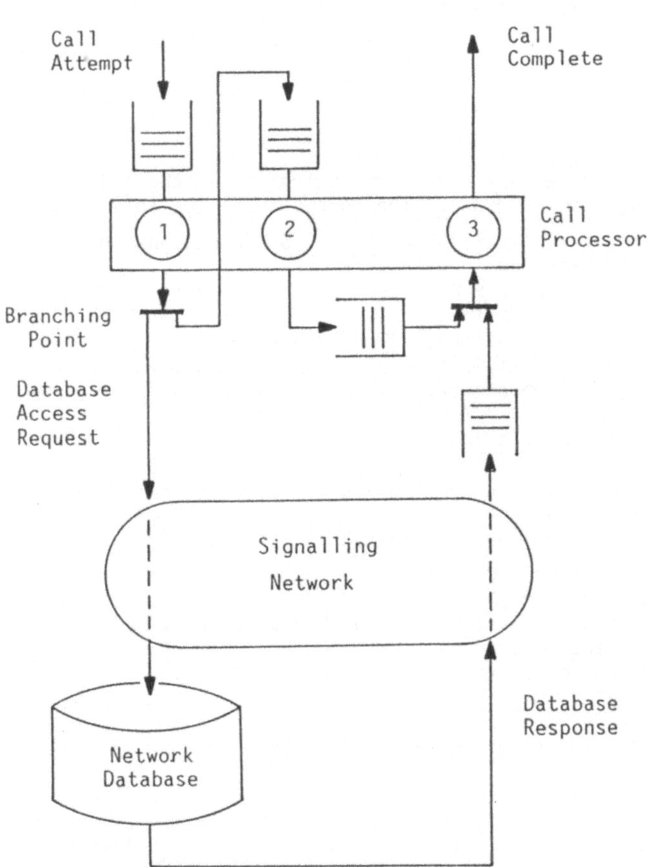

FIGURE 20
Feature Call Handling in ISDN Through Network Database Support

Models of the corresponding internetworking units as Bridges, Routers or Gateways reflect features as

- protocol layer of the relay function
- buffer structures for intermediate storage
- control procedures for transformations of connection types and data units
- flow control for back pressure functions, etc.

2.5. Source Traffic Models

For single service networks of the past, traffic has been modelled by a few parameters. Service integration leads to more complex traffic models with multiple parameters due to quite different service types, the separation of user information flow from control flow and due to variable bandwidth and throughput requirements. This may be illustrated by a few examples.

2.5.1 CS-Connections

Traffic source models for CS-calls for either service consist of

- Interarrival time distribution function
- Bandwidth requirement (number of B-channels)
- Holding time distribution function.

The simultaneous use of multiple services can be described by

- Joint probability of multi-service usage.

2.5.2 Connection Control

CS-connection set up/take downs are controlled in the ISDN using outband signalling through the D-channel. The source model for this channel consists of

- Requests for establishment of LAP D-connections for incoming/ outgoing B-channel-connections
- Holding time of LAP D-connections (as holding time in 2.5.1)
- Signalling information arrival processes from TE to ET and vice versa within each established LAP D-Connection
- Frame length distribution
- Flow control parameter.

2.5.3 PS-Virtual Connection

Packet Switched Virtual Connections (VC) may be established through D- or B-channels in the ISDN. According to the inherent inband signalling method of PS, there is only one source model which consists of

- VC establishment requests
- VC holding time
- Data packet arrival process within an established VC
- Packet length distribution
- Throughput class requirement
- Flow Control parameter.

2.6 Advanced Methods for Performance Analysis

The new traffic issues cannot be treated completely by the well-established traffic theoretic methods. Advanced Techniques of Performance Analysis are necessary since the complexity of the related traffic models has been increased considerably through

- Multi-layered functionalities of the protocol architectures
- Extensive use of microprocessor equipment
- Wide-spanning dependencies through protocol relations
- Introduction of new services.

2.6.1 Advanced Simulation Techniques

Simulation allows the analysis of arbitrarily complex relationships. The increase of discrete time events and the extension of data structures pose the problem of storage and execution time limitations which limits in turn the sufficient gathering of statistical data for comparatively rare events. New approaches direct to

- Hybrid Simulation Techniques
- Aggregated Submodel Simulation Techniques
- Environment Simulation Techniques
- Parallel Simulation Techniques.

2.6.2 Advanced Analysis Techniques

Network planning requires efficient analytical methods and tools. Most of these methods are approximate and have to be validated by simulation. For analytical treatment a major problem is the reduction of the complexity. Basic Methods and approaches are

- Aggregation of subsystems
- Decomposition of complex systems into subsystems
- Timed Petri Net analysis techniques for cooperating processes
- Merging/splitting of nonrenewal point processes
- System analysis under correlated input processes
- Nonstationary process analyses
- Flow time distribution analyses.

3. TRAFFIC ENGINEERING

Traffic Engineering refers to a set of issues which can be described by

- Grade of Service (GOS) Definition
- Traffic Measurements and Forecasting
- Traffic Characterization
- Traffic Analysis Procedures
- Interaction between System/Network Architecture and GOS
- Network Planning
- Network Management.

3.1 Grade of Service Definition

GOS definitions are subjected to international standardization by CCITT. Typical GOS measures are

- Probability of blocking
- Probability of loss
- Probability of delay
- Mean and 95-percentiles of major delays as
  dial tone delay
  post dialling delay
  releasing delay.

Such definitions are principally valid for CS-connections of a single service. They have to be completed with respect to

- multiple services GOS
- GOS under overload/breakdown conditions.

For PS virtual connections similar GOS criteria hold as

- VC establishment delay
- Data packet delay
- GOS for services under negotiation, as achievable maximum throughput of a VC

- GOS for particular network accesses as D-channel or B-channel packet network access.

The various GOS-criteria have to be related to standardized load conditions. For ISDN services a reference customer load should be defined; this question is also related to the busy hour definition for ISDN.

3.2 Traffic Measurements and Forecasting

Traffic measurements are taken for several reasons

- Monitoring of user behavior
- Monitoring of network load
- Monitoring of GOS
- Validation of analysis and planning tools.

New questions of traffic measurements are focused at

- Change of user behavior with respect to new features
- User behavior and traffic values for newly introduced services or new applications
- Busy Hours of individual services
- GOS for new services and features.

Forecasting of traffic and service developments is particularly difficult, since they are highly dependent on user acceptance and the development of applications, terminal systems and networks.

3.3 Interaction between System/Network Architecture and GOS

The development of System and Network Architectures has been largely driven by standardization bodies. Since these bodies work in parallel and are only loosely coordinated, the whole architecture may not always fit to an application or to a real time and performance-efficient implementation. It can be expected that tailored but still OSI-compatible solutions will be implemented for special applications. Traffic Engineering may therefore give an important momentum on future network architectures and their implementation.

3.4 Network Planning and Network Management

Network planning addresses the problem how to structure and how to size network resources according to load, GOS and cost objective functions. This classical issue gets new dimensions in the light of new services and ISDN. Problems related hereto are:

- Objective function definition
- Fixing of planning parameters
- Development of dimensioning methods
- Definition of reference loads and reference busy hours
- Sensitivity of solutions with respect to changes of parameters
- Multi-parameter optimization
- Development of Network Management procedures and practices.

CONCLUSION

With the introduction of digital networks and service integration a major step will be reached into an information society. The evolution of public and private networks requires large amounts of capital investment. Traffic Engineering will form a major part in the development, planning and operation of new networks and services. To play this part responsibly and constructively, much work is still to be done to penetrate and to understand the issues and to support development, planning and operation accordingly. This paper has been written in this spirit. It was not possible to go into details and to evaluate the various approaches more deeply, but it is hoped nevertheless to have contributed to the classification of the issues.

Der Einfluß des Wiederholrufeffekts auf die Leistungsgrößen von Verlustsystemen

U. Herzog, M. Paterok, C. Vogel *

Universität Erlangen-Nürnberg
Institut für mathematische Maschinen und Datenverarbeitung VII
Lehrstuhl für Rechnerarchitektur und -Verkehrstheorie
* DATEV Nürnberg

Zusammenfassung

Für die Planung von Systemen mit Rufabweisung sind Wiederholrufe ein schwer erfaßbares Phänomen. Sie entstehen durch das Verhalten abgewiesener Rufer, binnen kurzer Zeit weitere Versuche zur Belegung eines Kanals zu unternehmen.
In dieser Arbeit der Effekt von Wiederholrufen auf die Leistung von Verlustsystemen ermittelt. Hierzu werden Systeme, in denen Rufwiederholung auftritt, zum einen durch ein Modell ohne, zum anderen durch ein Modell mit expliziter Berücksichtigung der Wiederholrufe dargestellt. Dabei wird besonderen Wert darauf gelegt, daß beide Modellierungsansätze korrekt verfolgt werden. In diesem Fall läßt der Vergleich der Ergebnisse beider Modelle eine Abschätzung der Auswirkung von Wiederholrufen zu. Es wird gezeigt, daß dieser Effekt erheblich ist. Ferner wird eine starke Sensibilität der Ergebnisse bzgl. der Parameter, die das Wiederholerverhalten beschreiben, nachgewiesen. Die Konsequenzen der Ergebnisse für die Anwendung von Verlustmodellen in der Praxis werden kurz aufgezeigt.

Einführung

Reine Verlustsysteme mit mehreren Servern stellen ein wichtiges Element zur Modellierung von Systemen dar, in denen die fehlende Möglichkeit des Wartens zum Verlust eintreffender Aufträge bei voll belegten Bedieneinheiten führt.

Bereits 1918 hat Erlang mit seinen Formeln einen wichtigen Grundstock für die mathematische Behandlung solcher Systeme gelegt [ERLA18]. Sein Modell mit negativ exponentiell verteilten Ankunftsabständen eignet sich insbesondere zur Beschreibung von (einstufigen) Telefonsystemen, in denen man häufig von unabhängigen Rufen aus einem großen Reservoir potentiellen Anrufer ausgeht. Eine der besonderen Stärken dieses Modells liegt darin, daß es sich aufgrund seiner wenigen Modellparameter hervorragend zur Auswertung mittels Tabellen eignet - eine Eigenschaft, die angesichts der zunehmenden Möglichkeiten lokal verfügbarer Rechenkapazitäten nicht mehr die Bedeutung früherer Tage besitzt. Dennoch sind diese Tabellen auch heute noch eine der wichtigsten Hilfsmittel zur Dimensionierung von Telefonnetzen.

Die so ermittelten Leistungswerte basieren auf der Annahme der Unabhängigkeit der Rufe voneinander. Diese Eigenschaft gilt in Telefonsystemen in der Regel nicht mehr, wenn das Telefonsystem stark belastet ist. Es entsteht "unruhiger" Verkehr, zurückzuführen auf den sogenannten Wiederholrufeffekt : abgewiesene Anrufer starten innerhalb kurzer Zeit weitere Versuche zur Belegung eines freien Kanals. Als Folge davon schwillt der Ankunftsstrom bei stark belasteten Kanälen u. U. erheblich an. Für die Vermittlungseinrichtungen bedeutet jeder erfolglose Anwählversuch eine nutzlose Mehrbelastung, die sie in der Abarbeitung des anfallenden Verkehrs zusätzlich behindert. So kann sich die Wirkung dieses Effekts zu Zeiten besonders starken Verkehrs bis in die Nähe des Systemzusammenbruchs steigern.

Es sind zahlreiche Arbeiten veröffentlicht worden, die sich auf vielfältige Weise mit der Modellierung dieses Phänomens befassen. Eine Auswahl dieser Versuche findet sich in [BRET70, FALI87, GRIL79, JONI70, LEGA70, MACF79, MYSK76, STEP85, TRAN82]. Vielfach handelt es sich dabei um Grenzwertbetrachtungen zur Untersuchung verschiedener Extremfälle, deren praktische Umsetzung sehr erschwert ist. In einigen Modellen wird auch der Strom der Wiederholrufe einfach dem Strom der Rufe im Erlang'schen Modell hinzugefügt, wodurch sich im Modell mit Rufwiederholung eine sehr viel höhere Ankunftsrate ergibt als im Erlang'schen Modell. Aufgrunddessen weichen die ermittelten Leistungswerte um Größenordnungen voneinander ab.

Bei einer Modellierung derartiger Systeme sollte jedoch beachtet werden, daß das Erlang'sche Modell die wiederholten Rufe nicht gänzlich unberücksichtigt läßt. Tatsächlich integriert dieses Modell die Wiederholer in den negativ exponentiell verteilten Ankunftsstrom und unterschlägt somit also nur die Abhängigkeit solcher Rufe von der Vorgeschichte. Bei Modellen mit expliziter Rufwiederholung setzt sich der Ankunftsstrom dagegen aus den Erstrufern und den Wiederholrufern zusammen, wobei die Erstrufe mit konstanter Rate eintreffen und exponentiell verteilte Ankunftsabstände haben, während die Wiederholrufe einer lastabhängigen Verteilung unterliegen.

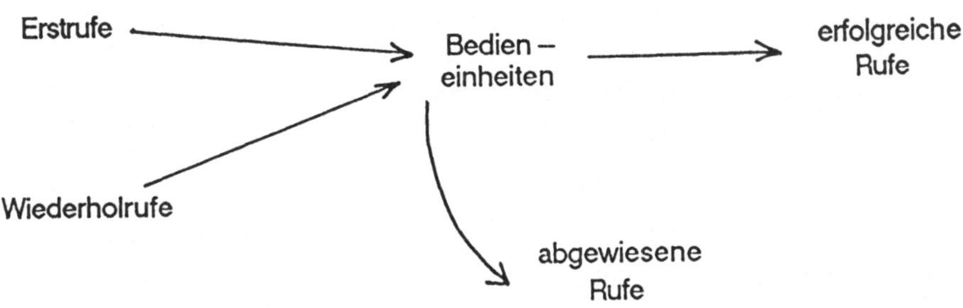

Abb. 1: Ankunfts- und Abgangsströme bei Rufwiederholung

Die Gesamtzahl der Belegungsversuche an den Kanälen darf also in beiden Modellen nicht voneinander abweichen. Aufgrund der verschiedenen Verteilungen der Ankunftsabstände werden sich die ermittelten Leistungswerte der Modelle mit bzw. ohne Rufwiederholung dennoch voneinander unterscheiden. Inhalt dieser Arbeit ist die systematische Untersuchung dieses Unterschieds.

Die praktische Relevanz dieser Untersuchung ist offensichtlich : sehr häufig werden auf Grundlage der gemessenen Kanalbelegungen die (nicht gemessenen) Abweiswahrscheinlichkeiten mit Hilfe von auf dem Erlang-Modell basierenden Tabellen berechnet. Diese Vorgehensweise führt in vielen Fällen zu starken Fehlbewertungen.

Notation

λ	Ankunftsrate der Rufe (im Erlang-Modell)
β	Ankunftsrate der Erstrufe
μ	Bedienrate einer Bedieneinheit
μ_0	Wiederholrate eines abgelehnten Rufes im Wiederholraum
$C = \mu_0/\mu$	normierte Wiederholrate
n	Anzahl der Bedieneinheiten
m	Anzahl der Plätze im Wiederholraum
Z_1	Rufwiederholwahrscheinlichkeit für abgewiesene Erstrufer
Z_2	Rufwiederholwahrscheinlichkeit für abgewiesene Wiederholrufer
Y	Belastung des Systems (mittlere Anzahl gleichzeitig belegter Bedieneinheiten)
P_c	Abweiswahrscheinlichkeit eines Rufes an den Bedieneinheiten
λ_s	Durchsatz durch die Bedieneinheiten

Ein Modell ohne explizite Rufwiederholung

Aus bereits zuvor geschilderten Gründen wurde aus dieser Klasse von Modellen das sogenannte Erlang-Modell M/G/n/0 [ERLA18] herangezogen, dessen Aufbau aus Abb. 2 zu ersehen ist.

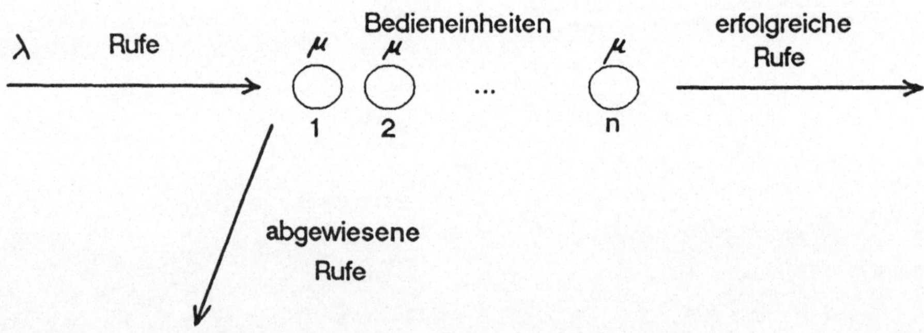

Abb. 2: Aufbau eines Erlang'schen Verlustmodells

Mit negativ exponentiell verteilten Zwischenankunftszeiten treffen Rufe mit Rate λ auf die n Bedieneinheiten, wo sie entweder mangels einer freien Bedieneinheit abgewiesen werden oder eine Einheit im Mittel für die Dauer $1/\mu$ belegen.

Unter diesen Annahmen sind wichtige Leistungsgrößen des Systems sehr kompakt beschreibbar. Da diese Formeln i. a. bekannt sind, sei nur auf die in diesem Zusammenhang wichtigsten Größen eingegangen [KLEI75].

Aufgrund der zustandsunabhängigen Ankunftsrate ist die Abweiswahrscheinlichkeit identisch mit der Wahrscheinlichkeit, daß alle Bedieneinheiten belegt sind :

$$p_n = \frac{(\lambda/\mu)^n/n!}{\sum_{k=0}^{n}(\lambda/\mu)^k/k!} = E_{1,n}(\lambda/\mu)$$

Der Durchsatz durch die Bedieneinheiten als Rate der erfolgreichen Rufe berechnet sich mit

$$\lambda_s = \lambda\,(1 - E_{1,n}(\lambda/\mu)) \quad ,$$

die Belastung als mittlere Anzahl belegter Bedieneinheiten mit

$$Y = \lambda_s \,/\, \mu$$

Sevast'yanow zeigte im Jahre 1957, daß diese von Erlang hergeleiteten Formeln auch für allgemein verteilte Bedienzeiten gültig sind [SEVA57].

Ein Modell mit expliziter Rufwiederholung

Das an dieser Stelle verwendete Modell muß besondere Eigenschaften bzgl. des Ergebnisvergleichs, dem Zweck dieser Untersuchung, aufweisen. Insbesondere im Hinblick auf die Auswertbarkeit und die Vergleichbarkeit mit den Ergebnissen des Modells von Erlang erwies sich dabei das Modell von Tran-Gia [TRAN82] als am vorteilhaftesten. Die in [TRAN82] gemachten Annahmen und Herleitungen mußten allerdings den Gegebenheiten angepaßt werden (näheres siehe [VOGE87]).

Das so abgeleitete Modell läßt sich wie in Abb. 3 gezeigt darstellen.

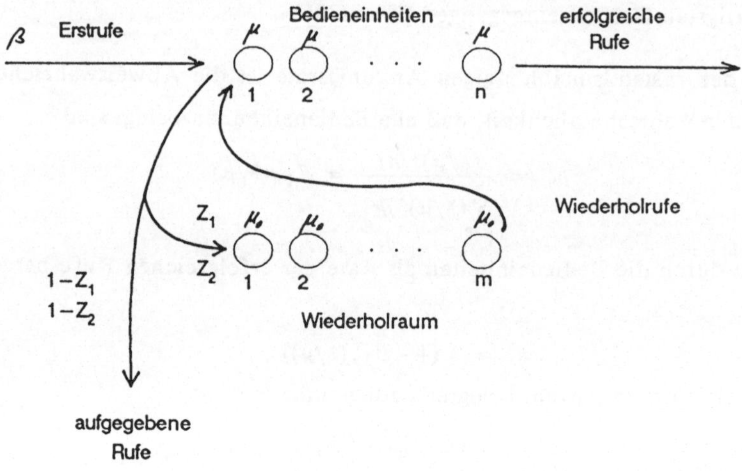

Abb. 3: Modell mit Rufwiederholung

Die Erstrufe treffen mit Rate β an den n Bedieneinheiten an, wo sie entweder mit Rate μ bedient oder aber abgewiesen werden. Abgewiesene Erstrufe verlassen mit der Wahrscheinlichkeit $1-Z_1$ das System oder begeben sich mit komplementärer Wahrscheinlichkeit in den Wiederholerraum, wo sie im Mittel für die Zeit $1/\mu_0$ verbleiben. Danach unternehmen sie als Wiederholrufe einen erneuten Versuch, eine Bedieneinheit zu belegen. Sind sie dabei erneut erfolglos, so verlassen sie mit Wahrscheinlichkeit $1-Z_2$ das System oder begeben sich mit Wahrscheinlichkeit Z_2 nochmals in den Wiederholerraum. Abgewiesene Wiederholrufe verhalten sich wie abgewiesene Erstrufe. Es wird also zwischen Wiederholruf- und Erstrufwiederholwahrscheinlichkeit unterschieden [BRET70]. Alle Zeiten gehorchen der exponentiellen Verteilung.

Der Wiederholerraum wird aus Gründen der Auswertbarkeit als endlich angenommen, erwies sich aber für hohe Genauigkeiten als hinreichend groß [VOGE87] (vgl. auch [STEP84]). Bei vollem Wiederholerraum abgewiesene Aufträge verlassen das System.

Zur Berechnung der Leistungsgrößen wird der Zustandsraum des Systems aufgestellt, der (m + 1) * (n + 1) Zustände umfaßt. Mit Hilfe eines rekursiven Verfahrens [HERZ75, TRAN82, VOGE87] können die Zustandswahrscheinlichkeiten im Gleichgewicht auf effiziente Art berechnet werden. Die Verkehrsströme im Modell lassen sich aus diesen Werten auf einfache Weise herleiten. Die Rate der Belegungsversuche an den

Bedieneinheiten ist die Summe aus den Raten der Versuche der Erst- und der Wiederholrufer. Die Abweiswahrscheinlichkeit ist der Quotient aus dem Strom der abgewiesenen Versuche und dem der Versuche überhaupt. Eine genaue Ableitung dieser Größen wird im Rahmen dieses Beitrags nicht dargestellt und kann in [VOGE87] nachgelesen werden.

Die Einschränkung dieses Modells auf negativ exponentiell verteilte Bedienzeiten stellt unseres Erachtens die Aussagekraft des Vergleichs im wesentlichen nicht in Frage, da die Verteilungen gemessener Belegungszeiten häufig nicht allzu stark von der Exponentialverteilung abweichen.

Nunmehr können die Ergebnisse der beiden Modelle miteinander verglichen werden.

Das Konzept des Vergleichs

Wie bereits geschildert, muß für einen Vergleich die Anzahl der Belegungsversuche in beiden Modellen gleich sein. Es bietet sich nun an, die Gleichheit dieser Größen über eine Regulation der Ankunftsrate der Erstrufe im System mit Rufwiederholung zu erreichen und dann die Ergebnisse der beiden Modelle miteinander zu vergleichen. Da in der Praxis die Anzahl der Versuche jedoch häufig nicht gemessen wird, sondern Anzahl und Dauer der Belegungen, wurde der Durchsatz durch die Bedieneinheiten (bzw. deren Belastung) als gleichzusetzende Größe ausgewählt.

Das Vergleichsverfahren gliedert sich demnach wie folgt :

1) Einstellung des Modells ohne Rufwiederholung auf eine vogegebene Belastung durch Regulation des Angebots.

2) Einstellung des Modells mit Rufwiederholung auf dieselbe Belastung durch Regulation des Angebots an Erstrufern.

3) Vergleich der Leistungsgrößen der beiden Modelle.

Systeme ohne Rufwiederholung werden vom Erlang'schen Modell sehr gut abgebildet. Andererseits liefert das Modell von Tran-Gia gute Resultate für Systeme mit Rufwiederholung. Der Ergebnisvergleich liefert also sowohl Aussagen über den Effekt der Rufwiderholung als auch über die Unzulänglichkeit des Erlang'schen Modells hinsichtlich des Rufwiederholungseffekts.

Ergebnisvergleich (für Standard-Benutzerverhalten)

Der Ergebnisvergleich erfolgt nach der Regulation der Rate der erfolgreichen Rufe in beiden Modellen auf den gleichen numerischen Wert. Dieser Wert liegt häufig neben der Anzahl der Kanäle und ihrer mittleren Belegungsdauer pro Ruf als Meßergebnis vor. Wir können damit von unterschiedlich genauen Modellen desselben Systems sprechen. Das Modell mit expliziter Rufwiderholung sollte in diesem Zusammenhang als genauer angesehen werden. Dies gilt umso mehr, als das Erlang'sche Verlustmodell ein Spezialfall dieses Modells ist (Wiederholwahrscheinlichkeiten auf Null setzen). Aufgrund der zuvor erwähnten Insensibilität bzgl. höherer Momente der Bedienzeitverteilung [SEVA57] werden in diesem Fall keine exponentiell verteilten Bedienzeiten mehr verausgesetzt.

Besonders geeignet ist dabei der Vergleich der Abweiswahrscheinlichkeiten in beiden Modellen als gemeinhin interessanteste Leistungsgröße und der Vergleich der Raten der Belegungsversuche an den Bedieneinheiten als Abschätzung der zusätzlichen Mehrbelastung im realen System.

Hierzu werden zwei Vergleichsgrößen definiert :

$$K_P = \frac{\text{Abweiswahrscheinlichkeit im Modell mit Rufwiederholung}}{\text{Abweiswahrscheinlichkeit im Modell ohne Rufwiederholung}}$$

$$K_A = \frac{\text{Rate der Belegungsversuche im Modell mit Rufwiederholung}}{\text{Rate der Belegungsversuche im Modell ohne Rufwiederholung}}$$

Die Definition dieser Größen macht sie als Korrekturfaktoren für die Werte von Erlang-Tabellen geeignet.

Die folgende Abb. 4 zeigt die Entwicklung von K_P über der Systembelastung. Die Belastung ist bzgl. der Anzahl der Kanäle normiert. Das Benutzerverhalten wurde in den Größen C, der normierten Wiederholrate, und den Wiederholwahrscheinlichkeiten für Erstrufe Z_1 und für Wiederholrufe Z_2 parametrisiert. Die hier aufgeführten Größen resultieren teils aus Messungen [VOGE87], teils sind sie geschätzt. Durch Variieren der Anzahl der Bedieneinheiten ergibt sich eine Kurvenschar.

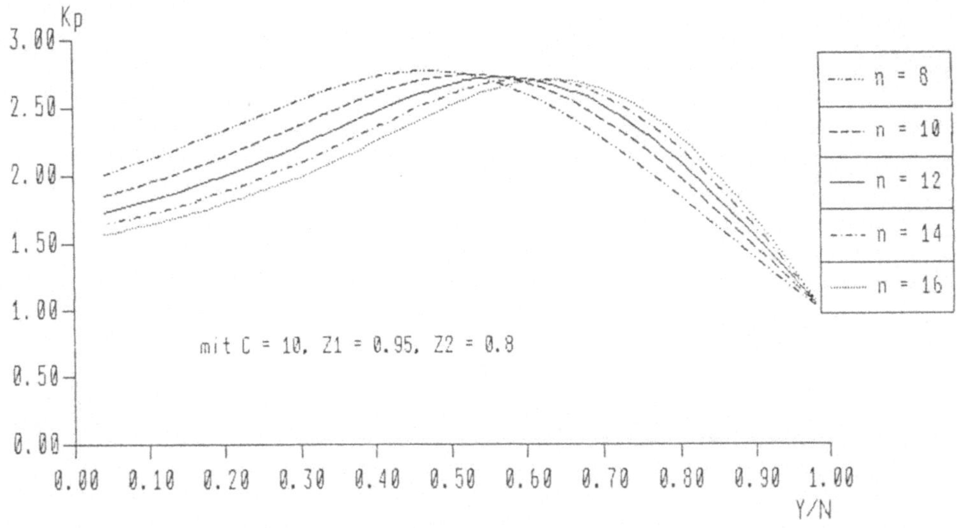

Abb. 4: Entwicklung der Abweiswahrscheinlichkeit über der Belastung

Die Kurven zeigen ein sehr ähnliches Verhalten. Sie sind bei höherer Serverzahl erwartungsgemäß etwas robuster in Bezug auf den Anstieg, nicht jedoch auf die Höhe der Abweichung. Der maximale Korrekturfaktor liegt bei 2.75. In diesem Fall beträgt der Wert nach Erlang nur 36% des Wertes bei expliziter Rufwiederholung. Wichtig ist, daß der maximale Fehler im Bereich von 0.45 - 0.65 normierter Belastung auftritt, also unter bei solchen Systemen häufig anzutreffender Betriebs- oder Spitzenlast.

Die für kleine Last bereits hohen Fehler sind die Quotienten sehr kleiner Werte, der Abfall bei hoher Last rührt daher, daß sich die numerischen Werte der Abweiswahrscheinlichkeiten in beiden Modellen gegen 1 entwickeln.

Abb. 5 zeigt den Fehler für die Anzahl der Belegungsversuche unter den bei Abb. 4 beschriebenen Bedingungen. Bereits bei der praktisch relevanten Belastung von 0.6 ergibt sich ein um 25% erhöhter Anfrageverkehr an den Bedieneinheiten aufgrund zusätzlicher abgewiesener Aufträge. Der Abfall bei sehr hoher Last rührt von der Division sehr großer Werte her (die Differenz der Ergebnisse beider Modelle steigt weiterhin an) [VOGE87]. In diesem Zusammenhang muß allerdings darauf hingewiesen werden, daß die Mehrbelastung der Kanäle durch abgewiesene Belegungsversuche im hier vorgestellten Modell nicht explizit erfaßt wird.

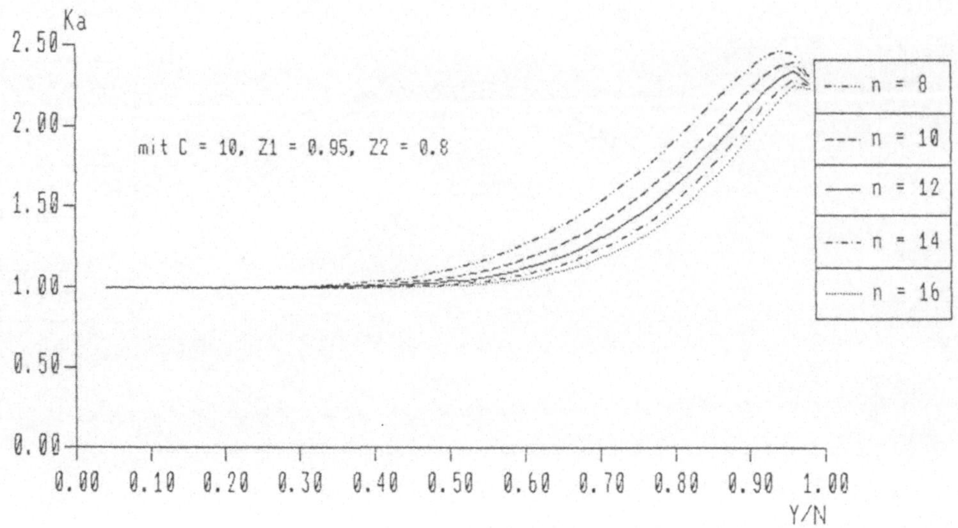

Abb. 5: Entwicklung der Anzahl der Belegungsversuche über der Belastung

Insgesamt gesehen sind die Abweichungen auch bei Standard-Benutzerverhalten bereits beträchtlich. Ferner sind diese Differenzen ohne systematische Untersuchung kaum abzuschätzen. Dies wird auch das folgende Kapitel zeigen. Eine explizite Modellierung des Wiederholrufeffekts muß in diesem Rahmen nicht als verzichtbare Ergänzung vorhandener Modelle, sondern als notwendiger Bestandteil anwendbarer Modelle gesehen werden.

Sensibilität bezüglich des Benutzerverhaltens

Die zuvor gezeigten Ergebnisse beruhen auf der Annahme eines genau ermittelten Benutzerverhaltens. Eine Untersuchung des Systems unter verschiedenen Benutzerverhalten erscheint insbesondere aus zwei Gründen wichtig :

- Die Parameter, die das Benutzerverhalten beschreiben, sind in der Regel schwer zu ermitteln, da in den Vermittlungseinrichtungen keine Wiederholrufe erkannt werden können.
  Die Sensibilität hinsichtlich Parameterschätzfehler muß also stark beachtet werden.
- Das Benutzerverhalten kann sich aus verschiedenen Gründen mit der Zeit ändern und somit verschiedene Prognose-Szenarien nötig machen.

Einfluß der Wiederholwahrscheinlichkeiten

In Abb. 6 ist der Ergebnisvergleich für die Abweiswahrscheinlichkeiten bei verschiedenen Annahmen über die Wiederholwahrscheinlichkeiten festgehalten.

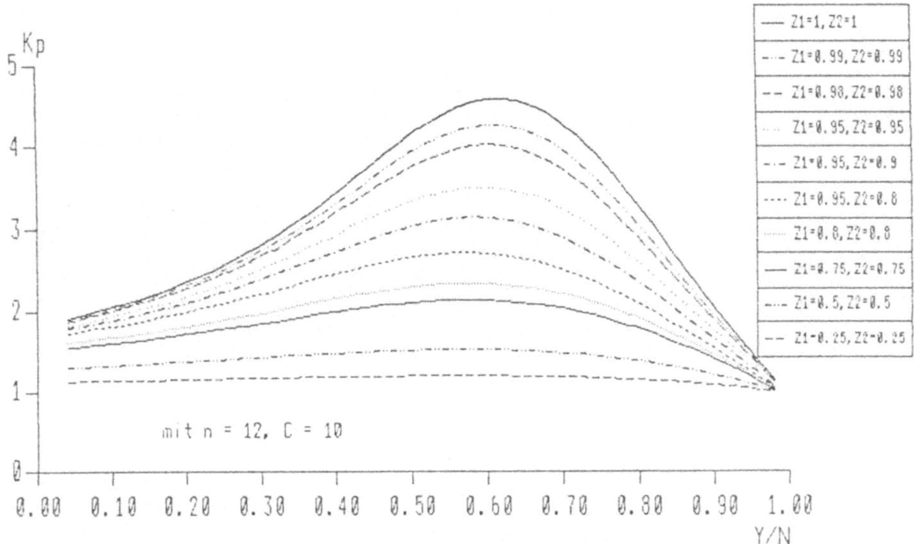

Abb. 6: Abweiswahrscheinlichkeiten bei verschiedenen Wiederholwahrscheinlichkeiten

Deutlich erkennbar ist die sehr starke Sensibilität bei hohen Wiederholwahrscheinlichkeiten. Ob ein Anrufer immer oder fast immer wiederholt, macht einen beträchtlichen Unterschied. Mit sinkender Wiederholwahrscheinlichkeit nimmt die Sensibilität ab. Der Spitzenfaktor 4.5 bei nicht aufgebendem Anrufer ist beachtenswert.

Eine derartig starke Sensibilität wirft erhebliche praktische Probleme auf, da die Ergebnisse bei unsicher geschätzten Parametern erheblichen Schwankungen unterliegen. Daß diese Parameter zumeist nicht genauer bekannt sind, kann zusätzliche Messungen notwendig machen, bei denen man aufgrund gemessener Abweiswahrscheinlichkeiten auf das Benutzerverhalten rückschließt. Das setzt allerdings die Richtigkeit des Modells voraus, läßt also keine Validierung des Modells an sich zu.

Einfluß der Wiederholrate

Die Kurvenschar in Abb. 7 zeigt die Vergleiche der Abweiswahrscheinlichkeiten für verschiedene normierte Wiederholraten. Die Anzahl der Bedieneinheiten und die Wiederholwahrscheinlichkeiten sind in allen Kurven gleich.

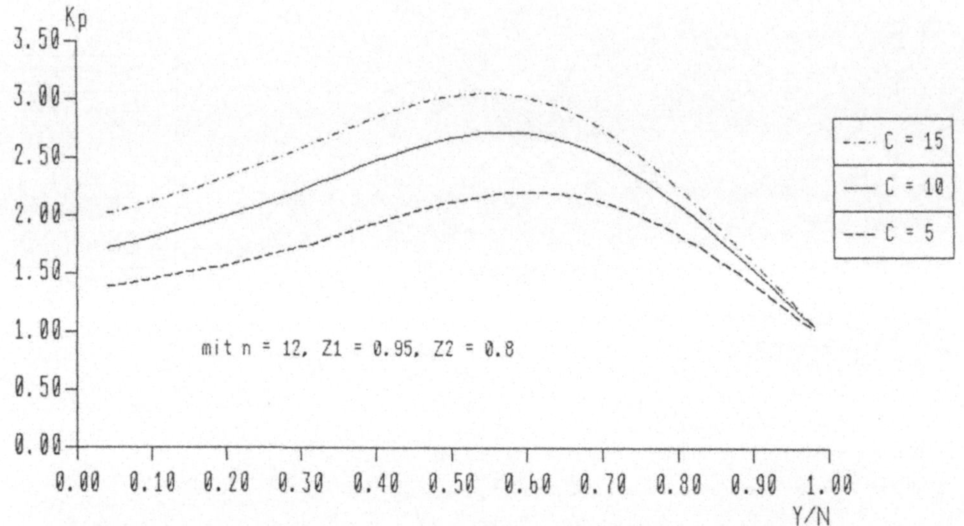

Abb. 7: Abweiswahrscheinlichkeiten bei verschiedenen Wiederholraten

Zwar ist der Einfluß der Wiederholrate (wachsendes C bedeutet kürzere Wieuerholabstände) auf die Ergebnisse nicht so stark wie der der Wiederholwahrscheinlichkeiten, doch ist er immer noch groß. Zudem können neue Techniken zur automatischen Rufwiederholung in Komforttelefonen und angeschlossenen Personal-Computern die Wiederholrate in gefährliche Höhen treiben. Daß derartige Steigerungen erhebliche Probleme aufwerfen, ist auch aus Abb. 8, in der eine andere Darstellung gewählt wurde, ersichtlich.

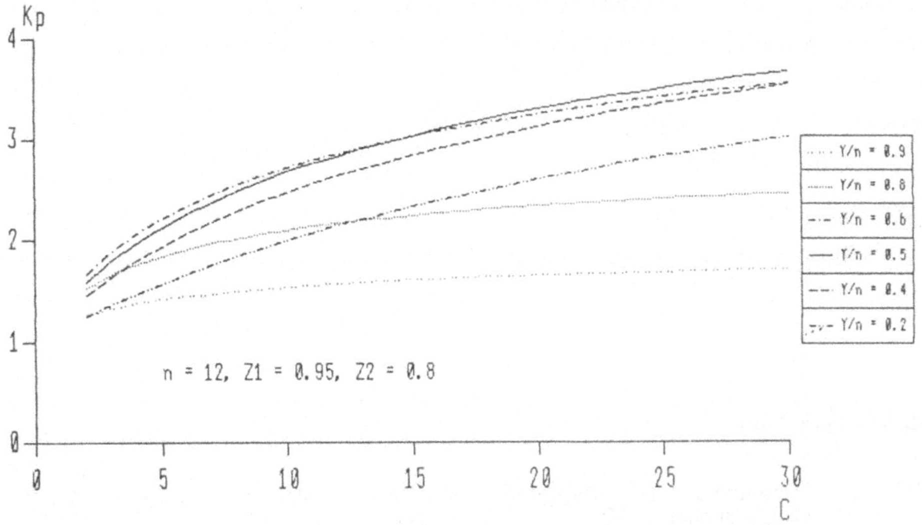

Abb. 8: Entwicklung der Abweiswahrscheinlichkeiten über der Wiederholrate

Hier ist K_P als Funktion über die Wiederholrate aufgetragen. Die Kurven unterscheiden sich in der gewählten Belastung.

Diese Größe ist nicht so schwer zu schätzen wie die Wiederholwahrscheinlichkeiten, aber dennoch mit einem Unsicherheitsfaktor behaftet. Dieser variiert natürlich mit den jeweils gewählten Wiederholwahrscheinlichkeiten: z. B. ergeben sich ohne aufgebende Rufer wesentlich höhere Faktoren. Die Sensibilität bzgl. des gesamten Benutzerverhaltens als Vereinigung der beiden Parameter ist somit als sehr stark einzustufen. Für die praktische Anwendungen erweist sich dies als sehr hinderlich.

Zusammenfassung

Ein systematischer Ansatz zur Bewertung des Einflusses von Wiederholrufen auf die Leistungsgrößen von Verlustsystemen wurde vorgestellt. Hierfür wurde gezeigt, daß sich die Ergebnisse von Modellen mit und ohne expliziter Modellierung von Wiederholrufen erheblich unterscheiden. Die dabei auftretenden Fehlerfaktoren sind so beträchtlich, daß von einer Modellierung von Systemen mit Wiederholrufeffekt ohne explizite Berücksichtigung dieses Effekts gewarnt werden muß. Angesichts der heutigen Rechnerkapazitäten erscheint daher die Aufstellung neuer Tabellenwerke unter Einbeziehung von parametrisiertem Wiederholerverhalten angebracht.

Es konnte ferner nachgewiesen werden, daß die Ergebnisse sehr stark von den Parametern abhängen, die das Wiederholerverhalten beschreiben. Daß es zumeist sehr schwierig ist, diese Parameter zu bestimmen, macht die Anwendung von Modellen mit Rufwiederholung problematisch. Andererseits bedeuten die gezeigten Ergebnisse, daß der weitverbreitete Gebrauch von Erlang-Tabellen zu beträchtlichen Fehlbewertungen führen kann. Fehlerfaktoren können diese Fehler abschätzen, weshalb die vorgestellten Graphiken auch als Fehlerdiagramme für Erlang-Tabellen verwendet werden können.

Literaturverzeichnis

[BRET70] Bretschneider, G.: Repeated Calls with Limited Repetition Probability, Proc. 6th ITC, München, 1970.

[ERLA18] Erlang, A.K.: Lösung einiger Probleme der Wahrscheinlichkeitsrechnung von Bedeutung für die selbsttätigen Fernsprechämter, Elektrotechnische Zeitung, 39, 504-508.

[FALI80] Falin, G. I.: Communication Systems with Repeated Calls, Probl. Peredachi Inform., 16, 2, 1980, 83-91.

[FALI87] Falin, G. I.: Multichannel Queueing Systems with Repeated Calls Under High Intensity of Repetition, Journal of Information Processing and Cybernetics, EIK 23, 1, 1987, 37-47.

[GRIL79] Grille, D.: Telephone Network Behaviour in Repeated Attempts Environment - A Simulation Analysis, Proc. 9th ITC, Torremolinos, 1979.

[HERZ75] Herzog, U.; Woo, L.; Chandy, K. M.: Solution of Queuing Problems by a Recursive Technique, IBM Journal of Research and Development, 19, May 1975, 295-300.

[JONI70] Jonin, G. L.; Sedol, J. J.: Telephone Systems with Repeated Calls, Proc. 6th ITC, München, 1970.

[KLEI75] Kleinrock, L.: Queueing Systems Vol. 1, John Wiley & Sons, New York, 1975.

[LEGA70] LeGall, P.: Sur l'influence des repetitions d'appels dans l'ecoulement du trafic telephonique, Proc. 6th ITC, München 1970.

[MACF79] Macfayden, N. W.: Statistical Observation of Repeated Attempts in the Arrival Process, Proc. 9th ITC, Torremolinos, 1979.

[MYSK76] Myskja, A.; Aagesen, F. A.: On the Interaction between Subscribers and a Telephone System, Proc. 8th ITC, Melbourne, 1976.

[SEVA57] Sevast'yanov, B. A.: An Ergodic Theorem for Markov Processes and its Application to Telephone Systems with Refusals, Theory of Probability and its Applications, II, 1, 1957, 104-112.

[STEP84] Stepanow, S. N.: Numerical Calculation Accuracy of Communication Models with Repeated Calls, Problems of Control and Information Theory, 13, 6, 1984, 371-381.

[STEP85] Stepanow, S. N.; Tsitovich, I. I.: The Model of a Full-Available Group with Repeated Calls and Waiting Positions in the Case of Extreme Load, Problems of Information and Control Theory, 14, 1, 1985, 25-32.

[TRAN82] Tran-Gia, P.: Überlastprobleme in rechnergesteuerten Fernsprechvermittlungs-systemen - Modellbildung und Analyse, Dissertation, University of Siegen, 1982.

[VOGE87] Vogel, C.: Belegtwahrscheinlichkeiten an Sammelanschlüssen, Diplomarbeit, Universität Erlangen-Nürnberg, IMMD VII, 1987.

ZÄHLMONITOR 4 :
Ein Monitorsystem für das Hardware- und Hybrid-Monitoring von Multiprozessor- und Multicomputer-Systemen

Richard Hofmann
Rainer Klar
Norbert Luttenberger
Bernd Mohr

Universität Erlangen-Nürnberg
IMMD 7
Martensstr. 3
D-8520 Erlangen
Federal Republic of Germany

Zusammenfassung

Die Beobachtung (das Monitoring) der inneren Abläufe in Multiprozessor- und Multicomputer-Systemen ist ein hervorragendes Werkzeug, um die Ursachen für die extern festgestellte Gesamtleistung solcher Systeme zu analysieren und damit die Voraussetzungen für eine Leistungsverbesserung zu schaffen. Im vorliegenden Papier wird ein Monitorsystem vorgestellt, das speziell für (auch räumlich verteilte) Multiprozessor- und Multicomputer-Systeme entworfen wurde. Seine wesentlichen Eigenschaften sind die Kombination von verteilter und zentraler Instrumentierung des Objektsystems, die systemweite Ermittlung von Ereignisreihenfolgen und die quellbezogene Auswertung von Ereignisspuren. Neben dem globalen Konzept werden vertiefende Untersuchungen dargestellt, und es wird von ersten Meßerfahrungen berichtet.

Schlüsselwörter

Monitoring, Instrumentierung, Leistungsanalyse, Multiprozessor- und Multicomputer-Systeme, verteilte Systeme, Ereignisspuren, Quellbezug, Beschreibungssprachen, VLSI-gestütztes Monitoring.

1. Einleitung

Die Leistung von Multiprozessor- und Multicomputer-Systemen ist in einem hohen Maß durch die Konfiguration des Systems, die Verteilung der Last auf die einzelnen Prozessoren bzw. Computer und nicht zuletzt durch den Entwurf, die Implementierung und die Nutzung der Kommunikationsdienste bestimmt.

Um die Leistung solcher Systeme zu bestimmen, geht man zunächst von pauschalen Leistungsmaßen wie Durchsatz oder Antwortzeit aus. Dieses Vorgehen liefert Leistungsaussagen, ist jedoch insoweit nicht befriedigend, als solche Maße kaum in der Lage sind, die dem beobachteten Verhalten zugrundeliegenden Ursachen aufzudecken. In den allermeisten Fällen ist man jedoch gerade hieran interessiert, um z.B. ein existierendes System zu verbessern oder eine Aussage für den Entwurf eines neuen Systems zu gewinnen.

Will man eine solche Einsicht in das innere Verhalten eines Multicomputer-Systems oder eines Multiprozessor-Systems gewinnen, liegt der meistversprechende Weg darin, die Realzeit-Folge der Aktivitäten der in den jeweiligen Prozessoren bzw. Computern aktiven Prozesse und ihrer Wechselwirkung aufzudecken. Dadurch ergibt sich für den Betreiber bzw. Entwerfer des Systems in den allermeisten Fällen ein direkter Hinweis, warum sich das System so verhält, wie es sich verhält. Kennt man aber die Ursachen des beobachteten Systemverhaltens, so weiß man oft schon, wie es verbessert werden kann.

Ein hervorragendes Werkzeug, um solche Einsicht in den inneren Ablauf von Rechensystemen zu gewinnen, ist das Monitoring, sprich die Beobachtung von Rechensystemen. In diesem Aufsatz soll ein kombiniertes Hardware-/Hybrid-Monitorsystem vorgestellt werden, das unter dieser Zielperspektive ("Gewinn von Einsicht") insbesondere für das Monitoring von Multiprozessor- und Multicomputer-Systemen entwickelt wurde und wird.

Bevor dieses von uns als Zählmonitor 4 (ZM4) bezeichnete Monitorsystem im dritten Kapitel in seiner Konzeption und im vierten Kapitel in Teilaspekten (Zeitbasis, Auswertung, dedizierte Meßelektronik) beschrieben wird, sollen im folgenden zweiten Kapitel einige grundsätzliche Ausführungen zum Monitoring von Multiprozessor- und Multicomputer-Systemen gemacht werden. Im fünften Kapitel schließlich wird von ersten Meßerfahrungen berichtet.

2. Monitoring von Multicomputer- und Multiprozessor-Systemen

Sind die Monitoring-Objekte Multiprozessor- und Multicomputer-Systeme, so treten einige neue Probleme auf, die beim Monitoring von zentralen Einprozessor-Systemen nicht gelöst werden müssen, und einige "alte" Fragestellungen des Monitoring erscheinen in einem neuen Licht. Dies wird erst in voller Schärfe deutlich, wenn man von räumlich verteilten Objektsystemen ausgeht, weshalb die im folgenden betrachteten Systeme auch immer verteilte Systeme sein sollen.

2.1 Knoten- und Netz-Monitoring

Denkt man sich vereinfacht ein verteiltes System aus den (Verarbeitungs- und ggf. Durchschalte-) Knoten und dem Verbindungssystem zwischen diesen Knoten aufgebaut, so gibt es naheliegenderweise zwei Möglichkeiten zum Monitoring eines solchen Systems (Bild 1):

- Zum einen kann man alle interessierenden Knoten des Objektsystems einzeln mit einem *Knotenmonitor* instrumentieren (Bild 1 A).
- Zum anderen kann man alle Abschnitte des Verbindungssystems mit *Verbindungsmonitoren* instrumentieren (Bild 1 B). Falls das Verbindungssystem eine einfache Bus-, Stern- oder Ring-Topologie mit Broadcast-Charakteristik hat, so kann das gesamte System mit einem einzigen Verbindungsmonitor, der an einem zentralen "Abhörpunkt" anzusetzen ist, instrumentiert wer-

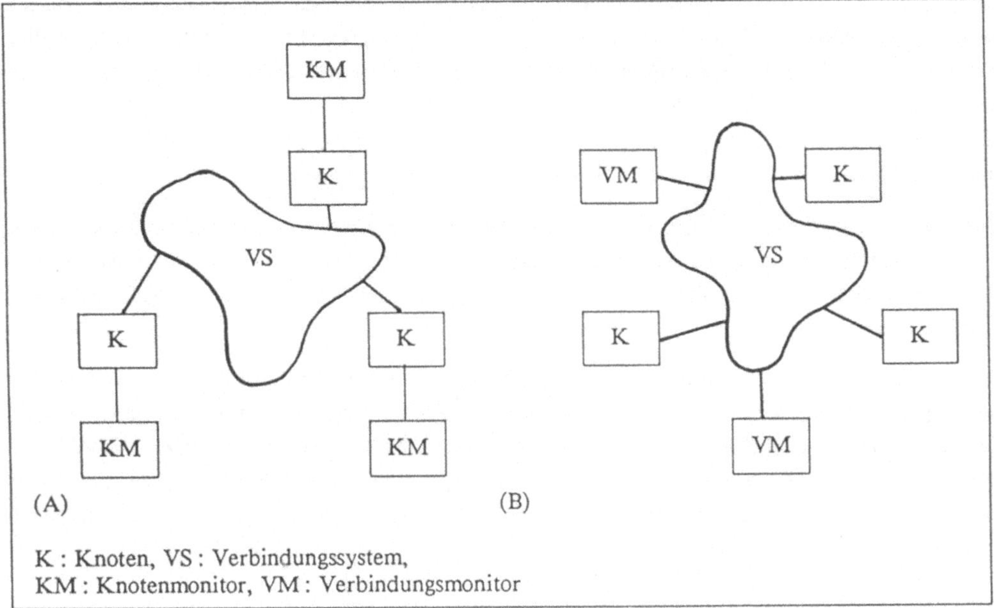

Bild 1. Knoteninstrumentierung (A) und Verbindungsinstrumentierung (B)

den.

Knoten- und Verbindungsmonitoring zeichnen sich durch jeweils spezifische Vorteile aus:

- Knotenmonitore gestatten nicht nur die Beobachtung des äußeren Knotenverhaltens, das sich an der Schnittstelle des Knotens zum Verbindungssystem zeigt, sondern sie gestatten auch die Beobachtung des "internen" Knotenverhaltens, aus dem sich oft interessante Aspekte für die Systemanalyse ergeben.
- Der Vorteil von Verbindungsmonitoren dagegen liegt in der einfachen Implementierbarkeit und in der fast immer möglichen rückwirkungsfreien Betriebsweise: Als Verbindungsmonitore werden meist zusätzliche Knoten eingesetzt, die den gesamten Datenverkehr auf dem Verbindungssystem "abhören"; für dieses Verfahren braucht oftmals keine eigene Monitor-Hardware entwickelt werden, und im Abhörbetrieb ergeben sich keine Rückwirkungen auf das beobachtete System. Es läßt sich aber damit das "interne" Verhalten eines Knotens nicht beobachten.

Besonders sinnvoll sind gemischte Instrumentierungen mit Knoten- *und* Verbindungsmonitoren, da sich so die Aussagen beider Instrumentierungstechniken miteinander verbinden lassen. Beim Entwurf des ZM4 sind wir entsprechend von einer solchen gemischten Instrumentierungstechnik ausgegangen, um zu einem möglichst allgemeinen Entwurf zu gelangen (vgl. Bild 2!). In ähnlicher Weise wird ja auch von Burkhart/Millen [BuMi86] betont, daß nur der gemeinsame Einsatz verschiedener Monitortechniken bei Multiprozessor-Systemen erfolgversprechend ist.

2.2 Monitoring verteilter Systeme = verteiltes Monitoring

Wie aus dem vorangehenden Abschnitt schon deutlich geworden ist, hat man es beim Monitoring verteilter Systeme in der Regel nicht mit nur einem Monitor zu tun, sondern mit einem *Monitor-*

system aus einer Menge von *verteilten Monitoren*, die für unterschiedliche Meßaufgaben an unterschiedlichen Stellen des Objektsystems eingesetzt werden. Für räumlich verteilte Systeme ist dies unmittelbar einsichtig; es ist jedoch auch für räumlich konzentrierte, aber funktionell verteilte Systeme gültig: Ein einzelner Monitor kann nur eine begrenzte Anzahl von unabhängigen Ereignisströmen behandeln, eine Anzahl, die von der Zeitauflösung, den Ereignisabständen, der zulässigen Monitorrückwirkung und bestimmten Anwenderanforderungen abhängt.

Um aus den Einzelbeobachtungen der verteilten Monitore eine Gesamtaussage über das Verhalten des beobachteten Systems zu gewinnen, müssen diese Monitore als *verteiltes Monitorsystem* zusammenarbeiten. Die Architektur des Objektsystems spiegelt sich also in der Architektur des Monitorsystems wider.

Da die Struktur eines Objektsystems im Einzelfall zwar fest ist, man mit einem Monitorsystem aber Messungen an höchst unterschiedlich konfigurierten Objektsystemen durchführen möchte, braucht man zum Messen ein ausgesprochen flexibles verteiltes Monitorsystem mit einer Vielzahl von Meßmethoden und -instrumenten und vor allem einer sehr guten *Rekonfigurierbarkeit*, um es entsprechend an eine möglichst große Zahl von Objektsystemen einfach anpassen und anschließen zu können.

Unabhängig von Art und Anzahl der verteilten Monitore wird es in jedem verteilten Monitorsystem jedoch zusätzlich eine *zentrale Instanz* geben, die die Aufgabe hat, die Einzelbeobachtungen zu einer Systembeobachtung zusammenzufassen. Dadurch ergibt sich in der Regel eine Master-Slave-Konfiguration mit einer Menge von verteilten Monitoren als Slaves und einem steuernden und auswertenden Master.

Diese notwendige Kombination von verteilter und zentraler Instrumentierung hat entscheidende Folgen für den Entwurf der verteilten Monitore: Die verteilten Monitore sind nicht als separate Einzelgeräte zu entwerfen, sondern als kommunikationsfähige Einheiten. Dies bedeutet praktisch, daß die verteilten Monitore neben der eigentlichen auf das Objektsystem ausgerichteten Monitorelektronik über einen Rechnerkern (Prozessor, Speicher), über einen Meßdatenpuffer (Haupt- und ggf. Peripherspeicher) und eine Kommunikationsschnittstelle zur Kommunikation mit dem Master verfügen müssen. Diese Zweiteilung wird von uns im folgenden in den Begriffen "dedizierte Monitorsonde" (engl.: Dedicated Probe Unit, DPU) für die Monitorelektronik und "verteilter Monitorcomputer" (Distributed Monitor Computer, DMC) für das Ensemble Rechnerkern, Datenpuffer und Kommunikationsschnittstelle reflektiert.

2.3 Rekonstruktion von Ereignisreihenfolgen

Bei Messungen an Rechensystemen steht man vor der grundlegenden Alternative, ob man eine ereignisgesteuerte oder eine zeitgesteuerte Meßmethode [Klar85] anwenden will.

Beim ereignisgesteuerten Verfahren wählt man bestimmte Vorgänge im Objektsystem aus, die man für signifikant hinsichtlich des Beobachtungsziels hält, und definiert sie als Meßereignisse. Jedes Auftreten eines solchen Ereignisses aktiviert den Datenerfassungsmechanismus des Monitors, der dann einen das Ereignis beschreibenden Datensatz, der insbesondere auch die Ereigniszeit umfaßt, erzeugt, diesen speichert und/oder an nachgeschaltete Monitorinstanzen weiterreicht. Ein solcher Datensatz wird von uns im folgenden als Ereignis-Record (engl.: event-record) oder kurz: E-Record bezeichnet, eine Folge von E-Records als Ereignisspur (event trace).

Beim zeitgesteuerten Verfahren wird die Datenerfassung eines Monitors in regelmäßigen zeitlichen Abständen aktiv, greift auf signifikante Statusinformationen im Objektsystem zu und stellt diese den nachgeschalteten Instanzen zur Verfügung.

Aufgrund des globalen Beobachtungsziels "Gewinn von Einsicht in den inneren Ablauf von Rechensystemen" ist die ereignisgesteuerte Methode die geeignetere, da sie die vollständige Rekonstruktion dieses Ablaufs ermöglicht, wogegen mit zeitgesteuerten Methoden in der Regel nur pauschale statistische Aussagen möglich sind. Im folgenden wird also stets eine ereignisgesteuerte Methode unterstellt.

Will man durch das Monitoring eines verteilten Systems Aussagen über den *kausalen Zusammenhang* zwischen den beobachteten Prozeßabläufen treffen, so muß man die *zeitliche Reihenfolge* solcher Ereignisse, die den Ablauf der Prozesse beeinflussen, eindeutig bestimmen. Werden solche Ereignisse von unabhängigen verteilten Monitoren erfaßt, so können zeitliche Reihenfolgen nur bei Ereignissen vom Typ SEND/RECEIVE logisch rekonstruiert werden. Basiert die Kommunikation im beobachteten System jedoch nicht auf dem message-passing-Konzept, sondern auf dem Konzept gemeinsamer Variablen im gemeinsamen Speicher (shared-variable-Konzept), so ist die Reihenfolge der Operationen, die auf diesen Variablen ausgeführt werden, nicht a priori bestimmbar. Hier können Reihenfolgeaussagen nur über die Zuordnung einer *globalen Zeitinformation* zu jedem einzelnen Ereignis rekonstruiert werden. Anders ausgedrückt: Um eine konsistente Sicht des Verhaltens des gesamten verteilten Systems zu gewinnen, ist es im allgemeinen erforderlich, die von den einzelnen (Knoten- und/oder Verbindungs-) Monitoren unabhängig voneinander erfaßten Ereignisströme auf der Basis einer globlaen Zeitinformation zu einem "Gesamt-Ereignisstrom" zusammenzufassen, in dem die beobachteten Ereignisse in aufsteigender zeitlicher Reihenfolge geordnet sind.

Dieses Ziel setzt voraus, daß alle Monitore, die beim Monitoring eines verteilten Systems zusammenwirken, sich einer gemeinsamen Zeitbasis bedienen, die zudem über eine genügend hohe zeitliche Auflösung verfügt, um die gewünschten Reihenfolgeaussagen treffen zu können (bei den anvisierten Beobachtungszielen ca. 100 ns - 10 μs Auflösung).

Dies hat unserer Meinung nach drei entscheidende Konsequenzen:

1. Die Forderung nach einer gemeinsamen Zeitbasis kann im allgemeinen nicht durch isoliert betriebene lokale Uhren erfüllt werden, auch wenn diese hochgenau sind. Zum einen könnte nicht sichergestellt werden, daß diese Uhren zu exakt demselben Zeitpunkt gestartet werden, und zum anderen könnten sie im Verlauf einer längeren Messung über das Maß der Uhrauflösung auseinanderdriften, so daß globale Reihenfolgeaussagen nicht mehr sicher getroffen werden können. Lokale Uhren können in den verteilten Monitoren nur dann verwendet werden, wenn sie nicht isoliert betrieben werden, sondern über eine Synchronisationseinrichtung miteinander verbunden sind. Über diese Synchronisationseinrichtung müssen die lokalen Uhren zum gleichen absoluten Zeitpunkt gestartet werden, und ihre Frequenz muß laufend synchronisiert werden.

2. Wegen der Forderung, die Reihenfolge von unabhängig erkannten Ereignissen anhand einer von der "Monitoruhr" gelieferten Zeitinformation eindeutig bestimmen zu können, müssen alle lokalen Uhren in jedem "Tick" (entsprechend der Uhrauflösung) synchron sein. Um diese

Forderung zu erfüllen, müssen die zusammenwirkenden Monitore über eine *hardwaremäßig* ausgeführte Synchronisationseinrichtung miteinander verbunden sein. Software-Synchronisationsalgorithmen (z.B. [GuZa83], [GoHT87]), wie sie in lokalen Netzen angewendet werden, scheiden wegen der praktisch erreichbaren Synchronisationsgenauigkeit von nur ca. 20 ms aus.

3. "Reines" Software-Monitoring, das nicht mindestens durch eine solche Synchronisationseinrichtung unterstützt wird, scheidet ebenfalls aus, da sich damit nur für bestimmte Klassen von Ereignissen (z.B. SEND/RECEIVE) überhaupt Reihenfolgeaussagen treffen lassen (im Bsp.: Eine Nachricht kann immer nur nach dem Senden empfangen werden). Diese Problematik wird in einigen Software-Monitor-Entwürfen für verteilte Systeme zwar erkannt, aber entweder nicht gelöst (z.B. [Snod82]) oder umgangen, indem man sich auf Ereignisse vom Typ SEND/RECEIVE zurückzieht (z.B. [MiMS86]).

Ein Synchronisationsproblem auf einer anderen Ebene tritt auf, wenn das Monitoring mit zeitlichen Rückwirkungen auf das beobachtete System verbunden ist: Solche zeitlichen Rückwirkungen, die in der Größenordnung weniger Befehlslaufzeiten je Ereignis z.B. für die Ausgabe einer Ereignisidentifikation über eine EA-Schnittstelle beim Hybrid-Monitoring bis hin zu einigen -zig Befehlslaufzeiten je Ereignis beim Software-Monitoring liegen, können dazu führen, daß im instrumentierten System andere Ereignisreihenfolgen beobachtet werden als sie im nicht instrumentierten System tatsächlich auftreten würden. Berg et al. [BeFW82] und Kluge [Klug82] schlagen für diesen Fall vor, mit dem Beginn einer Rückwirkungsphase auf einem Prozessor gleichzeitig alle Prozessoren und die Kommunikationseinrichtungen des beobachteten Systems anzuhalten. Nur so werde garantiert, daß Ereignisfolgen zeitlich unverfälscht beobachtet werden könnten.

Abgesehen von dem zusätzlichen Aufwand, den es bedeutet, ein "globales Anhalten" in räumlich verteilten Systemen zu realisieren, führt dieses Vorgehen nur unter zwei Voraussetzungen zu einer Lösung des dargestellten Synchronisationsproblems:

1. Das beobachtete System arbeitet nicht mit "natürlicher" Last unter Echtzeitbedingungen. Durch das globale Anhalten würden sich zum einen die Rückwirkungen vervielfachen, zum anderen müßten zeitgebundene (z.B. zyklische) Aufträge (wie sie z.B. für Prozeßrechensysteme typisch sind) ja nach wie vor in Echtzeit ausgeführt werden, so daß ein Anhalten des Systemzeitgebers nicht in Frage kommt. Damit würden sich die zeitlichen Bedingungen für das Objektsystem aber ähnlich verfälschen wie ohne globales Anhalten. Entsprechend wird von Berg et al. das globale Anhalten auch nur für Laboruntersuchungen in einem "Testbed" vorgeschlagen, in dem eine vollständige Kontrolle über das beobachtete Rechensystem und seine Umgebung hergestellt werden kann.

2. Es dürfen keine rotierenden Speicher im Objektsystem betrieben werden. Das vollständige Anhalten eines Systems ist auch in einem Testbed nicht möglich, wenn sich im System Geräte mit rotierenden Speichermedien (Platten-, Bandlaufwerke etc.) befinden: Diese Systeme können nicht schlagartig angehalten und ebenso schlagartig wieder gestartet werden, weshalb sich hier in jedem Fall eine Verfälschung des zeitlichen Verhaltens des beobachteten Systems durch Monitor-Rückwirkungen ergibt.

Die "Lösung" dieses Synchronisationsproblems kann nur darin liegen, mit den unvermeidlichen Rückwirkungen "zu leben", ein Weg, den auch Hinrichsen [Hinr86] vorschlägt.

2.4 Auswertesystem

Besondere Anforderungen richten sich nun an ein auf dem zentralen Steuer- und Auswerterechner zu installierendes Auswertesystem:

1. Es soll unabhängig von der Konfiguration des beobachteten Objektsystems sein; d.h., Annahmen über Konfiguration und Betriebsweise des Objektsystems dürfen nicht ins Auswertesystem "einprogrammiert" sein.
2. Das gleiche gilt für das verteilte Monitorsystem: Weder Art und Anzahl der verwendeten Monitore, noch die Struktur der von ihnen abgelieferten Daten und die interne Datendarstellung dürfen durch das Auswertesystem vorbestimmt sein.
3. Auswertungen sollen quellbezogen erfolgen; d.h., der Benutzer soll unter Bezug auf die ihm geläufigen Namen von Hard- und Softwareobjekten über die Dynamik des Systems informiert werden.

Diesen letzten Aspekt vertreten insbesondere auch Bemmerl [Bemm86] und Burkhart/Millen [BuMi86], indem sie eine enge Verknüpfung des Monitorsystems mit dem Objekt-Programmiersystem herstellen.

Im folgenden Kapitel wird nun der von uns auf der Basis dieser Überlegungen teils entwickelte, teils in Entwicklung befindliche "Zählmonitor 4" vorgestellt.

3. Der Zählmonitor 4

Bei der Entwicklung des ZM4 konnten wir uns auf Erfahrungen mit vorangehenden Monitoring-Projekten abstützen: Die Zählmonitore 1 und 2 [Klar71], [Klar75], [Schr78] waren ausgerichtet auf die Beobachtung von zentralen Eincomputer-Systemen und hatten als wesentliches Ergebnis die Technik des automatischen Komparators für die Erkennung von Meßereignissen, auf den im Pkt. 4.3 noch einmal eingegangen wird. Der Zählmonitor 3 [From83], [Klar81] wurde für Messungen an räumlichen konzentrierten Multiprozessor-Systemen eingesetzt; die eindeutige Rekonstruktion der Reihenfolge von Ereignissen aus unabhängig voneinander beobachteten Ereignisströmen konnte dabei durch eine FIFO-Schaltung gewährleistet werden, in die die einzelnen Monitore die von ihnen erfaßten Ereignis-beschreibenden Datensätze konkurrierend einspeisten.

3.1 Hardware-Architektur des ZM4

Der ZM4 ist eine Master-Slave-Konfiguration aus einem zentralen Steuer- und Auswerterechner und einer variablen Anzahl von verteilten Monitoren als Slaves (Bild 2).

In diesem Konzept stellen der zentrale Steuer- und Auswerterechner, das Meßnetz und die verteilten Monitorcomputer die für das Monitoring von verteilten Systemen notwendige Infrastruktur dar; das eigentliche Monitoring wird von dedizierten Monitorsonden durchgeführt.

Naheliegenderweise wurde die Monitoring-Infrastruktur aus käuflichen Standardrechnern aufgebaut: Der Steuer- und Auswerterechner ist ein 32-bit-Minicomputer (hp 9000/500), die verteilten

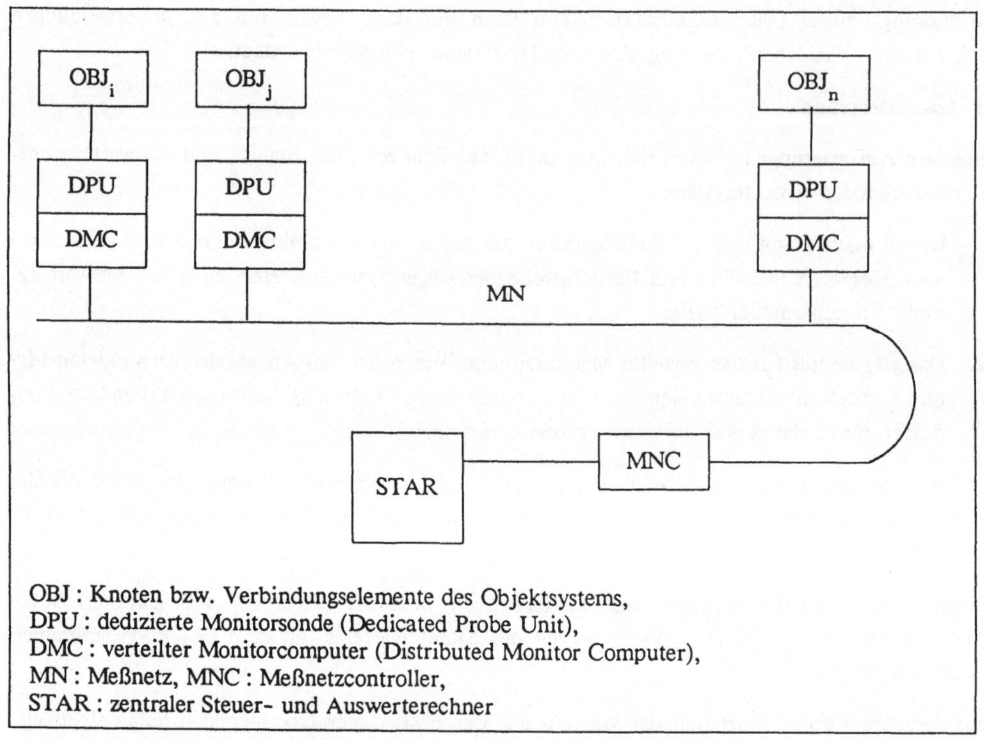

OBJ : Knoten bzw. Verbindungselemente des Objektsystems,
DPU : dedizierte Monitorsonde (Dedicated Probe Unit),
DMC : verteilter Monitorcomputer (Distributed Monitor Computer),
MN : Meßnetz, MNC : Meßnetzcontroller,
STAR : zentraler Steuer- und Auswerterechner

Bild 2. ZM4: Hardware-Architektur

Monitorcomputer sind PC's. Nicht so einfach ist der Aufbau des Meßnetzes: Es besteht aus einem Datenkanal zur Übertragung von Steuerinformationen und Meßdaten zwischen dem Steuer- und Auswerterechner und den verteilten Monitorcomputern und einem Zeitkanal zur Synchronisation der lokalen Uhren der verteilten Monitorcomputer. Der Datenkanal ist wiederum eine käufliche Standardkomponente (Netzwerk nach IEEE 802.3 mit Koaxialkabel), während der Zeitkanal eine Eigenentwicklung ist (s. nächstes Kapitel!).

3.2 Software-Architektur des ZM4

Die Schnittstelle des ZM4-Softwaresystems zum Objektsystem ist das Objekt-Programmiersystem. Die Software-Architektur des ZM4 ist entsprechend darauf ausgelegt, durch sorgfältig definierte Schnittstellen einen möglichst einfachen Anschluß an verschiedene Objekt-Programmiersysteme zu gewährleisten. Bild 3 zeigt ein entsprechendes Szenario.

Der Anwender benutzt das Objekt-Programmiersystem sowohl zur Erstellung von Software für das Objektsystem, als auch zur Formulierung von Monitoring-Anweisungen an den ZM4, mit deren Hilfe das dynamische Verhalten der erstellten Software transparent gemacht werden soll. Die Monitoring-Anweisungen führen einerseits zur Parametrierung des ZM4, andererseits wird im Programmiersystem ein "Schlüssel" zur Dekodierung und quellbezogenen Darstellung der von den verteilten Monitoren unabhängig gewonnenen Ereignisspuren und der vom ZM4 daraus durch einen Mischprozeß generierten "Systemspur" erzeugt. Die Auswertekomponente des ZM4 erzeugt mit Hilfe dieses Schlüssels Auswertungen, die, zum Programmiersystem zurückgeleitet, dem Benutzer den gewünschten Aufschluß geben.

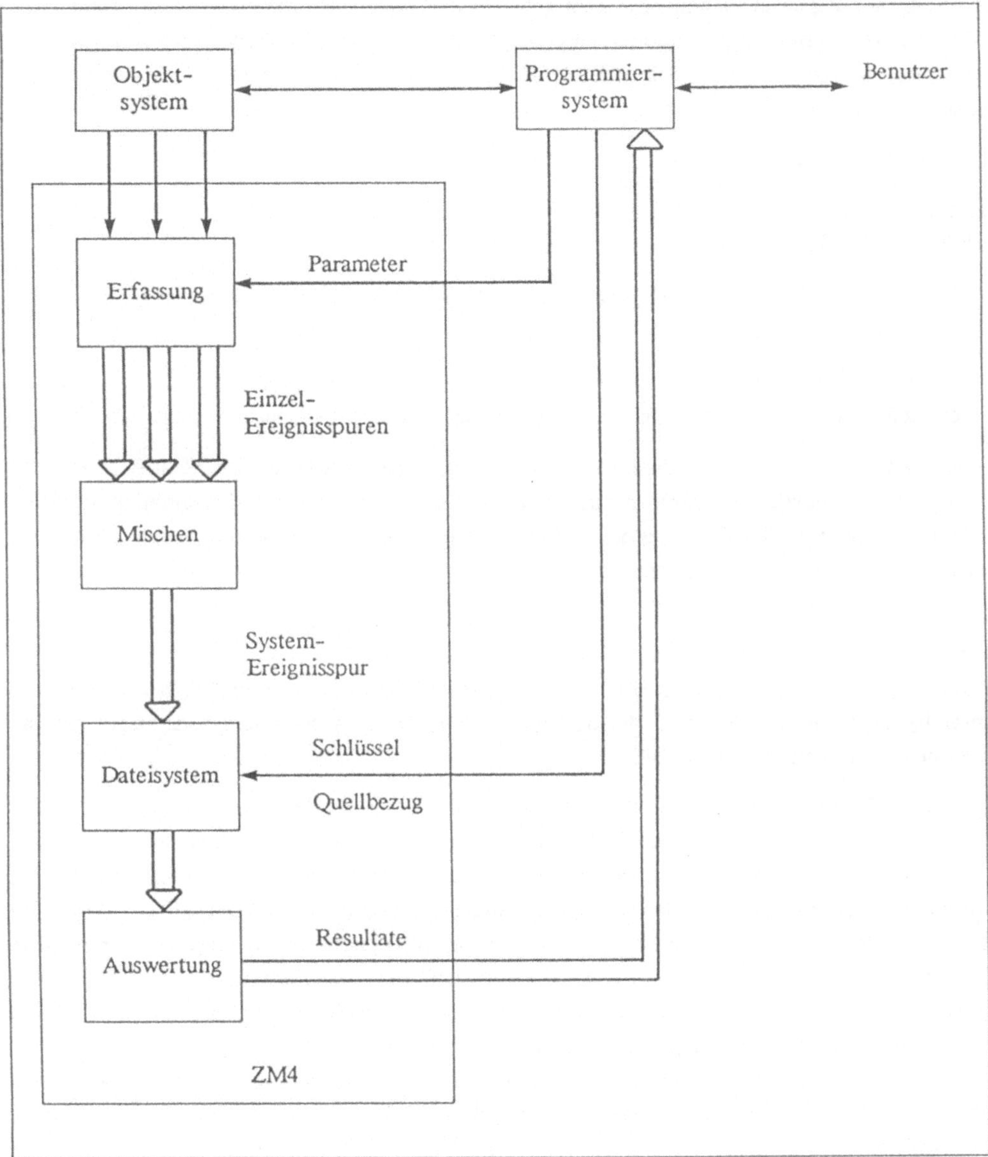

Bild 3. ZM4: Software-Architektur

4. Vertiefende Darstellung von drei ZM4-Komponenten

Ein so großes Projekt wie die Entwicklung des ZM4 läßt sich sinnvollerweise nur in einer vernünftigen Kombination von "horizontaler" und "vertikaler" Arbeitsweise bewältigen. Gemeint ist damit, daß zum einen möglichst weitgespannte Konzepte entworfen werden müssen, die aber nicht notwendigerweise in die Tiefe gehen müssen, sondern eher die Funktion einer "Klammer" haben, um das Projekt zusammenzuhalten; zum anderen müssen bereits in der Frühphase an Stellen von zentraler Wichtigkeit gründliche Detailuntersuchungen unternommen werden, um die realen Schwierigkeiten besser abschätzen zu können.

Das vorliegende Papier stellt nun nach dem globalen ZM4-Konzept drei solche Detailaspekte vor, die in der Hardware-Synchronisationseinrichtung für die lokalen Monitor-Uhren, in den die Auswertung unterstützenden Tools EDL/POET und in der dedizierten VLSI-Monitorsonde "CRAC-Baustein" liegen.

4.1 Die globale Zeitbasis

Die für systemweite Reihenfolgeaussagen notwendige globale Zeitbasis wird in der ZM4-Architektur wie folgt realisiert:

- Es existiert eine einzige hochgenaue Zentraluhr (Master Clock), die für die Bildung des globalen Zeitrasters maßgebend ist.
- Jeder verteilte Monitor verfügt über eine eigene lokale Uhr (Slave Clock), die extern synchronisiert werden kann. Die lokale Uhr kann lokal rückgesetzt, aber nur global gestartet werden.
- Die lokalen Uhren der verteilten Monitore sind über den Zeitkanal des Meßnetzes mit der Zentraluhr verbunden und werden mit der Zentraluhr durch bestimmte Synchronisationsinformationen, die über den Zeitkanal des Meßnetzes übertragen werden, synchronisiert. Diese Synchronisationsinformationen steuern zum einen den synchronen Start der lokalen Uhren und halten zum anderen die Synchronität der Uhren während der Messung aufrecht.

Bei räumlich verteilten Systemen mit einer großen Anzahl von Komponenten ist die Wahrscheinlichkeit von Übertragungsfehlern auf dem Meßnetz nicht mehr vernachlässigbar, so daß es notwendig ist, auch den Zeitkanal des Meßnetzes einerseits zu sichern und andererseits gewisse Fehlertoleranzmaßnahmen vorzusehen.

Übertragungsfehler werden in der Synchronisierschaltung der lokalen Uhren durch eine PLL-Filterung erkannt und in den meisten Fällen unterdrückt. Das Wiederaufsetzen einer Messung nach einem Fehler, der nicht ausgefiltert werden konnte und daher falsche Ereigniszeiten liefern würde, wird durch Synchronisation der Uhren in einem zweiten, gröberen Intervall ermöglicht. Dazu wird die von der Master Clock gebildete Zeit zunächst in gleich lange Abschnitte eingeteilt. Der Beginn eines Abschnittes wird von der Zentraluhr aus auf dem Meßnetz markiert, wobei diese Markierungen von der Zentraluhr zusammen mit den Synchronisationsinformationen über den Zeitkanal des Meßnetzes übertragen werden. Diese Zeitabschnitt-Marken werden in den verteilten Monitoren zusammen mit der dort gebildeten lokalen Zeit und dem Zustand der Synchronisierschaltung (synchron oder nicht) gemeinsam mit den erfaßten Ereignissen in der Ereignisspur abgespeichert.

Diese Dokumentation der Zeitabschnitt-Marken in der Ereignisspur in Verbindung mit den aufgezeichneten Meßereignissen - die ja ebenfalls zeitbehaftet sind -, liefert ein Hilfsmittel, die die registrierten Ereigniszeitpunkte systemglobal hinsichtlich ihrer Verläßlichkeit zu bewerten und ggf. sogar Fehler in der Zeiterfassung nach einer Messung in der Auswertung zu korrigieren.

4.2 Das zentrale Auswertesystem

Die wichtigste Forderung an ein zentrales Auswertesystem ist die Forderung nach der *Integration verschiedener Meßmethoden*, um so das Verhalten eines Objektsystems quasi "gleichzeitig aus verschiedenen Perspektiven" betrachten zu können. Die Forderung wurde erfüllt durch die

Erstellung einer Gesamtauswertung aus den von den verteilten Monitoren unabhängig erzeugten Meßdaten. Dazu muß das Auswertesystem die Struktur der von den unterschiedlichen Monitoren abgelieferten Meßdaten, ihre Bedeutung (Quellbezug) und die Konfigurationen von Objekt- und Monitorsystem kennen. Diese drei Punkte werden im folgenden diskutiert.

4.2.1 Monitorunabhängige Meßdatenauswertung

Die vom Auswertesystem auszuwertenden Meßdaten sind die von uns bereits im Pkt. 2.3 erwähnten Ereignisspuren. Um das Auswertesystem unabhängig von deren Struktur und von verschiedenen Datendarstellungen zu machen, soll es ausschließlich über eine einheitliche Prozedurschnittstelle darauf zugreifen können (s. [Mohr87]!). Dazu wird im ZM4-Konzept eine Ereignisspur als *abstrakte generische Datenstruktur* verstanden.

Eine *abstrakte* Datenstruktur besteht aus einem (meist sehr komplexen) Objekt und Operationen, die auf diesem Objekt oder seinen Komponenten definiert sind. Es kann *nur* über diese Operationen auf das Objekt zugegriffen werden. Sie bilden damit eine Schnittstelle zum Objekt, die möglichst einfach, vollständig und orthogonal sein sollte. Interne Details eines Objektes bleiben dem Benutzer verborgen. *Generisch* bedeutet hier, daß mehrere verschiedene sog. *Instanzen* eines Objektes existieren können. Jede Instanz kann durchaus andere Eigenschaften haben und verschieden parametriert sein; das Erscheinungsbild nach außen hin, also die Operationenschnittstelle, ist jedoch für alle gleich. Das bedeutet insbesonders, daß die *gleiche* Operation bei jeder Instanz die *gleiche* Wirkung hat.

Auf das gestellte Problem angewendet, bedeutet dies, daß jede konkrete Ereignisspur als Instanz der generischen abstrakten Datenstruktur "Ereignisspur" gesehen wird, auf die nur über einen Satz von Prozeduren zugegriffen werden kann.

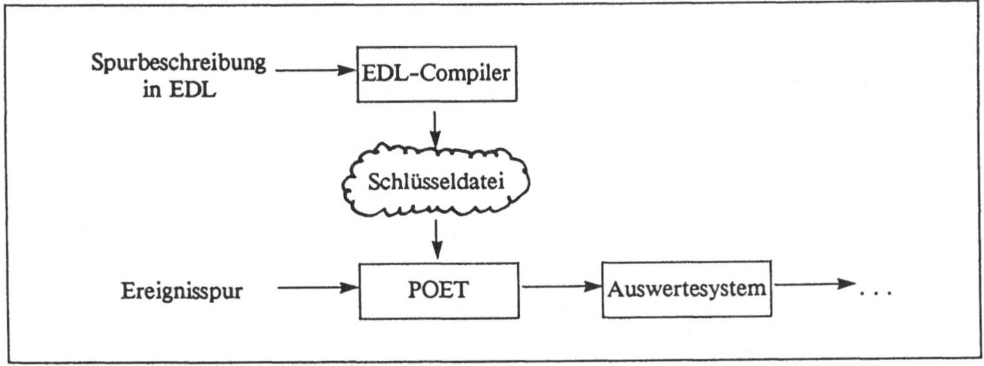

Bild 4. Ereignisspur-Auswertung mit EDL und POET

Die Zugriffsfunktionen für Ereignisspuren, zusammengefaßt unter dem Namen **POET** (Problem Oriented Event Trace interface), benötigen natürlich eine vollständige Beschreibung der Ereignisspur, die in einer Datei, dem sog. *Schlüssel*, zusammengefaßt wird (Bild 4). Ein solcher Schlüssel wird in einem Übersetzungsvorgang aus einer Ereignisspur-Beschreibung mit Hilfe einer Datenbeschreibungssprache gewonnen. Zu diesem Zweck wurde eine speziell für Meßdaten geeignete Beschreibungssprache entworfen, die "Event trace Description Language" **EDL**, mit der sich die Meßdatenformate aller gängigen Monitore beschreiben lassen. EDL wird im nächsten Abschnitt in

Verbindung mit dem Thema "Quellbezug" genauer dargestellt.

4.2.2 Herstellung des Quellbezugs

Beim Entwurf der Datenbeschreibungssprache EDL war es nicht nur das Ziel, Datenstrukturen und -darstellungen von Ereignisspuren beschreibbar zu machen, sondern es sollte darüberhinaus in einem gewissen Umfang möglich sein, einem Auswertesystem auch die Semantik der einzelnen Komponenten eines E-Records mitzuteilen, sodaß eine quellbezogene Auswertung erstellt werden kann. Wie dies erreicht wird, soll im folgenden nach einer kurzen Darstellung der EDL-Beschreibung eines E-Records erläutert (s. auch Bild 5!) werden.

```
EVENT RECORD :
  TAG IS #FF00;
  TOKEN :
     NAME IS EVENT;
     LENGTH IS 2 BYTE;
     VALUES ARE [ 0 .. 3 ];
     INTERPRETATION
       0 = 'Programmstart',
       1 = 'Aufruf Lesen',
       2 = 'Aufruf Schreiben',
       3 = 'Programmende';
  TIME :
     NAME IS ACQUISITION;
     FORMAT IS ( UNSIGNED * 4, s ), ( UNSIGNED * 2, ms );
     MODE IS POINT;
  DATA :
     NAME IS Text_1;
     IN RUCKSACK, LENGTH IN 2 BYTE;
  DATA :
     NAME IS Param_6;
     FORMAT IS UNSIGNED * 2;
```

Bild 5. EDL-Beschreibung eines E-Records

Ein E-Record ist eine Datenstruktur mit einer beliebigen Anzahl von Komponenten, wobei die Komponenten vom Typ TOKEN, TIME, DATA, RUCKSACK, TAG, FILLER oder LENGTH sein können.

TOKEN sind Aufzählungstypen mit endlichem Wertevorrat. Sie werden in EDL benutzt, um E-Record-Komponenten zu beschreiben, in denen der entsprechende Monitor Codierungen für z.B. die Art des jeweils erkannten Ereignisses, den auslösenden Prozessor, den beobachteten Kommunikationskanal etc. ablegt. Der Typ TIME wird benutzt, um die im E-Record abgelegten Zeitinformationen zu beschreiben. DATA und RUCKSACK sind Datentypen mit quasi unendlichem Wertevorrat, wobei der Typ DATA in einer festen Anzahl von Bytes dargestellt ist, der Typ

RUCKSACK in einer variablen Anzahl (er kann nur an einen E-Record angehängt werden, daher der Name). DATA und RUCKSACK werden benutzt, um E-Record-Komponenten zu beschreiben, die der Monitor aus dem beobachteten System beim Erkennen eines Ereignisses extrahiert hat. TAG, FILLER und LENGTH schließlich sind Datentypen, mit deren Hilfe sich die Struktur eines E-Records selbst beschreiben läßt.

Ein Quellbezug wird nun in zwei Stufen hergestellt: Zum einen kann den Komponenten eines E-Records ein *Name* zugeordnet werden, der z.B. im Falle eines DATA-Feldes identisch ist mit Namen der Variablen, deren Wert vom Monitor ausgelesen wird. Die Liste aller verwendeten Namen kann mit POET-Funktionen abgefragt werden, und E-Record-Komponenten können per POET über ihren Namen referenziert werden. Zum zweiten können für alle Werte eines TOKEN-Typs *Interpretationen* hinterlegt werden, sodaß das Auswertesystem die vom Monitor verwendeten Codierungen autonom in die Anwender-definierte Bedeutung rückcodieren kann.

Das Werkzeug-Paar EDL/POET wurde mit Hilfe der UNIX-Tools lex und yacc in C entwickelt und ist unter UNIX System V ablauffähig. Für die weitere Entwicklung des ZM4 ist geplant, EDL-Beschreibungen automatisch schon bei der Spezifikation der Monitoring-Objekte zu erstellen, die z.B. in Interaktion mit der Programmierumgebung des Objektsystems erfolgen kann. Die Mächtigkeit der EDL-Schnittstelle wird dabei auch den Zugang zu solchen Monitoring-Objekten eröffnen, die nicht explizit im Programmiersystem "vorhanden" sind: Beispiele sind Gerätenamen, Rechneradressen, Netzbezeichnungen etc.

4.2.3 Objektsystembeschreibung

Ebenso wie die Konfiguration der Hardware eines Monitorsystems an die Konfiguration des Objektsystems angepaßt sein muß, so muß es auch die Auswertesoftware sein. Dabei kann es nicht das richtige Vorgehen sein, für jedes Objektsystem eine neue Auswertesoftware zu schreiben; stattdessen darf die aktuelle Konfiguration des jeweiligen Objektsystems "nur" als Parameter in die Auswertung eingehen.

Daher wird die Entwicklung eines Werkzeugs angestrebt, mit dem sich die Konfiguration von Multiprozessor- und Multicomputer-Systemen für das Monitoring graphisch interaktiv modellieren läßt. Das Vorgehen wird dabei ähnlich wie beim Entwurf elektrischer Schaltpläne sein: Aus der Verknüpfung vorgefertigter Symbole für Prozessoren, Speicher, Geräte, Kanäle etc. wird ein graphisches Modell erstellt, das in einem nachfolgenden Compilationsvorgang in eine rechnerinterne Darstellung übersetzt wird, die dann im oben beschriebenen Sinn als Auswerteparameter dient.

Die entsprechenden Arbeiten sind noch nicht abgeschlossen.

4.3 Eine dedizierte Monitorsonde in VLSI-Technik: der CRAC-Baustein

Wie in der Einleitung dargelegt, bildet die Beobachtung von *Kommunikationsaktivitäten* einen Schwerpunkt bei der Leistungsmessung von Multiprozessor- und Multicomputer-Systemen. Speziell für eine Klasse von Objektsystemen, nämlich speichergekoppelte Multi-Mikrocomputer-Systeme (MMC-Systeme) wurde deshalb im Rahmen der ZM4-Entwicklung eine eigene dedizierte Monitorsonde entworfen und implementiert, der Chained Reference Address Comparator- (CRAC-) Baustein. Im folgenden werden nun zuerst grundlegende Mechanismen für die Kommunikation in speichergekoppelten Systemen betrachtet und daraufhin die entsprechenden Beobachtungtechniken des CRAC-Bausteins.

Der Grundmechanismus für die Kommunikation in speichergekoppelten Systemen ist die Kommunikation über *gemeinsame Variable*, d.h. Variable, die im gemeinsamen Speicher allokiert sind (in [Spie85] als indirekte oder prozedurorientierte Kommunikation bezeichnet). Die Kommunikation über gemeinsame Variable besteht in ihrem Kern aus Zugriffen auf den Wert dieser Variablen (*Wertzugriffen*), die, um die Konsistenz der entsprechenden Variablen zu erhalten, unter gegenseitigem Ausschluß durchgeführt werden. Als Kommunikationsvorgänge müssen entsprechend Schreib- und Leseoperationen, die auf diese Variablen ausgeführt werden, beobachtet werden.

Aufbauend auf dem Grundmechanismus "gemeinsame Variable" läßt sich in speichergekoppelten Systemen aber auch die Kommunikation über *Nachrichten* realisieren (in [Spie85] als direkte oder nachrichtenorientierte Kommunikation bezeichnet). Nachrichten werden in speichergekoppelten Systemen durch die Elemente von dynamischen Datenstrukturen repräsentiert, wobei jedes Datenelement als "Nachrichtenbehälter" fungiert. Die gesamte Datenstruktur läßt sich entsprechend als Kommunikationskanal interpretieren. Bei nachrichtenorientierter Kommunikation sind die Operationen Senden und Empfangen von Nachrichten zu beobachten. Diese Operationen fallen in speichergekoppelten Systemen mit dem Hinzufügen bzw. Entfernen von Datenelementen zur bzw. aus der den Nachrichtenkanal repräsentierenden Datenstruktur zusammen. Solche Zugriffe werden im folgenden zusammenfassend als *Strukturzugriffe* bezeichnet.

In speichergekoppelten Systemen ist die Basistechnik für die Beobachtung der Kommunikationsaktivitäten eines Prozessors die Überwachung der von diesem Prozessor auf den Bus geschalteten Adressen bzw. Adreßfolgen. Für eine solche Überwachung kommen wegen der oft angewendeten prozessorinternen Befehlspufferung nur *Daten*adressen (und nicht Befehlsadressen) in Frage, so daß aus einer Folge von Datenadressen auf einen Kommunikationsvorgang geschlossen werden muß.

Beim Zugriff auf eine gemeinsame Variable oder eine Nachricht müssen ggf. mehrere Adreßsubstitutionen und Indexrechnungen durchgeführt werden. Diesen Berechnungsvorgang bezeichnet man als Dereferenzierung. Dabei ergeben sich Folgen von Datenadressen, für deren Beobachtung im CRAC-Baustein geeignete Adreßvergleichstechniken zur Verfügung stehen müssen. Wir halten die folgenden drei für wesentlich:

Zugriffe auf den Wert *unstrukturierter Datentypen* können in der Regel mit Hilfe eines *einfachen Adreßvergleichs* auf die Variablenadresse erkannt werden. Zugriffe auf Elemente *statischer Datenstrukturen* können bei der meist angewendeten sequentiellen Allokation der Elemente trotz der zur Dereferenzierung erforderlichen Indexrechnung einfach mit Hilfe eines *Feldgrenzenvergleichs* erkannt werden.

Speziell für die Beobachtung von Wert- und Strukturzugriffen auf dynamische Datenstrukturen wurde die Technik des *geketteten Adreßvergleichs* entwickelt. Dabei wird die Tatsache ausgenutzt, daß das Objektsystem bei jedem Dereferenzierungsschritt über seinen Datenbus eine Adreßinformation zwischen Speicher und Rechenwerk transportiert. Solche Transporte finden beim Zugriff auf die zur Dereferenzierung verwendeten Pointer- und/oder Indexvariablen statt.

Beim geketteten Adreßvergleich nun wird dem CRAC-Baustein vor einem Beobachtungslauf vom zugeordneten Monitorcomputer ein Beobachtungsprogramm eingegeben, das den vom Objektsystem beim Zugriff auf das zu beobachtende Datenobjekt angewendeten Dereferenzierungsalgorithmus beschreibt. Sodann wird vom CRAC-Baustein, ausgehend von der "Wurzelvariablen" (einer Pointer- oder Indexvariablen mit statisch bekannter Adresse), der Zugriff auf die zur Dereferenzierung der beobachteten dynamischen Datenstruktur verwendeten Pointer- und/oder Indexvariablen per Adreßvergleich überwacht, und bei jedem entsprechenden Treffersignal wird die transportierte Adreßinformation "on the fly" vom Datenbus des Objektsystems gelesen. Der CRAC-Baustein bildet daraus entprechend dem hinterlegten Beobachtungsprogramm autonom eine neue Referenzadresse für den nächsten Dereferenzierungsschritt. Auf diese Art und Weise werden dynamisch die Adressen ermittelt, unter denen Wert- oder Strukturzugriffe stattfinden sollen. Strukturzugriffe lassen sich dabei an einer schreibenden Veränderung der zur Referenzierung verwendeten Pointer- und Indexvariablen erkennen.

Die dargestellte Komplexität des CRAC-Beobachtungsverfahrens ist Ausdruck der Tatsache, daß Mikroprozessoren wegen ihrer hohen Integrationsdichte und der im Verhältnis dazu geringen Anzahl nach außen führender Anschlüsse ("Pin-count-Limitation") nur wenige Zustandssignale für einen externen Hardware-Monitor anbieten. Da in MMC-Systemen die CRAC-Elektronik zudem für jeden Knoten des Systems vorhanden sein muß, liegt es nahe, sie selbst in Form eines hochintegrierten Schaltkreises aufzubauen: Es ergibt sich die paradox erscheinende Situation, daß sich die "Beobachtungsfeindlichkeit" von VLSI nur durch VLSI selbst wieder ausgleichen läßt.

Um alle vom beobachteten Prozessor generierten Adressen verfolgen zu können, wird der CRAC-Baustein direkt an dessen Adreß-/Datenbus angeschlossen. Dies bedingt, daß er nur für einen Prozessortyp, in unserem Fall den Intel Prozessor 8086, eingesetzt werden kann.

Aufgrund der zu erwartenden geringen Stückzahlen für einen solchen Beobachtungsbaustein versteht sich von selbst, daß er nicht in "full-custom"-, sondern in "semi-custom"-Technik (und zwar in Standardzellentechnik) realisiert wurde. Der CRAC-Baustein hat ca. 4000 Gatteräquivalente und eine Chipfläche von 87,5 mm^2 (3-μm-CMOS-Technologie mit 2-Lagen-Aluminium-Verdrahtung). Der CRAC-Baustein wurde 1987 gefertigt und befindet sich derzeit in der Test- und Erprobungsphase.

5. Erste Meßerfahrungen

Die modulare Architektur des Monitorsystems ZM4 erlaubt es, Ereignisspuren mit Monitoren unterschiedlichster Bauart zu erfassen. So war es schon vor der Fertigstellung der in Entwicklung befindlichen dedizierten Meßsonden (DPU) möglich, erste Messungen auszuführen. Dazu wurde ein handelsüblicher Logikanalysator als DPU eingesetzt, der Ereignisströme abschnittsweise sehr effizient erfassen kann, jedoch zwischen zwei Meßabschnitten recht lange Erfassungspausen hat, da die erfaßten Daten nicht parallel zur Erfassung an den Steuer- und Auswerterechner übertragen werden können (keine Wechselpufferung).

Das Objekt der Messung war ein Ring aus vier DIRMU-Moduln (DIstributed Reconfigurable MUltiprocessor) des DIRMU 25-Baukastens [Maeh86]. Ein solcher DIRMU-Modul besteht aus einem Prozessor mit seinem privaten Speicher für Code und lokale Daten und einem Multiportspeicher für gemeinsame Daten. Prozessor und Multiportspeicher sind miteinander über jeweils einen Port verbunden und stellen darüberhinaus jeweils sieben weitere Ports für Verbindungen zu anderen Multiportspeichern bzw. Prozessoren zur Verfügung. Bild 6 zeigt die beobachtete Ringkonfiguration, wobei aus Gründen der Übersichtlichkeit die Verbindung der

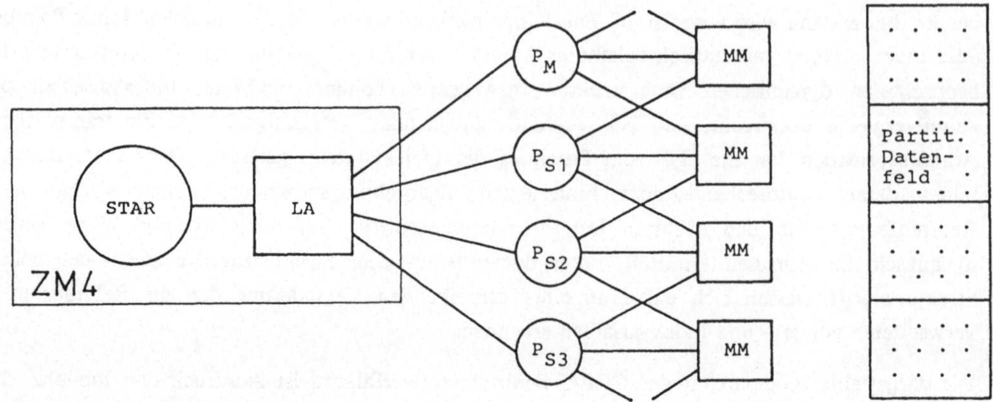

P : Prozessor- und Privatspeicher eines DIRMU-Moduls, MM : Multiport-Speicher,
Index M : Master, Index S1, S2, S3 : Slave 1, 2, 3,
LA : Logikanalysator, STAR : Steuer- und Auswerterechner

Bild 5. Beobachtung der Arbeit einer DIRMU-Ringkonfiguration auf einem partitionierten Gitter von Daten

Randmoduln miteinander nur angedeutet ist.

Die Beobachtung des DIRMU-Rings wurde als Hybridmessung ausgeführt. Dazu wurde jeder der vier beobachteten DIRMU-Moduln an einer parallelen EA-Schnittstelle mit einer 8 bit-breiten Meßsonde des Logikanalysators (LA) instrumentiert. Die Meßsonden empfingen an diesen Schnittstellen Ereigniskennungen, die vom Anwenderprogramm aus durch einen einfachen Ausgabebefehl (mit ca. 5 µs Befehlslaufzeit) ausgesendet wurden. Der Logikanalysator repräsentiert hier also sowohl vier verteilte Monitorcomputer DMC als auch vier dedizierte Monitorsonden DPU. Der LA lieferte in dieser Messung auch die gemeinsame Zeitbasis für alle vier Ereignisströme.

Die Last, unter der gemessen werden sollte, war ein asynchroner paralleler Gauß-Seidel-Algorithmus zur Lösung der Laplace'schen partiellen Differentialgleichung. Dabei bearbeitet jeder der vier DIRMU-Moduln (genannt "Master" und "Slave1,2,3") gemäß Bild 6 einen Streifen des gesamten Gitters der zu bestimmenden Werte. Die entsprechenden Daten sind in den Multiportspeichern der DIRMU-Module aufgehoben. Bei diesem Algorithmus synchronisieren sich die einzelnen Prozessoren nur am Ende der Berechnung, nicht jedoch nach der Berechnung der einzelnen Gitterpunkte. Der Master spielt in diesem Algorithmus nur insofern eine besondere Rolle, als er allein über die Terminierung des gesamten Rechenvorgangs entscheidet.

Uns interessierte bei dieser Messung insbesondere,

— wie sich der Algorithmus lokal auf jedem einzelnen Prozessor bezüglich des Erreichens einer gegebenen Schranke für den minimal zu erreichenden Iterationsfortschritt je Iteration verhält, und
— wie der Algorithmus global terminiert.

Bild 7 zeigt in vereinfachter Form die dazu vorgenommene Programmmarkierung: Von außen nach innen gehend sind Start- und Endepunkt des gesamten Algorithmus, Start- und Endepunkt der einzelnen Iterationschleife und die daran anschließende Meldung des Erreichens der lokalen Terminierungsbedingung an den Master und schließlich innerhalb der Iterationschleife Start- und Endepunkt der eigentlichen Iteration und die daran anschließende elementweise Überprüfung des Iterationsfortschritts markiert (•). Bei jedem Durchlaufen dieser Marken wird eine Ereigniskennung an die Meßschnittstelle ausgesandt und vom LA als Ereignis erfaßt.

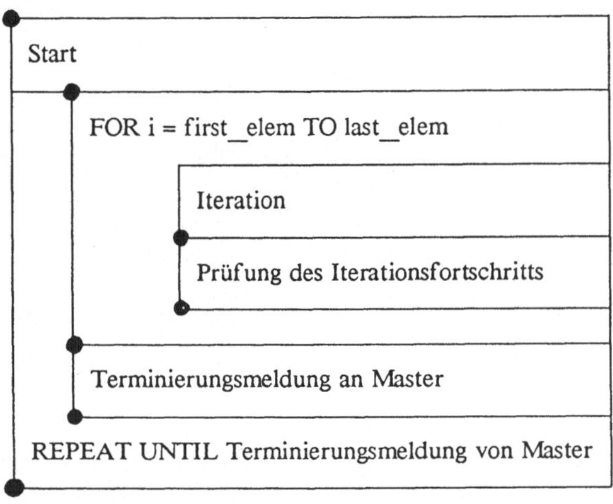

Bild 7. Markierung des parallelen asynchronen Gauß-Seidel-Algorithmus

Die Ergebnisse einer solchen Messung sind etwa 1 Kbit lange Abschnitte von (hier: vier) parallelen Ereignisspuren. Jede dieser Ereignisspuren gibt Auskunft darüber, wann ein Prozessor den entsprechenden Punkt im Algorithmus erreicht hat. Innerhalb jedes Meßabschnittes sind exakte Aussagen über die Reihenfolge von Ereignissen sowohl innerhalb einer Ereignisspur, wie auch zwischen verschiedenen Ereignisspuren möglich. Zwischen zwei Meßabschnitten allerdings gehen wegen der eingeschobenen Meßpause die Kausalzusammenhänge verloren.

Bild 8 illustriert das Verhalten der einzelnen Prozessoren: Die Diagramme zeigen je Prozessor in chronologischer Reihenfolge die Dauern der einzelnen Iterationsschleifen, wobei die Summe der Zeitanteile für die elementweise Überprüfung des Iterationsfortschritts durch die Strecke oberhalb der Markierung (*) in jedem Balken dargestellt ist. Aufgrund des gewählten Algorithmus und aufgrund von Compilereigenschaften ist die Größe dieses Zeitanteils proportional zur Anzahl der Gitterpunkte, für die kein genügend großer Iterationsfortschritt mehr erzielt werden konnte, d.h., daß eine lokale Terminierungsbedingung erreicht worden ist. Man sieht daraus, wie (z.B. beim Master) auch nach einem Erreichen dieser lokalen Terminierungsbedingung (in der Iteration-Nr. 12) eine erneute Iteration notwendig werden kann, und zwar dann, wenn sich die Werte der Gitterpunkte am Rande des Nachbarfeldes, die ja zur eigenen Iteration mitbenutzt werden, vom Nachbar-Prozessor entsprechend geändert wurden.

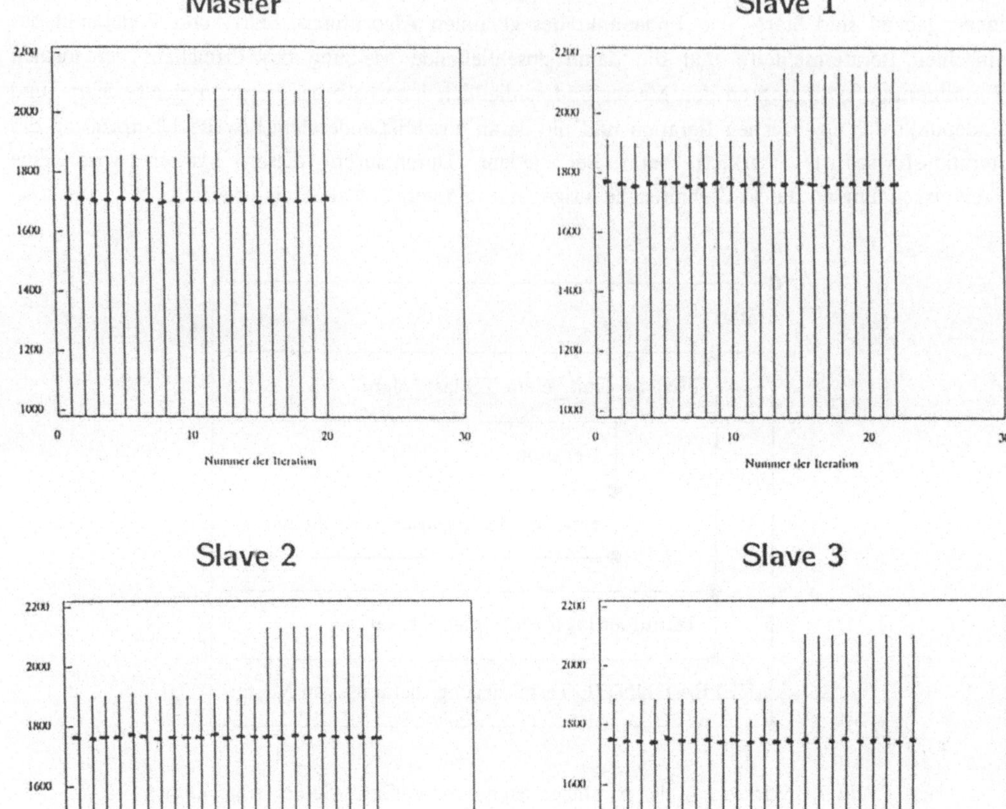

Bild 8. Iterationsdauern auf den vier Prozessoren (Zeit in µs)

Nach der 15. Iteration ist bei allen vier Prozessoren die lokale Terminierungsbedingung stabil erfüllt. An den Master wird dies durch ein ständig umlaufendes "Token" gemeldet, das jeder Prozessor von seinem oberen Nachbarn liest, in das er seinen eigenen Zustand einträgt, und das er dann an seinen unteren Nachbarn weiterschickt. Der Master erkennt die globale Terminierung daran, daß dieses Token zweimal nacheinander anzeigt, daß alle Prozessoren fertig sind. Diese beiden Token-Umläufe sind im Gantt-Diagramm des Bildes 9 mit ① und ② markiert. Bedingt durch Zugriffskonflikte kann das Token erst nach der 16. bzw. 17. Iteration zum ersten Mal dem Master von allen Prozessoren das Erreichen der lokalen Terminierungsbedingung melden. Nach diesem zweiten Token-Umlauf fordert der Master alle Slaves auf, nun zu terminieren (Umlauf ③). Bis dahin sind weitere Iterationen durchgeführt worden.

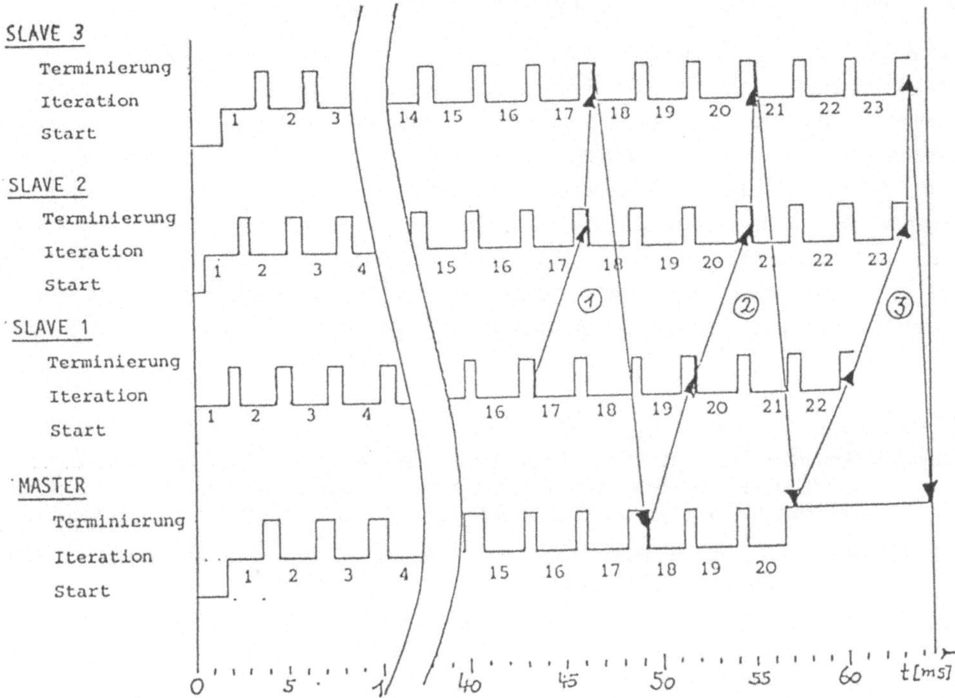

Bild 9. Globale Terminierung durch dreifachen Token-Umlauf

Dieses als erste Erprobungsmessung des neuen ZM4 angelegte Experiment bestätigte unser Entwicklungsziel, eine Monitor für die verteilte Erfassung von Ereignisspuren zu entwickeln, sehr gut: Aussagen über das Verhalten des dargestellten Algorithmus wäre ohne den Bezug der erfaßten Ereignisse auf eine gemeinsame Zeitbasis nicht möglich gewesen. Darüberhinaus ergab diese Messung, daß sich die Fernsteuerung des LA vom zentralen Steuerrechner STAR aus bewährte, die Datenübertragung vom LA zum STAR funktionierte [Quic87] und somit die Messung der Abläufe an räumlich eng benachbarten Mehrprozessorsystemen mit einem integrierten System Logikanalysator - Steuer- und Auswerterechner möglich ist.

6. Zusammenfassung

Der Zählmonitor 4 (ZM4) ist ein Monitor für räumlich und/oder funktionell verteilte Multiprozessor- und Multicomputer-Systeme. In seinem Anspruch, systemweit gültige Aussagen über die Reihenfolge von unabhängig beobachteten Ereignissen treffen zu können, grenzt sich das ZM4-Konzept von vorangehenden Monitorkonzepten für verteilte Systeme ab: Der Ansatz von Morgan et al. [MBGK75] etwa sieht nur eine Zusammenschau von Einzelauswertungen vor, die von den verteilten Monitoren "vor Ort" erstellt wurden, wobei in diesen Einzelauswertungen die einzelnen Ereignisse in einer statistischen Aussage aber bereits "verschwunden" sind. Der von uns zuvor entwickelte Zählmonitor 3 war zwar in der Lage, mehrere unabhängige Ereignisfolgen global aufzuzeichnen, dies allerdings nur bei räumlich eng benachbarten Multiprozessor-Systemen. Bei bestimmten Software-Monitoren für verteilte Systeme schließlich begnügt man sich damit, ausschließlich die Reihenfolge von SEND/RECEIVE-Ereignissen zu rekonstruieren [MiMS86].

Entsprechend der so formulierten Ziele liegt das Schwergewicht unseres Konzepts auf der Entwicklung eines flexibel rekonfigurierbaren Monitorsystems, das einerseits aus einer Menge von verteilten Monitoren und andererseits aus einem zentralen Steuer- und Auswerterechner besteht, die untereinander durch Netze zum Datenaustausch und zur Synchronisierung der lokalen Monitor-Uhren verbunden sind. Die Verarbeitung der von den verschiedenen Monitoren gewonnenen Ereignisströme mit im allgemeinen unterschiedlicher Ereignisdarstellung wird auf dem zentralen Steuer- und Auswerterechner durch die Werkzeuge EDL und POET sehr vereinfacht. Schließlich wird in diesem Papier ein spezieller, hochintegrierter Ereignisdetektor, der CRAC-Baustein, vorgestellt, und es wird von einem ersten Meßeinsatz am DIRMU-Multiprozessor-System berichtet.

Literatur

[Bemm86] **Bemmerl, T.** : Realtime High Level Debugging in Host/Target Environments. EUROMICRO Venedig 1986. Proceedings in: Microprocessing and Microprogramming, vol. 18, no. 1-5, 1986. Elsevier Science Publishers B.V. (North Holland) Amsterdam.

[BeFW82] **Berg, H.K., Franta, W.R., Wood, W.T.** : Issues and Approaches to Distributed Testbed Instrumentation. IEEE Computer, vol. 15, no. 10, Okt. 1982.

[BuMi86] **Burkhart, H., Millen, R.** : Performance Measurement Tools in a Multiprocessor Environment. Eidgenössische Technische Hochschule Zürich, Institut für Elektronik 86/10, Nov. 1986.

[Flep87] **Flepsen, G.** : Hardware-Messung des Speicherzugriffsverhaltens in einem Ring von DIRMU-Modulen. Studienarbeit am Lehrstuhl für Rechnerarchitektur und Verkehrstheorie des Instituts für Mathematische Maschinen und Datenverarbeitung der Universität Erlangen-Nürnberg, 1987.

[From83] **Fromm, H., Hercksen, U., Herzog, U., John, K.-H., Klar, R., Kleinöder, W.** : Experiences with Performance Measurement and Modeling of a Processor Array. IEEE Transactions on Computers, vol. C-32, no. 1, 1983.

[GoHT86] **Gora, W., Herzog, U., Tripathi, S.K.** : Clock Synchronization on the Factory Floor. in: Rosenthal, R. (Ed.) : Proc. Workshop on Factory Comunications, March 1987. National bureau of Standards, Gaithersburg, 1987.

[GuZa83] **Gusella, R., Zatti, S.** : TEMPO Time Services for the Berkeley Local Network. University of California, Berkeley, Computer Science Division, Report No. UCB/CSD 83/163, Berkeley 1983.

[Hinr86] **Hinrichsen, U.** : Ein Monitor für ein autonomes Datenübertragungssystem. Dissertation. Technische Universität Braunschweig, Informatik-Bericht 86-05, Braunschweig 1986.

[Klar71] **Klar, R.** : Messungen von Rechneraktivitäten. Dissertation. Arbeitsberichte des Instituts für Mathematische Maschinen und Datenverarbeitung der Universität Erlangen-Nürnberg, Bd. 4, Nr. 2, Erlangen 1971.

[Klar75] **Klar, R., Schreiber, H., Widjaja, H.C.** : Messungen mit dem Zählmonitor 2. Arbeitsberichte des Instituts für Mathematische Maschinen und Datenverarbeitung der Universität Erlangen-Nürnberg, Bd. 8, Nr. 9, Erlangen 1975.

[Klar81] **Klar, R.** : Hardware Measurements and their Application on Performance Evaluation in a Processor-Array. Computing, Suppl. 3, Springer-Verlag, Wien New York, 1981.

[Klar85] **Klar, R.** : Hardware-/Software-Monitoring. Informatik-Spektrum, Bd. 8, Heft 1, Feb. 1985. Springer Verlag, Berlin Heidelberg New York Tokyo.

[KlLu86] **Klar, R., Luttenberger, N.** : VLSI-based Monitoring of the Inter-Process-Communication in Multi-Microcomputer Systems with Shared Memory. EUROMICRO Venedig 1986. Proceedings in: Microprocessing and Microprogramming, vol. 18, no. 1-5, 1986. Elsevier Science Publishers B.V. (North Holland) Amsterdam.

[Klug82] **Kluge, W.** : Zur Problematik des Messens in verteilten Systemen. in: Glässer, K.H. (Hrsg.): Verteilte Systeme, Workshop in der GMD am 4.11.1981. R. Oldenbourg Verlag, München Wien, 1982.

[Maeh86] **Maehle, E., Wirl, K., Jäpel, D.** : Experiments with Parallel Programs on the DIRMU Multiprocessor Kit. in: Feilmeier, M., Joubert, G., Schendel, U. (Eds.) : Parallel Computing 85, Int. Conf. Berlin 1985. Elsevier Science Publishers B.V. (North Holland) Amsterdam, 1986.

[MiMS86] **Miller, B.P., Macrander, C., Sechrest, S.** : A Distributed Programs Monitor for Berkeley UNIX. Software - Practice and Experience, vol. 16, no. 2, Feb. 1986.

[Mohr87] **Mohr, B.**: Entwurf und Implementierung eines Systems zur Entschlüsslung von Monitordaten. Diplomarbeit am Lehrstuhl für Rechnerarchitektur und Verkehrstheorie des Instituts für Mathematische Maschinen und Datenverarbeitung der Universität Erlangen-Nürnberg, 1987.

[MBGK75] **Morgan, D.E., Banks, W., Goodspeed, D.P., Kolanko, R.** : A Computer Network Monitoring System. IEEE Transactions on Software Engineering, vol. SE-1, no. 3, Sep. 1975.

[Quic87] **Quick, A.** : Programmgesteuerte Hardwaremessung mit einem Logikanalysator. Studienarbeit am Lehrstuhl für Rechnerarchitektur und Verkehrstheorie des Instituts für Mathematische Maschinen und Datenverarbeitung der Universität Erlangen-Nürnberg, 1987.

[Schr78] **Schreiber, H.** : Hardware-Messung und Analyse des Ablaufgeschehens in Rechnerkernen. Dissertation. Arbeitsberichte des Instituts für Mathematische Maschinen und Datenverarbeitung der Universität Erlangen-Nürnberg, Bd. 11, Nr. 7, Erlangen 1978.

[Snod82] **Snodgrass, R.** : Monitoring Distributed Systems: A Relational Approach. PhD Dissertation. Carnegie-Mellon University, Department of Computer Science, Report-No. CMU-CS-82-154, Pittsburgh 1982.

[Spie85] **Spies, P.P.**: No-wait-send/Rendezvous. Informatik Spektrum, Bd. 8, Heft 5, Okt. 1985. Springer Verlag, Berlin Heidelberg New York Tokyo.

Ein universeller Datenmonitor zur Netzdiagnose und Leistungsbewertung

*W. Gora; H. Körzdörfer**

Universität Erlangen-Nürnberg
Lehrstuhl für Rechnerarchitektur und -Verkehrstheorie
IMMD VII
Martensstr. 3
8520 Erlangen, F.R.G.

Zusammenfassung:

Mit der Zahl der Rechnernetze steigt auch die Notwendigkeit für "intelligente" Systeme, die die komplexen und undurchsichtigen Vorgänge in einem Netz offenlegen und analysieren können. Aufgrund der Weiterentwicklung in Technologie und Standardisierung darf ein derartiges Diagnosesystem allerdings nicht nur auf ein spezielles Kommunikationssystem beschränkt sein, sondern muß vom Benutzer an unterschiedliche Kommunikationsprotokolle und -hierarchien anpaßbar sein. Im folgenden wird die Konzeption und Realisierung eines Datenmonitors vorgestellt, der es ermöglicht, "Protokollwissen" benutzerfreundlich zu integrieren, insbesondere Wissen über den Aufbau und die Struktur der einzelnen Protokolldateneinheiten. Hierzu wurde die Sprache FAN.1 (*Frame Analyzing Notation One*) in Anlehnung an die Notation ASN.1 (*Abstract Syntax Notation One*) der internationalen Standardisierungsorganisation (ISO) entworfen und für die universelle Analyse von Paketstrukturen und Protokollparametern implementiert. Auf FAN.1 aufbauend wurden komplexe Meß-, Auswertungs- und Analysefunktionen realisiert, die es, durch Angabe beliebiger Prädikate über die definierten Paketstrukturen, ermöglichen, das dynamische Ablaufgeschehen in einem großen verteilten System transparent zu machen. Der hier beschriebene Datenmonitor ist Bestandteil des LAN Protocol Testers B5100 der Firma Siemens.

1. Einleitung

Der konzipierte und realisierte Datenmonitor hat die Aufgabe, Entwickler von Kommunikationssoftware und Netz-Administratoren bei der Inbetriebnahme, Analyse, Wartung und Pflege eines verteilten Systems zu unterstützen. Darüberhinaus ist die Diagnose und Analyse

* Derzeitige Anschrift: Siemens AG, ESTE 36, 8520 Erlangen

der Abläufe in einem Rechnernetz die Voraussetzung für eine Bewertung der in einer konkreten Umgebung verwendeten Kommunikationsprotokolle.

Eine wesentliche Forderung an ein derartiges Werkzeug ist die vom Benutzer vorgebbare Beschreibung eines realen Kommunikationssystems mit gleichzeitiger Abstrahierung von vorhandenen Hardware-Möglichkeiten. Die Abhängigkeit vom konkreten Medium (Ethernet, Token Bus, Token Ring, etc.) wird beim vorzustellenden Datenmonitor auf leistungsfähige Hardware-Komponenten beschränkt, welche die Netzaktivitäten aufzeichnen und in Dateien ablegen können. Diese Dateien sind die Grundlage für die anschließenden Auswertungen und Analysen, womit die Unabhängigkeit von den physikalischen Gegebenheiten erreicht wird.

Der Begriff "Datenmonitor" ist im Gegensatz zu einem Protokollmonitor zu sehen, der in der nächsten Ausbaustufe geplant ist. Daten-Monitoring bezieht sich auf die Paketstrukturen und hat deren Dekodierung und Analyse, sowie die Überwachung bestimmter Rahmenbedingungen, beispielsweise die Einhaltung von Wertebereichen, zum Ziel. Als Protokollmonitor soll dagegen ein System verstanden werden, welches aufbauend auf einen Datenmonitor die internen Zustände und den prozeduralen Ablauf eines Protokolls kennt, sowie weitergehendes Wissen über das Netz und die verwendeten Kommunikationsprotokolle integrieren kann. Derartiges Wissen können unter anderem spezielle Zuverlässigkeits- oder Leistungsaspekte sein.

Der im folgenden erläuterte Datenmonitor wurde im Rahmen von [Körz86] an der Universität Erlangen-Nürnberg vorentwickelt und anschließend in den LAN Protocol Tester B5100 der Firma Siemens als ein wesentlicher Baustein integriert. Der Konzipierung des universellen Datenmonitors ging zunächst die Analyse und Bewertung des Funktionsumfangs und der Struktur einiger typischer Vertreter von kommerziell erhältlichen Netzdiagnose-Geräten voraus. Ein wesentlicher Nachteil aller untersuchten Systeme war es, daß diese nur für bestimmte Protokollkonfigurationen eingesetzt werden können und dem Anwender meist keine Auswahl zwischen verschiedenartigen Protokoll-Analysen lassen.

Um dieses Benutzer-Wissen über die verwendeten Kommunikationsprotokolle und -architekturen in ein universelles Netzdiagnose-System zu integrieren, wurde die Beschreibungs- und Dekodierungssprache FAN.1 (*Frame Analyzing Notation One*) zur Analyse und Auswertung von beliebigen Paketen bzw. Datenstrukturen entwickelt und implementiert. Mit Hilfe von FAN.1 können verschiedenartige Strukturen und Protokollparameter erkannt und über standardisierte Schnittstellen an höherwertige Auswertefunktionen übergeben werden. Dadurch wird für das Auswertesystem eine Unabhängigkeit bezüglich des realen Kommunikationssystems und der verwendeten Protokolle erreicht. Momentan existieren Standard-Definitionen basierend auf FAN.1 für ISO-Protokolle der Ebenen 2-4 gemäß dem OSI-Referenzmodell (*Open Systems Interconnection*) [ISO7498], für die TCP/IP-Architektur sowie dem Xerox Network Service (XNS). Der Benutzer des Datenmonitors kann jedoch darüberhinaus beliebige Protokolldefinitionen selbst erstellen und in das System integrieren.

Aufgrund der theoretischen Mächtigkeit der Notation FAN.1 ist deren Anwendungsbereich nicht nur auf das Monitoring von Netzen beschränkt, sondern es können beispielsweise auch Datenstrukturen, die in anderen Bereichen anfallen, beschrieben, analysiert und ausgewertet werden.

2. Anforderungen an einen universellen Datenmonitor

Die Netzdiagnose, d.h. die Analyse der Kommunikation, das Erkennen von Fehlern und das Aufzeigen von Engpässen, ist ein wichtiger Bestandteil eines Netzmanagement-Systems, welches auf der Basis der aufgezeichneten Informationen ein "Tuning" des Netzes und seiner Komponenten vornehmen kann. Die Anforderungen an einen "universellen" Datenmonitor lassen sich folgendermaßen zusammenfassen:

- Einsetzbarkeit bzw. Anpassungsfähigkeit an unterschiedliche Kommunikationssysteme, -architekturen und -protokolle
- Umfangreiche Funktionalität bezüglich der oben aufgeführten Aufgabenbereiche
- Strukturierung des Systems zur Beherrschung der Komplexität
- Integration von Wissen in prozeduraler und nicht-prozeduraler Form über das Ablaufgeschehen im zu beobachtenden Kommunikationssystem
- Einfache Bedienbarkeit mit gleichzeitiger Mächtigkeit der Benutzeroberfläche

Die in [Körz86] untersuchten kommerziell erhältlichen Netzdiagnose-Systeme für den Bereich der lokalen Netze (*Local Area Networks* LAN) konnten zwar eine hohe Leistungsfähigkeit im Bereich der Aufzeichnung des Netzverkehrs aufweisen und besitzen auch einen gewissen Umfang an Auswertefunktionen, jedoch sind diese LAN-Diagnose-Systeme fast ohne Ausnahme nur auf eine oder mehrere bestimmte Protokollkonfigurationen abgestimmt.

Die gleichzeitig von den Herstellern hervorgehobene Protokollunabhängigkeit wird meist dadurch erreicht, daß die gesammelten Netzdaten bezüglich der höheren Protokolle uninterpretiert gelassen werden. Dies ist angesichts der Vielzahl unterschiedlichster Protokolle und Normen nicht verwunderlich, denn selbst bei einer Einschränkung auf ISO-Normen existieren bereits zu den niedrigen Schichten im LAN-Bereich viele Varianten [From86]. Die Headerdaten der einzelnen Protokolle sind sehr individuell und kompliziert strukturiert, so daß die Interpretation der gesammelten Daten bei den untersuchten Geräten nicht unterstützt, sondern ganz dem Benutzer überlassen wird.

Ein Hauptziel der eigenen Untersuchungen und Entwicklungen für ein universelles Netzdiagnose-System war es, Wege aufzuzeigen, um die skizzierten Einschränkungen in der Unterstützung von Analysen der höheren Protokollschichten zu beheben. Der im folgenden beschriebene Datenmonitor soll durch intelligente Software bestehende Systeme ergänzen und gleichzeitig soviel Wissen wie möglich vom Menschen in die Maschine verlagern.

Zur Realisierung bzw. Umsetzung dieser Anforderungen in ein konkretes Netzdiagnosegerät wurde in einem ersten Schritt die Paketbeschreibungssprache FAN.1 entwickelt, mit der das Kriterium der Einsetzbarkeit und Anpassungfähigkeit in verschiedenen Kommunikationsumgebungen erreicht wird. Auf dieser Notation basierend wurden Auswerte-Funktionen konzipiert, prototypisch erprobt und in der Komponente "Datenmonitor" des neuentwickelten Netzdiagnose- und Monitoring-Systems LAN Protocol Tester B5100 der Firma Siemens größtenteils bereits auch realisiert. Mit Hilfe dieser Auswertungen kann der Benutzer den Netzverkehr analysieren und die gewonnenen Daten zur Leistungsbewertung seiner konkreten Kommunikationsumgebung verwenden.

Neben dem Wissen über die Struktur der Pakete oder zum Wertebereich der Protokollparameter, muß ein universelles Netzdiagnose-System jedoch auch Wissen über die Dynamik und die Zusammenhänge des gesamten Ablaufgeschehens in einem Netz integrieren können. So sollten beispielsweise nicht nur Fehler im Hinblick auf die Verletzung von Protokollnormen, sondern auch ein "Fehlverhalten" in der Kommunikation diagnostiziert und analysiert werden können.

Als Fehlverhalten im Kommunikationssystem kann beispielsweise ein zu häufiger Verbindungsaufbau und -abbau zwischen zwei Partnern in dem beobachteten Zeitintervall verstanden werden. Zwar kann die gesamte Kommunikation fehlerfrei im Sinne eines "echten" Protokollfehlers bzw. einer Verletzung von Protokollparametern sein, jedoch ist intuitiv klar, daß das beschriebene Verhalten das Kommunikationssystem unter Umständen unnötig belastet.

Derartiges Wissen über die Abläufe im einem Rechnernetz bzw. Meta-Wissen über die Protokolle läßt sich beispielsweise mit Methoden aus dem Bereich der künstlichen Intelligenz repräsentieren und vearbeiten [Haye83] [Mich83]. Die Beschreibungssprache FAN.1 stellt hier den ersten Schritt zu einem wissensbasierten Netzdiagnose-System dar. Weitere Arbeiten auf diesem Gebiet, insbesondere unter Einbeziehung von regelbasierten Wissenskomponenten, sind in der Entwicklung.

3. Architektur des universellen Datenmonitors

Das Konzept zur Netzdiagnose ist auf komplexe und deshalb relativ zeitaufwendige *off-line* Auswertungen und Analysen von gesammelten Netzdaten ausgerichtet. Verschiedene einfache Auswertungen und Vorverarbeitungen von Netzdaten werden jedoch auch in Realzeit unterstützt. Dem Datenmonitor erscheint das Rechnernetz als spezielle Datei, die er über leistungsfähige Hardware und geeignete Treibersoftware direkt oberhalb der physikalischen Schicht ansprechen kann.

Abbildung 1 zeigt die Software-Architektur und die einzelnen Module des Datenmonitors. Die Module kommunizieren im allgemeinen über Prozedurschnittstellen. Für besondere Aufgaben, die zeitparallel ausgeführt werden sollen, können auch

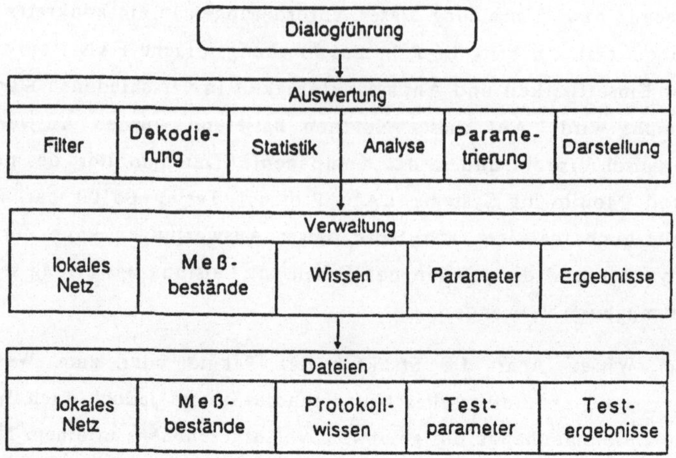

Abbildung 1. Architektur des Datenmonitors

Prozeßschnittstellen definiert werden. Beispielsweise kann die *LAN-Verwaltung* Netzdaten sammeln und puffern, während die *Meßdatenverwaltung* gleichzeitig die Daten in eine Datei schreibt. Im folgenden wird auf die Struktur der einzelnen Ebenen kurz eingegangen.

3.1 Dateiebene

Insgesamt operiert der Datenmonitor auf fünf Typen von Informationsbeständen. In einem Dateityp ist das Wissen über die Paketstrukturen mit Hilfe der Beschreibungssprache FAN.1 festgehalten. Ein weiterer Dateityp definiert die Test- und Auswerteparameter, beispielsweise bestimmte Filterbelegungen, Statistikparameter, Prädikate oder Testfolgen aus Paketen und vom Benutzer konfigurierte Parametrierungen. Weitere Dateien enthalten Test- und Auswerte-Ergebnisse, z.B. berechnete Statistiken.

3.2 Verwaltungs-Ebene

Die Ebene 2 der Software-Architektur des Datenmonitors enthält Module zur Verwaltung der darunterliegenden Informationsbestände, deren komplexe Struktur der darüberliegenden Auswerte-Schicht und auch dem Benutzer verborgen bleiben sollen. Die *Meßdatenverwaltung* stellt Funktionen zum Öffnen und Schließen von Meßbeständen zur Verfügung, sowie zum paketorientierten Lesen, Schreiben und zum Suchen nach Paketen, die benutzerspezifizierbare Kriterien erfüllen.

Das Modul *Parameter- und Ergebnisverwaltung* unterstützt zusammen mit dem darüberliegenden Modul *Parametrierung* den Benutzer bei der Konfiguration von Tests und Auswertungen. Zusammen mit dem Modul *Darstellung* wird durch die *Parameter- und Ergebnisverwaltung* eine anschauliche und benutzerfreundliche Aufbereitung und Darstellung der Ergebnisse am Bildschirm oder über den Drucker erreicht. Das Modul *Wissensverwaltung und Anwendung* ist konzipiert für verschiedene Wissensbasen über die Struktur der Protokolldateneinheiten, über den prozeduralen Ablauf der Protokolle, sowie Wissen über Symptome von "Fehlverhalten" und dessen Ursachen.

3.3 Auswertungs-Ebene

Die Funktionalität der Auswertungsebene gliedert sich in die vier Teile *Filter*, *Dekodierung*, *Statistik*, *Analyse*, *Parametrierung* und *Darstellung*. Auf die Auswertungsebene wird in Kapitel 5 detaillierter eingegangen.

3.4 Dialog-Schicht

Der Benutzer wird bei seinen Arbeiten mit dem Datenmonitor von einem Dialogmodul unterstützt, der Menütechniken und Softkeys verwendet. In einer weiteren Ausbaustufe ist geplant, auch Wissen über Benutzerklassen, sowie das Verhalten einzelner Benutzer mit zu integrieren.

4. Die Paket- und Datenbeschreibungssprache FAN.1

4.1 Anforderungen an eine Paketbeschreibungssprache

Die grundlegendste Forderung an ein universell einsetzbares Netzdiagnose-System ist die freie Konfigurierbarkeit der Monitoring-Einheit bezüglich der Kodierungsvorschriften verschiedenster Protokolle. Flexibilität hinsichtlich unterschiedlicher Protokolle und Protokollarchitekturen kann jedoch nur dann erreicht werden, wenn ein genügend mächtiges Werkzeug zur Verfügung steht, das es erlaubt, einheitlich die unterschiedlichen Paketstrukturen zu beschreiben.

Im folgenden wird die Beschreibungssprache FAN.1 (*Frame Analyzing Notation One*) vorgestellt, die auf der ISO-Notation ASN.1 basiert und folgenden Anforderungen genügt:

- Theoretische Mächtigkeit
- Kompaktheit und Anschaulichkeit der Darstellung von Strukturen
- Effektive maschinelle Nutzung der Beschreibung

Auf das Kriterium der theoretischen Mächtigkeit soll an dieser Stelle nicht im Detail eingegangen werden, in [Körz86] wird jedoch gezeigt, daß FAN.1 sich in die Klasse der kontextsensitiven Sprachen einordnen läßt.

Die wesentlichsten Charakteristika bezüglich Syntax und Semantik von FAN.1 werden im folgenden anhand von Beispieldefinitionen bezüglich der ISO-Protokolle für die Ebenen 2 bis 4 kurz vorgestellt. Der genaue Sprachumfang bezüglich Syntax und Semantik von FAN.1, die vom Parsergenerator *YACC* [John78] akzeptierte Grammatik, sowie Aufbau und Funktionsweise des FAN.1-Compilers sind in [Körz86] beschrieben.

4.2 Aufbau der Beschreibungssprache FAN.1

FAN.1 orientiert sich an der ISO-Notation ASN.1 aufgrund der Tatsache, daß diese Syntax verstärkt bei der Definition von Normprotokollen eingesetzt wird, wie beispielsweise bei den *Common Application Service Elements* [ISO8650] oder dem *Manufacturing Message Service* [ISO9506]. Ziel dieser Orientierung an ASN.1 ist es, dem Benutzer eine weitestgehend einheitliche Sprachoberfläche für alle Definitionen im Kommunikationssystem anzubieten und somit den Aufwand für das Erlernen verschiedener Notationen zu verringern.

ASN.1 hat jedoch eine Reihe von Nachteilen und Einschränkungen [Gora87a], die zu Erweiterungen bei FAN.1 führten. So wurde ASN.1 hauptsächlich zur abstrakten Syntaxspezifikation entworfen, während die Umsetzung in eine konkrete Transfersyntax, d.h. die Transformation der abstrakten Strukturen gemäß eindeutiger Kodierungs- und Dekodierungsvorschriften, mit Hilfe der *Basic Encoding Rules for Abstract Syntax Notation One* [ISO8825] definiert wird.

```
-- Definition eines OSI-Kommunikationssystems

    PACKET ::= SEQUENCE OF LENGTH(li_2 + 14)
    {
        link            LinkLayer,
        network         NetworkLayer,
        transport       TransportLayer,
        ...
    }

    LinkLayer ::= SEQUENCE          -- Typdefinition der Ebene 2
    {
        destination     Dest,
        source          Source,
        li_2            Li_2,
                        L_dsap,
                        L_ssap,
                        Control
    }
```

Abbildung 2. Beispiel für die Typdefinition einer OSI-Umgebung

Um jedoch auch Daten entschlüsseln zu können, die nach anderen Regeln als die von [ISO8825] kodiert wurden (z.B. die Protokolldaten der unteren Schichten des OSI-Referenzmodells), wurde FAN.1 erweitert um die Beschreibung von einfachen Datentypen, sowie um Sprachelemente, die speziell für die Dekodierung der Netzpakete von Bedeutung sind und beispielsweise die Länge eines Feldes definieren. Dadurch wird neben der Strukturbeschreibung bzw. der abstrakten Syntax von Datentypen, auch die Beschreibung der Kodierung und der Repräsentation dieser Datentypen in das Sprachkonzept von FAN.1 integriert.

Zur Interpretation und Dekodierung eines Paketes müssen sämtliche Datenstrukturen, die auftreten können, definiert werden. Die Definition einer Protokollarchitektur auf der Basis von FAN.1 besteht aus einer Folge von *Typdefinitionen* (siehe Abbildung 2) und *Wertdefinitionen* (siehe Abbildung 3). Typdefinitionen beschreiben die Datenstruktur der von dem jeweiligen Protokoll verwendeten Protokolldateneinheiten, während Wertdefinitionen Instanzen, d.h. konkrete Ausprägungen, solcher Protokolldateneinheiten darstellen.

Eine Typdefinition besteht aus einer Typreferenz (z.B. *LinkLayer*) und einer Liste von weiteren FAN.1-Typen, wobei diese Liste gemäß der Angabe eines speziellen Konstruktionsmechanismus (z.B. *SEQUENCE*) interpretiert wird. Jedem einzelnen Typ, aus denen sich ein konstruierter Typ zusammensetzt, kann ein sogenannter Identifier zugeordnet werden. Dieser Identifier wird bei der Darstellung des dekodierten Paketes, sowie bei den Auswerte- und Analysefunktionen verwendet. Eine Typreferenz muß mit einem Großbuchstaben beginnen, während ein Identifier durch einen Kleinbuchstaben am Anfang gekennzeichnet wird.

Die erste Typdefinition in FAN.1 wird durch das Schlüsselwort PACKET identifiziert. Hierbei kann der Benutzer verschiedene Protokollarchitekturen mit unterschiedlichen Schichtprotokollen angeben (z.B. TCP/IP, ISO/OSI). Abbildung 2 zeigt dies am Beispiel des ISO/OSI-Referenzmodell. Vergleicht man beispielsweise die *LinkLayer*-Typdefinition mit einer Grammatikregel, so bedeutet dies, daß aus dem Symbol *LinkLayer* die Folge der Symbole *Dest*, *Source*, *Li_2*, *L_dsap*, *L_ssap* und *Control* ableitbar ist.

Eine Wertdefinition (s. Abbildung 3) besteht aus einer Wertreferenz (z.B. *Dest*), dem Typ (BYTE_HEX) des zu dieser Wertreferenz zugeordneten Wertes, der Länge des Wertes (LENGTH), sowie Angaben zum Wertebereich (RANGE) oder der Aufzählung aller möglichen Wertzustände (s.a. Symbol *Reason* in Abbildung 3). Bei der Auswertung wird auf diese Definition zurückgegriffen, um ein Paket zu identifizieren und den Protokollparametern (*destination*, *source*, ...) den aktuellen Wert innerhalb dieses Paketes zuordnen zu können.

4.2.1 Einfache FAN.1-Typen

Von FAN.1 werden drei einfache Datentypen unterstützt, Hexadezimalstrings (BIT_HEX und BYTE_HEX), Binärstrings (BIT_BIN und BYTE_BIN) und Dezimalzahlen (BIT_DEC und BYTE_DEC). Der Benutzer kann, von diesen einfachen Datentypen ausgehend, weitere zusammengesetzte Datentypen mit Hilfe der nachfolgenden Konstrukte definieren.

```
-- Wertdefinitionen der Ebene 2

Dest   ::= BYTE_HEX { LENGTH(6) }
Source ::= BYTE_HEX { LENGTH(6) }
Li_2   ::= BYTE_DEC
{
        LENGTH(2)
        VALUE(RANGE(0, 1518)
}
. . .

Reason ::= BYTE_HEX   -- Fehlermeldungen der Transportebene
{
        LENGTH (1)
        COMMENT
        (
          ['00' H] "Reason not specified",
          ['01' H] "Congestion at TSAP",
            . . .
        )
}
```

Abbildung 3. Beispiele für Wertdefinitionen

4.2.2 Zusammengesetzte FAN.1-Typen

Besteht ein Paket oder eine Datenstruktur aus einer Folge mehrerer Komponenten, die verschiedene Typen besitzen (s.a. Abbildung 2), kann dies durch einen *SEQUENCE*-Typ ausgedrückt werden falls die Reihenfolge der Komponenten relevant ist, ansonsten durch einen *SET*-Typ. Beim *SET*-Typ werden zwei Alternativen unterschieden, der *SET UNTIL*-, sowie der *SET OF LENGTH*-Typ.

Die *SET OF LENGTH*-Konstruktion führt einen neuen Datentyp ein, der aus einer ungeordneten Menge vorhandener Typen besteht, wobei diesen Typen auch Identifier zugeordnet werden können. Die Länge in Bytes der Repräsentation eines Wertes dieses Typs ist aus dem in runden Klammern folgenden arithmetischen Ausdruck berechenbar. Aus Abbildung 4 ist ersichtlich, daß über den Identifier *li_4* Bezug auf den Wert eines anderen Datentyps genommen wird, d.h. FAN.1 besitzt Eigenschaften einer attributierten Grammatik.

Mit der *SET UNTIL*-Konstruktion wird ebenfalls ein neuer Datentyp eingeführt, der aus einer ungeordneten Menge vorhandener Typen besteht. Die Mächtigkeit der Menge wird hier über ein Ende-Kennzeichen bzw. ein entsprechendes *tag* bestimmt. Das Ende der Repräsentation ist gefunden, falls der angegebene Wert an der aktuellen Position im Paket erscheint.

```
Transport ::= CHOICE        -- Pakete der Transportebene
{
        [9, '1110' B]   Connect_Request,
        [9, '1101' B]   Connect_Confirm,
        [9, '1101' B]   Disconnect_Request,
        . . .
}

Dr_vars ::= SET OF LENGTH (li_4 - 6)
{
        [1, 'e0' H]     Dr_vare0,
        [1, 'c3' H]     Dr_varc3
}

Tcp_vars ::= SET UNTIL ('00' H)
{
        [1, '01' H]     Typ1,
        [1, '02' H]     Typ2,
        [1, '03' H]     Typ3
}
```

Abbildung 4. Beispieldefinition von *CHOICE*- und *SET*-Typen

Die Auswahl aus vorhandenen Typen wird mit Hilfe der *CHOICE*-Konstruktion unterstützt. Jedem Typ muß hierbei analog dem *SET*-Typ ein *tag* vorangestellt werden. Bei der Dekodierung wird der Typ der Auswahlliste genommen, dessen *tag* im Paket an der entsprechenden Position gefunden wird.

4.2.3 Weitere Sprachelemente

Ein Benutzerkommentar, der nicht weiter interpretiert wird, kann an allen Stellen der Definition durch die Zeichen "--" begonnen werden. Der Kommentar endet durch erneute Angabe dieser Zeichen, spätestens jedoch am Ende einer Zeile. Bei der Wertdefinition können Kommentare zu bestimmten Werten, die bei der Auswertung am Bildschirm erscheinen sollen, durch das Schlüsselwort *COMMENT* eingeleitet werden (s.a. Abbildung 3). Einem Wert kann hierbei eine Zeichenkette zugeordnet werden, die bei der Dekodierung am Bildschirm erscheint, um neben dem Identifier noch weitere Informationen (z.B. "maximale TPDU-Größe") ausdrücken zu können.

Durch die Angabe des Schlüsselwortes LENGTH sowie einem in runden Klammern eingeschlossenen Ausdruck kann die Länge der zu dekodierenden Information angegeben und beschränkt werden. Hierbei sind folgende Ausdrücke möglich:

- Angabe einer numerischen Konstante, z.B. LENGTH(5)
- Angabe eines Identifiers, dem ein Typ eindeutig zugeordnet ist. Bei der Dekodierung dieses Paketes wird der aktuelle Wert dieser Typreferenz ermittelt, beispielsweise bei Parametern zur Längenidentifikation eines Pakets (s.a. Symbol *li_2* in Abbildung 2)
- Angabe eines arithmetischen Ausdrucks, der aus den arithmetischen Operatoren "-", "+" und "*", sowie Identifier und numerischen Konstanten gebildet werden kann.

Vor dem Schlüsselwort *LENGTH* ist bei dezimalen Typdeklarationen (*BYTE_DEC*, *BIT_HEX*) die Angabe von REVERSE möglich, dadurch wird im Gegensatz zur normalen Gewichtung der Binärstellen (von links nach rechts abnehmend), eine entsprechend umgekehrte Gewichtung angenommen.

Die Kennzeichnung bestimmter Positionen und Werte in einem Paket oder einer Datenstruktur ist durch die Angabe sogenannter *tags* möglich. Ein *tag* ist die Folge einer Dezimalzahl und eines binären (B) oder hexadezimalen (H) Wertes, deren Folge durch eckige Klammern eingeschlossen wird (s.a. Abbildung 4). Die Semantik der *tags* ist, daß der Typ durch die Bit- bzw. Bytefolge des Wertes, ab der angegebenen dezimalen Position von Beginn der Repräsentation des Gesamtwertes an, bestimmt wird.

Durch *NONDEC* können Paket- oder Datenstrukturen definiert werden, die bei der Dekodierung nicht auf dem Bildschirm ausgegeben werden sollen, beispielsweise Längenidentifikationen (*li_2*), die für den Benutzer keine relevanten Informationen mit sich führen.

4.3 Übersetzung der FAN.1-Definition

Mit Hilfe eines eigenes entwickelten FAN.1-Compilers wird eine definierte Paketspezifikation in hierarchisch angeordnete Datenstrukturen, sogenannte H-Graphen [Körz86], umgewandelt, die einfach und effizient implementiert werden können.

Auf der obersten Stufe des H-Graphen wird genau ein Knoten (Startknoten) vorausgesetzt, der dem Datentyp PACKET entspricht. Wendet man die sog. Inhaltsfunktion auf den Startknoten an, so erhält man einen Graphen, der die Grobstruktur eines Netzpaketes wiederspiegelt. Auf jeden Knoten dieses Graphen kann wiederum diese Inhaltsfunktion angewendet werden, wobei hierdurch die Struktur immer mehr verfeinert wird.

In terminalen Knoten ist die Information zum jeweiligen einfachen Datentyp gespeichert. Dazu zählen Angaben über den Typ des Wertes (hexadezimal, binär, dezimal), den Typ der Repräsentation (Bit, Byte) und verschiedene interne Kennzeichen. Abgespeichert sind weiterhin eine Liste mit den zulässigen Werten für den Datentyp, sowie die mit *tags* versehenen Kommentare. In nichtterminalen Knoten sind der Typ des Knotens (*SEQUENCE*, *CHOICE*, *SET*), Verweise auf die Unterknoten und die zugehörigen Identifier enthalten.

5. Auswertungen mit Hilfe des universellen Datenmonitors

5.1 Übersicht

In einem ersten Schritt beim Aufruf des Datenmonitors muß eine FAN.1-Definitionsdatei angegeben werden, die die Spezifikation der im Kommunikationssystem auftretenden Datenstrukturen enthält. Nach dem Laden dieser FAN.1-Definitionen (siehe voriges Kapitel) und dem Aufruf einer der nachfolgend beschriebenen Auswertefunktionen, liest der Datenmonitor die zu betrachtenden Pakete aus dem Meßbestand, oder falls in Realzeit gewünscht, vom Netz.

Jedes einzelne Paket wird gemäß der FAN.1-Definition analysiert, wobei rekursive Abarbeitungen teilweise notwendig sind, da beispielsweise in einer OSI-Umgebung variable Header-Längen auftreten und somit die Grenzen der einzelnen Schichten im Paket nicht statisch erkennbar sind. Hierbei wird zunächst das Paket gemäß den *LinkLayer*-Definitionen (Ebene 2) überprüft, wobei die Paketsegmente der höheren Protokolle als Nutzdaten betrachtet werden. Diese Vorgehensweise wird anschließend für die Ebene 3 (*NetworkLayer*) vorgenommen und solange wiederholt, bis das ganze Paket abgearbeitet wurde und somit eindeutig die Strukturen den Benutzerdefinitionen zugeordnet werden konnten, oder im Fehlerfall, bis eine Datenstruktur nicht in der FAN.1-Definitionsdatei spezifiziert ist.

Die in den nachfolgenden Kapiteln aufgeführten Abbildungen sind das Ergebnis konkreter Netzanalysen mit dem LAN Protocol Tester B5100 im PAP-Netz (*Projekt für flexibel automatisierte Produktionssysteme*) der Universität Erlangen-Nürnberg. Als Kommunikationsprotokolle werden im PAP-Projekt die MAP-Protokolle (*Manufacturing Automation Protocols*) [MAP86] [Gora87b] eingesetzt, von denen die Ebenen 5-7 gemäß dem OSI-Referenzmodell selbst implementiert wurden.

Der LAN Protocol Tester B5100 der Firma Siemens enthält als einen Bestandteil den hier beschriebenen Datenmonitor mit praktisch allen im folgenden skizzierten Auswertungsfunktionen. Dadurch ergeben sich umfassende Test-, Analyse- und Bewertungsmöglichkeiten der im Rahmen des PAP-Projektes eingesetzten Kommunikationsprotokolle.

5.2 Prädikate

Die Funktionalität des Datenmonitors ist insbesondere auf Ereignisse in den höheren Protokollschichten ausgelegt. Um bestimmte Ereignisse bzw. Pakete mit bestimmten Eigenschaften charakterisieren zu können, werden in allen Moduln des Datenmonitors boolsche Prädikate über Protokollparameter oder Datenstrukturen verwendet. Die boolschen Prädikate des Monitors sind induktiv definiert:

(1) ein einfaches Prädikat hat die Form *Identifier Operator Wert*,

(2) falls *P* ein Prädikat ist, dann ist auch *NOT P* ein Prädikat,

(3) falls *P* ein Prädikat ist, dann ist auch *(P)* ein Prädikat,

(4) falls *P1* und *P2* Prädikate sind, dann ist auch *P1 AND P2* ein Prädikat,

(5) falls *P1* und *P2* Prädikate sind, dann ist auch *P1 OR P2* ein Prädikat.

Auf die Werte von gesendeten Protokollparametern wird über die *Identifier* zugegriffen. Als *Operator* sind die Elemente "==" (gleich), "!=" (ungleich), ">" (größer), "<" (kleiner), ">=" (größer-gleich) und "<=" (kleiner-gleich) zugelassen.

Ein *Wert* wird als Hexadezimalstring, Binärstring oder Dezimalzahl angegeben. In Hexadezimalstrings und Binärstrings darf an beliebigen Stellen das Zeichen "x" für "don't care" verwendet werden. Der Typ des Wertes muß mit dem Typ des Protokollparameters übereinstimmen, dem der *Identifier* in der mit FAN.1 definierten Paketspezifikation zugeordnet wurde.

5.3 Filterfunktionen

Das Filtermodul stellt dem Benutzer Funktionen zur Verfügung, die es gestatten, Meßbestände in reduzierte Meßbestände überzuführen, die nur Pakete von besonderem Interesse enthalten. Das lokale Rechnernetz selbst wird als spezieller, lesbarer Meßbestand betrachtet. Zur Filterung können verschiedene Kriterien angegeben werden, wie Selektion

(1) nach *bestimmten* Zeitintervallen

(2) nach *Paketen*, die mit Fehlern bestimmter Art behaftet sind

(3) nach *Schichten*, d.h. der Zielmeßbestand enthält nur die Protokolldaten der Schicht *n* nach Norm *m*

(4) nach *Normen*, wobei mehrere Unterteilungen möglich sind, wie beispielsweise, daß der Zielmeßbestand nur Pakete enthält, die bis zur Schicht *n* nach Norm *m* dekodierbar sind, oder daß der Quellmeßbestand in *i* Meßbestände aufgeteilt und ein Paket im Meßbestand *j* abgelegt wird, falls es nach einer bestimmten Norm dekodierbar ist

(5) nach *Prädikaten*, d.h. der Zielmeßbestand enthält nur Pakete, die ein bestimmtes boolsches Prädikat bezüglich einer Norm erfüllen, oder allgemeiner, mindestens eines von *n* gegebenen boolschen Prädikaten erfüllen (z.B.: (source == '00aa00001824'H AND tpdu_code == 'f0'H))

(6) nach *Kombinationen* der obigen Filterkriterien. Diese können mit AND, OR und NOT verküpft werden

(7) nach *Ereignissen*. Die Pakete eines Meßbestandes sind zeitlich geordnet. Daraus ergibt sich die Möglichkeit Netzereignisse zu definieren, die die Aufnahme von Paketen in dem Zielmeßbestand ein- und ausschalten. Netzereignisse können

Kombinationen der obigen Kriterien oder andere Ereignisse sein, wie z.B. Kollisionen in einem CSMA/CD-Netz.

5.4 Dekodierfunktionen

Mit Hilfe von Dekodierfunktionen werden die Protokoll- und Nutzdaten benutzerfreundlich dargestellt. Hierbei sind umfangreiche Möglichkeiten zum Positionieren innerhalb des Meßbestandes, zur Auswahl von Schichten und Protokollen, sowie verschiedene Ausgabemodi realisiert.

```
A1HINWEIS A2INFO    A3STATUS  A4HARDCOPYA5BEGINN  A6ENDE    A7ABBRUCH A8FERTIG
SIEMENS LANTESTER B5100                                      15/06/1987 15:50:31
------------------------------ 42 DATENMONITOR ---------------------------------

DATEI     :DEMO.DAT         |PAKET_POS :1                  |NORM      :ISO.NRM
DATUM     :15/06/87 15:21   |PAKET_LNG :46                 |EBENE     :3
PAKETE    :4765             |ZEIT_STMP : 0: 0:37:712: 50   |DATEN AB  :4
AUFZ_LNG  :82               |ZEIT_AUFL :10 usec            |FEHLER    :0000000000

---------------------------------- transport ----------------------------------
li_4                        :D 27                          :
tpdu_code                   :H e                           :Connection Request
cdt                         :D 0                           :
dst_ref                     :H 0000                        :
src_ref                     :H c7af                        :
class                       :D 4                           :
format                      :B 1                           :Extented
flow_control                :B 0                           :No use of flow control
------------------------------ variable_parameters -----------------------------

---F1------F2--------F3---------F4--------F5--------F6--------F7--------F8------
VORWARTS |ZURUCK   |PAKET_POS |ZEIT+    |ZEIT-    |SUCHEN   |         |>WEITER<
link     |network  |transport |         |         |         |         |DATEN
```

Abbildung 5. Dekodierung eines Transport-Paketes

Abbildung 5 zeigt die Bildschirmausgabe eines dekodierten Netzpaketes, welches Protokoll-Teile der Ebenen 2-4 enthält (*link, network, transport*). In Spalte 1 wird der vom Benutzer definierte Identifier ausgegeben, Spalte 2 enthält den aktuellen Wert dieses Identifiers. Falls ein Kommentar zu einem bestimmten Wert mit Hilfe des Schlüsselwortes COMMENT angegeben wurde, wird dieser in Spalte 3 ausgegeben. Beispielsweise erscheint beim Wert für den Typ des Paketes auf der Transportebene (*tpdu_code*) der Benutzerkommentar "Connection Request" am Bildschirm.

5.5 Statistikfunktionen

Das *Statistikmodul* unterstützt Statistiken von Netzereignissen, die vom Darstellungsmodul graphisch oder alphanumerisch über das jeweilige Endgerät ausgegeben werden. Welche Statistiken möglich und sinnvoll sind, ist sehr stark vom zugrundeliegenden Rechnernetz und den Protokollen abhängig. Daher werden neben Standardstatistiken, die vor allem Ereignisse in den niedrigeren Protokollschichten unterstützen, auch vom Benutzer frei konfigurierbare Statistiken über höhere Protokollschichten angeboten. Beispiele für derartige

Standardstatistiken sind:

(1) Analyse von Paketlängen

(2) Analyse von Fehlerarten

(3) Messung der Netzauslastung

(4) Ausgabe einer Kommunikationsmatrix

Bei einer Kommunikationsmatrix repräsentiert der Eintrag in der Komponente *(i,j)* den relativen Datenfluß von Station *i* zur Station *j* in Relation zum Gesamtverkehr. Die Prozentzahlen werden entweder über die Paket- oder die Byteanzahl ermittelt. Aus den einzelnen Einträgen in der Kommunikationsmatrix ist ersichtlich, inwieweit eine Station das Netz belastet.

```
A1HINWEIS A2INFO     A3STATUS   A4HARDCOPYA5BEGINN  A6ENDE       A7ABBRUCH ABFERTIG
SIEMENS LANTESTER B5100                                          15/06/1987 15:36:48
──────────────────────────────── 42 DATENMONITOR ──────────────────────────────────

MESSDATEI :WOLV1.DAT    |BEGINN_POS :1          |PRADIKAT  :KEINES
PAKET_ANZ :315          |ENDE_POS   :315        |GEFUNDEN  :315

 NR |tpdu_code          |KOMMENTAR                    |PAKETE |ANTEIL
 1  |e                  |Connection Request           |    15 |  4.8%
 2  |d                  |Connection Confirm           |    15 |  4.8%
 3  |60                 |Acknowledge                  |   172 | 54.6%
 4  |f0                 |Data                         |    57 | 18.1%
 5  |10                 |Expedited Data               |    14 |  4.4%
 6  |20                 |Expedited Acknowledge        |    14 |  4.4%
 7  |80                 |Disconnection Request        |    14 |  4.4%
 8  |c0                 |Disconnection Confirm        |    14 |  4.4%

───F1──────F2──────F3──────F4──────F5──────F6──────F7──────F8───
DIAGRAMM |DRUCKEN |       |       |       |       |       |
```

Abbildung 6. Auswertung der relativen Häufigkeit der Pakete auf der ISO-Transportschicht

Beispiele für vom Benutzer konfigurierbare Statistiken sind:

(1) Analyse des Protokoll-Overheads aufgrund des durchschnittlichen Verhältnisses von Nutz- zu Protokolldaten für verschiedene Protokolle und Schichten

(2) Ausgabe aller Werte, die ein Protokollparameter in einem Meßbestand annimmt

(3) Auswertung von Prädikaten, d.h. der Benutzer definiert Eigenschaften und in einem Diagramm wird gezeigt, wieviele Pakete des Meßbestandes die jeweilige Eigenschaft erfüllen.

In Abbildung 6 wird am Beispiel der ISO-Transportschicht gezeigt, wie die Darstellung auf dem Bildschirm erfolgt. Der Meßbestand besteht hierbei aus 315 Paketen und es wird

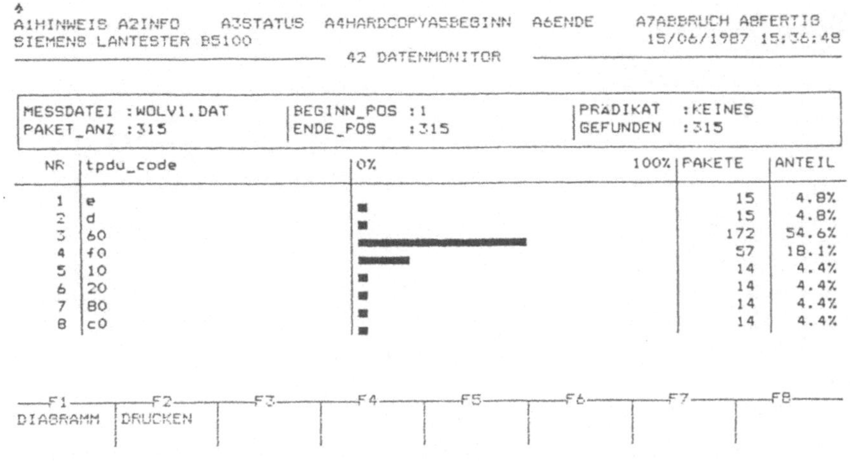

Abbildung 7. Histogramm der Daten von Abbildung 6

ausgewertet, mit welcher Verteilung die einzelnen Paket-Typen (*Connection Request, Connection Confirm, ...*) auftreten. Die Identifizierung dieser Pakettypen erfolgt hierbei durch den Identifier *tpdu_code*, wobei jedem möglichen Wert vom Benutzer mit Hilfe des Schlüsselworts COMMENT ein Kommentar zugeordnet wurde.

5.6 Analysefunktionen

Die bisher vorgestellten Funktionen erleichtern es dem Benutzer, die Aktivitäten von Kommunikationseinheiten zu untersuchen und zu bewerten. Beispielsweise kann bei zu häufigen Übertragungswiederholungen angenommen werden, daß ungünstig eingestellte Timer vorliegen oder daß Pufferspeicher überlastet sind. Ein Beispiel zur Leistungsbewertung des ISO-Transportprotokolls [ISO8073] ist in Abbildung 8 dargestellt. Hierzu wurde mit Hilfe der Prädikat-Angabe

tpdu_code == 'e'H OR tpdu_code == 'd'H

der ursprüngliche Datenbestand derartig reduziert, daß nur noch "Connect Request"- (CR) und "Connect Confirm"-Pakete (CC) enthalten waren. Abbildung 8 zeigt die mit Hilfe der Funktion "Abstandsverteilung" gewonnene Verteilung der Zwischenankunftszeiten.

Allerdings ist hier zu beachten, daß auch der Abstand zwischen CC- und CR-Paketen einfließt, d.h. die Bestätigung der (n).ten und die Anforderung der (n+1).ten Transport-Verbindung. Dieser Zeitabstand liegt jedoch im Sekunden- oder Minutenbereich und kann durch geeignete Einstellung der Zeitskala (Funktionstaste F5) ausgeblendet werden.

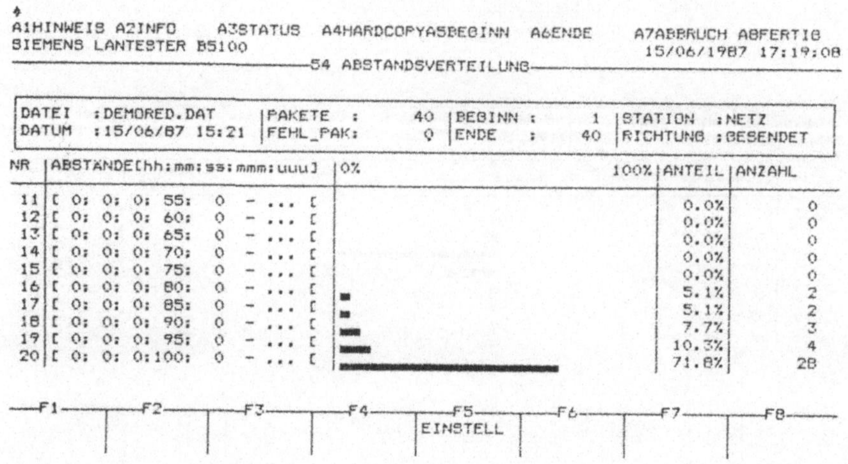

Abbildung 8. Verteilung der Folge CR- und CC-Pakete auf der Transportebene

6. Ausblick

Die Protokollparameter der im Netz beobachteten Pakete sind Ursachen, Wirkungen oder Reaktionen auf externe und interne Ereignisse in den Kommunikationseinheiten der Rechner und stehen in enger Beziehung zueinander. Bei der Diagnose und Analyse komplexer dynamischer Abläufe, beispielsweise wenn die Einhaltung von ausgehandelten Protokollparametern überwacht werden soll, wird aber der Benutzer vom bisher realisierten System noch wenig unterstützt. Folgende wichtigen Anforderungen sind deshalb Gegenstand weiterer Untersuchungen und Entwicklungen:

- Überwachung der Paketreihenfolgen aufgrund der vom Protokoll vorgegebenen Abläufe
- Feststellen, in welchen Zuständen sich Kommunikationseinheiten zu bestimmten Zeitpunkten aufgrund der Netzbeobachtungen befinden können
- Kodieren von Paketen aufgrund der Paket- und Protokollspezifikation und Senden dieser Pakete als Reaktion auf Netzereignisse

Derartige Funktionen sind ein wesentlicher Bestandteil für ein umfassendes Netzmanagent-System, welches über die Netzdiagnose hinaus auch weitere Bereiche, wie die automatische Reaktion auf bestimmte Ereignisse, zu unterstützen hat.

Referenzen

[From86] Fromm, I.: "Standardisierung Lokaler Netze (LAN)", Informationstechnik, 28. Jahrgang, Heft 1/1986, S. 30-37

[Gora87a] Gora, W.; Speyerer, R.: "Abstract Syntax Notation One", Datacom, Heft 4, April 1987, S. 78-86

[Gora87b] Gora, W.: "Anwendungsfunktionen und -protokolle in MAP", in "Automatisierungssystem MAP", Bauer, M. (Hrsg.), Weidler-Verlag, Berlin, 1987, S. 47-70

[Haye83] Hayes-Roth, F.; Waterman, D.A.; Lenat, D.B.: "Building Expert Systems", Addison-Wesley, 1983

[ISO7498] ISO IS 7498: "Information Processing Systems - Open Systems Interconnection - Basic Reference Model", November 1983

[ISO8073] ISO IS 8073: "Connection oriented Transport Protocol Specification", 1985

[ISO8650] ISO DP 8650 "Common Application Service Elements Protocol Specification", 1985

[ISO8824] ISO DIS 8824: "Specification of Abstract Syntax Notation One (ASN.1)", 1986

[ISO8825] ISO DIS 8825: "Specification of Basic Encoding Rules for Abstract Syntax Notation One (ASN.1)", 1986

[ISO9506] ISO DP 9506: "Manufacturing Message Service for Bidirectional Transfer of Digitally Encoded Information", Part 1: Service Definition, Part 2: Protocol Definition, May 1987

[John78] Johnson, S. C.: "YACC: Yet Another Compiler Compiler", in "UNIX Programmer's Manual", Bell Laboratories, Murray Hill, 1978

[Körz86] Körzdörfer, H.: "Konzipierung und Implementierung eines Systems zur Netzdiagnose", Diplomarbeit am IMMD VII, Universität Erlangen, September 1986

[MAP86] General Motors: MAP Specification, Version 2.1.A and 2.2, August 1986

[Mich83] Michalski, R.S.; Carbonell, J.G.; Mitchell, T.M.: "Machine Learning - An Artificial Intelligence Approach", Tioga Publ. Co., 1983

Approximation von empirischen Verteilungsfunktionen mit Erlangmischverteilungen und Coxverteilungen

Leonhard Schmickler *

Juli 1987

Zusammenfassung

Messungen der Verteilungsfunktionen stochastischer Prozesse werden für die Modellierung und Analyse von Datenkommunikationssystemen benötigt. Die empirischen Verteilungsfunktionen müssen in eine für die Verkehrstheorie geeignete mathematische Beschreibung überführt werden, die die Anwendung bekannter Analysemethoden erlaubt. Im folgenden werden Anforderungen an ein Approximationsverfahren hergeleitet und zwei bereits bekannte Verfahren erläutert. Dann wird das neue Approximationsverfahren MEDA[1] vorgestellt, das diese verkehrstheoretischen Anforderungen erfüllt. Es verwendet eine spezielle Erlangmischverteilung als Approximierende. Die ersten drei empirischen Momente werden exakt abgeglichen und alle vorhandenen Stützstellen zur Näherung des Verteilungsfunktionsverlaufs herangezogen.

Anschließend wird ein Umrechnungsalgorithmus hergeleitet, mit dem die Erlangmischverteilung in eine äquivalente Coxverteilung überführt werden kann. Dadurch wird sie bereits etablierten Analysewerkzeugen zugänglich.

Sowohl MEDA als auch der letztgenannte Algorithmus wurden in PASCAL realisiert.

1 Einführung

Die Modellierung und Bewertung von Datenkommunikationssystemen beruht neben vorgegebenen Systemparametern im wesentlichen auf der Einbeziehung von Messungen der stochastischen Prozesse, wie Zwischenankunftszeiten oder Bedienzeiten. In den letzten Jahren wurde eine Fülle von Daten in Telefonnetzen, Terminalnetzen, lokalen Rechnernetzen oder an Datenverarbeitungsanlagen selbst gemessen ([10], [11], [15], [17], ...). Aus solchen Messungen können Stichproben von N Realisationen einer Zufallsvariablen X gewonnen werden

$$X_N = \{x_1, x_2, ..., x_N\}. \tag{1}$$

Zur Gewinnung der empirischen Verteilungsfunktion wird die Abszisse in m Intervalle I_i unterteilt, die üblicherweise die gleiche Breite Δx im Bereich $0 \le x < m\ \Delta x$ haben. Wenn wir aus der Stichprobe die Zahl der Werte y_u ermittelt haben, die in das Intervall $(u-1)\Delta x \le x < u\Delta x$ fallen, gilt für die empirische Verteilungsfunktion (Summenhäufigkeitsfunktion [9]) in diesem Intervall

$$F_E(x) = \frac{1}{N}\sum_{i=1}^{u} y_i, \qquad (u-1)\Delta x \le x < u\Delta x. \tag{2}$$

Die empirischen Momente j-ter Ordnung werden nach

$$M_E(j) = \frac{1}{N}\sum_{i=1}^{N} x_i^j, \qquad j = 1, 2, .., N \tag{3}$$

*Lehrstuhl für Allgemeine Elekrotechnik und Datenfernverarbeitung, RWTH Aachen

[1]Mixed Erlang Distributions for Approximation

aus der Stichprobe berechnet. In dem häufigen Fall, daß die Verteilungsfunktion $F_E(x)$ keine signifikanten Sprünge aufweist, kann sie mit einem linearen Verlauf zwischen den Stützstellen dargestellt werden. Für die meistens sehr großen Stichproben ist der Fehler, der dadurch gemacht wird, vernachlässigbar klein. Die so gewonnenen empirischen Verteilungsfunktionen sollen bei der Modellierung und Leistungbewertung von Datenkommunikationsnetzen geeignet eingesetzt werden.

Es läßt sich daraus z.B. ein Tabellenzufallsgenerator für ein vorhandenes Simulationssystem erstellen. Alle Eigenschaften der empirischen Verteilungsfunktion werden dabei berücksichtigt [2]. In vielen Fällen wird jedoch eine mathematische Berechnung des Modells angestrebt und daher nach einer möglichst einfachen, aber dennoch ausreichend detaillierten mathematischen Beschreibung der empirischen Verteilung gesucht. Zuerst sollen einige Anforderungen an diese formuliert werden.

1.1 Anforderungen an ein Approximationsverfahren

Wie Cox [5], Schassberger [13], Bux [4] und andere gezeigt haben, eignet sich die Beschreibung eines Zufallsprozesses durch eine Phasenverteilung besonders für die mathematische Analyse. Sie ermöglicht durch die Verwendung exponentieller Phasen eine Beschreibung eines stochastischen Prozesses als Markov-Prozeß. Dadurch können bei der Analyse Standardverfahren eingesetzt werden. Man versucht daher, die empirische Verteilung mit einer Phasenverteilung möglichst genau zu approximieren.

Neben der Verwendung einer Phasenverteilung ist besonders die Berücksichtigung der niederen empirischen Momente von großer Bedeutung [3]. Wesentliche Leistungsmerkmale (wie mittlere Wartezeit, Belegungszeit) hängen in vielen Wartesystemen ausschließlich von diesen ab [7]. Drei Momente der Bediendauerverteilung reichen z.B. im System M/G/1 aus, um Mittelwert und Streuung der Warte- und Belegungszeit zu berechnen. Darüberhinaus empfiehlt sich die Näherung des Verteilungsfunktionsverlaufes zur Einbeziehung der Einflüsse der höheren Momente [4].

Die Anforderungen an ein Approximationsverfahren sind also:

- Verwendung einer Phasenverteilung mit wenigen Parametern;
- Abgleich der niederen Momente $M_E(j)$ der empirischen Verteilung;
- Berücksichtigung der höheren Momente durch ausreichende Näherung des gemessenen Verteilungfunktionsverlaufes in allen Stützstellen x_i.

In Kapitel 2 wird gezeigt, daß das Approximationsverfahren MEDA[1] diese Anforderungen erfüllt.

Auf der Basis einer von Iversen [8] entwickelten Umwandlungsvorschrift wird in Kapitel 3 gezeigt, wie aus einer beliebigen Erlangmischverteilung eine äquivalente Coxverteilung bestimmt werden kann. Dadurch ist es möglich, die der Coxverteilung zugeschriebenen Vorteile bei der Modellanalyse mit denen der Erlangmischverteilung zu verbinden, die für eine Approximation besser geeignet ist.

1.2 Coxverteilung und Erlangmischverteilung

Die Coxverteilung [5] besteht aus einer Reihenschaltung von exponentiellen Phasen mit im allgemeinen unterschiedlicher Intensität λ_i (Abb. 1a). Mit der Wahrscheinlichkeit α_i wird die Reihe nach dem Durchlaufen der i-ten Phase verlassen. Jeder dieser Wahrscheinlichkeiten kann also die Reihe der ersten i Generatoren zugeordnet werden. Eine solche Reihe, die vom Typ einer allgemeinen Erlang-k-Verteilung ist, wird im folgenden als Variante i mit der Variantenwahrscheinlichkeit α_i bezeichnet. Bildet man alle Varianten der Coxverteilung, so liegt eine allgemeine Erlangmischverteilung mit i Generatoren im i-ten Zweig vor, der mit der Zweigwahrscheinlichkeit α_i durchlaufen wird (s. Abb 1a).

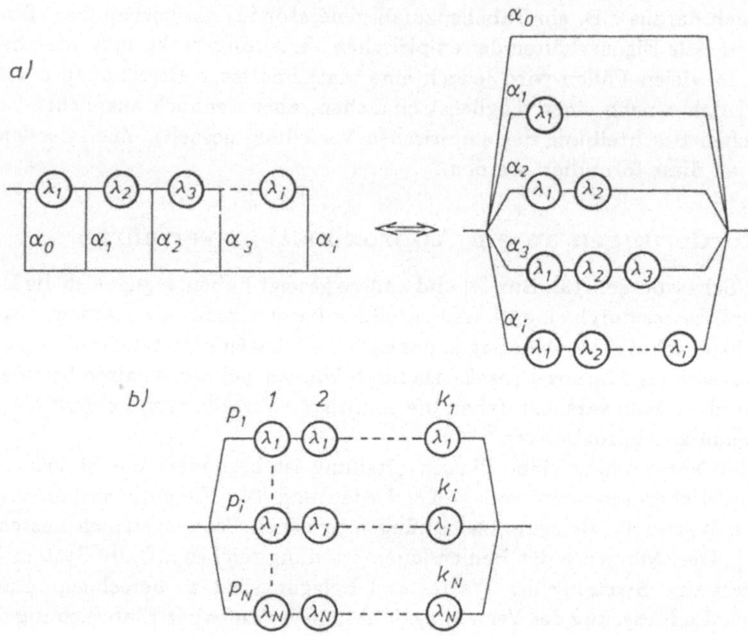

Die Laplace-Transformierte der Coxverteilung lautet

$$L(s) = \alpha_0 + \sum_{i=1}^{N} \alpha_i \prod_{j=1}^{i} \frac{\lambda_j}{\lambda_j + s}. \tag{4}$$

Verschiedene Approximationsverfahren (z.B. [4], s. Kap. 1.3) verwenden eine spezielle Coxverteilung, deren Intensitäten λ_i alle gleich sind ($\lambda_i = \lambda$). Die von MEDA verwendete spezielle Erlangmischverteilung hat jeweils in jedem Zweig ein konstantes λ_i (s. Abb. 1b). Sie hat die Laplace-Transformierte

$$L(s) = \sum_{i=1}^{N} p_i \left(\frac{\lambda_i}{\lambda_i + s} \right)^{k_i}. \tag{5}$$

Beide Phasenverteilungen sind dazu geeignet, jede Verteilung mit einer rationalen Laplace-Transformierten beliebig genau zu nähern [5]. Sie können also beide für Approximationen verwendet werden. Bei der Erlangmischverteilung ist jedoch die Zahl der notwendigen Parameter und damit auch der Rechenaufwand wesentlich geringer.

An dieser Stelle sei bemerkt, daß die Überführung der Coxverteilung in die äquivalente Erlangmischverteilung zwar trivial ist, die direkte Überführung einer Erlangmischverteilung in eine äquivalente Coxverteilung jedoch relativ schwierig. In Kapitel 3 wird ein Verfahren beschrieben, das auf der Umwandlung einer einzelnen Exponentialverteilung in eine äquivalente Coxverteilung beruht.

1.3 Andere Approximationsverfahren

Es gibt eine ganze Reihe von bekannten Approximationsverfahren, die eine mathematische Beschreibung einer nur in Stützstellen vorliegenden empirischen Verteilungsfunktion liefern. Sie sind jedoch meist nicht für verkehrstheoretische Anwendungen geeignet, weil sie keine Phasenverteilung als Approximierende verwenden. Zwei Verfahren, die mit Phasenverteilungen approximieren, werden hier verglichen und bewertet.

Bux und Herzog [4] verwenden die spezielle Coxverteilung (Kap. 1.2). Für vorgegebene Phasenzahl r und Phasenintensität μ werden die Variantenwahrscheinlichkeiten α_i durch ein lineares Optimierungsverfahren bestimmt. Der optimale Wert für μ ergibt sich als Minimum einer nichtlinearen Funktion von μ. Die Approximation wird beendet, wenn dieses Minimum gleich Null ist, andernfalls wird die Phasenzahl r um eins erhöht und die Parameterbestimmung erneut durchgeführt. Die ersten zwei empirischen Momente werden exakt abgeglichen. Der Verlauf der gemessenen Verteilungsfunktion wird in einigen Stützstellen, deren Zahl wegen des hohen Rechenaufwandes auf maximal 10 begrenzt ist, bis auf eine vorgegebene Abweichung genähert. Der von Bux in [3] nachgewiesenen Bedeutung des dritten empirischen Moments wird nicht Rechnung getragen. Als Gütekriterium wird die maximale Abweichung in den Stützstellen verwendet. Prinzipiell kann jede empirische Verteilung mit beliebiger Genauigkeit genähert werden. Eine hohe Genauigkeit erfordert jedoch in der Regel eine sehr große Phasenzahl ($r > 100$).

Das von Pawlita in [12] vorgeschlagene Zufallssuchverfahren optimiert die Parameter einer N-zweigigen Erlangmischverteilung. Die Parameter der Approximierenden werden als Zufallszahlen generiert. Zu Beginn sind die Zufallszahlen im gesamten Wertebereich gleichverteilt, im Laufe der Approximation wird dieser jedoch eingeschränkt und die Parameter um das bisher erzielte Optimum normalverteilt variiert. Bei der Unterschreitung einer vorgegebenen Fehlerschranke des mittleren Fehlerquadrates der Klassenhäufigkeiten wird der Algorithmus beendet. Es werden keine empirischen Momente abgeglichen. Die Zahl der berücksichtigten Stützstellen der empirischen Verteilungsfunktion ist mit maximal 60 sehr begrenzt. Die erreichbare Übereinstimmung im Verteilungsfunktionsverlauf ist trotz eines hohen Rechenzeitbedarfs gering.

Auf Grund der Nachteile dieser und anderer Verfahren wurde das neue Approximationsverfahren MEDA entwickelt.

2 Das Approximationsverfahren MEDA

Das Approximationsverfahren MEDA [14] nähert eine linearisierte empirische Verteilungsfunktion durch eine Erlangmischverteilung. Es gleicht die ersten drei empirischen Momente exakt ab und berücksichtigt alle (in den folgenden Beispielen Größenordnung ≈ 1000) Stützstellen bei der Näherung der Verteilungsfunktion. Die Eigenschaften der Erlangmischverteilung (Kap. 2.1) und das verwendete Gütekriterium (Kap. 2.2) haben einen großen Einfluß auf den Algorithmus. Die Approximation wird zuerst mit einer 2-zweigigen Erlangmischverteilung durchgeführt, die nur fünf freie Parameter hat (Kap. 2.3). Durch die Hinzunahme eines dritten Zweiges (Kap. 2.4) wird die Möglichkeit der Approximierenden, sich dem Verlauf der empirischen Verteilungsfunktion anzupassen, weiter verbessert.

2.1 Eigenschaften der Erlangmischverteilung

Zunächst etwas zu den besonderen Eigenschaften der Erlangmischverteilung, die einen einfachen Optimierungsalgorithmus ermöglichen. Während bei der Coxverteilung die Abhängigkeit des Verteilungsfunktionsverlaufes von einem einzelnen Parameter nicht mehr nachvollziehbar ist, stellt sich diese Abhängigkeit bei der Erlangmischverteilung sehr einfach dar [14]. Wir betrachten dazu eine einzelne Erlang-k-Verteilung. Sie hat als Parameter die Phasenzahl k und die Phasenintensität λ. Der Mittelwert ist $M(1) = k/\lambda$. Die Dichtefunktion der Erlang-k-Verteilung hat ein einzelnes ausgeprägtes Maximum in der Nähe des Mittelwertes. Eine gleichzeitige Vergrößerung der Phasenzahl k und der Intensität λ (Mittelwert $M(1) = k/\lambda = const.$) bewirkt, daß die Verteilungsfunktion $F(x)$ um den Funktionswert $F(\overline{x})$ am Ort des Mittelwerts $\overline{x} = M(1)$ "gedreht" wird, d.h. der Übergang vom Bereich, in dem die Funktion noch fast Null ist, in den Bereich des steilen Anstiegs wird abrupter (Abb. 2a). Wird λ variiert ($k = const.$), so erfährt die Verteilungsfunktion eine Stauchung bzw. Streckung in x-Richtung (Abb. 2b). Als Teil einer Erlangmischverteilung ist der einzelne Erlang-k-Zweig mit der Zweigwahrscheinlichkeit p_i gewichtet. Dieser Parameter bewirkt eine Stauchung bzw. Streckung der Verteilungsfunktion in y-Richtung. Diese Abhängigkeiten können in einem Optimierungsverfahrens dazu benutzt werden, jeden Parameter einzeln zu optimieren. Die Komplexität des Algorithmus kann dadurch erheblich eingeschränkt werden.

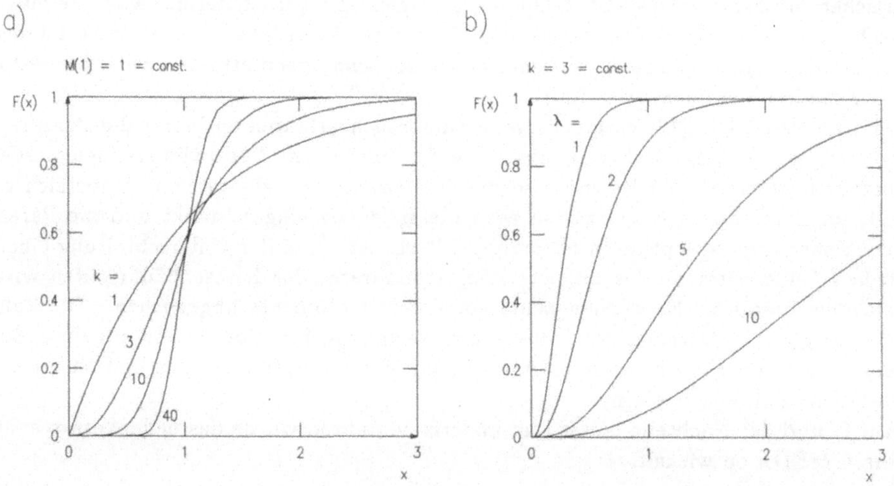

Abb. 2: **Abhängigkeit des Verteilungsfunktionsverlaufs einer Erlang-k-Verteilung von a) der Ordnung k ($M(1) = k/\lambda = const$) und b) der Phasenintensität λ**

2.2 Das Gütekriterium

Als Gütekriterium wird die auf den Mittelwert bezogene Fläche Δ zwischen dem Verlauf der empirischen Verteilungsfunktion $F_E(x)$ und dem der approximierenden Erlangmischverteilung $F_A(x)$ verwendet (Abb. 3):

$$\Delta = \frac{1}{M_E(1)} \int_0^\infty |F_E(x) - F_A(x)| dx. \tag{6}$$

Bei ausreichend hoher Stützstellenzahl können die Flächenstücke durch Trapeze bzw. Dreiecke genähert werden.

Die Differenzfläche berücksichtigt nicht nur die Abweichungen δ in den Stützstellen, sondern auch die Schnittpunkte der beiden Kurven. Der Bezug auf den Mittelwert $M_E(1)$, der

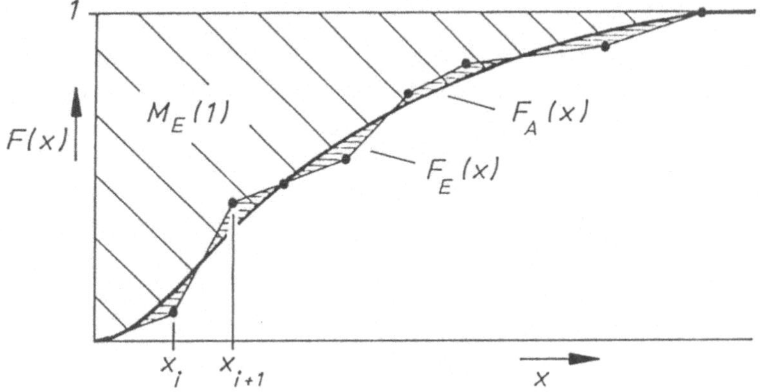

Abb. 3: **Definition der Differenzfläche**

gleich der Fläche oberhalb der Verteilungsfunktion ist [6]

$$M_E(1) = \int_0^\infty (1 - F(x))dx, \tag{7}$$

erlaubt außerdem den Vergleich der Approximationen verschiedener Verteilungen. Ein lokales Kriterium, wie die maximale Abweichung $\delta_{max} = \max(|F_E(x_i) - F_A(x_i)|)$, ist für die Optimierung nicht so gut geeignet. Es führt in der Regel zu einem instabilen Zustand des Algorithmus, da oft viele Nebenminima einer solchen Größe existieren. Zur Beurteilung der erreichten Güte und zum Vergleich verschiedener Approximationen ist die maximale Abweichung jedoch durchaus zweckmäßig.

2.3 Approximation mit zwei Erlangmischzweigen

Der Abgleich der ersten drei empirischen Momente erfolgt durch die freien Parameter p_2, λ_1 und λ_2 der ersten zwei Zweige der Erlangmischverteilung; $p_1 = 1 - p_2$. Für die Momente einer Erlangmischverteilung mit 2 Zweigen gilt

$$M(j) = \sum_{i=1}^{2} p_i \begin{pmatrix} k_i - 1 + j \\ j \end{pmatrix} \frac{j!}{\lambda_i^j}. \tag{8}$$

Für beliebige Phasenzahlen $k_1 \geq 1$ und $k_2 \geq 1$ entsteht aus dem Gleichungssystem der ersten drei Momente $M(j)$, $j = 1, 2, 3$, ein Polynom sechsten Grades, das mit Hilfe eines numerisch algebraischen Verfahrens gelöst wird [1]. In der Regel gibt es eine oder zwei gültige Lösungen für die gesuchten Parameter.

Zunächst wird die empirische Verteilung mit einer 2-zweigigen Erlangmischverteilung approximiert. Die Optimierung ist sehr einfach, da die Fläche Δ bei einem vorgegebenen k_1 eine monotone Funktion von k_2 ist. Es müssen nur wenige k_1, k_2-Kombinationen durchsucht werden, um daraus eindeutig das Optimum für Δ zu gewinnen. Zu Beginn wird $k_1 = k_2 = 1$ gesetzt und k_2 hochgezählt bis ein lokales Minimum von Δ durchlaufen wurde (in Abb. 4 durch $\circ$ markiert). Dann wird k_1 um eins erhöht und wieder k_2 von $k_2 = k_1$ an erhöht bis auch hier ein lokales Minimum von Δ gefunden wurde (Abb. 4).

Ergibt der Vergleich zweier aufeinander folgender lokaler Minima, daß keine Verbesserung erzielt wurde, so sind die Parameter für den approximierenden 2-zweigigen Erlangmischgenerator gefunden.

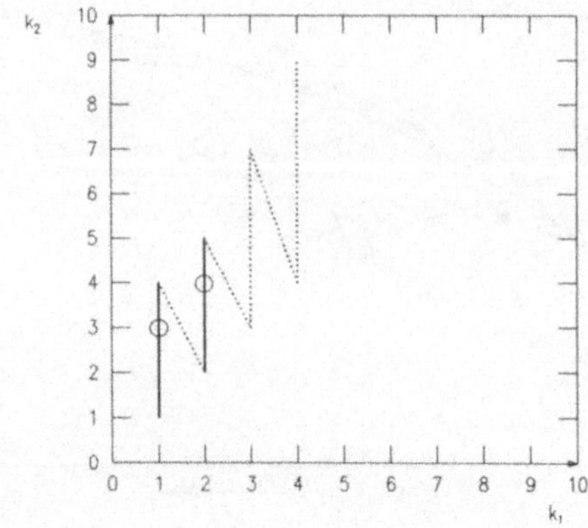

Abb. 4: **Variation der Phasenzahlen k_1 und k_2**

Abb. 5 und 6a zeigen das Ergebnis der Approximation einer Gleichverteilung (vergl. [4]) und der empirischen Verteilung "EWW" (Zwischenankunftszeiten im Rechensystem des EinWohnerWesens in Westberlin [15]).

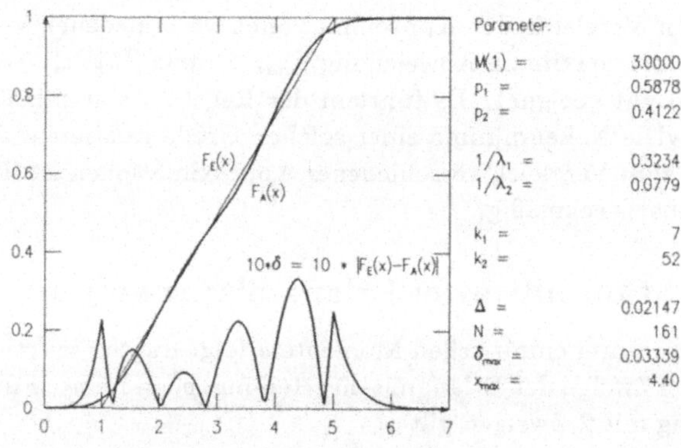

Abb. 5: **Approximation einer im Bereich $0 \leq x \leq 5$ gleichverteilten Verteilungsfunktion $F_E(x)$ mit zwei Zweigen**

In vielen Fällen reicht diese einfache und schnelle Methode aus (im Mittel 20 Verteilungsfunktionsvergleiche), um eine erste Näherung durchzuführen. Bei einigen Verteilungen ist die Güte schon im angestrebten Bereich ($\delta_{max} < 0.05$, bzw. $\Delta < 0.05$).

2.4 Approximation mit drei Erlangmischzweigen

Um die Anpassungsfähigkeit der approximierenden Erlangmischverteilung zu verbessern, kann die Zahl ihrer Parameter durch die Hinzunahme eines dritten Zweiges erhöht werden. Dann gibt es acht freie Parameter p_2, p_3, k_1, k_2, k_3, λ_1, λ_2 und λ_3; $p_1 = 1 - p_2 - p_3$. Drei (p_2, λ_1, λ_2) liegen durch den Momentenabgleich fest, der wieder mit den ersten beiden Zweigen durchgeführt wird. Die Momente $M_3(j)$ der Erlangverteilung im dritten Zweig der Approxi-

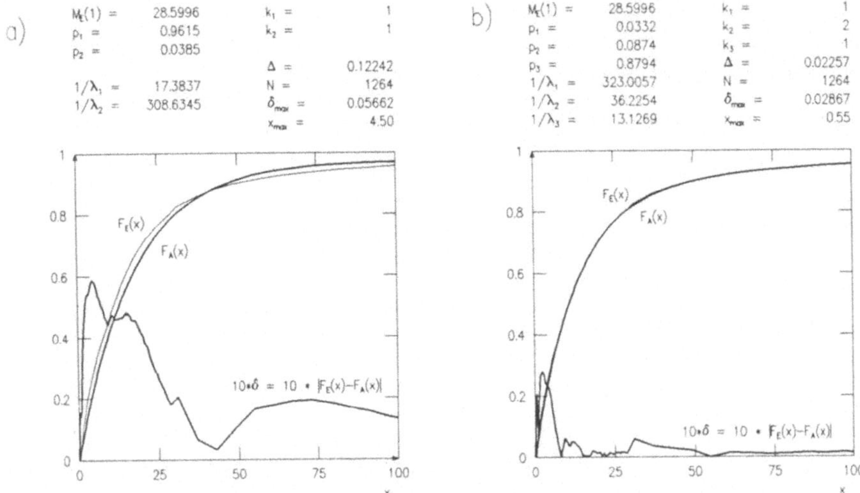

Abb. 6: **Approximation empirischen Verteilungsfunktion $F_E(x)$ der Zykluszeiten im System EWW; a) mit 2 Zweigen, b) mit 3 Zweigen**

mierenden werden von den empirischen Momenten $M_E(j)$ abgezogen und auf $(1-p_3)$ normiert.

$$M(j)' = \frac{M_E(j) - M_3(j)}{(1 - p_3)} \tag{9}$$

Die Momente $M(j)'$ werden an den in Kapitel 2.3 erwähnten Abgleichalgorithmus übergeben. Die resultierenden Zweigwahrscheinlichkeiten p_1' und p_2' müssen durch Multiplikation mit $(1 - p_3)$ auf die 3-zweigige Erlangmischverteilung umgerechnet werden.

2.4.1 Ermittlung des Startvektors für den dritten Zweig

Für den dritten Zweig wird aus einer Linearisierung der empirischen Verteilungsfunktion ein Startvektor abgeleitet (Abb. 7).

Das Maximum der ersten Ableitung $f(x)$ dieser Linearisierung mit dem kleinsten Abszissenwert wird zur Konstruktion einer Geraden $F_L(x)$ mit dieser Steigung verwendet. Diese Gerade nähert auf einfache Weise den ersten Anstieg der empirischen Verteilungsfunktion. Aus den Daten der Näherung von Erlang-k-Verteilungen durch drei Geradenstücke [14] kann auf heuristische Weise von der gewonnenen Geraden auf die Parameter p_3, k_3 und λ_3 des dritten Zweiges geschlossen werden. Dieser soll den ersten Abschnitt der empirischen Verteilungsfunktion nähern. Aus dem Schnittpunkt x_1 der Geraden $F_L(x)$ mit der Abszisse kann über den Abstand $1.3x_1$ der Geraden von der empirischen Verteilungsfunktion $F_E(x)$ auf p_3 geschlossen werden (Abb. 7). Eine Schätzung für den Mittelwert $\overline{x} \approx M_3(1)$ des dritten Zweiges kann aus der Beziehung $F_L(\overline{x}) = 0.7p_3$ ermittelt werden. Dem Wert x_1 wird über eine Tabelle der Erlang-k-Linearisierungen ein k_3 zugeordnet. Damit liegt auch ein Schätzwert für λ_3 vor: $\lambda_3 = k_3/\overline{x}$.

Dieser heuristische Ansatz bedeutet eine starke Vereinfachung des Optimierungsverfahrens. Er hat sich bei der großen Zahl der bereits durchgeführten Approximationen sehr bewährt.

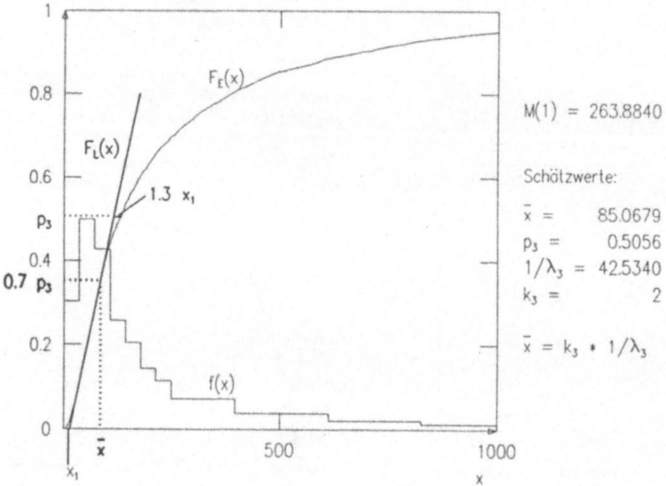

Abb. 7: **Ermittlung des Startvektors aus der empirischen Verteilungsfunktion $F_E(x)$**

2.4.2 Variation der Parameter

Anschließend werden die Parameter k_1, k_2, k_3, p_3 und λ_3 abwechselnd einzeln variiert bis keine Verbesserung mehr möglich ist. Dazu wird nach der Startvektorgenerierung mit der Suche nach einer k_1, k_2-Kombination begonnen, die so abläuft wie bei der Optimierung des 2-zweigigen Generators.

Die beste Phasenzahl des dritten Zweiges k_3 wird durch einfaches Hochzählen von $k_3 = 1$ an ermittelt. Der Zweigmittelwert des dritten Zweiges $M_{13} = k_3/\lambda_3$ wird dabei konstant gehalten (vergl. Abb. 2a). Nach dem Durchlaufen eines lokalen Minimums der Differenzfläche wird die Variation von k_3 abgebrochen.

Die Parameter p_3 und λ_3 werden iterativ verändert, bis mit der kleinsten Schrittweite keine Verbesserung der Differenzfläche mehr möglich ist.

In der ersten Phase der Approximation folgt auf die Variation eines kontinuierlichen Parameters die Optimierung der ganzzahligen Parameter. Nach der zehnten Variation eines Parameters werden zur Verbesserung der Genauigkeit der Approximation nur noch p_3 und λ_3 verändert. Bei der Variation eines ganzzahligen Parameters sind die Veränderungen am Verlauf der Approximierenden in diesem Zustand des Algorithmus schon zu grob. Eine Änderung der vorgegebenen Reihenfolge der variierten Parameter ist möglich, bringt jedoch nur in Ausnahmefällen eine Verbesserung der Approximationsgüte.

Die Optimierung wird beendet, wenn das lokale Minimum der bezogenen Differenzfläche Δ, das mit der Variation eines Parameters erreicht wurde, durch Veränderung des nächsten Parameters um weniger als 0.05% verbessert werden konnte. Es können auch Grenzen für die bezogenen Differenzfläche Δ oder die maximale Abweichung δ_{max} vorgegeben werden.

Das Ergebnis für die bereits oben mit zwei Zweigen approximierte Verteilung der Zykluszeiten im System EWW, jetzt mit drei Zweigen, zeigt Abb. 6b.

Die Zahl der für verschiedene Approximationen benötigten Iterationen, d.h. Verteilungsfunktionsvergleiche, liegt in der Regel zwischen 100 und 1300. Eine große Anzahl von Vergleichen ist nur notwendig, wenn das Minimum des Gütekriteriums sehr schwach ausgeprägt ist.

2.5 Zusammenfassung

Damit liegt ein Verfahren vor, das mit wenigen Parametern eine für die Datenverkehrstheorie geeignete Beschreibung von gemessenen Zufallsprozessen liefert. Der Abgleich der ersten drei empirischen Momente und die Näherung des Verteilungsfunktionsverlaufes, der zusätzlich die Einflüsse der höheren Momente berücksichtigt, reicht aus, um die für eine mathematische Analyse wesentlichen Kennwerte einer Verteilung wiederzugeben. Empirische Verteilungsfunktionen mit nur einem Maximum in der ersten Ableitung können bis auf eine maximale Abweichung $\delta_{max} < 0.05$ bzw. bezogene Differenzfläche $\Delta < 0.05$ genähert werden (vgl. Abb. 6). Um die Genauigkeit für allgemeine Verteilungen weiter zu erhöhen, ist eine Erweiterung des approximierenden Generators auf 4 - 5 Zweige in Vorbereitung, die an mögliche weitere Maxima der ersten Ableitung der empirischen Verteilungsfunktion angepaßt werden soll.

Signifikante Sprünge der empirischen Verteilungsfunktion, die in der mathematischen Beschreibung berücksichtigt werden müssen, werden vor der Übergabe der Verteilungsfunktion an das Approximationsverfahren subtrahiert und nach der Approximation als zusätzliche Zweige in die Erlangmischverteilung aufgenommen (z.B. als Erlang-k-Verteilung der Ordnung unendlich oder als Sprungfunktion). Die Vollständigkeitsbedingung $\sum p_i = 1$ wird durch entsprechende Transformationen immer erfüllt.

3 Bestimmung des äquivalenten Coxgenerators aus einer Erlangmischverteilung

In der Analyse von Kommunikationssystemen ist die Beschreibung von stochastischen Prozessen mittels Coxverteilungen verbreitet (vergl. [3], [5], [8]). Dabei wird in der Regel die spezielle Coxverteilung verwendet, die eine konstante Intensität in allen Phasen aufweist. Sie vereinfacht die mathematische Analyse sehr, erfordert aber unter Umständen eine hohe Phasenzahl. Jede allgemeine Coxverteilung kann in eine spezielle Coxverteilung mit im allgemeinen unendlich vielen Phasen überführt werden.

Grundsätzlich ist die in MEDA verwendete Erlangmischverteilung einfacher zu beschreiben, da sie mit einer festen Zahl von 5 bzw. 8 Parametern auskommt. Das Modell ist nur in der Zahl der Zweige und der Intensitäten beschränkt, nicht jedoch in der Zahl der Phasen. Das Approximationsverfahren soll in vorhandene Analysesysteme eingebracht werden. Daher soll die mit MEDA gewonnene Erlangmischverteilung in eine Coxverteilung transformiert werden. Die in den Kapiteln 3.2 ff. hergeleitete Überführungsvorschrift basiert auf einer von Iversen [8] entwickelten Äquivalenzbeziehung zwischen einer Exponential- und einer zweiphasigen Coxverteilung.

3.1 Exponentialprozeß als Coxverteilung nach Iversen

Gegeben sei ein Exponentialgenerator mit der Rate μ (s. Abb. 8a). Er kann in zwei Exponentialgeneratoren zerlegt werden. Einer hat eine beliebige Rate $\lambda \geq \mu$, der andere dieselbe Rate μ wie der Ausgangsgenerator. Die beiden Prozesse sind zu einem Mischprozeß nach Cox zusammengeschaltet (s. Abb. 8b).

Die Identität der Laplace-Transformierten ist für $p = \mu/\lambda$ und $q = 1 - p = 1 - \mu/\lambda$ leicht zu zeigen:

$$L_a(s) = \frac{\mu}{\mu + s} \tag{10.1}$$

$$L_b(s) = p\,\frac{\lambda}{\lambda + s} + q\,\frac{\lambda}{\lambda + s}\,\frac{\mu}{\mu + s} \tag{10.2}$$

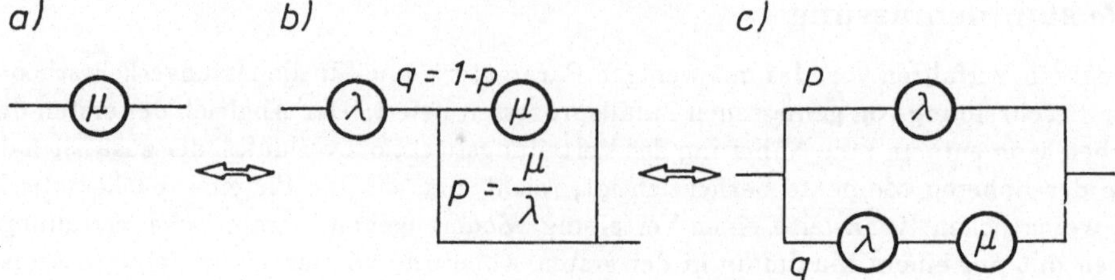

Abb. 8: **Zerlegung nach Iversen: a) Exponentialgenerator; b) äquivalenter Coxgenerator; c) Variantendarstellung**

$$= \frac{\mu}{\mu+s}\left(\frac{\mu+s}{\mu}\frac{\mu}{\lambda}\frac{\lambda}{\lambda+s} + (1-\frac{\mu}{\lambda})\frac{\lambda}{\lambda+s}\right) = \frac{\mu}{\mu+s} = L_a(s) \quad (10.3)$$

Abbildung 8c zeigt die beiden Coxvarianten.

Die Identität der Laplace-Transformierten gilt auch dann, wenn der Exponentialgenerator Bestandteil eines Summenzufallsprozesses ist. Der zitierte Zerlegungsmechanismus soll nun dazu verwendet werden, aus einer Erlangmischverteilung die Coxverteilung mit der identischen Laplace-Transformierten zu gewinnen.

3.2 Erlangmischverteilung mit zwei Zweigen als Coxverteilung

Wir wollen dazu mit einer 2-zweigigen Erlangmischverteilung beginnen und schließlich die allgemeine Formel für einen N-zweigigen Mischgenerator herleiten. Es wird angenommen, daß die Intensitäten λ_i in jedem Zweig konstant sind. Die Zweige der Mischverteilung seien so geordnet, daß $\lambda_i \geq \lambda_{i+1}$. Nehmen wir an, daß für die Phasenzahlen k_1 und k_2 gilt $k_1 \leq k_2$. Dies ist der einfachere Fall. Betrachten wir ein Beispiel:

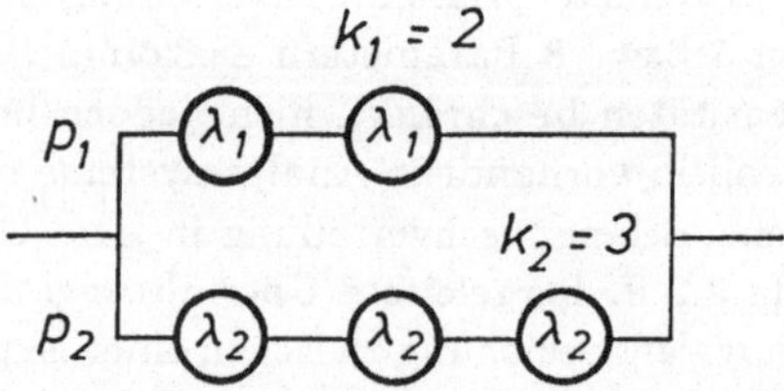

Abb. 9: **Beispiel einer Erlangmischverteilung mit zwei Zweigen und $k_1 \leq k_2$**

Die aus der Zerlegung des i-ten Zweiges errechnete Wahrscheinlichkeit, mit der die Variante j durchlaufen wird, wird als $p_{Ci}(j)$ geschrieben. Die endgültigen Variantenwahrscheinlichkeiten berechnen sich dann aus der Summe über die Zerlegungen, da gleiche Reihenfolgen von Exponentialgeneratoren entstehen können:

$$p_C(j) = \sum_{i=1}^{2} p_{Ci}(j). \quad (11)$$

Der erste Zweig kann bereits als Coxvariante angesehen werden (Abb. 10)

$$p_{C1}(j) = \begin{cases} p_1 & j = k_1, \\ 0 & \text{sonst.} \end{cases} \quad (12)$$

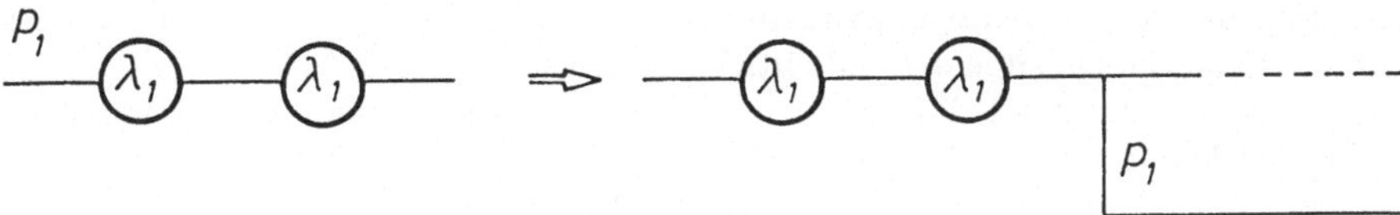

Abb. 10: **Der erste Zweig als Coxvariante**

Der zweite Zweig muß sich in diesen bereits festliegenden Teil des Coxmodells einfügen. Den Generatoren λ_2 müssen also zwei Generatoren λ_1 vorgeschaltet werden. Dies geschieht schrittweise wie Abb. 11 zeigt. Es gilt $p = \lambda_2/\lambda_1$ und $q = 1 - \lambda_2/\lambda_1$.

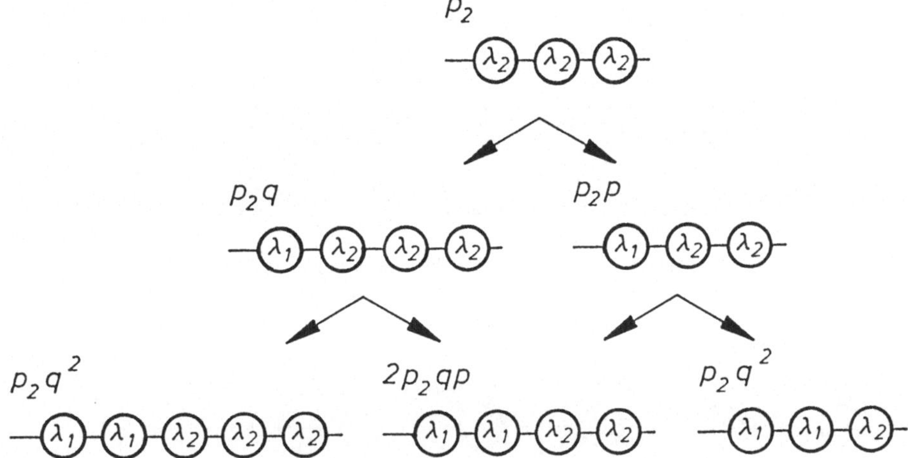

Abb. 11: **Zerlegung des zweiten Zweiges in Coxvarianten**

Es entsteht auf diese Weise ein Pascalsches Dreieck, aus dessen untersten Zeile die Coxvarianten und ihre Wahrscheinlichkeiten bestimmt werden können. Dieses Dreieck hat k_1 Ebenen, da k_1 Zerlegungen zum Vorschalten von k_1 Generatoren der Intensität λ_1 führen. Das resultierende Coxmodell zeigt Abb. 12.

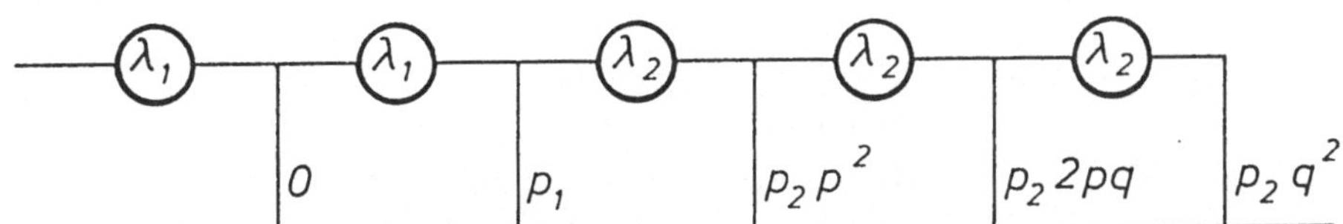

Abb. 12: **Das äquivalente Coxmodell**

Wir finden die Zweige der Erlangmischverteilung hintereinandergesetzt vor. Die Zahl der Phasen der Coxverteilung ist $k_1 + k_2$. Die Zerlegung des zweiten Zweiges ergibt folgende verallgemeinerte Gleichung für die resultierenden Variantenwahrscheinlichkeiten:

$$p_{C2}(k_1 + k_2 - i) = p_2 \binom{k_1}{i} q^{k_1 - i} p^i, \qquad 0 \leq i \leq k_1,\ k_1 \leq k_2. \tag{13}$$

Die Addition der beiden Zerlegungsergebnisse $p_{C1}(j)$ und $p_{C2}(j)$ ergibt die resultierenden Variantenwahrscheinlichkeiten der endgültigen Coxverteilung (Gl. 11).

Betrachten wir nun den Fall $k_1 > k_2$. Die Zweige der Erlangmischverteilung seien bereits nach fallenden Intensitäten $\lambda_i \geq \lambda_{i+1}$ geordnet. Es sei $k_1 = 4$ und $k_2 = 3$.

Der erste Zweig kann auch hier mit Gleichung (12) übernommen werden. Das aus der Zerlegung des zweiten Zweiges hervorgehende Pascalsche Dreieck ist jetzt nicht mehr vollständig (s. Abb. 13). Das äquivalente Coxmodell zeigt Abb. 14.

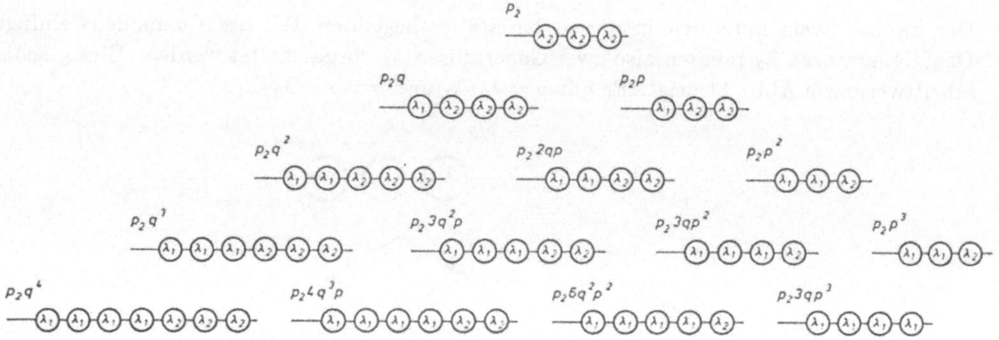

Abb. 13: **Zerlegung des zweiten Zweiges der Erlangmischverteilung; $k_1 > k_2$**

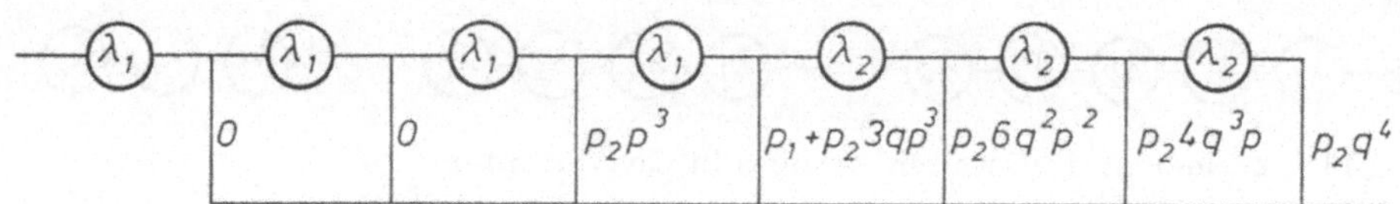

Abb. 14: **Äquivalentes Coxmodell**

In der allgemeinen Gleichung muß jetzt eine Unterscheidung vorgenommen werden. Die Glieder $i \geq k_2$, die in der rechtsseitigen Diagonalen liegen, haben die Koeffizienten der Elemente links oberhalb von ihnen im Dreieck.

$$p_2(k_1 + k_2 - i) = \begin{cases} p_2 \binom{k_1}{i} q^{k_1 - i} p^i & 0 \leq i < k_2, \\ p_2 \binom{k_1 + k_2 - i - 1}{k_2 - 1} q^{k_1 - i} p^{k_2} & k_2 \leq i \leq k_1. \end{cases} \tag{14}$$

Die endgültigen Variantenwahrscheinlichkeiten ergeben sich wieder nach Gleichung (11).

Es kann also jetzt jede beliebige 2-zweigige Erlangmischverteilung in die äquivalente Coxverteilung überführt werden. Der Beweis der Identität läßt sich über die Laplace-Transformierten leicht erbringen, da alle Zerlegungen auf die in Kap. 3.1 bewiesene Umformung zurückgeführt werden können.

3.3 Umwandlung einer speziellen Erlangmischverteilung mit N Zweigen in eine äquivalente allgemeine Coxverteilung

Die Gewinnung der äquivalenten Coxverteilung aus einer Erlangmischverteilung mit mehr als zwei Zweigen geht in mehreren Schritten vor sich. Der erste Zweig kann nach Gleichung (12) sofort in das Coxmodell aufgenommen werden. Dem zweiten Zweig müssen die Generatoren des ersten Zweiges vorgeschaltet werden (Kap. 3.2). Dem dritten Zweig wird zuerst der erste, dann der zweite Zweig vorgeschaltet usw.. Bei jeder Zerlegung entsteht ein Pascalsches Dreieck der Variantenwahrscheinlichkeiten. Die Fußpunkte des Dreiecks nach dem Vorschalten eines Zweiges bilden die Spitzen der Pyramiden der darunterliegenden Dreiecke, die sich durchdringend die Vorschaltung des nächsten Zweiges repräsentieren.

Es folgt die allgemeine Gleichung, mit der rekursiv die einzelnen Zweige in das Coxmodell umgerechnet werden. In Gleichung (15.6) muß, wie in Gleichung (14), berücksichtigt werden, daß die Pascalschen Dreiecke für bestimmte Werte der k_i unvollständig sein können. Die Gleichung kann leicht in ein Programm überführt werden.

3.3.1 Allgemeine Gleichung für die äquivalente Coxverteilung einer speziellen Erlangmischverteilung mit N Zweigen

Zuerst definieren wir die allgemeinen Zerlegungswahrscheinlichkeiten p_{nz}, q_{nz} für $2 \leq n \leq N$, $1 \leq z < n$

$$p_{nz} = \frac{\lambda_n}{\lambda_z}, \tag{15.1}$$

$$q_{nz} = 1 - p_{nz} \tag{15.2}$$

und die Phasensummen m_j

$$m_j = \sum_{l=0}^{j} k_l, \qquad k_0 = 0, \quad 0 \leq j \leq N. \tag{15.3}$$

Die Variantenwahrscheinlichkeiten $p_{Cnz}(i)$ gehen aus der Vorschaltung der k_z Generatoren des z-ten Zweiges vor den n-ten Zweig hervor. Sie können rekursiv berechnet werden. Die ersten Glieder sind für $1 \leq n \leq N$

$$p_{Cn0}(i) = \begin{cases} p_n & i = m_n, \\ 0 & \text{sonst.} \end{cases} \tag{15.4}$$

Die Rekursion setzt sich wieder aus zwei Teilschritten zusammen. In den folgenden Gleichungen (15.5+6) gilt $0 \leq i \leq m_z$, $2 \leq n \leq N$ und $1 \leq z < n$. Bereits fertige Varianten aus der letzten Zerlegung werden übernommen

$$p_{Cnz}(m_{n-1} - k_z - i) = \begin{cases} p_{Cnz-1}(m_{n-1} - i) & 0 \leq i \leq m_{z-1} - n,\ k_1 \geq k_n,\ z > 1, \\ 0 & \text{sonst,} \end{cases} \tag{15.5}$$

und zu den neuen Zerlegungsergebnissen addiert

$$p_{Cnz}(m_n - i) = p_{Cnz}(m_n - i) +$$

$$+ \begin{cases} \sum\limits_{j=0}^{m_{z-1}} p_{Cnz-1}(m_n - j) \binom{k_z}{i-j} q_{n,z}^{k_z-(i-j)} p_{n,z}^{i-j} & 0 \leq i < k_n, \\ \sum\limits_{j=0}^{min(m_{z-1}, k_n - 1)} p_{Cnz-1}(m_n - j) \binom{k_n + k_z - i - 1}{k_n - j - 1} q_{n,z}^{k_z-(i-j)} p_{n,z}^{k_n - j} & k_n \leq i < k_z + k_n. \end{cases} \tag{15.6}$$

Die Summe über die Ergebnisse der letzten Zerlegungen ergibt die resultierenden Coxvariantenwahrscheinlichkeiten

$$\alpha_i = p_C(i) = \sum_{j=1}^{N} p_{Cnn-1}(i), \qquad 1 \leq i \leq \sum_{l=1}^{N} k_l. \tag{15.7}$$

Für die Intensitäten λ_{Ci} der äquivalenten Coxverteilung gilt:

$$\lambda_{Ci} = \begin{cases} \lambda_1 & 1 \leq i \leq k_1, \\ \lambda_2 & k_1 < i \leq k_1 + k_2, \\ \dots & \dots \\ \lambda_N & m_{N-1} < i \leq m_N. \end{cases} \tag{15.8}$$

Mit dieser allgemeinen Gleichung (15) kann jede spezielle Erlangmischverteilung, deren Intensitäten λ_i im i-ten Zweig konstant sind, in eine allgemeine Coxverteilung umgerechnet werden. Der äquivalente Coxgenerator besteht aus einer Reihenschaltung der nach fallendem λ_i sortierten Erlangmischzweige. Die Summe der Generatoren und ihre Intensitäten bleiben also bei dieser Überführung konstant. Aus den Verhältnissen der λ_i, der Zahl k_i der exponentiellen Phasen in den einzelnen Erlang-k-Zweigen und den Zweigwahrscheinlichkeiten p_i können die Variantenwahrscheinlichkeiten $\alpha_i = p_C(i)$ berechnet werden. Die beiden Laplace-Transformierten sind auf Grund der in Kapitel 3.1 geführten Beweise gleich.

Literaturverzeichnis

[1] Basermann, A., *Nullstellenbestimmung von Polynomen zum Momentenabgleich im Approximationsverfahren MEDA.* Studienarbeit am Lehrstuhl für Allgemeine Elektrotechnik und Datenfernverarbeitung, RWTH Aachen, Aachen Juli 1987.

[2] Bratley, P., Fox, B., Schrage, L., *A Guide to Simulation.* Springer-Verlag, New York 1983.

[3] Bux, W., *Single Server Queues with General Interarrival and Phase-Type Service Time Distributions - Computational Algorithms.* Proc. 9. ITC, Torremolinos 1979, paper 413.

[4] Bux, W., Herzog, U., *The Phase Concept: Approximation of Measured Data and Performance Analysis.* Proc. Int. Symp. on Computer Performance, Measurements, and Evaluation, Elsevier North-Holland, New York, Amsterdam 1977.

[5] Cox, D.R., *A Use of Complex Propabilities in the Theory of Stochastic Processes.* Proc. Camb. Phil. Soc., 51 (1955), pp. 313-319.

[6] Feller, W., *An Introduction to Probability Theory an Its Applications, Volume I.* Wiley, New York, London 1957.

[7] Halfin, S., *Delays in Queues, Properties and Approximations.* Proc. 11. ITC, Kyoto 1985, paper 1.4-3.

[8] Iversen, V.B., Nielsen, B.F., *Some Properties of Coxian Distributions with Applications.* Int. Conference Modelling Techniques and Tools for Performance Analysis, Paris 1985.

[9] Kreyszig, E., *Statistische Methoden und ihre Anwendungen.* Vandenhoeck und Ruprecht, Göttingen 1968.

[10] Marshall, W.T., Morgan, S.P., *Statistics of Mixed Data Traffic on a Local Area Network.* Proc. 11. ITC, Kyoto 1985, paper 4.1A-1.

[11] Pawlita, P., *Traffic Measurements in Data Networks, Recent Measurement Results, and Some Implications.* Proc. of IEEE 4 (1981), 525-535.

[12] Pawlita, P., *Messungen und Analyse des Ein-/Ausgabeverkehrs von Datenstationen in Fernverarbeitungssystemen.* Dissertation, RWTH Aachen, Aachen 1977.

[13] Schassberger, R., *Warteschlangen.* Springer-Verlag, Wien, New York 1973.

[14] Schmickler, L., *Approximation gemessener Ankunftsprozesse und ihre Auswirkung auf das Verhalten des Warteraumes GI/M/1.* Diplomarbeit am Lehrstuhl für Allgemeine Elektrotechnik und Datenfernverarbeitung der RWTH Aachen, Aachen 1985.

[15] Südhofen, H.-D., *Modelle für Benutzergruppen zur Beschreibung von Datenverkehrsflüssen in Kommunikationssystemen.* Dissertation, RWTH Aachen, Aachen 1986.

[16] Südhofen, H.-D., Pawlita, P., *Modeling of Compound Traffic Streams in Computer Communication Networks.*Proc. 11. ITC, Kyoto 1985, paper 3.2A-3.

[17] Welzel, T., *Analyse und Messung der Workload einer technisch-wissenschaftlichen Programmentwicklungsumgebung als Grundlage der Leistungsbewertung Lokaler Netze.* Informatik Fachberichte: Kommunikation in verteilten Systemen, GI/NTG Fachtagung, Aachen, Springer-Verlag, Berlin 1987.

Simulated Time and the Ada Rendezvous

Werner Pohlmann
Technische Universität München
Institut für Informatik

Postfach 202420, D-8000 München

Abstract:
Choosing Ada as a base language for the process view of simulation, one would like to adopt the Ada rendezvous for process interaction. This paper investigates the implications of the rendezvous in the context of simulated time. To avoid the danger of deadlock (as exhibited by some proposed systems), a very thorough control of the rendezvous is shown to be necessary. A suitable simulation mechanism is presented and proven. But judged by the ensuing costs the rendezvous is considered to be no attractive choice for general simulation purposes.
Considering methods, this paper uses and contributes to the "space-time view" (L.Lamport) of processes: processes are sequences of actions, and simulated time is a mapping of actions to numerical values.

1. Introduction

There is a natural interest in Ada as a base language for discrete event simulation. The reasons include Ada's software engineering qualities as well as the expected dissemination; Ada's provisions for concurrent programming attract a special and twofold attention. First, having work done in parallel may speed up simulation runs. Second, Ada tasks are eligible for the process view of simulation, i.e. a plausibly structured representation of the object system (cf. [9]).

Both topics present open questions; this paper exclusively deals with the second one. To be more specific: we assume that a modeller wants to depict real processes and their interactions by Ada tasks and rendezvous, and we investigate the technical implications. Real processes evolve in real time; the shift to Ada tasks omits this essential aspect. So we must add a mechanism for simulated time. But Ada tasks bear a temporal structure of their own which is not altogether trivial. So if we want to superimpose simulated time we must take care to clarify our aims and to avoid insecurities that might result from incompatible temporal concepts.

For an illustration, but equally necessary in itself, we start with a critical discussion of some published Ada based systems (§2). Next we state basic facts and requirements for the combination of Ada tasks and simulated time (§3) and develop a suitable simulation mechanism (§4). So far our concern is correctness; in the concluding chapter (§5) we sum up and try an opinion on the choice of Ada for process oriented simulation.

2. Insecurities of some proposed systems

The proposals discussed in this chapter wish to exploit Ada's potential for parallelism or at least accept Ada's rendezvous for process interaction. For simulated time they adapt the classical event list mechanism. Model tasks (i.e. tasks representing real processes) suspend themselves for specified intervals by calling a hold procedure, which enters their resumption times into the event list. Using this list, a special scheduler task reactivates suspended model tasks in the proper order and advances the simulation clock. For the technical problem of how to put an Ada task to sleep and wake it up again we refer the reader to the literature; we concentrate on a more logical difficulty. The scheduler must wake up the next model task when and only when all current model activity has ended - but the rendezvous might obscure this situation.

2.1 The author of [5] decided to let model tasks proceed in a strict coroutine fashion, seemingly avoiding all problems with concurrency:

> "To avoid data synchronization problems and to follow standard process oriented practice, we will need a mechanism to make sure that only one task is running at a time. This is the purpose of our simulation package." [5],p.117

The scheduler therefore acts as follows, [5],p.121 :

```
loop
    according to first notice in event list
    reactivate model task and advance simulation time;
    accept next;
```

```
        eliminate first notice from event list;
    endloop
```

The entry next is called whenever a model task suspends itself via **hold**.

Unfortunately, the author employs the normal Ada rendezvous for process interaction. This is a conflict of concepts: there is just one active model task at a time, but it takes two to bring about a rendezvous. Before a rendezvous the currently active task will have to wait for its inactive partner, but will not be able to instruct the scheduler accordingly. After a rendezvous both tasks will be active, and prevent the scheduler from establishing the intended temporal order.

To illustrate the problem, let us look at the author's M/M/1 example, consisting of a service process and an arrival process and a queue accessed by both, [5],p.119:

```
  arrival:
     loop
          hold( );
          create new job;
          if queue is empty
               then put job in queue;
                    service.wakeup;
               else put job in queue; endif;
     endloop

  service:
     loop
          if queue is empty
               then accept wakeup;
               else hold( );
                    take job out of queue; endif;
     endloop
```

Now suppose that service executes accept wakeup while arrival is suspended via hold. Then there is a deadlock:

service waits for rendezvous with arrival;
arrival waits to be reactivated by the scheduler;
scheduler waits for service to call next.

If, however, a rendezvous did take place, two concurrent calls of next (via hold) would follow. The scheduler had to treat them serially, incorrectly advancing the simulation clock in between.

(For a pure coroutine system, the reader may consult [13].)

2.2 With a view to possible gains in speed the authors of [10] and [11] allow their model tasks to operate in parallel as far as they operate at the same simulated time:

> "Task scheduler, ..., will resume at one time all tasks which are scheduled for activation at the same simulation time. These tasks in turn will need to schedule themselves for subsequent resumptions, but since both they and the scheduler are running in parallel, the scheduler could resume one out of order. ...
> The problem is solved by having the scheduler keep a count of all active tasks. Then no tasks are reactivated until the currently active tasks have completed their scheduling."
>
> [10],p.480

This scheme may be viewed as a generalization of the arrangement in 2.1: instead of relying on the one-process-at-a-time invariant, the scheduler now relies on the known number of active tasks. This knowledge is maintained as follows, [10],p.486:

```
loop select
           accept next; index:=index+1;
        or accept wait; number_active:=number_active-1;
        or accept spawn; number_active:=number_active+1;
     endselect;
     if index=number_active then exit; endif;
endloop
```

Exiting from this loop the scheduler reactivates all tasks due for the next moment of simulated time. Their number is assigned to number_active, and index is set to zero. The scheduler then reenters the loop above. Entry next is called whenever a model task calls hold. Wait and spawn are used to announce model task interaction; they tell the scheduler that a task starts to wait for a rendezvous or is set free again.

For an illustration of this method (there are no precise rules), we can again look at a M/M/l program by the authors, [10],p.487:

```
arrival:
        loop
            create new job;
            if queue is empty and not job_in_service
               then put job into queue;
                    scheduler.spawn;
                    service.wakeup;
               else put job into queue; endif;
            hold(  );
        endloop

service:
        loop
            if queue is empty
               then scheduler.wait;
                    accept wakeup; endif;
            take job from queue;
            job_in_service:=true;
            hold(  );
            job_in_service:=false;
        endloop
```

This program is not correct (but a modeller might defend it with stochastic arguments):
Suppose that arrival evaluates its condition after service has emptied the queue but before job_in_service is set to true. Next, arrival executes its then-alternative, but service gets busy and calls hold. So there is a deadlock:

```
arrival waits for rendezvous with service;
service waits for reactivation by scheduler;
scheduler waits for number_active to be decreased
                (since it was increased by arrival's spawn).
```

More important, a simple counting scheme like this is in general not adequate for the Ada rendezvous. The rendezvous is handshaking communication and to that extent symmetric: it is not a priori clear which partner will have to wait for the other. Imagine e.g. an unbuffered producer-consumer relationship:

```
producer:                          consumer:
  loop                               loop
       hold( );                           accept deliver;
       consumer.deliver;                  hold( );
  endloop                            endloop
```

Wait and spawn provide no straightforward way of informing the scheduler about the delays that result from this coupling of tasks. But the modeller must supply calls of wait and spawn and therefore try to secure their intended effect by additional means of synchronization, e.g. status variables; he will, as in the M/M/1 example, end up with solutions that are awkward and unreliable and do not represent good Ada style.

2.3 In contrast to the proposals of 2.1 and 2.2, the authors of [7] and [8] and the author of [4] choose to supply simulation specific constructs for process interaction. There are provisions for a model task to e.g.

- acquire or release a resource
- activate another model task.

The general meaning of these operations is the traditional one. But in accordance with 2.2, simultaneous activity is allowed to happen in parallel, and there is a similar counting scheme to direct the scheduler. Two comments are necessary:

First. The rendezvous is, of course, used to program the proposed higher level interaction primitives, but this specialized use does not necessarily constitute a difficulty: a simple counting mechanism suffices to watch over the model activity. We must and can arrange matters so that a model task will never spend simulated time waiting for such a rendezvous. For instance, a resource may be represented by a task that is always ready to engage in a rendezvous with a user task, though not always ready to grant what this user is asking for.

Some care is nevertheless indicated. Suppose that a model task p reactivates a model task q by e.g. releasing a resource q is waiting for. Then the program must guarantee that the counter for active tasks is increased on behalf of q before p or q gets a chance to suspend itself and consequently decrease the counter. The program given in [4] clearly violates this rule and risks an erroneous advancement of simulation time.

Second. If a modeller attempts to use the rendezvous in addition to the supplied facilities for process interaction, he will experience the difficulties discussed in 2.2. The authors of [7] and [8] claim that their system is a "natural extension of Ada" ([8],p.54) or, more specific:

> "This method of implementing processes allows the tasks representing the entities to communicate directly by means of the normal Ada rendezvous." [7],p.70

This claim is unjustified and misleading. As illustrated in 2.2, hidden waiting for a rendezvous will lead to a deadlock.

(For a basically similar, but more radical solution the reader may consult [15]: the authors provide a user interface that tries to completely hide that the model program is in effect a system of Ada tasks.)

3. Sequences of actions and simulated time

This chapter studies the rendezvous in the context of simulated time. To simplify the exposition, we ignore task creation and termination and restrict our attention to a given collection of coexisting tasks. We reduce the rendezvous to its simple form, the exchange of information, and forbid shared variables.

Time, with its usual attributes like duration, serves as a popular frame of reference in informal explanations of the Ada rendezvous. But more often than not it is inadequate, unnecessary and misleading to apply this notion of time to the design of programs, and there is an additional source of confusion if we have to reason about simulated time too. (Prompted by some odd experience, I do not hesitate to state the obvious: program execution time and simulated time are as different as apples and pears and not to be added, subtracted or otherwise combined; and an attempt to use Ada's real time clock for simulated delays will end up with a process spending some unpredictable 10 units of time in random number generation just to learn that he should wait for 2 units of time.) We therefore fall back on a more modest concept, as advocated in [12]:

A task gives rise to a sequence of actions, naturally endowed with a total happened-before relation, and a system is a collection of disjoint sequences of actions. Actions are denoted by x, y, and z; in general, we need not be more specific about them. But we have to consider entry call and entry accept actions, denoted by c and a, which may constitute a rendezvous c~a. This relation means mutual communication and synchronization of tasks, so that the past of one task will add to the past of the other since it may influence its future. We therefore generalize the local happened-before relation:

3.1 The relation $\rightarrow$ on the set of actions of the system is the smallest transitive relation such that:
- x happens before y => $x \rightarrow y$
- c~a => (c happens before x => $a \rightarrow x$) &
  (a happens before y => $c \rightarrow y$) .

We will use terms like "happened before", "predecessor", etc. for $\rightarrow$ as well. The relation $\succ$ has to be irreflexive, since it is meant to describe the evolution of a system, and Ada adds some obvious constraints. A c or a action may engage in just one rendezvous c~a, with the right entry, and it cannot have a successor without engaging in a rendezvous.

Next we introduce simulated time. We define it to be a function $x \mapsto T(x)$ assigning numerical values to actions. So actions occur instantaneously, at the moment denoted by T; if the modeller wishes to represent duration, he can use a pair of actions.

A plausible representation of time must satisfy irreversibility:

3.2 $x \succ y \Rightarrow T(x) \leq T(y)$.

Time must not go backwards in an evolving system; but since the modeller may wish to e.g. divide into several steps what was an atomic action in reality, we allow the same value of T for consecutive actions.

Applying 3.2 to 3.1 we learn that simulated time may advance arbitrarily in the course of actions, but at least to max(T(c),T(a)) if there was a rendezvous c~a. The modeller needs a meaningful standard. For the rendezvous we choose the least possible increment, which may be interpreted as the time one partner spends waiting for the other. For all other situations we supply the modeller with a special hold action which he may use to advance time by increments of his own choice:

3.3 Let y be the immediate successor of x in a task. Then:
T(y)=max(T(c),T(a)) if x is the c or a of some c~a,
T(y)=T(x)+t if x is some hold(t),
T(y)=T(x) otherwise.

In Ada, several tasks may call the same entry. Such calls need not be related to each other in any way, but the Ada runtime system imposes an ordering on them since they are accepted serially if they are accepted at all. Let us for the moment denote this imposed ordering by $\succ_{im}$; we can partly infer it from the effected rendezvous. Namely, for calls and accepts of the same entry:

3.4 $c \sim a \ \& \ \neg\exists a': c' \sim a' \ \& \ a' \succ a \quad \Rightarrow \quad c \succ_{im} c'$.

Establishing $\succ_{im}$ the Ada machine will of course not respect simulated time; so an entry call issued later in simulated time could be accepted in precedence to an earlier one. This could lead to phenomena of time consumption or causal relationship that in no way reflect a possible behaviour of the real system. We therefore extend irreversibility 3.2 of time to $\succ_{im}$; its application to 3.4 yields:

3.5 $c \sim a \ \& \ T(c') < T(c) \quad \Rightarrow \quad \exists a': a' \succ a \ \& \ c' \sim a'$.

It is by this rule that simulated time becomes a determinant of model behaviour. (For a more complete treatment of Ada tasking, e.g. task creation or the select form of the rendezvous, we had to add similar rules.) 3.5 looks like FCFS, with T for FC, $\succ$ for FS and indeterminate result for T(c)=T(c'). But please note the rationale for this rule: the rendezvous shall enable a task to interact with the rest of the model in a temporally well defined and understandable manner. Neither do we intend to mimic Ada's FCFS rule (which is defined in terms of the implementation) nor is there any bias towards FCFS service in resource-client models.

Next we turn to the task of implementing the rules we found necessary. 3.3 suggests that we should furnish each model task with its own clock variable, which is incremented by hold actions. For a rendezvous individual clock readings must be exchanged and updated in accordance with 3.3.

Rule 3.5 is more difficult to establish. A programmer cannot directly control $\succ_{im}$, but fortunately 3.5 does not require us to do exactly this. In the next chapter we show how to restrict concurrent behaviour sufficiently.

4. The scheduler

Suppose an observer wants to verify 3.5 for a rendezvous c~a. - A true disciple of the empirical method, the observer does not rely on insight but on a set M of observed actions. M is an initial part of system history: if x is in M so is y if y≻x. Clearly, the observer must require M to include, in addition to c and a, all calls c' of the same entry if T(c')<T(c). How to define a suitable set M is less clear as he will need to recognize that he has really gathered all the facts he wants. More formally: let M={x ¦ E(x) }; then, knowing just M, the observer must be able to conclude that no possible system future will produce an additional x with E(x).

It is not difficult to show (but please remember that we abstract from task termination; otherwise we should help the observer by special terminate actions) that a sufficient and not over-cautious choice is

4.1 $M=\{x \mid \forall y \succ x: T(y)<T(c) \text{ or } y \succ a \text{ or } y \succ c \}$

so that for maximal elements m of M

4.2 $T(m) \geq T(c)$ or
    m is some c" & $\neg\exists$a" in M: c"~a" & T(a")<T(c) or
    m is some a" & $\neg\exists$c" in M: c"~a" & T(c")<T(c).

We are convinced that, in the absence of any a priori information about system behaviour, no algorithm to direct the rendezvous of model tasks can do better than our observer. So we turn 4.2 into a rule of operation, roughly reading as follows:

4.3 An entry call c may be granted a rendezvous when all other model tasks either have reached or passed T(c) in their simulated time or cannot advance because they themselves want a rendezvous.

We replace our observer by a scheduler task (its job could, at the price of more communication overhead, be placed upon the model tasks). The amount of present information may be reduced to:

4.4 The scheduler maintains the system state, i.e. a set of records $\langle p,t,x,e\rangle$, one for each model task p, denoting

t = simulated time,
x = activity, i.e. r = running or
c = entry call or
a = entry accept,
e = entry (and owner) in the case of c or a.

Unlike the observer, the scheduler cannot look at the model tasks but has to be told what is going on; he furthermore wishes to interfere with their desires. Scheduler and model tasks interact as follows (given the obvious initialization):

4.5 *model tasks:*

before hold action:
    tell scheduler about new t;

before entry call or entry accept action:
    tell scheduler about c or a and e;
    continue when allowed by scheduler;

4.6 *scheduler:*

loop
  receive information from model task
  and update system state accordingly;
  while there is $\langle p_1,t_1,c,e_1\rangle,\langle p_2,t_2,a,e_2\rangle$ such that
      $e_1=e_2$ and
      for all $\langle q,t,x,e\rangle$: $t<t_1 \Rightarrow$
        x=a or
        (x=c and there is no $\langle q',t',a,e'\rangle$: e=e')
  do
    choose such a pair p_1 and p_2;
    change their state description to $\langle p_i,\max(t_1,t_2),r,-\rangle$;
    allow p_1 and p_2 to go on, informing them about $\max(t_1,t_2)$;
  endwhile
endloop

We now show that this mechanism is correct w.r.t. 3.5 and not subject to the kind of deadlock described in §2.

4.7 Theorem:

a) Suppose the scheduler reactivates tasks p_1 and p_2 with $\langle p_1,t_1,c,e_1\rangle$, $\langle p_2,t_2,a,e_2\rangle$, $e_1=e_2$. Then neither the present system state nor any future one does contain a record $\langle q,t,c,e\rangle$ with $e=e_1$ and $t<t_1$.

b) Suppose that all model tasks are in state c or a but the scheduler cannot activate any. Then there is no matching pair $\langle p_1,t_1,c,e_1\rangle$, $\langle p_2,t_2,a,e_2\rangle$ with $e_1=e_2$.

Proof:

a) Define A to be the subset $\{\langle q,x,t,e\rangle \mid t<t_1\}$ of the system state. By 4.6, these assertions hold for the present state and are invariant under transitions:

i. $\neg\exists\langle q,t,c,e\rangle$ in A: $e=e_1$

ii. $\neg\exists\langle q,t,c,e\rangle$, $\langle q',t',a,e'\rangle$ in A: $e=e'$

iii. $\neg\exists\langle q,t,r,-\rangle$ in A.

b) If there were such pairs, the pair with minimal t_1 would be allowed to go on.

This mechanism restricts synchronization of model tasks to the case of rendezvous; our treatment of hold actions involves no waiting for each other. For conceptual or technical reasons, e.g. to model preemptive service, one might prefer simulated time to be a common standard to all model tasks. It is easy to adjust the considerations of this chapter to this purpose, essentially recasting rule 4.3 as

4.8 A model task may advance its simulated time to the value t when all model tasks either wish to advance their time to at least t or cannot advance because they want to rendezvous.

But please note that this variant does not obviate the need of supervising the rendezvous.

5. Conclusion

We found that to direct the temporal behaviour of the model tasks, the scheduler must control their rendezvous; one may think of the scheduler as reproducing the rendezvous related job of the Ada runtime system, with simulated time instead of real time.

It is neither easy nor efficient to program this job in Ada, and the result will mean some extra burden to the modeller since any rendezvous in the model must be preceded by an interaction with the scheduler. First, there is the problem of denoting and communicating the entries and tasks in question. Second, the interplay of model tasks and scheduler is an example of the general resource management problem: model tasks apply for a specified piece of progress and get permission according to the logic of simulated time. A single rendezvous does not suffice to program such a situation; for a general and safe solution we need, as is explained e.g. in chapter 10 of [6], agent tasks and three rendezvous (in advance to the really intended entry call).

So to set up an Ada simulation system one should choose the rendezvous not just because it is offered by the base language. Since it is expensive to accomodate the rendezvous to simulated time, one may as well decide to implement an interaction mechanism closer to one's own needs or fancy. A very special point in favour of the rendezvous is that a simulation program may serve as a prototype for production software that is to be written in Ada.

Programming language elements shall enable the user to express his ideas with ease and clarity. In the context of modelling, there is little experience with the rendezvous, but it is a priori reasonable to provide the modeller, he may be a skilfull Ada programmer or not, with ready-made constructs for typical situations. This can be easily done using Ada packages and generic units; for examples see [4],[7],[13],[15]. A well chosen set of such higher level constructs may answer to most needs; it may even allow an easier implementation since, in contrast to our exposition in §4, it may come along with some restriction (cf. 2.3 in §2) of the possible interaction in the model. But if, with a view to this facilitation, the time mechanism is programmed to rely on the use of these higher level constructs, the modeller will hardly be able to move beyond their scope if he has to do so.

In response to some inadequate solutions, this paper was meant to cope with the logical aspect of simulated time and the rendezvous and to outline what a correct simulation mechanism must achieve. Doing so, I was, and am, convinced that Ada tasking can easily match the parallel nature of the real systems under study and thus help to construct clear and convincing model programs. But in concurrent programming it is not sufficient to take the language for a suitable abstraction of the machine: here, the utility of the algorithm strongly depends on the machine architecture employed. I did not investigate the problem of performance and machines. But since we found that the use of the rendezvous for model interaction implies a considerable overhead (especially: additional rendezvous), I do not expect the modeller to be rewarded by quick program execution if he turns to the Ada machine at hand, which will be a one processor system. A classical coroutine structured program might do much better. I do not know, and therefore will not subscribe to the optimism shown in some of the quoted papers, whether the future shift to a modest form of true parallelism will automatically compensate those ineffiencies. If, on the other hand, one envisaged very specialized hardware and every process as having a processor of his own, the problems to solve are closer to simulation methodology in general than to programming in a given language.

Acknowledgement:
For the main part of our investigation, the essential source of ideas is L.Lamport's paper [12]. The same holds for a method described in [3], which in addition to a central simulation clock uses local clocks for the synchronizing aspect of the rendezvous. As I understand the sketchy presentation, this method fails to distinguish among different possible entries.

References:

[1] H.H.Adelsberger:
ASSE - Ada Simulation Support Environment.
Proc. Winter Simulation Conf., 1982

[2] ANSI:
The Programming Language Ada Reference Manual.
1983

[3] C.J.Antonelli, R.A.Vok, T.N.Mudge:
Hierarchical Decomposition and Simulation of Manufacturing Cells using Ada.
Simulation, Apr.1986

[4] G.Bruno:
Rationale for the Introduction of Discrete Event Primitives in Ada.
Simulation in Strongly Typed Languages, ed.R.M.Bryant, 1984

[5] R.M.Bryant:
Discrete Event Simulation in Ada.
Simulation,Oct.1982

[6] A.Burns:
Concurrent Programming in Ada.
1985

[7] V.A.Downes, R.Tellaeche Bosch:
Discrete Event Simulation with Ada.
Proc. UKSC Conf. on Computer Simulation, 1984

[8] V.A,Downes, R.Tellaeche Bosch:
Discrete Event Modelling in Ada.
Proc. Joint Ada Europe/ Ada Tec Conf. 1984

[9] W.R.Franta:
The Process View of Simulation.
1977

[10] P.Friel, S.Sheppard:
Implications of the Ada Environment for Simulation Studies.
Proc. Winter Simulation Conf. 1984

[11] P.Friel, D.Reese, S.Sheppard:
Simulation in Ada: an Implementation of two World Views.
Simulation in Strongly Typed Languages, ed.R.M.Bryant, 1984

[12] L.Lamport:
Time, Clocks, and the Ordering of Events in a Distributed System.
CACM, July 1978

[13] G.Lomow, B.Unger:
The Process View of Simulation in Ada.
Proc. Winter Simulation Conf. 1982

[14] R.Pooley:
Languages for Discrete Event Simulation.
Proc. Simula Users' Conf. 1985

[15] S.A.Steele, R.Beeby:
A Process Simulation Package concealing Multi-tasking.
Proc. Ada Europe Conf. 1986

<u>Anmerkung der Herausgeber</u>

Dieser Artikel ist auch im Tagungsband des 4. Symposiums "Simulationstechnik", das vom 9. - 11.9. in Zürich stattfand, enthalten.

Analysis of Reversible and Nonreversible Queueing Networks with Rejection Blocking

I. F. Akyildiz

School of Information and Computer Science
Georgia Institute of Technology
Atlanta, Georgia 30332
U. S. A.

ABSTRACT

Queueing networks which contain finite capacities have proved useful in modeling actual computer systems and communication networks. The finite capacity of stations introduces blocking events which should be considered in the performance evaluation. Specifically, we shall examine the effects of rejection blocking upon queueing networks. Rejection blocking is defined in the following manner. Upon completion of its service of a particular station's server, a job attempts to proceed to its next station. If, at that moment, its destination station is full, the job is rejected. The job goes back to the server of the source station and immediately receives a new service. This is repeated until the next station releases a job and a place becomes available. In the first part of this work the well known exact product form solution for the equilibrium state probabilities is presented for closed rejection blocking networks which have reversible routing. An algorithm is given for computation of performance measures in reversible networks with rejection blocking. In the second part nonreversible networks with rejection blocking are analyzed. The analysis is based on the transformation of the state space of a blocking queueing network into an equivalent state space of a nonblocking network with infinite station capacities. It is shown that the state spaces of both systems are isomorphic under a given condition. Markov processes describing the evolution of both networks over time have the same structure. This leads to the product form solution for blocking networks. Based on product form solution new formulae are given for the exact computation of performance measures.

"Key Words:" Performance Evaluation, Queueing Networks, Blocking, Equilibrium State Probabilities

This work is supported in part by the Airforce Office of the Scientific Research under Grant No. AFOSR-87-0160.

1. Introduction

Queueing networks have received a special attention in the last fifteen years in performance evaluation and analysis of complex multiprocess, multiresource computer systems. A queueing network is composed of a collection of stations (devices with queues) in which jobs/processes proceed from one station to another in order to satisfy their service requirements. The basic results of queueing network theory were given by Jackson and Gordon/Newell [JACK63,

GORD67a]. They showed that open and closed queueing networks with a single job class, with exponential arrival and service time distributions, and FCFS queueing disciplines at each station have a *product form solution.* The product form solution states that the equilibrium state probabilities consist of a product of terms where each term represents a state of the queues. Their result implies that the individual stations behave as if they were separate queueing systems. Baskett, Chandy, Muntz and Palacios [BCMP75] extended the results of [JACK63, GORD67a] to obtain product form solutions for open, closed and mixed queueing networks with different job classes, non-exponential service time distributions and different queueing disciplines such as FCFS, Processor Sharing (PS) and Last Come First Served Pre-emptive Resume (LCFS-PR). Several algorithms have been introduced for effective computation of performance measures for queueing networks [BUZE71, CHAN75, REIS75, REIS80, SAUE81].

Product form networks (also known as BCMP or separable networks) have proved invaluable for the modeling of a variety of computer and communication systems. They are sufficiently flexible as to adequately represent the features arising in such applications. They have not, however, been able to provide proper insight into the phenomenon of *blocking*. This is because product form networks assume that each station in the network has an infinite capacity. Since in actual systems the resources have a finite capacity, queueing networks with blocking must be used for performance analysis. Blocking arises because of the limitations imposed by the capacity of these stations.

In recent years there has been a growing interest in the development of computational methods to analyze queueing networks with blocking. Researchers from various areas such as Computer Performance Analysis, Operations Research, and Electrical Engineering Telecommunication Systems have studied blocking networks. Several papers have been published dealing with various types of blocking types.

Formally, we distinguish between three types of blocking: *"Transfer Blocking"*, *"Service Blocking"* and *"Rejection Blocking"*.

In the *"Transfer Blocking"* case, the blocking event occurs when a job completing service at station i cannot proceed to station j because station j is full. The job resides in station i's server, which stops processing until station j releases a job. This type of blocking has been used to model systems such as production systems and disk I/O subsystems. [AKYL87a,b,c,d, PERR81, PERR86, ONVU87, PERR87, SURI86, TAKA80].

In the *"Service Blocking"* case, blocking occurs when a job in front of queue at station i declares its destination station j before it starts its service in station i's server. If the destination station j is full, the i-th server becomes blocked, i.e., it can not serve jobs. When a departure occurs from destination station j, the i-th server becomes unblocked and the job begins receiving service. This blocking type has been used to model systems such as telecommunication systems and production systems. [BOXM81, GORD67b, SURI84].

In the *"Rejection Blocking"* case, blocking occurs when a job completes service at station i's server and wants to join station j, whose capacity is full. The job is rejected by station j. That job goes back with a certain probability (rejection probability) to station i's server and receives a new service with the same mean service time. This activity is repeated until station j releases a job, and a place becomes available.

The "rejection blocking" type has been used to model systems such as communication networks, computer systems with limited multiprogramming, production lines and flexible manufacturing systems. Most of the previous work was done on the "rejection" blocking in both open and closed queueing networks. Within this category, the studies fall into two groups. The first group provides exact results for both open queueing networks [KONH76, KONH77] and closed queueing networks [BALS83, HORD81, PITT79, YAO85]. Konheim/Reiser [KONH76, KONH77] propose an algorithm for the solution of an open network with two single server stations exhibiting exponential service time distributions. It also permits feedback by allowing some departures from the second station to proceed back to the first station's queue. Balsamo/Iazeoalla [BALS83], Hordijk/VanDijk [HORD81], Pittel [PITT79] and Yao/Buzacott [YAO85] have shown the existence of product form solutions for closed queueing networks which satisfy one of the following three conditions:

i) The network routing matrix is *reversible*.

ii) The probability of blocking is constant; that is, independent of the number of jobs in the station causing the blocking event.

iii) The service rate of each station is constant provided that it is impossible to have empty station at any time.

The second group of studies is characterized by the specific solution method [AKYL85, CASE79, LABE80, SURI84]. Caseau/Pujolle [CASE79] studied a blocking queueing network consisting of two or more stations in tandem in an effort to obtain an approximate expression of the maximum throughput.

Since the "service blocking" is identical to "rejection blocking" in case of tandem networks, we consider the work of Suri/Diehl [SURI84] also in the group of studies for rejection blocking networks. The Suri/Diehl [SURI84] study examined closed tandem queueing networks with finite station capacities in which the first queue has a capacity greater than the number of jobs in the system. By an application of Norton's Theorem [CHAN75], they reduce each two-station to a single station with a variable size queue capacity that is easily analyzed. An approximation algorithm is derived for the total mean residence time of the network, assuming exponentially distributed service times. The major disadvantage to this technique is that only the total throughput and the total mean residence time of a network can be determined. Performance measures for individual stations are impossible to compute. Another disadvantage is that it is restricted to networks with small populations (computation of marginal probabilities) and with serially switched stations. In addition the capacity of the first station must be infinite.

Several other investigators in recent years have published results on queueing networks with rejection as well as transfer blocking. A bibliography of studies about queueing network models with all types of blocking is given by Perros [PERR84].

In this work we present a computational algorithm for analyzing closed queueing networks with rejection blocking and reversible routing. As mentioned before, an exact product form solution exists for closed rejection blocking networks which have a reversible routing. However, no algorithm has been proposed for the computation of the normalization constant. In order to compute the equilibrium state probabilities in the product form solution, the normalization constant must be determined for closed networks. A naive technique to compute the normalization constant is to enumerate all states and compute their relative probabilities. Absolute probabilities can then be determined from the relative probabilities by normalizing their sum to one. This is of course feasible only for small networks, because for larger networks, the number of states grows rapidly.

We consider closed queueing networks with N stations and K jobs which constitute a single job class. Each station has single server and each server has an exponentially distributed service time with mean value $1/\mu_i$ (for $i = 1,2,...,N$). Each station has a fixed finite capacity, M_i, where M_i = *queue capacity* + 1. A job which is serviced by station i proceeds to station j with the transition probability p_{ij} (for $i,j = 1,2,...,N$), if station j is not full. In other words, the number of jobs in station j, k_j, is less than M_j. Otherwise, the job will be rejected from the station j, and it will return to the server of station i and receive another round of service. This is repeated until a place is available in station j. Furthermore, we assume that

$$K < \sum_{i=1}^{N} M_i$$

which means that the total number of jobs, K, in the network may not exceed the total capacity of the entire network. The service discipline of each station is First-Come-First-Served.

Section 2 explores the product form solution for reversible networks given by Hordijk and van Dijk [HORD81]. In section 3 we introduce an algorithm for the computation of $G(K)$ and derive formulae for performance measures for reversible networks. Section 4 and 5 contain the analysis of nonreversible networks.

2. Product Form Solution for Reversible Networks with Rejection Blocking

A queueing network is "reversible" [KELL79, MELA82], if the following condition is satisfied:

$$e_i \, p_{ij} = e_j \, p_{ji} \quad \text{for } all \; i,j = 1,2,..,N. \tag{1}$$

This equation states that the rate at which jobs arrive at station j from station i equals the rate at which jobs leave station j to return to station i. Simple examples for reversible networks are two-station networks and central server models.

It is well known that there exists a positive solution for e_i:

$$e_i = \sum_{j=1}^{N} e_j \, p_{ji} \quad \text{for } i,j = 1,2,...,N \tag{2}$$

The relative utilization (also called loadings) of station i is denoted by x_i and is computed by

$$x_i = e_i / \mu_i \qquad \text{for } i = 1, 2, \cdots, N \tag{3}$$

We define the vector, $\underline{k} = (k_1, k_2, \ldots, k_N)$, as a state of the system where k_i denotes the number of jobs in the station i. We say that a state, $\underline{k} = (k_1, k_2, \ldots, k_N)$, is "feasible" if all k_i's are less than or equal to their respective M_i's, the capacity of the station i. Otherwise, the state is said to be "infeasible". In an infeasible state, at least one of the stations in the system violates its capacity restrictions. The following global balance equation is derived for the proposed model.

$$[\sum_{i=1}^{N} \sum_{j=1}^{N} \mu_i \; p_{ij} \; \delta_i(k_i)] \; p(k_1, k_2, \ldots, k_N) = \tag{4}$$

$$= \sum_{i=1}^{N} \sum_{j=1}^{N} \mu_j \; p_{ji} \; \delta_j(k_j - 1) \;\; p(k_1, \cdots, k_i + 1, \cdots, k_j - 1, \ldots, k_N)$$

Informally, the left-hand side of equation (4) denotes the stream out of the state $(k_1, k_2, \cdots, k_N)$, and the right-hand side denotes the stream into that state.

The binary function, δ_i, eliminates the infeasible states.

$$\delta_i(k_i) = \begin{cases} 0 & \text{if} \quad k_i > M_i \\ 1 & \text{if} \quad \textit{otherwise} \end{cases}$$

Hordijk and VanDijk [HORD81] have shown that in a closed queueing network with rejection blocking that satisfies the reversibility condition (1), the equilibrium state probability solution for equation (4) for a feasible state $(k_1, k_2, \cdots, k_N)$ is represented by the following product form of marginal probabilities:

$$p(k_1, k_2, \cdots, k_N) = \frac{1}{G(K)} \prod_{i=1}^{N} x_i^{k_i} \;\; \delta_i(k_i) \tag{5}$$

where

$$\sum_{i=1}^{N} k_i = K$$

$G(K)$ is the normalization constant, which adjusts all the probabilities of "feasible" states so that they sum to one. For formal proof of the theorem, see [HORD81].

Although Hordijk/VanDijk [HORD81] have shown that a product form solution exists for networks with rejection blocking, they do not provide an algorithm for an efficient computation of the normalization constant, $G(K)$, as well as other performance measures. In the next section we give a convolution algorithm for the computation of the normalization constant and as well as formulae for performance measures in such queueing networks.

3. Performance Measures for Reversible Networks with Rejection Blocking

In order to compute the normalization constant for "rejection" blocking networks, we use the "convolution" algorithm. The normalization constant, G, is computed by the convolution of N vectors, G_i:

$$G = G_1 \otimes G_2 \otimes \ldots\ldots \otimes G_N$$

where G_i (for $i = 1,2,..,N$) is a $(K+1)$ dimensional vector with

$$G_i = \begin{bmatrix} g_i(0) \\ g_i(1) \\ g_i(2) \\ \cdots \\ \cdots \\ g_i(K) \end{bmatrix} \tag{6}$$

and

$$g_i(k) = \begin{cases} 1 & \text{if} \quad k = 0 \\ \dfrac{e_i \; g_i(k-1)}{\mu_i} & \text{if} \quad k \le M_i \\ 0 & \text{if} \quad k > M_i \end{cases} \tag{7}$$

with $\otimes$ as the convolution operation

Informally, if the number of jobs in the station i exceeds its capacity, M_i, the component $g_i(k)$ will be set to 0. This eliminates all infeasible states. Once $G(K)$ is computed, the other performance measures can easily be obtained using the formulae which are given in the following.

The marginal probability, $p_i(n)$, which denotes the probability that there are n jobs in the station i, is obtained using the following equation:

$$p_i(n) = \sum_{\sum_{i=1}^{N} k_i = K \;\&\; k_i = n} p(\underline{k}) \tag{8}$$

By substituting the solution for $p(\underline{k})$ in equation (5) into equation (8) we get

$$p_i(n) = \frac{g_i(n)}{G(K)} \; G_{i^-}(K-n) \qquad \text{for} \quad i = 1,2,\ldots,N \text{ and } n = 1,2,\ldots,M_i \tag{9}$$

where $g_i(n)$ is computed by equation (7) and G_{i^-} is the normalization constant calculated without considering station i.

$$G_{i^-} = G_1 \otimes G_2 \otimes \ldots \otimes G_{(i-1)} \otimes G_{(i+1)} \otimes \ldots\ldots \otimes G_N$$

The *mean number of jobs* in each station is computed using the following formula:

$$\bar{k}_i(K) = \sum_{n=1}^{M_i} n \;\; p_i(n)$$

Substituting the value for marginal probability, $p_i(n)$, equation (9) into this formula, we obtain

$$\bar{k}_i(K) = \sum_{n=1}^{M_i} n \; \frac{g_i(n)}{G(K)} \; G_{i^-}(K-n) \tag{10}$$

The *utilization* of each station is determined by the following equation:

$$\rho_i(K) = \sum_{n=1}^{M_i} p_i(n) \tag{11}$$

By substituting the values for $p_i(n)$ in equation (11) we can find a direct solution (i.e., using the normalization constant) for utilization.

$$\rho_i(K) = \sum_{n=1}^{M_i} \frac{g_i(n)}{G(K)} \; G_{i^-}(K-n) \tag{12}$$

4. Product Form Solution for Nonreversible Networks with Rejection Blocking

Our concept is based on finding an equivalent non-blocking network which has the same number of states and the same state space structure as the blocking network. To solve this problem we use the concept of "holes" (as introduced by Gordon/Newell [GORD67b]). Note that Gordon/Newell [GORD67b] investigated closed networks with serially connected stations and *service blocking* where the blocked job stays at the head of the queue and resides there until a space becomes available in the destination station. A "hole" is the number of available places in the non-blocking network. We assume that the "holes" in the nonblocking network are moving in the opposite direction of the jobs in the blocking network. For the sake of simplicity in the following we will use the notations Γ for the blocking network, Φ for the nonblocking network, "holes" as jobs in Φ.

We assume that each station in Γ must have capacity equal or larger than n, the number of "holes" in Φ, such that the state spaces of both networks Γ and Φ are isomorphic.

$$M_i \geq n \quad \text{for } all \;\; i = 1, 2, \cdots, N. \tag{13}$$

where n is computed by:

$$n = \sum_{i=1}^{N} M_i - K \tag{14}$$

Theorem. A closed queueing network with rejection blocking satisfying the condition, equation (13), has the following product form solution for the equilibrium probability distribution of feasible states:

$$p(\underline{k}) = \frac{1}{G(n)} \prod_{i=1}^{N} x'_i{}^{(M_i - k_i)} \tag{15}$$

where

$G(n)$ represents the normalization constant with n jobs computed by equation (14). The normalization constant $G(n)$ can be obtained by the convolution algorithm [BUZE71, CHAN75] or mean value analysis [REIS80].

$x'_i = e'_i / \mu'_i$ is the relative utilization of the i-th station in Φ.

e'_i is computed by equation (2). Note that the transition probabilities of "holes" in Φ are computed by considering the fact that the "holes" move in the opposite direction in Φ than in Γ:

$$p'_{ij} = \frac{\mu_j \, p_{ji}}{\sum\limits_{1 \le t \le N} \mu_t \, p_{ti}} \tag{17}$$

μ'_i is the service rate of the i-th station in Φ and is computed by:

$$\mu'_i = \sum_{1 \le j \le N} \mu_j \, p_{ji} \tag{18}$$

Note also that equation (18) is derived from the fact that the job flow in Φ is in the opposite direction than in Γ.

Proof.

The network Φ has the same number of stations as Γ, $N' = N$. The difference is that the station capacities are unlimited, hence no blocking occurs and the total number of "holes" in Φ is $n \neq K$. Another difference is that the "holes" in Φ move in the opposite direction than in Γ as mentioned above. The service times in Φ are also exponentially distributed with rates μ'_i, computed by equation (18) and the transition probabilities in Φ are determined by equation (17).

The behavior of Γ can be modeled by a Markov process $X(t)$. The transition structure of $X(t)$ can be described by the global balance equation for Γ:

$$\{\sum_{i=1}^{N} \sum_{j=1}^{N} \mu_i \; p_{ij} \; \varepsilon_i(k_i) \; \delta_j(k_j)\} \; p(\underline{k}) = \sum_{i=1}^{N} \sum_{j=1}^{N} \mu_j \; p_{ji} \; \varepsilon_j(k_j) \; \delta_i(k_i) \; p(k_1, \ldots, k_i + 1, \cdots, k_j - 1, \cdots, k_N)$$

where the binary functions ε and δ express the impossibility of jobs departing from a station that is empty and entering a station that is full:

$$\varepsilon_i(k_i) = \begin{cases} 0 & \text{if} \quad k_i = 0 \\ 1 & \text{if} \quad k_i > 0 \end{cases}$$

$$\delta_i(k_i) = \begin{cases} 0 & \text{if} \quad k_i > M_i \\ 1 & \text{if} \quad otherwise \end{cases}$$

The Markov process $X(t)$ has the following state space:

$$S = \{ \underline{k} \ / \ \forall \ i \ \ (0 \le k_i \le M_i \ ) \ \& \ \sum_{i=1}^{N} k_i = K \ \}$$

Similarly, we define a Markov process $X'(t)$ whose transition structure is described by the following global balance equation for Φ :

$$\{\sum_{i=1}^{N} \sum_{j=1}^{N} \mu'_i p'_{ij} \ \varepsilon_i(k'_i)\} \ p'(\underline{k}') = \sum_{i=1}^{N} \sum_{j=1}^{N} \mu'_j \ p'_{ji} \ \varepsilon_{(j)}(k'_j) \ p'(k'_1 , ..., k'_i + 1 , \cdots , k'_j - 1 , \cdots , k'_N)$$

The Markov process $X'(t)$ has the following state space:

$$S' = \{\underline{k}' \ / \forall \ i \ \ (0 \le k'_i \le n \ ) \ \& \ \sum_{i=1}^{N} k'_i = n\}$$

We assert that $S = S'$

i) The number of jobs in Φ is defined in the following range:

$$0 \le k'_i \le n$$

Replacing the values for n we get

$$0 \le k'_i \le \sum_{i=1}^{N} M_i - K$$

Substituting $k'_i = (M_i - k_i)$ and considering $K = \sum_{i=1}^{N} k_i$ we get

$$0 \le (M_i - k_i) \le \sum_{i=1}^{N} M_i - \sum_{i=1}^{N} k_i$$

Rewriting

$$0 \le \sum_{i=1}^{N} k_i - k_i \le \sum_{i=1}^{N} M_i - M_i$$

we obtain

$$0 \le \sum_{i=2}^{N} k_i \le \sum_{i=2}^{N} M_i$$

which provides

$$0 \le k_i \le M_i$$

ii) Substituting the value of n we obtain

$$\sum_{i=1}^{N} k'_i = \sum_{i=1}^{N} M_i - K$$

Rewriting

$$\sum_{i=1}^{N} (M_i - k'_i) = K$$

and substituting $k_i = (M_i - k'_i)$ we get

$$\sum_{i=1}^{N} k_i = K$$

This implies that the equilibrium state probability $p(\underline{k})$ of Γ is equivalent to the equilibrium state probability $p'(\underline{k}')$ of Φ :

$$p(\underline{k}) = p'(\underline{k}')$$

Substituting the value $\underline{k}' = (\underline{M} - \underline{k})$ we obtain

$$p(\underline{k}) = p'(\underline{M} - \underline{k})$$

Since Φ has product form solution, the $p'(\underline{M} - \underline{k})$ values are obtained from the Gordon/Newell Theorem [GORD67a] which provides equation (15).

Remark. As mentioned in the introduction Balsamo/Iazeoalla [BALS83] and Hordijk/VanDijk [HORD81] investigate queueing networks with rejection blocking. However, they have the following conditions which are not required in our concept:

i) The capacity of each station must be equal to the total number of jobs with one less job divided by the total number of stations with one less station, $M_i = \dfrac{K-1}{N-1}$ for $all\ i$.

ii) The total number of jobs must be greater than the total number of stations, $K > N$. This implies that no station is allowed to be empty.

5. Performance Measures for Nonreversible Networks with Rejection Blocking

Corollary. Since the Markov processes in Γ with K jobs have the same structure as the Markov processes in Φ with n "holes" computed by equation (14), the throughput of Γ with K jobs satisfying the condition, equation (13), is equal to the throughput of Φ with n jobs:

$$\lambda^{\Gamma}(K) = \lambda^{\Phi}(n) \tag{19}$$

Since Φ with n jobs has product form solution, any exact algorithm such as mean value analysis [REIS80] can be applied for the computation of $\lambda^{\Phi}(n)$.

Each station's throughput in Γ is then computed by:

$$\lambda_i(K) = e'_i \cdot \lambda^{\Gamma}(K) \qquad \text{for } i = 1, \cdots, N. \tag{20}$$

The *mean number of jobs in the i-th station of* Γ is computed by:

$$\bar{k}_i(K) = \sum_{\min\{b\} \text{ in station } i}^{M_i} b \; p_i(b) \qquad \text{for } i = 1, \cdots, N. \tag{21}$$

where $p_i(b)$ is the *marginal probability* that there are b jobs in the i-th station which are obtained from the equilibrium state probabilities, equation (15):

$$p_i(k) = \sum_{\text{feasible } \underline{k}} p(\underline{k}) \qquad \text{for } i = 1, \cdots, N. \tag{22}$$

Using Little's law the *mean residence time* of jobs at the i-th station in Γ is determined by:

$$\bar{t}_i(k) = \frac{\bar{k}_i(k)}{\lambda_i(k)} \qquad \text{for } i = 1, \cdots, N. \tag{23}$$

6. Conclusion

We have presented an algorithm for the computation of the normalization constant and formulae for other performance measures in queueing networks with rejection blocking and reversibility. This permits effective analysis of such networks. As generally known the set of reversible networks is contained in the set of non-reversible networks. However, the exact analysis of non-reversible networks is possible under the condition, equation (13), which weakens the set of the non-reversible networks. We are in the process of finding an approximate solution for cases where the condition, equation (13), is not satisfied. It is also interesting to investigate the cases where the stations have generally distributed service times and FCFS scheduling disciplines. H. von Brand [VONB87] gives an exact product form solution for reversible open, closed and mixed queueing networks with rejection blocking. He also introduces an algorithm for the computation of performance measures.

References

AKYL85 I. F. Akyildiz, "Approximate Product Form Solution of Blocking Queueing Networks", (in German), Elektronische Rechenanlagen Journal, Dec. 1985, pp. 333-343.

AKYL87a I. F. Akyildiz, "Exact Product Form Solution for Queueing Networks with Blocking", *IEEE Transactions on Computers, Vol. 1, No. 1, January 1987, pp. 121-126.*

AKYL87b I. F. Akyildiz "On the Exact and Approximate Throughput Analysis of Closed Queueing Networks with Blocking", *to appear in IEEE Transactions on Software Engineering.*

AKYL87c I. F. Akyildiz, "Mean Value Analysis for Closed Queueing Networks with Blocking", *to appear in IEEE Transactions on Software Engineering.*

AKYL87d I. F. Akyildiz, "Product Form Approximations for Closed Queueing Networks with Multiple Servers and Blocking", to appear in IEEE Transactions on Computers.

AKVO87 I. F. Akyildiz and H. von Brand, "Duality in Open and Closed Markovian Queueing Networks with Rejection Blocking", *Technical Report of Louisiana State University,* TR-87-011, March 1987.

BALS83 S. Balsamo and G. Iazeolla, "Some Equivalence Properties for Queueing Networks with and without Blocking", *Proceedings of Performance 83 Conference,* editors A. K. Agrawala and S. Tripathi, North Holland Publ. Co., 1983, pp. 351-360.

BCMP75 F. Baskett, K. M. Chandy, R. R. Muntz and G. Palacios, "Open, Closed and Mixed Network of Queues with Different Classes of Customers", *Journal of the ACM,* Vol. 22, Nr. 2, Apr. 1975, pp.248-260.

BOXM81 O. I. Boxma and A. G. Konheim, "Approximate Analysis of Exponential Queueing Systems with Blocking," *Acta Informatica,* vol. 15, January 1981, pp. 19-66.

BUZE71 J. P. Buzen "Queueing Network Models of Multiprogramming" *PhD Thesis, Div. Eng. and Applied Sciences, Harvard Univ., Cambridge,* Mass., Aug. 1971.

CASE79 P. Caseau and G. Pujolle, "Throughput Capacity of a Sequence of Queues with Blocking due to Finite Waiting Room", *IEEE Transactions on Software Engineering,* Vol. SE-5, No. 6, November 1979, pp. 631-642.

CHAN75 K. M. Chandy, U. Herzog and L. Woo, "Parametric Analysis of Queueing Network Models, *IBM Journal Res. Dev.,* Vol. 19, Nr. 1, Jan. 1975, pp.43-49.

GORD67a W. J. Gordon and G. F. Newell, "Closed Queueing Systems with Exponential Servers", *Operations Research,* 15, 1967, pp. 254-265.

GORD67b W. J. Gordon and G. F. Newell, "Cyclic Queueing Systems with Restricted Queues", *Operations Research,* 15, Nr. 2, April 1967, pp. 266-277.

HORD81 A. Hordijk and N. van Dijk, "Networks of Queues with Blocking," *Proceedings,* 8th International Symposium on Computer Peformance Modelling, Measurement, and Evaluation, Amsterdam, November 4-6, 1981.

JACK63 J. J. Jackson, "Jobshop-like Queueing Systems", *Management Science,* 10, 1, 1963, pp. 131-142.

KELL79 F. P. Kelly, "Reversibility and Stochastic Networks", J. Wiley and Sons Publ. Co., New York, 1979.

KONH76 A. G. Konheim and M. Reiser, "A Queueing Model with Finite Waiting Room and Blocking," *Journal of the ACM,* vol. 23, Number 2, April 1976, pp. 328-341.

KONH77 A. G. Konheim and M. Reiser, "Finite Capacity Queueing Systems with Applications in Computer Modeling", SIAM Journal on Computing, Vol. 7, Number 2, Mai 1977, pp. 210-229.

LATE80 G. Latouche and M. Neuts, "Efficient Algorithmic Solutions to Exponential Tandem Queues with Blocking", *SIAM, Alg. Dis. Meth.*, Vol. 1, Nr. 1, March 1980.

MELA82 B. Melamed, "On the Reversibility of Queueing Networks", *Stochastic Processes and their Applications,* Vol 13, 1982, pp. 227-234.

ONVU86 R. O. Onvural and H. G. Perros, "On Equivalences of Blocking Mechanisms in Queueing Networks with Blocking", *Operations Research Letters,* Dec. 1986.

ONVU87 R. O. Onvural and H. G. Perros, "Some Exact Results for Closed Queueing Networks with Blocking", Technical Report of North Carolina State University, 1986.

PERR81 H. G. Perros, "A Symmetrical Exponential Open Queue Network with Blocking and Feedback", *IEEE Transactions on Software Engineering,* SE-7, 1981,pp. 395-402.

PERR84 H. G. Perros, "Queueing Networks with Blocking: A Bibliography", *ACM Sigmetrics Performance Evaluation Review,* August 1984.

PERR86 H. G. Perros and T. Altiok, "Approximate Analysis of Open Networks of Queues with Blocking: Tandem Configurations", *IEEE Transactions on Software Engineering,* Vol. SE-12, No. 3, March 1986, pp. 450-462.

PERR87 H. G. Perros, A. A. Nilsson and Y. C. Liu, "Approximate Analysis of Product Form Type Queueing Networks with Blocking and Deadlock", *to appear in Performance Evaluation.*

PITT79 B. Pittel, "Closed Exponential Networks of Queues with Saturation: The Jackson Type Stationary Distribution and Its Asymptotic Analysis," *Mathematics of Operations Research,* vol. 4, 1979, pp. 367-378.

REIS80 M. Reiser and S. S. Lavenberg, "Mean Value Analysis of Closed Multichain Queueing Networks", *Journal of the ACM, Vol. 27, No. 2, April 1980, pp. 313-322.*

SAUE81 C. H. Sauer and K. M. Chandy, "Computer Systems Performance Modeling" *Prentice Hall,* Englewood Cliffs, N. J., 1981.

SURI84 R. Suri and G. W. Diehl, "A New Building Block for Performance Evaluation of Queueing Networks with Finite Buffers", *ACM Sigmetrics Conference Proceedings,* Cambridge, Mass., Aug. 1984, pp.134-142.

SURI86 R. Suri and G. W. Diehl, "A Variable Buffer-Size Model and its Use in Analyzing Closed Queueing Networks with Blocking", *Management Science,* Vol. 32, No. 2, February 1986, pp. 206-225.

TAKA80 Y. Takahashi, H. Miyahara and T. Hasegawa, "An Approximation Method for Open Restricted Queueing Networks", *Operations Research,* Vol. 28, Nr. 3, May-June 1980, pp. 594-602

VONB87 Horst von Brand, "Queueing Networks with Blocking", PhD Dissertation, Computer Science Department of LSU, July 1987.

YAO85 D. D. Yao and J. A. Buzacott, "Modeling a Class of State-Dependent Routing in Flexible Manufacturing Systems", *Annals of Operations Research, Vol. 3, 1985, pp. 153-167.*

COMPUTATIONAL METHODS FOR MARKOV CHAINS OCCURRING IN QUEUEING THEORY

Manfred Kramer
Konstanz

Summary: An algorithmic method for computing the probability vector of finite irreducible Markov chains is developed. The block elimination scheme used is especially well suited for highly structured and/or sparse transition matrices. Special variants for block Hessenberg and tridiagonal matrices often occurring in queueing theory are derived. The algorithm is then applied to the embedded Markov chain describing the queue length in a discrete-time queue with state-dependent arrival rates.

0. INTRODUCTION

Many problems in computer and communication engineering lead to queueing models involving embedded Markov chains in discrete time. The computation of the invariant probability vector is then one of the most important steps towards the solution.

The components of the invariant vector normalized to unity are determined by a system of linear and homogeneous equations. One of these equations is redundant and usually replaced with the normalization condition. A more sophisticated method is to incorporate the normalizing equation into the system itself /6,p.37/. The resulting system is then solved by a standard algorithm, say by Gaussian elimination /2/. However, this approach entails several drawbacks such as the destruction of a natural block partitioning and the accumulation of rounding errors when the state space is large, as it is often the case in queueing models.

It is generally desirable to devise algorithmic methods which retain the probabilistic significance of intermediate results by expressing them in terms of inherently nonnegative probabilities. This approach leads not only to a more thorough understanding of the underlying stochastic processes but also to improved stability.

These guidelines were respected in two alternative computational methods closely related to the present paper. The first method adopts concepts known from the theory of regenerative processes and uses them to refine the standard Gaussian elimination procedure /3/. The second method is based on a matrix bordering approach to reduce the transition matrix /8/. After determining an invariant vector for the reduced transition matrix the stationary probability vector is computed in a backsubstitution step.

Another class of techniques for obtaining the stationary probability vector using generalized inverses of modified transition matrices is reviewed in /4,pp.96-105/. Aggregation methods lead also to efficient approximate solution algorithms exploiting the lumpability or a given sparsity pattern of large transition matrices /1/,/9/. Particular and efficient algorithms for block-structured Hessenberg or tridiagonal matrices were developed in /7/ and /11/.

In the following we derive a simple and numerically stable algorithm particularly well suited for Markov chains with a highly structured transition matrix. We apply the algorithm to the computation of queue lengths in a discrete-time queueing model with state-dependent arrival rates and give some examples.

1. THE ELIMINATION APPROACH

Consider a Markov chain in discrete time with a finite and irreducible transition matrix P. Let $P=(P_{i,j})_{i,j=0,\dots,K}$ be an arbitrary partitioning of this matrix. A main problem is then to find the stationary probability vector $\vec{p}$ of the Markov chain which satisfies the system $\vec{p}=\vec{p}P$ and whose components sum to unity. For $\vec{p}$ consistently partitioned as $\vec{p}=(\vec{p}_k)_{k=0,\dots,K}$ this may be written as

$$\vec{p}_k = \sum_{i=0}^{K} \vec{p}_i P_{i,k} \text{ , } k=0,\dots,K \text{ ,} \tag{1}$$

and

$$\sum_{k=0}^{K} \vec{p}_k \vec{1} = 1 \text{ .} \tag{2}$$

Here and henceforth the symbol $\vec{1}$ denotes a column vector of appropriate dimension with all its components equal to 1.

Our approach to solve (1) consists of eliminating the subvectors $\vec{p}_j$, j=K,...,k+1 successively from the right-hand side and of expressing $\vec{p}_k$ by the remaining ones. In doing so we have at step k the equivalent reduced system of linear equations

$$\vec{p}_j = \sum_{i=0}^{k} \vec{p}_i P_{i,j}^{(k)} \quad , \; j=0,\ldots,k \; , \tag{3}$$

and the subvector $\vec{p}_k$ represented by

$$\vec{p}_k = \sum_{i=0}^{k-1} \vec{p}_i Q_{i,k}^{(k)} \; . \tag{4}$$

The system (3) defines a transition matrix $P=(P_{i,j}^{(k)})_{i,j=0,\ldots,k}$. At the beginning the starting matrix P_K is identified with P. The next elimination step produces the system

$$\vec{p}_j = \sum_{i=0}^{k-1} \vec{p}_i P_{i,j}^{(k-1)} \quad , \; j=0,\ldots,k-1 \; . \tag{5}$$

Inserting $\vec{p}_k$ from equation (4) in (3) and comparing we obtain the first transformation rule

$$P_{i,j}^{(k-1)} = P_{i,j}^{(k)} + Q_{i,k}^{(k)} P_{k,j}^{(k)} \quad , \; i,j=0,\ldots,k-1 \; , \tag{6}$$

which can be rewritten in the expanded form

$$P_{i,j}^{(k-1)} = P_{i,j}^{(K)} + \sum_{l=k}^{K} Q_{i,l}^{(l)} P_{l,j}^{(l)} \quad , \; i,j=0,\ldots,k-1 \; , \tag{7}$$

by summing these equations for k=0,...,K. Note that only $Q_{i,k}^{(k)}$, i=0,..., k-1, and $P_{k,j}^{(k)}$, j=0,...,k-1 are needed to apply (7) in the subsequent elimination steps.

Suppose P_k is irreducible and stochastic. Since then there is at least one transition leading out of block k, $P_{k,k}^{(k)}\vec{1} < \vec{1}$ holds with a strict inequality for at least one row of $P_{k,k}^{(k)}$. From a basic theorem concerning nonnegative matrices /10, Theorem 4.35 and Corollary 2/ it follows that the matrix $(I-P_{k,k}^{(k)})^{-1}$ exists and has nonnegative entries only. (I represents an identity matrix whose size is clear by context) Thus equation (3) for j=k can be solved for $\vec{p}_k$. Comparing the resulting equation with (4) yields the second transformation rule

$$Q_{i,k}^{(k)} = P_{i,k}^{(k)} (I-P_{k,k}^{(k)})^{-1} \quad , \; i=0,\ldots,k-1 \tag{8}$$

Both rules (6) and (8) defining a basic elimination step are summarized in a comprehensive form as

$$P_{i,j}^{(k-1)} = P_{i,j}^{(k)} + P_{i,k}^{(k)} (I-P_{k,k}^{(k)})^{-1} P_{k,j}^{(k)} \quad , \; i,j=0,\ldots,k-1 \tag{9}$$

Suppose P_k is an irreducible transition matrix. Then each state in its index set can be reached from another state in a finite number of steps. Direct transitions between the states in blocks $0,\ldots,k-1$ are preserved in P_{k-1} by virtue of the first term $P_{i,j}^{(k)}$ in (9). In addition, for all possible transitions going through states in block k the second term introduces a new entry, because the matrix $(I-P_{k,k}^{(k)})^{-1}$ accounts for all transitions within block k. Thus P_{k-1} is irreducible too. Using

$$\sum_{j=0}^{k} P_{i,j}^{(k)}\vec{1} = \vec{1} \quad , \; i=0,\ldots,k \; , \tag{10}$$

it is easy to check by postmultiplying (9) by $\vec{1}$ and summing over $j=0,\ldots,k-1$ that

$$\sum_{j=0}^{k-1} P_{i,j}^{(k-1)}\vec{1} = \vec{1} \quad , \; i=0,\ldots,k-1 \; , \tag{11}$$

holds also.

Therefore the elimination procedure can be continued since the matrix P_{k-1} is irreducible and stochastic and these properties are passed from step to step to all the matrices arising.

After the reduction process is completed, the final system $\vec{p}_0=\vec{p}_0P_0$ is solved for $\vec{p}_0$ and the remaining components $\vec{p}_k$ are recovered by backsubstitution using (4). On the other hand, it is often desirable to express the vector $\vec{p}_k$ explicitly in terms of $\vec{p}_0$ by means of the representation

$$\vec{p}_k = \vec{p}_0 R_k \quad , \; k=1,\ldots,K \; . \tag{12}$$

Obviously the matrices R_k are determined by the recurrence relation

$$R_k = \sum_{i=0}^{k-1} R_i Q_{i,k}^{(k)} \quad , \; k=0,\ldots,K \; , \tag{13}$$

which is similar to (4).

The steady-state vector is finally obtained in both cases by normalizing $(\vec{p}_k)_{k=0,\ldots,K}$ according to (2).

2. MATRICES OCCURRING IN QUEUEING THEORY

The method allows further simplifications for the transition matrices of Markov chains related to the queueing models M/G/1 and G/M/1 with phase type services /6/. These matrices belong to the classes of (i) block Hessenberg and (ii) block tridiagonal matrices defined as follows:

(i) A matrix P has the upper (lower) Hessenberg block structure, if all submatrices below the block subdiagonal (above the block superdiagonal) have zero entries, that is $P_{i,j}=0$ for $i>j+1$ ($j>i+1$).

(ii) A matrix P has the tridiagonal block structure, if all submatrices except for those on the main block diagonal, the block sub- and superdiagonal have zero entries, that is $P_{i,j}=0$ for $i>j+1$ and $j>i+1$.

We show inductively that in these cases the reduced matrices P_{k-1}, $k=K,\dots,1$ bear the same block structure as the leading principal submatrices $(P_{i,j})_{i,j=0,\dots,k-1}$ of the starting matrix $P_K=P$.

(i) Suppose P_k is an upper (lower) block Hessenberg matrix. Then its last block row (column) is zero except for the last two submatrices, that is, $P^{(k)}_{k,j}=0$ for $j=0,\dots,k-2$ ($P^{(k)}_{i,k}=0$ for $i=0,\dots,k-2$) holds. Consequently the transformation rule (9) reduces to $P^{(k-1)}_{i,j}=P^{(k)}_{i,j}$ except for the last column (row) with index $i=k-1$ ($j=k-1$). This proves that P_{k-1} is also an upper (lower) block Hessenberg matrix.

(ii) Suppose P_k is a block tridiagonal matrix. Since these matrices form a subclass of both kinds of Hessenberg matrices, P_{k-1} is also block tridiagonal.

Since the reduced matrices exhibit the same sparsity pattern as the starting matrix P, the transformation rules (6) and (8) have to be applied only

(i) for $i=0,\dots,k-1$ and $j=k-1$ ($i=k-1$ and $j=0,\dots,k-1$) in the case of upper (lower) block Hessenberg matrices,

(ii) for $i=j=k-1$ in the case of block tridiagonal matrices.

The alternative equations (7) are identical with (6) in these cases, whereas the equations (4) and (13) for lower block Hessenberg matrices reduce further to

$$\vec{p}_k = \vec{p}_{k-1} Q^{(k)}_{k-1,k} \quad , \; k=1,\dots,K \; , \tag{14}$$

and

$$R_k = R_{k-1} Q^{(k)}_{k-1,k} \quad , \; k=1,\dots,K \; . \tag{15}$$

Formula (14) reminds of the matrix-geometric representation of the invariant probability vector for an infinite transition matrix of the G/M/1 type which is also a lower block Hessenberg matrix /7, Theorem 1.3.2/. In fact, the matrix R_k corresponds then to the kth power of the rate matrix and (15) may be used for computing truncated versions of these representations.

3. THE ALGORITHM

According to the preceding derivations the algorithmic procedure starts with $P_K=P$. It can be summarized as follows:

(i) For $k=K,\dots,1$ compute Q_k by (8) and P_{k-1} by (6) or by (7).

(ii) Solve $\vec{p}_0=\vec{p}_0P_0$ for $\vec{p}_0$.

(iii) For $k=1,\dots,K$ compute $\vec{p}_k$ by (4) or by (13).

In order to solve the reduced system in pass (ii) the same method can be applied recursively until a single scalar component remains which is arbitrarily set to unity.

From an inspection of the transformation rules (6) and (8) it becomes obvious that the submatrices $P_{i,j}^{(k-1)}$, $Q_{i,k}^{(k)}$ can occupy the same storage locations as the corresponding submatrices $P_{i,j}^{(k)}$ with the same pair of subscript indices. Consequently the storage locations for $P_{i,j}$ can be used throughout. Likewise, the superscripts indicating the actual stage of the elimination process are unnecessary if the actual submatrices are always denoted by $P_{i,j}$ for $i\geq j$ and by $Q_{i,j}$ for $i<j$.

The following diagram shows the storage array after pass (i).

$$
\begin{matrix}
P_{0,0} & Q_{0,1} & Q_{0,2} & \cdots & Q_{0,K-1} & Q_{0,K} \\
P_{1,0} & P_{1,1} & Q_{1,2} & & Q_{1,K-1} & Q_{1,K} \\
P_{2,0} & P_{2,1} & P_{2,2} & & Q_{2,K-1} & Q_{2,K} \\
\vdots & & & \ddots & & \vdots \\
P_{K-1,0} & P_{K-1,1} & P_{K-1,2} & & P_{K-1,K-1} & Q_{K-1,K} \\
P_{K,0} & P_{K,1} & P_{K,2} & \cdots & P_{K,K-1} & P_{K,K}
\end{matrix}
$$

The subvectors $\vec{p}_k$ (submatrices R_k) arising in pass (iii) can also be stored in the first (block) row of the storage array and no extra space is needed at all. Furthermore, if the entire array is retained on a mass storage device, only the matrices in the next computation step must be loaded into the main memory of the computer.

The basic computational procedure is exposed in the following

Algorithm 1

(i) {Reduction pass}

For k=K,...,1 do

For i=0,...,k-1 do

$$Q_{i,k} = P_{i,k}(I-P_{k,k})^{-1}$$

For j=0,...,k-1 do

$$P_{i,j} = P_{i,j} + Q_{i,k}P_{k,j}$$

(ii) {Solution pass}

Solve $\vec{p}_0 = \vec{p}_0 P_{00}$ for $\vec{p}_0$

(iii){Evaluation pass}

For k=1,...,K do

$$\vec{p}_k = \sum_{i=0}^{k-1} \vec{p}_i Q_{i,k}$$

Using the expanded transformation rule (7) the bordering submatrices $P_{i,k-1}$ and $P_{k-1,j}$ can be computed directly without storing the whole reduced matrix P_{k-1}. Thus a compact variant of the algorithm is implemented by replacing pass (i) with

(i) {Reduction pass}

For k=K,...,1 do

For i=0,...,k-1 do

$$Q_{i,k} = P_{i,k}(I-P_{k,k})^{-1}$$

$$P_{i,k-1} = P_{i,k-1} + \sum_{l=k}^{K} Q_{i,l}P_{l,k-1}$$

For j=0,...,k-2 do

$$P_{k-1,j} = P_{k-1,j} + \sum_{l=k}^{K} Q_{k-1,l}P_{l,j}$$

The matrices R_k are obtained with pass (iii) modified as follows

(iii){Evaluation pass}

For k=1,...,K do

$$R_k = \sum_{i=0}^{k-1} R_i Q_{i,k}$$

If all the matrices $P_{i,j}$ degenerate to scalars the algorithm reduces to the procedure derived in /3/. Steps (i) and (iii) can be simplified for block Hessenberg or tridiagonal matrices by restricting loop and summation ranges as described above. We display the resulting algorithm only for the case of tridiagonal matrices

Algorithm 2

(i) {Reduction pass}

For k=K,...,1 do

$$Q_{k-1,k} = P_{k-1,k}(I-P_{k,k})^{-1}$$

$$P_{k-1,k-1} = P_{k-1,k-1} + Q_{k-1,k}P_{k,k-1}$$

(ii) {Solution pass}

Solve $\vec{p}_0 = \vec{p}_0 P_{00}$ for $\vec{p}_0$

(iii){Evaluation pass}

For k=1,...,K do

$$\vec{p}_k = \vec{p}_{k-1} Q_{k-1,k}$$

It should be pointed out that only nonnegative quantities occur in the algorithmic processes, therefore the results are extremely insensitive to rounding errors. Moreover, the row sum criterion for the stochastic matrices P_k, k=K,...,0, can be checked at each elimination step to supervise the calculations.

Since step k of the reduction pass involves $O(k^2)$ operations on submatrices and this pass is the dominating one, the total number of these operations is $O(K^3)$. For block Hessenberg (tridiagonal) matrices however, the corresponding step involves only $O(k)$ ($O(1)$) operations on submatrices and the operation count goes down to $O(K^2)$ ($O(K)$) in that case.

4. AN APPLICATION

We demonstrate the algorithm by a numerical treatment of the discrete-time queueing system Geometric/G/1/K with state-dependent arrival rates. This system can be viewed as a simplified model of a message switching node with finite capacity buffer in a network operated under a load-dependent routing strategy /5/.

Assume a time axis divided into contiguous slots numbered 1,2,... and let the arrival and service processes be as follows:

(i) A unit arrives just prior to a slot mark and enters the system with probability $\lambda_k>0$ ($\lambda_k=0$), when there are $k<K$ ($k\geq K$) units in the system.
(ii) Services start or end at slot marks only and their length is determined by the probability vector $\vec{f}=(f_j)_{j=1,\dots,J}$.

The model can be called a late arrival system with delayed access because an arrival may occur just before the end of a time slot and a service may terminate at the beginning of a time slot. A similar version of this model with bulk arrivals and an unbounded queue was investigated in /4,pp.205-237/.

The process of the queue length observed at slot marks is rendered Markovian by the inclusion of the residual service time to be completed till the next departure. Let p_0 be the stationary probability that the system is empty at an arbitrary slot mark. Let $\vec{p}_k=(p_{k,j})_{j=0,\dots,J}$ be the subvector containing the probabilities $p_{k,j}$ that k units are in the system and a service is in progress with j slots yet to be completed. Further define the (column) unit vector $e=(1,0,\dots,0)^T$ and the shift matrix

$$E = \begin{vmatrix} 0 & 0 & \cdot & \cdot & \cdot & \cdot & \cdot & 0 \\ 1 & 0 & & & & & & 0 \\ 0 & 1 & \cdot & & & & & 0 \\ \cdot & & \cdot & \cdot & & & & \cdot \\ \cdot & & & \cdot & \cdot & & & \cdot \\ \cdot & & & & \cdot & \cdot & & \cdot \\ 0 & & & & & 1 & 0 & 0 \\ 0 & \cdot & \cdot & \cdot & \cdot & 0 & 1 & 0 \end{vmatrix}$$

The following recurrence relations can be deduced by elementary arguments concerning arrival and departure probabilities.

$$p_0 = (1-\lambda_0)p_0 + (1-\lambda_1)\vec{p}_1\vec{e} \tag{16}$$

$$\vec{p}_1 = \lambda_0 p_0 \vec{f} + \lambda_1\vec{p}_1\vec{e}\vec{f} + (1-\lambda_1)\vec{p}_1 E + (1-\lambda_2)\vec{p}_2\vec{e}\vec{f} \tag{17}$$

$$\vec{p}_k = \lambda_{k-1}\vec{p}_{k-1}E + \lambda_k\vec{p}_k\vec{e}\vec{f} + (1-\lambda_k)\vec{p}_k E + (1-\lambda_{k+1})\vec{p}_{k+1}\vec{e}\vec{f},\ k=2,\dots,K \tag{18}$$

It suffices to recall that postmultiplication by the dyadic vector product $\vec{e}\vec{f}$ corresponds to a departure followed by a new service. In the same way, postmultiplication by the shift matrix E corresponds to the decrease of the residual service time at slot marks without a departure.

The system of matrix equations can be rewritten in the form (1) hereby defining the block tridiagonal matrix

$$P = \begin{vmatrix} \vec{p}_{0,0} & \vec{p}_{0,1} & \vec{0} & \cdot & \cdot & \cdot & \cdot & \vec{0} \\ \vec{p}_{1,0} & P_{1,1} & P_{1,2} & & & & & 0 \\ \cdot & \cdot & \cdot & \cdot & & & & \cdot \\ \cdot & & \cdot & \cdot & \cdot & & & \cdot \\ \cdot & & & \cdot & \cdot & \cdot & & \cdot \\ \vec{0} & & & & \cdot & & P_{K-1,K-1} & P_{K-1,K} \\ \vec{0} & 0 & \cdot & \cdot & \cdot & & P_{K,K-1} & P_{K,K} \end{vmatrix}$$

with the scalar $p_{0,0}=(1-\lambda_0)$, the subvectors $\vec{p}_{0,1}=\lambda_0\vec{f}$, $\vec{p}_{1,0}=(1-\lambda_1)\vec{e}$ and the submatrices

$$P_{k,k-1} = (1-\lambda_k)\vec{e}\vec{f} \quad , \ k=2,\dots,K \ ,$$

$$P_{k,k} = \lambda_k\vec{e}\vec{f} + (1-\lambda_k)E \quad , \ k=1,\dots,K \ ,$$

$$P_{k,k+1} = \lambda_k E \quad , \ k=1,\dots,K-1 \ .$$

P is obviously irreducible and stochastic. Thus the invariant vector $\vec{p}$ can be computed by applying Algorithm 2. After normalization the stationary probabilities p_k are given by p_0 and $\vec{p}_k\vec{1}$ for k=1,...,K.

In order to illustrate the algorithm these probabilities are shown in the tables below for a model with K=7 waiting positions and service times determined by a shifted binomial distribution of order J=20 with parameter p=0.5. The arrival probabilities are linearly decreasing with the queue length in the first and constant in the second example. They are given by $\lambda_k=0.025(K-k)$ and $\lambda_k=0.1$ for k=0,...,K-1 respectively.

k	0	1	2	3	4	5	6	7
p_k	0.016	0.077	0.190	0.286	0.261	0.134	0.033	0.003

Tab. 1 The stationary queue length distribution for decreasing arrival rates

k	0	1	2	3	4	5	6	7
p_k	0.049	0.095	0.120	0.136	0.151	0.168	0.186	0.094

Tab. 2 The stationary queue length distribution for constant arrival rates

Although the overall arrival rate $\lambda=0.1$ is the same for both examples, the queue length distribution in the first case is more concentrated on the mean value than in the second case, as can be seen from the tables.

5. CONCLUSIONS

A computational algorithm for the invariant vector of a finite irreducible Markov chain has been developed. In a recursive elimination scheme the triangular decomposition of the matrix defined by the system of equilibrium equations is generated. The stationary vector is then computed in a backsubstitution pass. The feasibility of the algorithm is demonstrated by means of a nontrivial example concerning the queue length distribution in a discrete-time queueing model.

References

/1/ P.J. COURTOIS, Analysis of Large Markovian Models by Parts; Applications to Queueing Network Models
in: *H. Beilner (ed.), Messung, Modellierung und Bewertung von Rechensystemen*, 1-10
Springer, New York-Heidelberg-Berlin 1985

/2/ G.H. GOLUB, C.F. VAN LOAN, *Matrix Computations*
Johns Hopkins University Press, Baltimore 1983

/3/ W.K. GRASSMANN, M.I. TAKSAR, D.P. HEYMAN, Regenerative Analysis and Steady State Distributions for Markov Chains
Oper. Res. 33 (1985), 1107-1116

/4/ J.J. HUNTER, *Mathematical Techniques of Applied Probability Vol. 2 Discrete Time Models: Techniques and Applications*
Academic Press, New York-London 1983

/5/ H. KOBAYASHI, Discrete-Time Queueing Systems
in: *G. Louchard, G. Latouche (eds.), Probability Theory and Computer Science*, 53-84
Academic Press, New York-London 1983

/6/ M.F. NEUTS, *Matrix-Geometric Solutions in Stochastic Models*
Johns Hopkins University Press, Baltimore 1981

/7/ S.N. RAJU, U.N. BHAT, Recursive Relations in the Computation of the Equilibrium Results of Finite Queues
TIMS Studies in Man. Sci. 7 (1977), 247-270

/8/ T.J. SHESKIN, A Markov Chain Partitioning Algorithm for Computing Steady State Probabilities
Oper. Res. 33 (1985), 228-235

/9/ R. SCHASSBERGER, An Aggregation Principle for Computing Invariant Probability Vectors in Semi-Markovian Models
in: *G. Iazeolla (ed.), Mathematical Computer Performance and Reliability*, 259-272
North-Holland, Amsterdam-New York-Oxford 1984

/10/E. SENETA, *Non-Negative Matrices and Markov Chains*
Springer, New York-Heidelberg-Berlin 1981

/11/D. WIKARSKI, An Algorithm for the Solution of Linear Equation Systems with Block Structure
E.I.K. 16 (1980), 615-620

Synchronized Queueing Networks:
Concepts, Examples and Evaluation Techniques

Bruno Mueller-Clostermann, Guenter Rosentreter
Informatik IV, University of Dortmund, P.O. Box 50 05 00, 4600 Dortmund 50

Abstract

Synchronized queueing networks (SQN) are a new type of models, which allow the analysis of queueing systems subject to synchronization constraints. SQN-models originate from traditional queueing networks which have been enlarged by a new type of nodes. These nodes, called counters, can be viewed as synchronization primitives, which are used by the tasks moving through the network. Due to this combination of synchronization and queueing features SQN-models are a new approach towards the unified representation of areas which have been studied separately for a long time.

The usability of this approach is demonstrated by the evaluation of models for parallel task systems and for serialization delays due to the presence of critical regions. Evaluation of SQN-models have been performed with the numerical modelling tool NUMAS, which provides various facilities for model construction and solution. In particular the use of aggregation methods, either for speeding up the convergence of iterative algorithms or for an "a priori" off-line analysis of queueing subnetworks embedded in the SQN-model, are efficient techniques to reduce the costs of an quantitative evaluation.

Features for the evaluation of SQN-models have been integrated with the modelling environments of the computer manufacturers NIXDORF CAG and SIEMENS AG.

1. Introduction

Queueing network models have been extensively used in modelling and quantitative evaluation of computer systems. In particular when contention for shared resources is an important issue in the analysis, queueing network models are an appropriate tool for system design and analysis, see e.g. (LZGS83). A considerable amount of results does exist for separable networks possessing the computational advantageous product form. Further approximate algorithms for non-separable networks with less severe restrictions on the model's characteristic have been studied extensively.

Despite the progress in queueing network research, system features like process synchronization, multiple resource holding, concurrency and conflict phenomena are difficult to represent in traditional queueing network models. As pointed out in THBA83, traditional queueing models assume that "a program consists of a single process (task) which obtains service in a serial fashion from the devices in the queueing network model. Specifically, this implies that programs cannot hold more than one device at a time, and as such does not provide an accurate model for parallel processing". Evidently there is a growing interest to develop models which include, besides queueing phenomena, the features mentioned above.

In HETR82, HETR83 QN-based models for internal concurrency have been investigated; in THBA83 a task system model which comprehends the precedence relation between tasks is combined with a QN-model; a software blocking phenomenon due to critical sections has been investigated by combining stochastic Petri nets with a QN-model (BABG85). Especially the Petri net approach has received growing interest during the last years; a variety of Petri net-based models and modelling tools for the validation and evaluation of synchronization schemes has been developed (TURIN85). An important example is the well-kown GSPN-tool for the evaluation of Generalized Stochastic Petri Net models (BMCC84).

The concept of Synchronized Queueing Networks (SQN) introduced in this paper is based on the traditional queueing network model approach. A rather general class of QN-models, known from the numerical modelling tool NUMAS (MUEL84), was enlarged by a new type of resources which can be used for synchronization purposes by the tasks moving through the network. We believe that SQN-models are a new approach towards the unification of areas which have been studied separately for a long time. In our opinion, SQN present advantages from a pragmatic point of view making them more acceptable by practioners. The synchronization features of SQN are easier to understand than the very detailed mechanism of timed Petri-nets and they can be easily integrated with conventional queueing networks. The SQN-model world is introduced in chapter 2, chapter 3 is devoted to the evaluation of SQN and chapter 4 takes up two examples known from the literature and shows their formulation and evaluation. In chapter 5 we finally discuss some aspects of the future work.

2. Synchronized Queueing Network Models

We consider network models where stations (from now on called nodes) provide services which can be requested by tasks travelling from node to node.

The **tasks** belong to **classes**, which are grouped into disjoint **chains** describing the routing behaviour of the tasks; each possible transition of a class-r-task from node i to node j changing the task's class index from r to s is given by a **routing probability** p_{irjs}. For each chain we have a separate routing matrix.

In case of **closed chains** the population is constant i.e. tasks cannot enter or leave the system, whereas in case of **open chains** arrivals from outside occur with exponentially distributed interarrival times. In order to keep the model's state space finite the population of an open chain must be limited to a maximum value; as a consequence arrivals at a completely filled chain are lost.

Tasks have certain service requirements which can be satisfied at the nodes of the model. Different from traditional queueing network models we distinguish two different kinds of nodes, namely **servers** and **counters.** Servers are similar to (and partially include) the service centers known from e.g. product form networks, whereas counters are a new approach to integrate synchronization phenomena explicitly with traditional queueing network models. First we define servers; subsequently we discuss counters in more detail.

Servers consist of a number of waiting queues (one for each admissable class) and a number of identical processors. A task arriving at a server has to queue for one of the processors; if a processor is available the processing of the task can start. The duration of the service is determined by the amount of the task's service demand and the speed of the processor; the service demand of a task of class r at server i is given by a random variable which has a Coxian distribution function. Suppose that the mean service demand is m_{ir}; the mean service time is m_{ir}/s, where s denotes the speed of the processor.

As scheduling disciplines are available (mainly out of implementation restrictions) random, infinite server, processor sharing, preemptive and non-preeemptive priority. (Fcfs-scheduling is not admitted, but the use of random-scheduling leads to very good approximate performance measures in many cases, see MARU84.)

The **state** of a **server** is given by the number of tasks of class t in Coxian service phase q which are in service at server i and the number of tasks of class t which are passive (= waiting for service) at server i.

Note that for most applications the maximum number of phases is either 1 (exponential service time distribution) or 2 (e.g. Coxian service time distribution with arbitrary coefficient of variation > 1). The state variables for those classes t T which do not visit node i have a constant value of zero.

An important feature are task-mix dependent service speeds. For each task of class r visiting server i the processors' speed is given by a function $s_r(n_1,..,n_r,..,n_R)$, where n_r denotes the actual number of tasks of class r at server i.

There are some more features included in the current implementation of SQN, like limited waiting queues, processor failure and repair. We omit these features which are of minor interest in the given context.

Counters consist of a number of waiting queues (one for each class) and an integer **status** variable x, which can be changed (incremented or decremented) by tasks visiting the counter. The feasible interval for the status variable is defined by a lower bound l_i and an upper bound u_i; if a request c_{ir} of a task of class r at counter i can be satisfied, i.e. $l_i <= x_i + c_{ir} <= u_i$, the status changes instanteneously from x_i to $x_i + c_{ir}$ and the task leaves the counter without delay (and moves to the next node or leaves the system).

If a request cannot be satisfied, because the desired change would move the status x out of the interval (l,u), the requesting task must wait until x is changed by another task, say of class s, in a way that the waiting tasks' request can be satisfied. Because in general more than one class can visit the counter, there is a need for scheduling. Which one of several waiting tasks must be handled first is determined according to a priority or a random discipline.

The initial value of a counter's status variable can be essential for the model's behaviour. Unless otherwise specified the default initial value equals the upper bound u_i of the feasible range (l_i, u_i).

The state of a counter is given by the number of tasks which are "suspended" (= waiting that the requested change of the status can be performed) and the value of the integer status variable of the counter.

A simple example for a counter is a **binary semaphor.** The lower bound of the status variable is zero, the upper bound is one. Tasks which visit the counter decrement or increment the status variable by one (P or V-operation). The binary semaphor can be used for modelling critical sections (see fig. 1) where only one task can visit the critical section at the same time. We assume that all tasks do start in the left subnet and that the initial counter status is 1.

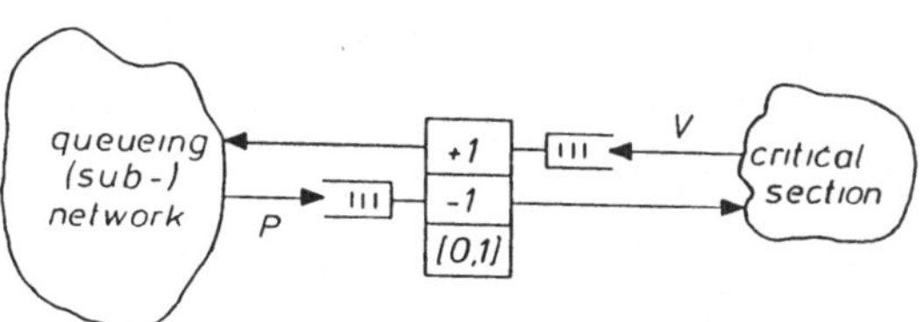

Fig. 1: Use of a counter as binary semaphor

The next example illustrates the use of **counters for synchronization in parallel systems;** this is important for modelling parallel execution in computer systems or in communication networks where messages are divided into packets which must be resequenced in the right order after transmission. Two counters and an additional chain are needed for synchronization. The range of the counter's status variable is (0,n), where n is the number of tasks to be synchronized; the initial counter status is also n. The additonal chain contains a single task which circulates among the two counters.

The model works as follows: Three tasks (or if you like "packets") travel through a queueing subnetwork (modelling e.g. a communication system) and do reach the first counter after a random delay. The first task arriving at the counter decrements the counter status from 3 to 2, the second task from 2 to 1 and the third task from 1 to 0; all tasks proceed to the second counter without any delay and try to increment the second counter's status variable. This situation immediately before synchronization is depicted in fig. 2; the latest task is leaving the queueing subnetwork, changes the status of the first counter from 1 to 0 and enters his queue at the second counter. The following avalanche effect happens in zero-time: the "synchronization token" waiting at the first counter changes the status from 0 to 3, moves to the second counter, changes the status variable from 3 to 0 and returns to the first counter, where it get stuck; now the requests of the suspended tasks are satisfied and the status of the second counter is set back to 3 again. The three tasks arrive at the queueing network in the same time instant, but note that they have an order which is determined by the scheduling policy of the second counter. For simplicity a priority discipline can be employed which arranges the tasks always in the same order. (A random discipline is much more intricate, especially with respect to implementation).

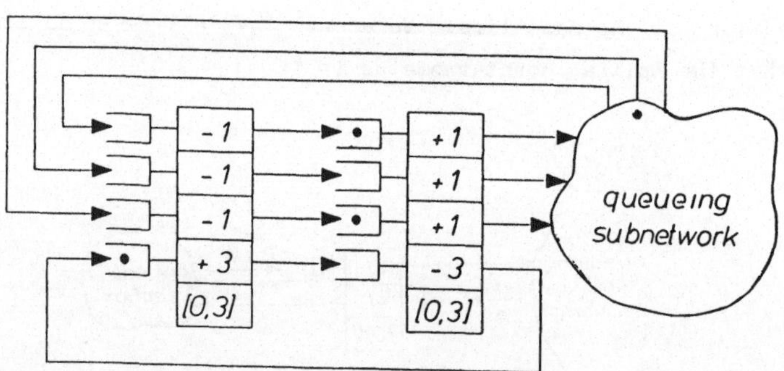

Fig. 2: Synchronization of tasks

3. Evaluation of SQN-Models

3.1 SQN as Markov Chains

SQN-models are defined in terms of nodes (servers and counters) and tasks circulating from node to node. The state z of a model is simply given by the states of all nodes. Of course the feasibility of model states is determined by certain conditions, like "maximum number of tasks of chain k" or "number of processors at server i", and by rules governing the transition behaviour of tasks, like scheduling disciplines at the nodes or the routing matrices for each chains.

Like in Timed Petri nets, see e.g. MBCO84, TURIN85, we have to distinguish between **tangible** states and **vanishing** states. Tangible states are states where the model spends a certain amount of time, whereas vanishing states are entered and left in zero time. An example for the occurence of a vanishing state is the binary semaphor sketched in the preceeding chapter: if the semaphor variable has the value 1 and a task wants to perform a P-operation (decrementing the state variable) this request is satisfied immediately (i.e. in our model: without any delay) and the model transits in zero-time into a successor state.

For an evaluation of the model we have to set up the transition matrix of the model. Two ways of establishing and storing the transition matrix can be distinguished.

1. The model's state space explicitely includes tangling **and** vanishing states and as a consequence time-less transitions out of (and into) vanishing states.

2. The vanishing states are eliminated during the process of matrix construction. Each state is a tangling state and each state transition z -> z' is associated with an exponential rate r (z -> z').

In case 2 the transition matrix is identical to the infinitesimal generator (or transition rate matrix) of the markov chain associated with the model under consideration. In case 1 a markov chain is embedded in the transition matrix and quantitative measures can be obtained by solution techniques which are based on the study of the reduced embedded markov chain that can be defined on the set of tangible states; this approach has been employed in the GSPN-tool (BMCC84).

3.2 On the Stationary Analysis of SQN-models

To ensure that the stationary probability distribution of the markov chain associated with the model can be computed by numerical techniques, the following assumptions must hold.

1. The model's state space (i.e. the set of tangible states) is finite. Due to the restrictions in the number of tasks which can belong to the chains of the model this condition is always met.

2. The markov chain must be time homogeneous, i.e. all transition rates are time independent. By definition this condition is always met.

3. The markov chain is irreducible; because of the finiteness of the state space this implies that all states are recurrent.

The third condition is **not** always met! Of course you can specify models having **transient states**, which are occupied with stationary probability zero or models with an **absorbing state** (and all other states transient), which has the stationary probability one. Moreover, it is possible that sets of **vanishing states** occur during the process of state space generation, which turn out to be a so called livelock or dynamic deadlock. Simple livelock-examples can result from models which contain constructions leading to infinite non-productive operation-cycles.

We do not consider non-irreducible markov chains in more detail, although the study of deadlocks, livelocks or other functional anomalies is an interesting subject in itself (see again the growing interest in Petri nets and related models). In the given context we consider the possible detection of such phenomena during an intended steady state analysis just as a by-product.

Now we turn to the numerical solution of SQN-models. If the markov chain satisfies the requirements sketched above, the vector of steady state probabilities can be obtained as the solution of a set of global balance equations. We summarize some of the most important solution algorithms which have been succesfully applied to markov models.

Direct algorithms, like Gauss elimination or LU-factorization, are quite favorable in cases where the generation of new matrix elements during the elimination process is not too excessive; this is the case for band matrices, which do occur (for example) as transition matrices of closed two-node networks or open single-node models. In most cases, the transition matrix is large and sparse; as a consequence iterative approaches, like Gauss-Seidel-iteration or SOR-techniques, are the only promising way to compute the solution vector (MUEL81, STGO85).

A technique which increases the speed of convergence in many cases is the combination of iteration and aggregation; the idea of this technique is to insert aggregation steps into the iterative procedure. In particular in case of "nearly completely decomposable" models the speed-up of the iteration can be enormous (MUEL81, SCHW83).

3.3 Hierarchical Analysis of SQN

Because the state description includes the status variables (of the counters) as well as the potentially very detailed description of the servers, the resulting state space explosion prevents a numerical evaluation of many (harmless looking!) models.

We present an approach which is based on the isolated analysis (off-line analysis) of model parts which can be considered as separable queueing network subsystems. Off-line analysis is a well known technique from queueing network analysis, which aggregates (transforms) a subnetwork into a flow-equivalent state dependent service center. Depending on the type of model in which such substitute representations of subnetworks are embedded, the (marginal) steady state distributions and consequently the derived performance measures are obtained exactly or approximately, see e.g. LZGS84.

This principle of subnetwork aggregation can be applied as well to SQN-models: A separable queueing network submodel embedded in a SQN can be analyzed (and transformed) by efficient product-form algorithms. The obtained state dependent server replaces the original subnetwork and the resulting more abstract SQN-model still shows the complete synchronization structure of the original model and is now - due to its reduced state space - amenable to numerical evaluation.

We consider the aggregation of separable queueing subnetworks in more detail. A task entering a subnetwork is routed from server to server, requiring and receiving services of a certain amounts. Of course the behaviour pattern of a class r task can be different from the behaviour pattern of class s; as a consequence we have to distinguish between the different behaviour patterns which are possible "inside" the subnetwork. In the following we refer to these different behaviour patterns as **services**, which are provided by the subnetwork and can be used by the tasks.

Assume that the subnet provides S services and that the actual load on the subnet is given by a population vector $(n_1,...,n_S)$; an off-line analysis of the subnet yields for every subnet service 1,..,S a load dependent service speed function g_s $(n_1,..n_s,..,n_S)$, such that the considered subnet can be replaced by a load dependent server which processes the service requests according the functions g_s. In particular, this load dependent server provides the same number of services like the original subnet and in case this equivalent server is embedded in a separable network ("the environment is separable") the mean delay of a task in the original subnet and in the substitute server are identical. For this reason a substitute representation is sometimes called an equivalent server.

In a non-separable environment the use of an equivalent server introduces approximation errors, but experience shows that the quality of the approximation is quite satisfactory; approximation is excellent if the considered subnet is "weakly coupled" with its environment, see e.g. COUR77, LZGS84. In connection with numerical solution techniques the main advantage of the equivalent server technique is the reduction of the state space size. Note that the state description of the equivalent server - in contrast to the detailed original subnet - does not take into account the location of a task; only the actual number of tasks requesting (using) service s, s = 1,...,S, is included in the state description.

For illustration we sketch the aggregation of a subnet consisting of three disk units, which can be accessed according the following behaviour patterns called "services", see fig. 3.

Service 1: a task requests one of the three disks with probability .4375, .28125 and .28125 respectively.

Service 2: the task requests service at disk 1

Service 3: the task requests service at disk 2

The equivalent server obtained by off-line analysis of the subnet also offers three services. Each request of service s, s = 1,2,3, is served with speed $g_s(n_1,n_2,n_3)$.

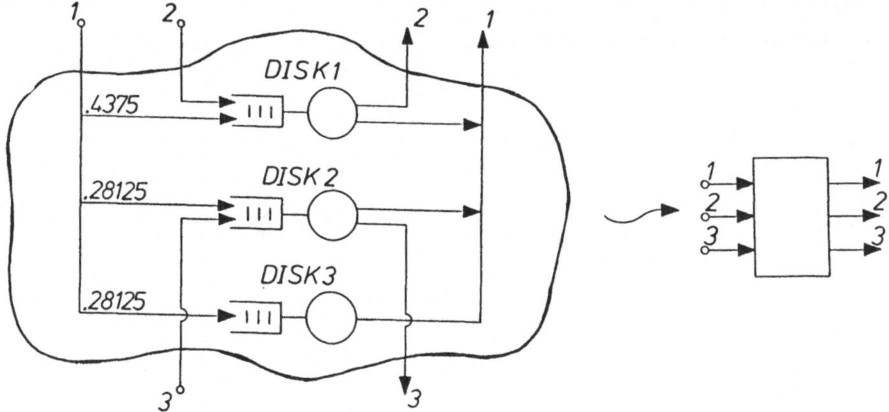

Fig. 3: Disk subsystem and equivalent server

4. Examples for SQN-Models

We illustrate the evaluation techniques described in the preceeding chapter by applying it to synchronization problems known from the literature. In the first example we consider a **serialization** phenomenon due to critical sections, which has recently be analyzed (AGBU83,BABG85). The second example deals with the **parallel execution of tasks.** We analyze a model which has been studied by Thomasian and Bay (THBA83). Furthermore the new SQN-features have been used for various modeling studies. In HAVE86 phenomena in an FDDI-token ring have been modelled with the help of counters. Another example is the modelling of spin locks and suspend locks (FRIC87).

4.1 Serialization Delays

The model under study is of the central server type. It consists of a cpu, three disks and two critical sections which are implementend with the help of counters. Jobs may process a non-serialized mode of execution or do enter one of two critical sections. For reasons of clarity figure 4a) - 4c) show the critical sections as separate instances. The model parameters have been taken from (AGBU83). The service times for each job per visit are: 0.02 time units at the CPU and 0.04 time units at the disks.

In the aggregated model the three disks are replaced by a load-dependent server (see fig. 5; the critical sections are analogous to those of fig. 4b and 4c).

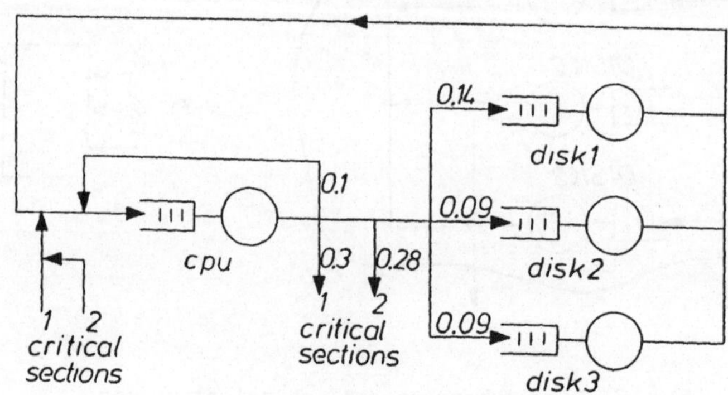

Fig. 4a: Non-critical section

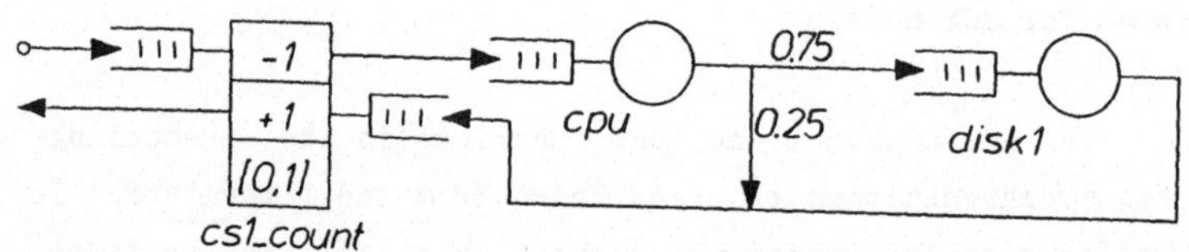

Fig. 4b: Critical section 1

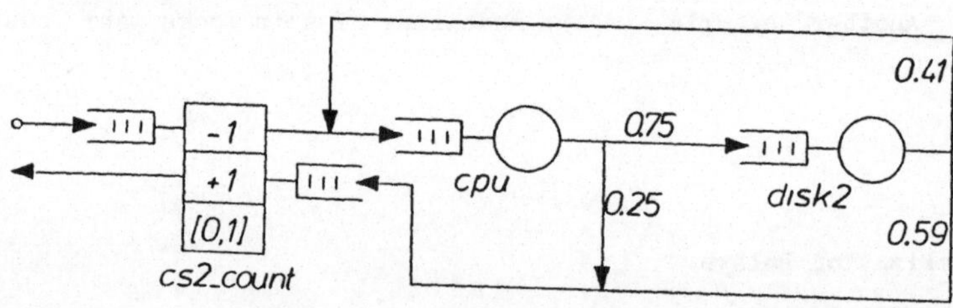

Fig. 4c: Critical section 2

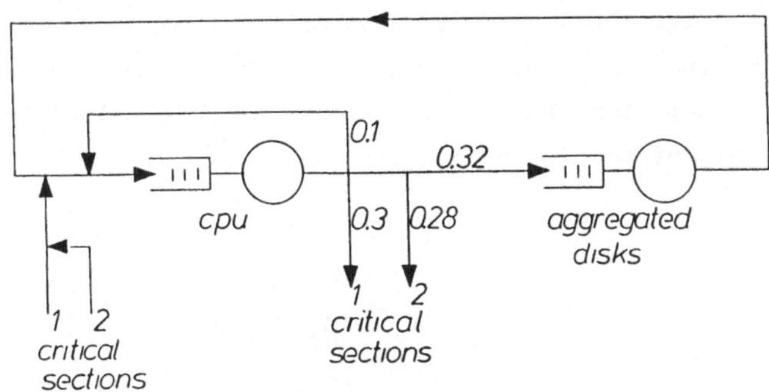

Fig. 5: The aggregated model

The approximate results obtained from the aggregated model show errors below 0.5% when compared with exact results obtained by numerical evaluation of the detailed original model. Moreover we display space and time consumption for the original and the aggregated model. All evaluations have been performed with the modelling tool NUMAS, see MUEL84.

The size of the aggregated model as well as the time for its solution is reduced essentially. (Space and time consumptions for the original and the aggregated model are displayed in appendix 1).

4.2 Parallel Execution of Tasks

In this section we use the counter to model synchronization aspects in parallel computing. We consider a task system with precedence relationships in their execution sequence, depicted in the precedence graph below. The problem and its parameters has been adopted from (THBA83).

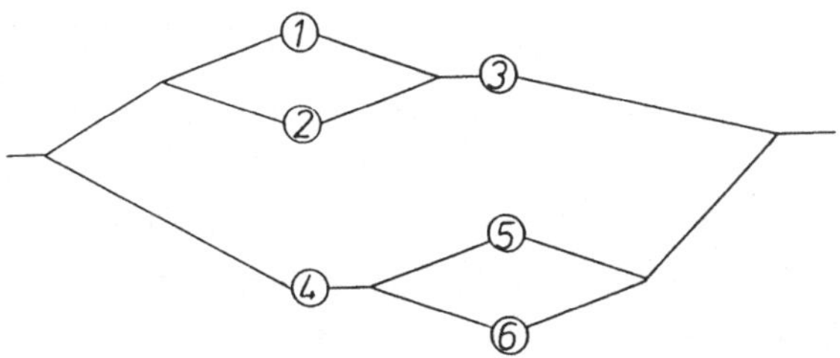

Fig. 6: Precedence graph of the task system

The model is shown in fig. 7. The black box "start synchronization" is a 'two counter subsystem' which works like described in chap. 2, cf. fig. 2, the "central server" is shown in fig. 8. It provides two different services, namely one for the tasks 1, 2, 5 and 6 (marked -) and another one for the tasks 3 and 4 (marked ---). In the aggregated model the central server is replaced by a load dependent equivalent server. The rest of the network was kept unchanged.

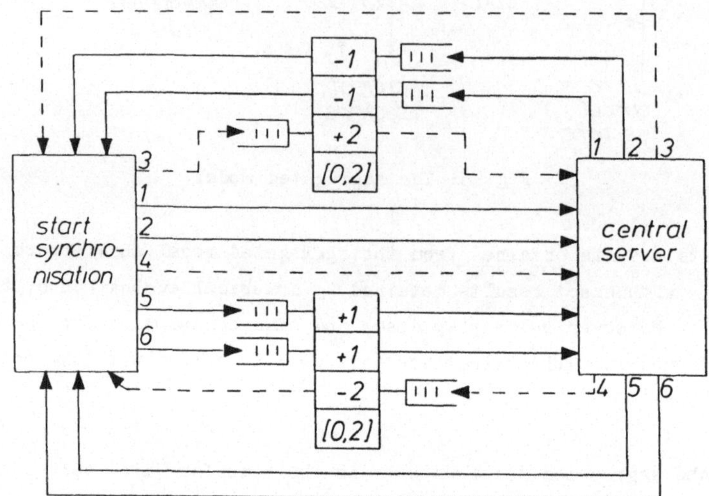

Fig. 7: SQN-model of a parallel task system

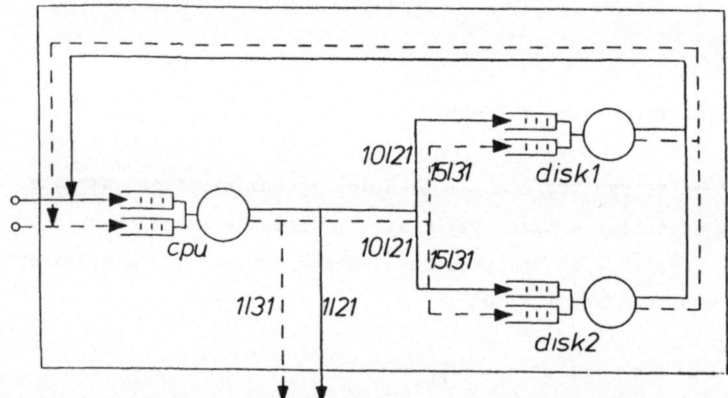

Fig. 8: The central server queueing subnetwork

The key performance measure of interest is the mean completion time for the task system; i.e. we have to determine the cycle time between two successive synchronizations. Like in the serialization example we construed and evaluated a detailed as well as an aggregated model. Due to the weakly coupled central server subnet the error of approximation was very small (about 0.6%); the exact value for the mean cycle time was 6.116 versus 6.081 seconds for the approximation. Again the technique of subnet aggregation yields a considerable gain in effiency. (For numbers see appendix 2).

5. Concluding Remarks

We introduced synchronized queueing networks as models for the evaluation of synchronization phenomena. Examples and applications of SQN-models show the practicability of the approach. Extending the scope of SQN is straightforward although theoretical investigations and comparisons with other specification techniques, e.g. in the area of stochastic Petri nets should preceed further implementation efforts. More stimulating effects on the further developement of SQN are expected to emerge from industrial applications.

The SQN-approach has been introduced first into the numerical modelling tool NUMAS, an early version of which has been described in MUEL84. The development of NUMAS has been supported by the "Minister fuer Wissenschaft und Forschung des Landes Nordrhein-Westfalen" in the years 1981-1983. Based on the universitary NUMAS-installation, features for SQN-modelling have been integrated with the modelling environment of computer manufacturing companies. For Nixdorf Computer Company the NUMAS-features have been integrated with the hierarchical performance evaluation tool HIT, see BESC85, BEST87. For Siemens Company, queueing network performance evaluation algorithms including solvers for separable networks as well as NUMAS-based numerical evaluation techniques have been implemented (in PASCAL-XT under BS2000). Graphic workstations are used for interactive model specification and fast evaluations, whereas the more time and space consuming evaluations are submitted to a host computer.

We acknowledge the many contributions of Michael Sczittnick to various NUMAS implementations. Last but not least we thank Iris Koch and Nathalie Muenter for typing this paper.

6. References

AGBU83 Agrawal, S.C.; Buzen, I.P.: "The Aggregate Server Method for Analyzing Serialization Delays", ACM Transactions On Computer Systems, 1, 2, May 1983, 116-143

BABG85 Balbo, G.; Bruell, S.C.; Ghanta, S.: "Combining Queueing Network and Generalized Stochastic Petri Net Models for the Analysis of Software Blocking Phenomenon", in: (TURIN85)

BESC85 Beilner, H; Scholten, H.: Strukturierte Modellbeschreibung und strukturierte Modellanalyse - Konzepte des Modellierungswerkzeugs HIT, 3. GI/NTG-Fachtagung "Messung, Modellierung und Bewertung von Rechensystemen", Dortmund, 1985

BEST87 Beilner, H.; Stewing, F.-J.: Concepts and Techniques of the Performance Modelling Tool HIT, European Simulation Multiconference, Vienna, July 1987

BMCC84 Balbo, G.; Marsan, A.; Ciardo, G.; Conte, G.: "A software tool for the automatic analyis of generalized stochastic Petri nets models", in: (POTI84)

COUR77 Courtois, P.J.: Decomposability - Queueing and Computer System Applications, Academic Press 1977

FRIC87 Fricke, B.: Projektdokumentation, Project on BS2000-modelling, Informatik IV, University of Dortmund

HAVE86 Haverkort, B.R.H.M.: "Performance Modelling of Network Gateways using USPE and NUMAS", Diploma Thesis, Technische Hogeschool Twente, Onderafdeling der Informatica, 1986

HETR82 Heidelberger, P.; Trivedi, K.S.: "Queueing network models for parallel processing with asynchronous tasks", IEEE Transact. Comp., vol. C-31, Nov 1982, pp. 1099-1109

HETR83 Heidelberger, P.; Trivedi, K.S.: "Analytic Queueing Models for Programs with Internal Concurrency", IEEE Trans.Comp., vol. C-32, Jan. 1983, pp. 73-82

LZGS84 Lazowska, E.D.; Zahorjan, J.; Graham, G.S.; Sevcik, K.C.: "Quantitative System Performance - Computer System Analysis Using Queueing Network Models", Prentice Hall, 1984

MARU84 Marie, R.; Rubino, G.: "An Approximation for a Multiclass ./M/1/FIFO Queue Imbedded in a Closed Queueing Network", Proc. of the Int. Workshop on Modelling and Performance Evaluation of Parallel Systems, Grenoble (France), December 13-18, 1984

MBCO84 Marsan, M.A.; Balbo, Conte, G.: "A Class of Generalized Stochastic Petri Nets for the Performance Evaluation of Multiprocessor Systems", ACM Transactions on Computer Systems (May 1984)

MUEL81 Mueller-Clostermann, B.: "Decomposition Methods in the Construction and Numerical Solution of Queueing Network Models", Proc. of Performance '81, Amsterdam (Netherlands), 1981, pp. 99-112

MUEL84 Mueller-Clostermann, B.: "NUMAS - A Tool for the Numerical Analysis of Computer Systems", in: (POTI84)

POTI84 Potier, D. (ed.): Proceedings of the International Conference on Modelling Techniques and Tools for Performance Analysis, May 16-18, 1984, Paris (North Holland Publishing Company)

SCHW83 Schweitzer, P.J.: "Aggregation Methods for Large Markov Chains", Proc. of the Int. Workshop on Mathematical Computer Performance and Reliability, Pisa (Italy), September 26-30, 1983

STGO85 Stewart, W.J.; Goyal, A.: "Matrix Methods in Large Dependability Models", IBM Research Report RC 11485, 1985

THBA83 Thomasian, A.; Bay, P.: "Queueing Network Models for Parallel Processing of Task Systems", in: Proceedings of the 1983 International Conference on Parallel Processing, Bellaire (Minnesota), August 23-26, 1983 (see also (TURIN85))

TURIN85 International Workshop on Timed Petri Nets, Turin, Italy, July 1-3, 1985, (IEEE Computer Society Press)

Appendix 1 : Results for Serialization Delays

The following tables show exact and approximate performance values as well as space and time consumption for the analysis. The model and its parameters are given in chap.4.1 and have been taken from AGBU83.

Table 1 shows the system throughput for different populations.

Population	Throughput	
	Exact	aggregated model
1	1.5438	1.5438
3	2.3661	2.3685
5	2.5590	2.5622
8	2.63	2.63

Table 1: Exact and approximate results

Table 2 shows the number of model states and the CPU-time used for solution.

Population	Number of states: exact model	Number of states: aggregated model	CPU-time for model construction and solution: exact model	CPU-time for model construction and solution: aggregated model
1	8	6	9.3	8.7
3	118	44	61	18.3
5	720	146	331	83
8	4887	489	-	308

Table 2: Space and time consumption

Appendix 2 : Evaluation of the Parallel Task Model

We display some figures concerning the evaluation of the detailed and the aggregated model described in chap. 4.2; the construction of the model's transition rate matrix consumed 120 CPU-seconds whereas the CPU-time spent for the solution of the global balance equations varies between 17 and 218 seconds, depending on the employed algorithm, see table 3.

Evaluation technique	#states	matrix construction	solution	total
LU-factorization	360	120 sec	218 sec	338 sec
Gauss-Seidel-iteration	dito	dito	73 sec	193 sec
Gauss-Seidel-iteration speeded-up by aggregation	dito	dito	17 sec	137 sec
Subnet-aggregation and LU-factorization	24	8 sec	3 sec	11 sec

Table 3: Comparison of evaluation techniques

- The insertion of aggregation steps into Gauss-Seidel-iteration reduces the time consumption from 73 seconds (195 iteration steps) to 17 seconds (7 iteration steps and 3 aggregations).

MODELING TOKEN RING NETWORKS – A SURVEY

Werner Bux
IBM Research Division, Zurich Research Laboratory
8803 Rüschlikon, Switzerland

This paper discusses single and multiple token rings, describes major aspects of their architecture, and discusses in detail the performance issues arising in the design of such local area networks. Following a survey of analytical queueing models to describe the basic token-ring operation, three topics are discussed in detail: 1) The IEEE 802.5 Token Ring and its performance; 2) performance aspects of the ANSI Fiber Distributed Data Interface (FDDI) token ring, and 3) performance questions arising in the interconnection of token-ring networks.

1. INTRODUCTION

Token rings, the subject of this paper, are a typical example of local-area networks (LAN's) that have emerged since the late 1970's |Kümmerle 87a|, |Schwartz 87|, |Stallings 84|. These networks are characterized by the following elements |Kümmerle 87b|, see Fig. 1:

a) A transmission medium shared among the participating stations (workstations, hosts, servers, etc.), and providing a broadcast capability.
b) A distributed protocol, the Medium Access Control (MAC) protocol which controls access to the medium and provides recovery mechanisms where necessary.
c) A set of cooperating adapters (also called LAN interfaces) through which stations attach to the network, and which execute the MAC protocol and interface with the attaching stations.

The groups concerned with LAN standardization, IEEE, ECMA, ANSI, and ISO have adopted a LAN-architecture model which describes the relationship between LAN architecture and the OSI Reference Model. As shown in Fig. 2, the OSI data link layer is split into two sublayers, the medium-dependent "Medium Access Control" (MAC) sublayer and the medium-independent "Logical Link Control" (LLC) sublayer. Peculiarities of the various local-network techniques are thus restricted to the medium, the physical layer, and the MAC sublayer.

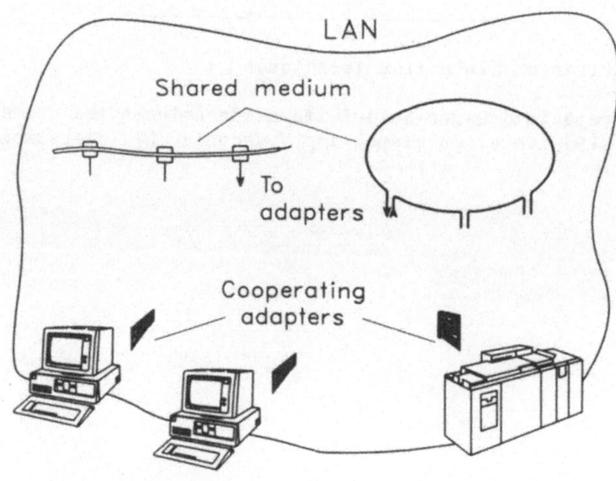

Fig. 1. LAN definition. From |Kümmerle 87b|. © 1987 by IEEE.

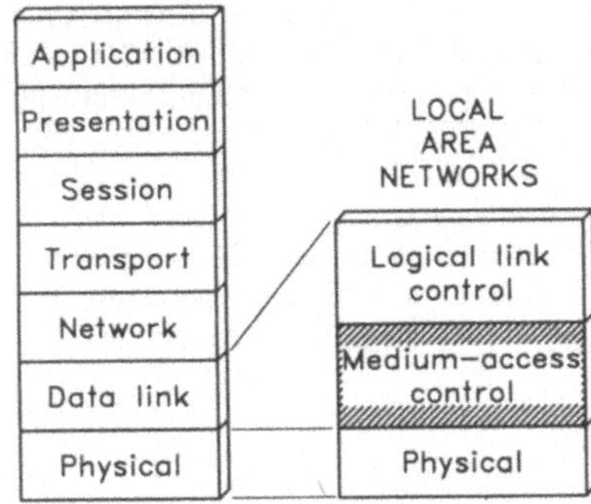

Fig. 2. LAN architecture reference model. From |Kümmerle 87b|. © 1987 by IEEE.

Compared to other LAN techniques, token rings have a relatively long technical history; experimental systems had shown the feasibility of the ring technique long before alternative methods, e.g., CSMA/CD or token-passing bus systems, were invented |Farber 73|, |Farmer 63|, |Penney 79|, however, they were not implemented on a broad basis. With the advent of local-area networks, the token-ring principle was re-examined and found to provide an attractive solution because of its favorable attributes in respect of wiring, transmission technology, performance, and the potential for low-cost implementation |Bux 81a|, |Bux 83b|, |Dixon 83|, |Penney 79|, |Saltzer 80|. Further work revealed that what had been considered a potential problem of token rings, namely, lack of reliability, can be overcome by an access protocol providing adequate recovery functions and a suitable wiring strategy |Bux 81a|, |Bux 83b|.

A token ring consists of a set of stations connected serially by a transmission medium, e.g., twisted-pair cable, see Fig. 3. Information is transferred sequentially from one active station to the next. A given station (the one having access to the medium) transfers information onto the ring. All other stations repeat each bit received. The addressed destination station copies the information as it passes. Finally, the station which transmitted the information removes it from the ring.

A station gains the right to transmit when it detects a token passing on the medium. The token is a control signal comprised of a unique signaling sequence that circulates on the medium following each information transfer. Any station, upon detection of a token, may capture the token by modifying it to a start of frame sequence, and then append appropriate control and address fields, the LLC-supplied data, the frame-check sequence, and a frame-ending delimiter. On completion of its information transfer, and after appropriate checking for proper operation, the station generates a new token which provides other stations the opportunity to gain access to the ring.

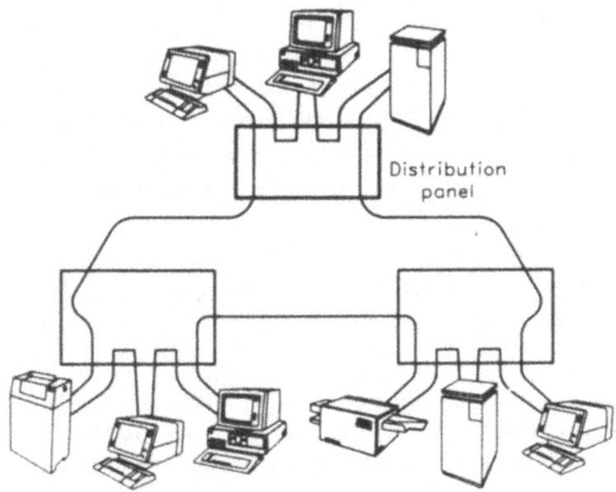

Fig. 3. Token-ring network. From |Kümmerle 87b|. © 1987 by IEEE.

The token-ring technology has been chosen by various manufacturers as the basis of their LAN offerings. It was also selected by standardization organizations such as IEEE, ECMA, ISO, and ANSI for their standardization of a major LAN technology. The two most prominent token-ring standards are the IEEE 802.5 Standard |IEEE 84b| and the ANSI FDDI Draft Standards |ANSI 85a,b,c,d|. The IEEE 802.5 Standard specifies format, MAC protocol, physical-layer functions, and the twisted-pair attachment of stations to the medium. The standardized speeds are 1 and 4 Mbps. The ANSI FDDI Draft Standards define a token-ring system that makes use of optical-fiber transmission, and operates at 100 Mbps. We shall discuss the MAC protocols and the performance of both standard systems in detail in Sections 3 and 4 of this paper, respectively.

2. MODELING THE BASIC TOKEN-RING OPERATION

Analytical models of token-ring operation have been developed to determine the basic performance measures, i.e., the delay packets or frames experience between their generation at one station and their reception at the destination station, and the maximum throughput achievable over a token ring with given characteristics.

The principle of token-ring operation can be described by a model as shown in Fig. 4. The active stations are represented by their transmit queues. These queues are serviced in a cyclic manner symbolized by the rotating switch which stands for the token.

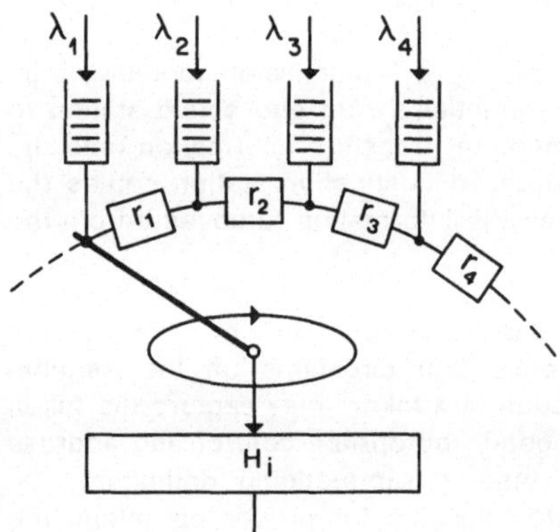

Fig. 4. Token-ring queueing model. From |Bux 83a|. © 1983, Elsevier Science Publishers B.V. (North Holland).

The time needed to pass the token from station i to station (i+1) is modeled by a constant delay r_i. On an actual ring, delay r_i corresponds to the propagation delay of the signals between stations i and (i+1) (approximately 5 μs per km cable) plus the latency caused within station i by the repeater and by actions such as alteration of the token bit. The station latency is usually in the order of one bit time. The traffic generated in station i is characterized by a packet arrival process with rate λ_i and a packet transmission-time distribution H(i) with mean h_i and second moment $h_i^{(2)}$.

In the standard token-ring protocols |IEEE 84b|, |ANSI 85a|, the length of time a station is allowed to transmit when in possession of the token is controlled by a so-called token-holding timer. Long token-holding timeouts will lead to an exhaustive type of service in which stations holding the token transmit all frames they have queued up. Short token-holding timeouts yield an operation similar to the nonexhaustive service discipline where only a single packet per token is transmitted. Therefore, both of these limiting cases are of interest in the performance evaluation of token rings.

A large number of variants of the cyclic-service model has been treated in the literature, see, e.g. the excellent monograph by Takagi |Takagi 86|. The models differ relative to the number of buffers, the service discipline, i.e., exhaustive or nonexhaustive service, and the assumptions of the arrival processes. We subsequently review the analytic results available for the models with infinite buffer size |Takagi 86|. Most of them are based on the assumption of Poisson arrival processes and generally distributed and independent service times.

The **exhaustive-service** model with an arbitrary number of stations has been analyzed in |Eisenberg 72|, |Hashida 72a|, |Brodetskii 73|, |Humblet 78|, and |Ferguson 85| (continuous time, asymmetrical traffic), in |Konheim 74| (discrete time, symmetrical traffic), and in |Swartz 80|, |de Moreas 81| and |Rubin 83| (discrete time, asymmetric traffic). Prior to these, simpler models with either two stations only or an arbitrary number of stations but zero token-passing overhead had been considered in |Avi-Itzhak 65|, |Neuts 68|, |Takacs 68| |Cooper 69|, |Cooper 70|, |Sykes 70|, and |Eisenberg 71|.

In case of different packet arrival rates at the stations/queues, and/or nonidentical service-time distributions, determining the mean queueing delays requires solution of a linear set of equations with N^3 variables. Ferguson and Aminetzah have illustrated a way to reduce the complexity to N^2 |Ferguson 85|.

An interesting general result which is also of practical use is the so-called pseudo-conservation law for exhaustive service which gives an explicit formula for the weighted sum of the station-specific waiting times w_i

$$\sum_{i=1}^{N} \frac{\rho_i}{\rho} w_i = \frac{R\left(\rho - \sum\limits_{i=1}^{N} \rho_i^2\right)}{2\rho(1-\rho)} + \frac{\sum\limits_{i=1}^{N} \lambda_i h_i^{(2)}}{2(1-\rho)} \tag{2.1}$$

with $\rho_i = \lambda_i h_i$; $\rho = \sum_{i=1}^{N} \rho_i$; $R = \sum_{i=1}^{N} r_i$.

If the mean service times at all stations are the same, the mean queueing delay of the entire system, i.e., the delay averaged over all stations, can obviously be easily computed from (2.1). The pseudo-conservation law was first conjectured in |Bux 83a| and later proven by Watson |Watson 84| and Ferguson and Aminetzah |Ferguson 85|. A discerning analysis and explanation of the pseudo-conservation law are given in |Boxma 86a|.

A corollary of (2.1) is the mean queueing delay in a completely symmetrical system, i.e., identical service-time distributions with mean h and second moment $h^{(2)}$ and equal arrival rates λ/N, |Eisenberg 72|, |Hashida 72a|, |Konheim 74|:

$$w_{Sym} = \frac{R(1-\rho/N)}{2(1-\rho)} + \frac{\lambda h^{(2)}}{2(1-\rho)} \quad . \tag{2.2}$$

Calculating mean queueing delays for systems with a large number of queues is hampered by the size of the system of equations N^2. For practical traffic-engineering purposes, an approximate result such as the one in |Bux 83a| is therefore very useful. It has been shown in |Bux 83a| that the relative error of this approximation is typically less than 10%. In practice, this error is presumably negligible compared to the inaccuracies introduced by the Poisson arrival-process assumption, not to mention other simplifications in the modeling of the real operation, see Sections 3 and 4.

The mean queueing delay at station (queue) i is approximated as

$$w_i = \frac{R(1-\rho_i)}{2(1-\rho)} + \frac{\lambda_i h_i^{(2)}}{2(1-\rho_i)} + \frac{1}{2(1-\rho)(1-\rho_i)} \sum_{\substack{k=1 \\ k\neq i}}^{N} \frac{\lambda_k h_k^{(2)}(1-\rho_i)^2 + \lambda_i h_i^{(2)} \rho_k^2}{1-\rho_i-\rho_k+2\rho_i\rho_k} \ . \quad (2.3)$$

The approximation was derived by making use of the exact pseudo-conservation law (2.1), the exact result for N = 2 queues, and a heuristic formula for the second moment of the cycle time distribution.

In the case of **nonexhaustive service**, a station receiving the token is allowed to transmit one frame and then has to pass the token along. This system is even more difficult to analyze than the exhaustive-service case. The model with two stations and nonzero token-passing overhead has been analyzed in |Eisenberg 79|, |Cohen 81|, and |Cohen 83|, the one with nonzero token-passing overhead in |Iisaku 81| and |Boxma 84|. A very useful exact and explicit result for the perfectly symmetrical case is available in |Nomura 78| and |Takagi 85|:

$$w_{sym} = \frac{R(1+\rho/N)}{2(1-\rho-\lambda R)} + \frac{\lambda h^{(2)}}{2(1-\rho-\lambda R)} \ . \quad (2.4)$$

Approximate analyses for an arbitrary number of stations and asymmetrical traffic are given in |Hashida 72b|, |Kuehn 79|, |Kurosawa 81|, |Arndt 84|, and |Boxma 86b|.

For the asymmetric case, Watson was able to derive a pseudo-conservation law, unfortunately a less straightforward relationship among the mean queueing times than in the exhaustive case |Watson 84|:

$$\sum_{i=1}^{N} \frac{\rho_i}{\rho}\left(1-\frac{\lambda_i R}{1-\rho}\right) w_i = \frac{R\left(\rho + \sum_{i=1}^{N} \rho_i^2\right)}{2\rho(1-\rho)} + \frac{\sum_{i=1}^{N} \lambda_i h_i^{(2)}}{2(1-\rho)} \ . \quad (2.5)$$

A practical value of this relationship lies in its possible use to construct approximations for the asymmetrical nonexhaustive-service model. In fact, Boxma and Meister |Boxma 86b| used an approach similar to |Bux 83a| to derive an approximate result for that model. Making use of (2.5) and employing a clever heuristic approach to estimate the second moment of the cycle time distribution, they derived the following approximation for the mean queueing delay at station i:

$$w_i = \frac{1-\rho+\rho_i}{1-\rho-\lambda_i R} \, \frac{1-\rho}{(1-\rho)\rho + \sum_{i=1}^{N} \rho_i^2} \left[\frac{\rho}{2(1-\rho)} \sum_{i=1}^{N} \lambda_i h_i^{(2)} + \frac{R}{2(1-\rho)} \sum_{i=1}^{N} \rho_i(1+\rho_i) \right] \ . \quad (2.6)$$

Through comparison with simulation, they showed an amazing accuracy of the approximation, although it turns out that finding an approximation with good accuracy covering a broad spectrum of parameters and traffic loads is more problematic than in the exhaustive-service case.

3. THE IEEE 802.5 MAC PROTOCOL AND ITS PERFORMANCE

MAC Protocol

The basics of token-ring operation have already been described in Section 2. The general format for transmitting information on a ring operating according to the IEEE 802.5 Standard is called a frame, see Fig. 5 |IEEE 84b|. The variable-length information field is preceded by a header, which includes an access-control field for controlling access to the transmission medium. Frames also include fields for frame status and ring management information, and contain the addresses of the source and destination nodes.

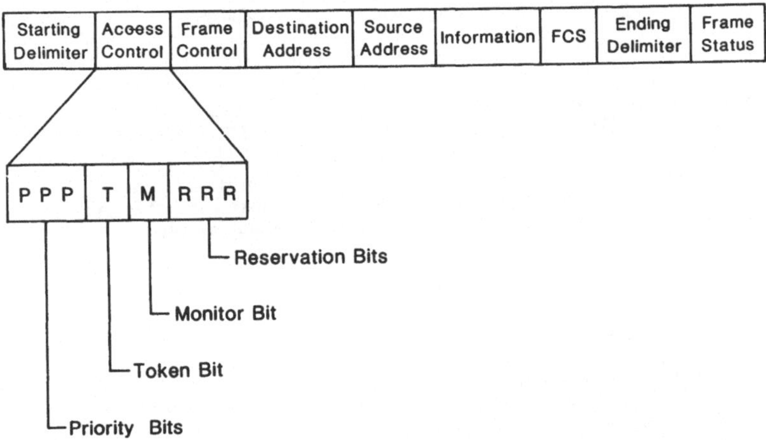

Fig. 5. IEEE 802.5 frame format.

The access-control field contains four fields: 1) The **priority bits** indicate the priority of a token and therefore which stations are allowed to use the token. 2) The **token bit** is a zero in a token and a one in a frame. When a station with a frame to transmit detects a token which has a priority equal to or less than the frame to be transmitted, it may change the token bit to one and transmit the frame. 3) The **monitor bit** is used to prevent a token whose priority is greater than zero or any frame from continuously circulating on the ring. 4) The **reservation bits** allow stations with high priority frames to request (in frames or tokens) that the next token be issued at the requested priority. A description of the priority scheme follows which is based on the principle described in |Bux 81a|.

If a station has a priority packet to send and no token is available for the station, it will make a reservation for a priority token in any passing frame, see Phase 1 of Fig. 6. When the transmitting station receives its frame with a priority reservation, the original token priority (called "normal" in the sequel) is suspended and a new token is issued at the priority requested, see Phase 2 in Fig. 6. This station must retain information on the level of the token suspended, in order to regenerate the suspended token after the priority token has circulated the ring. This is accomplished by writing the necessary information into a stack. When the time comes to transmit a token, each transmitting station will transmit one of the same priority it received (assuming no further reservations have been made). Thus, the high-priority token will progress around the ring, and the station requesting the priority will receive it, transmit its frame (Phase 3 in Fig. 6) followed by a priority token (Phase 4). When the station that originated the priority token receives it, indicating that the priority token has progressed all the way around the ring, it replaces the priority token with the suspended token by changing the priority bits accordingly (Phase 5).

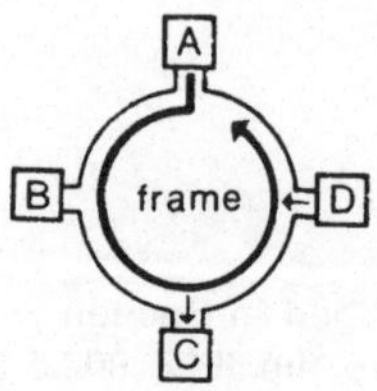

Phase 1: A transmits to C
D makes a reservation in the frame

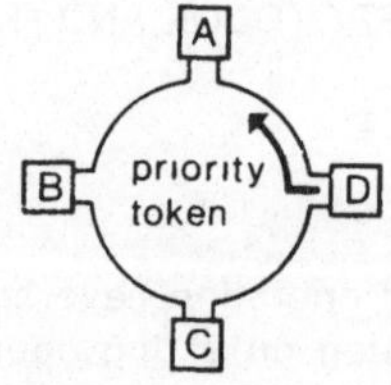

Phase 4 D issues a priority token

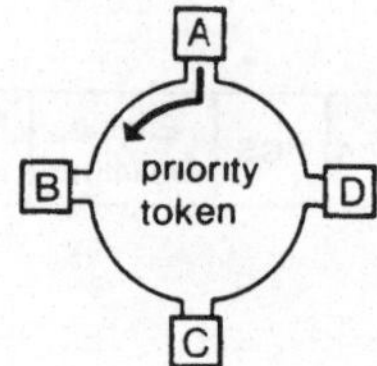

Phase 2 A issues a priority token and retains information on priority level

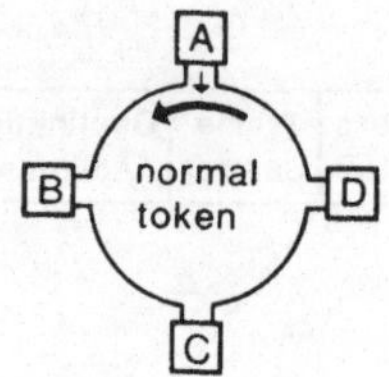

Phase 5. A detects the priority token and changes it to normal priority

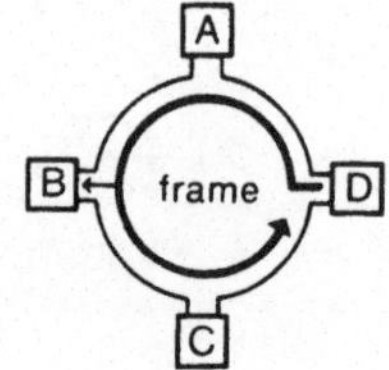

Phase 3: D uses priority token to transmit a frame to B

Fig. 6. IEEE 802.5 priority operation.

The station that suspended the token must be the station to restore it in order to maintain fairness within each priority level. Otherwise, some station may have multiple opportunities to use a token of a given priority before other stations have any. Assume, for example, station D in Fig. 6 to be a heavy user of high-priority bandwidth, and assume further that the priority protocol would work in such a way that the **user** of the priority token would be responsible for lowering the priority again. In this case, station A, the next station downstream from D would have a much better chance of obtaining a normal priority token than the stations further downstream. By using the priority stacking mechanism described above, the IEEE 802.5 token-ring protocol provides for regular progression of the tokens of all priorities around the ring, and thus guarantees fairness. We shall explain later in Section 4 that the FDDI protocol does not guarantee fairness in this sense.

The priority scheme requires stations not to issue a new token before having received back the header of their transmitted frame. While improving the reliability of the MAC protocol (each transmitting station can check the proper functioning of the ring), this "single-token operation" |Bux 81a,b| leads to inefficiencies in case of high speeds and long distances, as will be discussed in the next subsection.

Performance Results

We start our discussion of the IEEE 802.5 token-ring performance by mentioning some fundamental results for the delay-throughput characteristic obtained through simulation and analysis (where applicable).

A fundamental performance characteristic of any LAN medium-access protocol is its sensitivity to transmission speed and distance. Figures 7 and 8 show how token rings perform for various speeds and distances. Figure 7 illustrates the mean frame transfer delay as a function of the information throughput for 4 Mbps rings with 1 and 5 km cable lengths. It is assumed that all 100 stations generate the same amount of traffic. A further assumption is that frames are generated according to Poisson processes. Stations follow the single-token rule described above. Only one frame per access opportunity can be transmitted. It can be seen that increasing the ring length from 1 to 5 km has virtually no impact on the delay-throughput characteristic.

Under the same assumptions, except for a transmission rate of 16 Mbps, Fig. 8 exhibits the same performance measures as the previous one. Increasing the cable length from 1 to 5 km leads to more noticeable differences here, primarily because of the single-token rule.

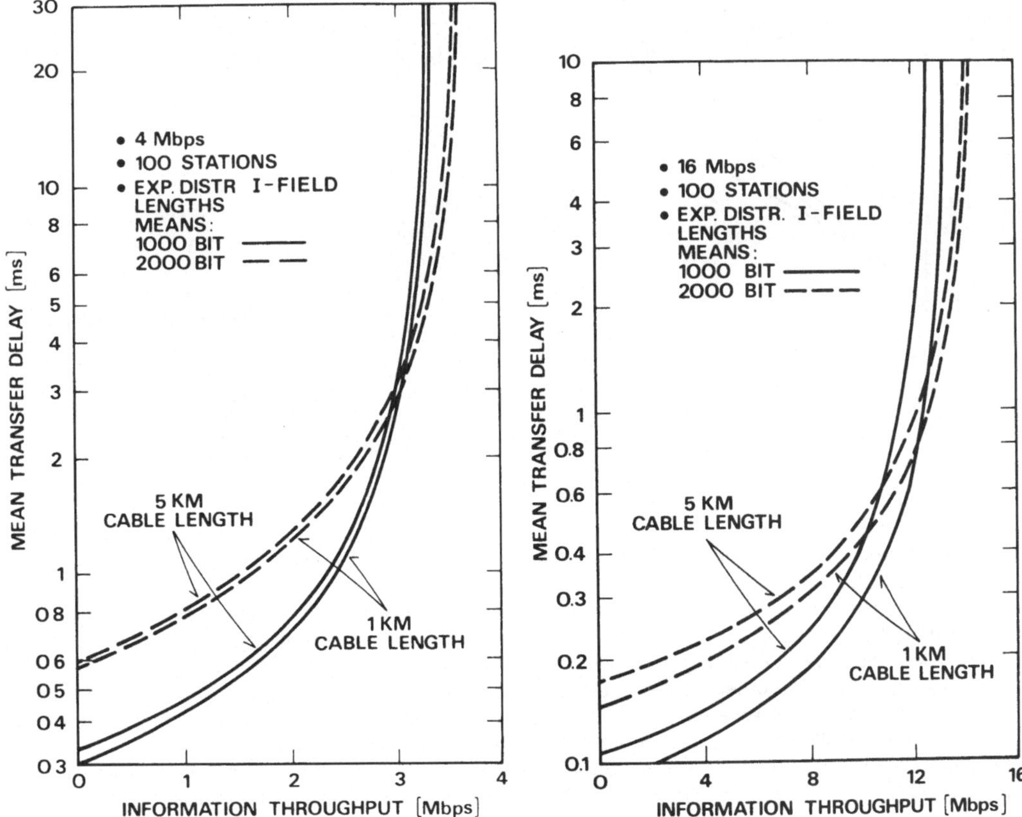

Fig. 7. Token ring delay-throughput characteristic. 4 Mbps transmission rate. Symmetrical traffic pattern. From |Bux 85a|. © by Springer-Verlag Berlin Heidelberg 1985.

Fig. 8. Token ring delay-throughput characteristic. 16 Mbps transmission rate. Symmetrical traffic pattern. From |Bux 85a|. © by Springer-Verlag Berlin Heidelberg 1985.

In these examples, the single-token rule had no or little impact on the ring performance. This changes when transmission speeds become higher and/or cable lengths longer. In Fig. 9, we show for rings with 100 stations and 5 km cable length, how the maximum throughput changes as a function of the ring transmission rate. Exponentially distributed information-field lengths with a mean of 1000 or 4000 bits have been assumed. We observe that for speeds up to roughly 20 Mbps, the single-token rule does not affect the ring efficiency even when the frames are relatively short. At higher speeds, stations spend a non-negligible fraction of their token-holding time waiting for the frame header to return. Hence, more and more bandwidth is lost as the transmission rate increases.

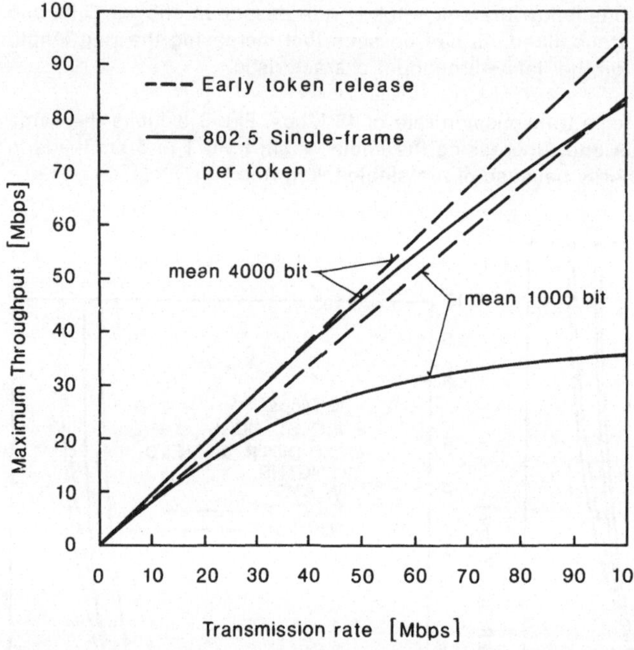

Fig. 9. Token-ring throughput. 5 km cable length. 100 stations. Exponentially distributed information field lengths.

There are two ways of overcoming this problem:

1) Permit stations to transmit more than one frame per token. Before issuing a new token, a transmitting station waits only until the header of its *first* packet has returned. This is sufficient for the priority-reservation scheme to work, and is the way the IEEE 802.5 protocol is specified |IEEE 84b|.
2) Release the token immediately after a frame has been completely transmitted. At first glance, this inhibits the use of the priority-reservation mechanism described above. It can be shown, however, that by modifiying the protocol suitably, the priority reservation can be made workable. Still another possibility of releasing the token early is not to use the priority-reservation mechanism but to use a timed-token protocol. The latter is the approach taken by the FDDI Standards Committee |ANSI 85a|, see also Section 4.

For comparison purposes, we have included the efficiency results of the early token release protocol in Fig. 9. As expected, this protocol is not sensitive to the speed/distance product.

Overall efficiency of an access protocol is the most basic performance property; a further important criterion is the quality of service given to individual stations, especially in the case of unbalanced traffic situations. This service can differ significantly, depending on the rule defining the time a station is allowed to transmit per access opportunity. As mentioned above, the IEEE 802.5 Standard specifies the use of a token-holding timer which limits the time a station is allowed to transmit continuously. To demonstrate the impact of this timer, we subsequently consider two extreme cases, a very short timer, such that stations can only transmit one frame per token (Fig. 10), and a very long timer, such that stations can always completely empty their transmit queues at each transmission opportunity (Fig. 11).

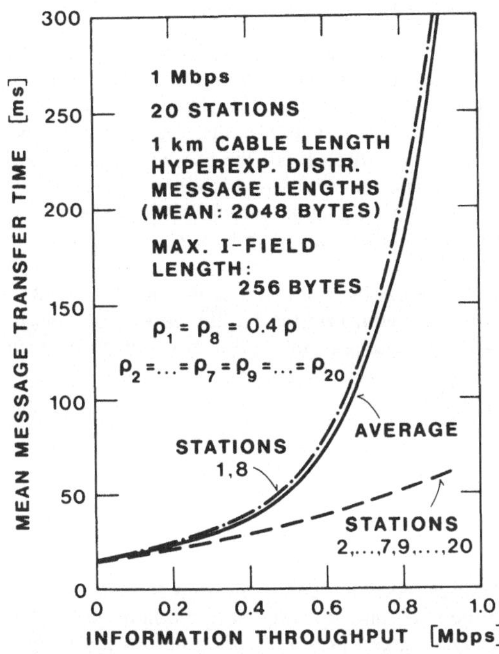

Fig. 10. IEEE 802.5 delay-throughput characteristic. 1 Mbps transmission rate. Asymmetrical traffic pattern. Short token-holding time-out. From |Bux 85a|. © by Springer-Verlag Berlin Heidelberg 1985.

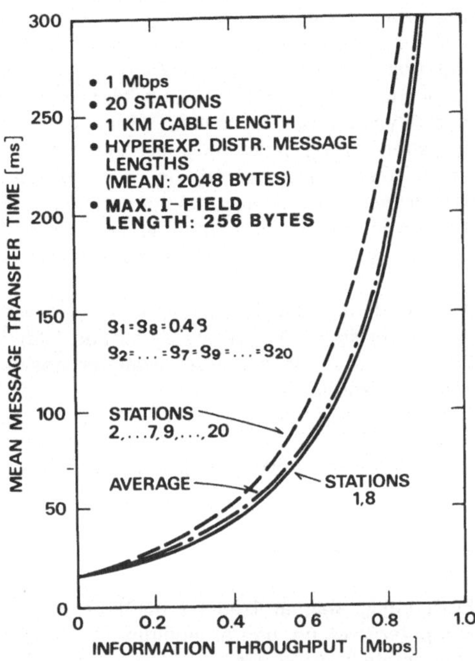

Fig. 11. IEEE 802.5 delay-throughput characteristic. 1 Mbps transmission rate. Asymmetrical traffic pattern. Long token-holding time-out. From |Bux 85a|. © by Springer-Verlag Berlin Heidelberg 1985.

For both examples, Poisson arrival processes have been assumed. However, the arriving data units are not single frames but entire messages, the lengths of which are distributed according to a hyperexponential distribution with a coefficient of variation equal to 2. In cases where a message is longer than the maximum information-field length of a frame (256 bytes), the message is segmented. In both examples, the traffic pattern assumed is very unbalanced: two of the 20 stations (Nos. 1 and 8) each generate 40% of the total traffic; each of the other 18 stations generates only 1.1% of the total traffic.

For the single-frame-per-token operation, Fig. 10 shows the mean transfer delay of the messages (not frames!) as a function of the total information throughput. Of course, the delay averaged over all stations increases with increasing ring throughput. The same is true for the delay of the messages transmitted by the heavy-traffic stations 1 and 8. On the other hand, the delay experienced by the light-traffic stations remains rather small even for very high utilizations. In this sense, the token-passing protocol combined with a single-frame-per-token operation, provides fair access to all users.

From Fig. 11, it can be seen that the relationship of the delay experienced by light and heavy users is reversed when the token-holding time is long. Here, the mean message-transfer delay of light-traffic stations is even higher than the one of heavy-traffic stations. This is due to the fact that messages generated at a heavy-traffic station have a relatively good chance of their station holding the token and, in this case, of becoming transmitted before frames waiting in other stations. These two examples demonstrate that the token-holding timer can be used to control the station-specific quality of service.

We wish to conclude this section by briefly commenting on two interesting applications of the token-ring priority mechanism.

The first is the integration of voice and data on a single token ring. Various techniques to accomplish this have been described in |Bux 81a| and |Bux 82|, all of them based on the concept of a "Synchronous Bandwidth Manager" responsible for switching the ring from asynchronous operation during which normal data frames are transmitted into synchronous operation during which voice is transmitted. In order to guarantee delay and throughput for the voice traffic, the Synchronous Bandwidth Manager makes use of the priority-reservation mechanism to interrupt the asynchronous data flow at regular intervals. Details and refinements of this technique together with a thorough and comprehensive performance analysis are given in |Zafiropulo 86|.

A second important application of priorities is in the area of multi-ring LANs interconnected through bridges. As will be described later in Section 5, giving bridges higher priority access to the rings than normal user stations has significant advantages both with respect to overall efficiency and fairness.

4. THE FDDI MAC PROTOCOL AND ITS PERFORMANCE

MAC Protocol

FDDI being standardized by ANSI provides a 100 Mbps communication system to interconnect computer and peripheral equipment using fiber optics as the transmission medium in a ring configuration |ANSI 85a,b,c,d|, |Ross 86|. FDDI employs a token protocol which allows the station transmitting to pass the token immediately after the end of frame transmission. As pointed out in Section 3, alternative token-passing strategies require the transmitting station to delay issuing a new token until some portion of the transmitted bits have returned.

The FDDI priority mechanism is designed to provide different classes of service that permit the network to support traffic simultaneously with a wide range of transmission requirements. The highest priority, "synchronous-access" class, guarantees bandwidth and response time, and is therefore suitable for applications such as voice or video.

All remaining traffic in FDDI belongs to the "asynchronous-access" class, which is further subdivided into restricted and nonrestricted subclasses. Both subclasses compete for access to the ring on an equal basis, however, once a token has been captured for use in transmitting traffic belonging to a restricted dialog (an extended exchange of frames between specific users), the bandwidth and response-time guarantees provided to synchronous traffic are extended to the particular asynchronous dialog. Applications that have periodic requirements for large amounts of bandwidth or guaranteed response time could be implemented using the asynchronous restricted-access class. For example, use of the restricted token mode would ensure meeting the response-time requirements of a data-transfer protocol between a channel and a device controller once the data transfer had been initiated.

The normal asynchronous-access class has no bandwidth or response-time guarantees. FDDI provides for up to eight different priority levels within this class in order to further distinguish levels of service. Therefore, interactive traffic could still receive better service (e.g., experience shorter average delays) than low-priority file transfers, within this class.

FDDI uses a so-called timed-token protocol in which the length of time the token may be held for transmitting frames of a given class of traffic (i.e., the token-holding time) depends in part on the time between successive arrivals of the token at the transmitting station (i.e., the token rotation time) |Grow 82|. We shall first describe the timers and variables used at each station to implement the protocol, and then outline the token-holding rules for each access class. For description purposes, in this paper we shall assume that all timers are initialized with zero, and expire when they have counted up to their target time.

As part of the ring initialization process, all stations negotiate a Target Token-Rotation Time (TTRT) which is the maximum average time required for the token to circulate completely around the ring (if the ring is not busy, then the token will circulate faster than this on average). The protocol guarantees that the maximum token rotation time will not exceed 2 * TTRT |Johnson 85|, |Sevcik 87|, so stations with strict response-time requirements must request a TTRT equal to one half of their required response time. At the end of the negotiation, the TTRT, which is equal to the shortest TTRT requested, becomes the operative TTRT, and is used to set the variable T_Opr identically in each station.

A Token-Rotation Timer (TRT) is used in each station to measure the time between successive arrivals of the token at that station. Normally, the TRT is reset each time the token has been received in order to time the next token rotation. The TRT will expire if it counts up to T_Opr before the token has arrived back at the station. When the TRT expires, Late_Ct, a counter which is initially zero, is incremented, TRT is reset to zero and continues timing. When the token arrives late at a station (Late_Ct = 1) the TRT is not reset, rather it is allowed to continue timing, thus accumulating the lateness of the current token rotation into the next token-rotation time. Late_Ct is reset to zero each time the token is received.

A Token-Holding Timer (THT) is used by each station to control the amount of time the token is held for transmitting asynchronous frames. The THT is loaded with the residual value of the TRT when the token is received on time (Late_Ct = 0) at a station. When the THT reaches the token-holding time threshold for a particular priority class, the token may no longer be used for transmitting frames of that class. T_Pri(i) (i = 1 to 8) defines the token-holding time threshold for asynchronous priority level i. The convention adopted is that the priority increases from 1 to 8. Numerically greater threshold values allow more time to elapse from the THT before the token must be passed, and therefore give the associated priority level a greater transmission window, and consequently higher priority relative to those priority levels with smaller token-holding time thresholds. The maximum threshold value for a priority level is T_Opr. The token-holding time threshold for restricted asynchronous traffic is T_Opr.

We now describe the token capture and holding rules for the synchronous and normal asynchronous-access classes. For a description of the asynchronous restricted class, the reader is referred to |ANSI 85a| or |Dykeman 87a,b|. Each station transmitting *synchronous* traffic must obtain an allocation of the synchronous bandwidth, expressed as a percentage of T_Opr, from the station management (SMT) entity. Each time the token is received by the station, it may transmit synchronous traffic for this allocated length of time. Since the average token-rotation time is less than or equal to T_Opr |Sevcik 87|, the synchronous-bandwidth allocation is actually a percentage of the total bandwidth of the ring. Owing to the guaranteed maximum token-rotation time of 2 * T_Opr, this class of traffic is guaranteed both response time and bandwidth.

A token may be captured for transmitting *normal* (i.e., nonrestricted) *asynchronous* traffic of priority i if the token is not restricted, and if the TRT has not expired (Late_Ct = 0) and is less than the token-holding time threshold for priority i [T_Pri(i)]. When the token has been successfully captured, the residual value of the TRT is loaded into the THT, and transmission of priority-i frames may continue until the THT has exceeded T_Pri(i). Transmissions already in progress when the THT expires are completed.

Performance Results

Given the rather short time since the FDDI MAC protocol was completely specified, it is not surprising that performance studies of its MAC protocol are rare. Some performance-related aspects of FDDI have been addressed in |Ulm 82|, |Johnson 85|, |Johnson 86|, |Sevcik 87|, and |Goyal 87|. We subsequently report on an investigation we recently completed, whose goal was to obtain a broader understanding of the performance characteristics of the FDDI MAC protocol |Dykeman 87a,b|.

We first examine the effects of ring latency (r_l), target token-rotation time (T_Opr), and the number of actively transmitting stations (N), on the maximum total throughput (γ_{max}) obtainable on an FDDI token ring. We assume that each actively transmitting station always has frames queued for transmission, and that any other stations connected to the ring are idle. In this analysis, we do not make a distinction between information and framing bits.

Our first result is an equation relating the ring characteristics to the maximum total throughput for an FDDI token ring with only one asynchronous priority level. The maximum total throughput is determined by examining frame-transmission scenarios for a ring with a small number of active stations. We assume that all transmissions beyond expiration of the token-holding timer, owing to frame transmissions in progress, are of equal length. Frame transmission times are assumed to be of constant length F. As shown in |Dykeman 87b|, the following expression for the maximum total throughput holds under the above assumptions:

$$\gamma_{max} = \frac{(N * tot\_tx\_time + N^2 * tx\_window) * v}{(N * tot\_tx\_time + N^2 * tx\_window + (N^2 + 2*N + 1) * r\_l)}, \tag{4.1}$$

where

v = the transmission rate

tot_tx_time = CEILING(tx_window / F) * F

tx_window = T_Opr - r_l.

If we take the limit of the above expression as the number of active stations N goes to infinity, we find that

$$\lim_{N \to \infty} \gamma_{max} = \frac{tx\_window * v}{tx\_window + r\_l} = \frac{T\_Opr - r\_l}{T\_Opr} * v \tag{4.2}$$

This limit is in agreement with the simple formula given in |Ulm 82| for the FDDI ring utilization.

The ring latency consists of propagation delay in the fiber, and the sum of the station latencies. We use a propagation delay of 5.085 microseconds per kilometer, and a station latency of 0.6 microseconds per station to obtain the results presented in this paper.

Figure 12 illustrates the effects of the target token-rotation time and the number of actively transmitting stations on the maximum total throughput, for a 50 kilometer ring with 30 connected stations and a frame length of 12660 bits. The results shown were obtained using Eq. (4.1), and verified using a simulation model (the differences in the results were always less than one percent). We also show results obtained using Ulm's formula |Ulm 82| which is an upper bound on the maximum throughput since it corresponds to the case of an infinite number of transmitting stations, see (4.2). These results indicate that Eq. (4.1) yields very good approximations for the maximum total throughput. Ulm's result is accurate when the number of stations is large, or when the target token-rotation time is large relative to the ring latency.

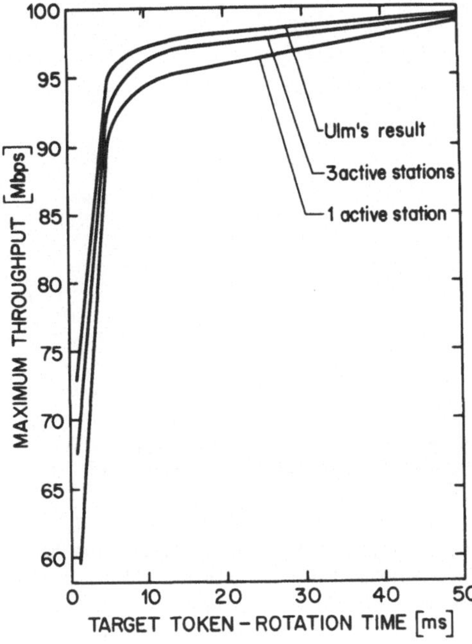

Fig. 12. FDDI maximum throughput versus target token-rotation time for n = 1,3 and Ulm's result (ring latency = 0.272 ms). From |Dykeman 87a|. © 1987 Information Gatekeepers, Inc., Boston, MA.

Increasing the target token-rotation time increases the maximum total throughput of the ring since there will be more frame transmissions on each token rotation. Therefore, the ratio of useful transmission time to token-rotation time (i.e., overhead) increases. To understand why increasing the number of active stations increases the maximum throughput, we shall consider a simplified transmission scenario in which there are no transmissions beyond the THT expiration. On the first token rotation in this system, assume that station 1 transmits frames for the maximum possible time (T_Opr - r_l). All other stations will pass the token since their TRT's will expire owing to the transmission by station 1. On the next token rotation, station 1 will pass the token since its TRT timed the previous transmission. Therefore, on the second token rotation, only station 2 will transmit (again for the maximum possible time) and so on. Finally, on the token rotation following transmission by station N (the final active station in our ordering), no stations will be allowed to transmit. The cycle will then repeat, beginning with station 1 on the next token rotation. Thus, by increasing the number of active stations, we increase the ratio of token rotations in which transmissions occur (N), to the total number of token rotations (N+1). Although real transmission scenarios are more complicated than this example, the same phenomenon can be observed.

Figure 13 illustrates the effects of ring latency (r_l) on the total maximum throughput. These results were obtained using Eq. (4.1). We consider a system with ten active stations, and a frame size of 12660 bits. The ring latency is varied from 0.011 millisec, representing a ring with ten connected stations and a 1 kilometer fiber, to 1.62 millisec, which represents a ring with 1000 connected stations and a 200 kilometer fiber. We can observe that with a target token-rotation time of 10 millisec, the maximum total throughput remains high, in this case over 82 Mbps, even when the ring latency is very large (1000 stations and 200 kilometer fiber length are FDDI maximum values). When the target token-rotation time is reduced to 5 millisec, the maximum total throughput corresponding to the largest ring latency drops to 65 Mbps.

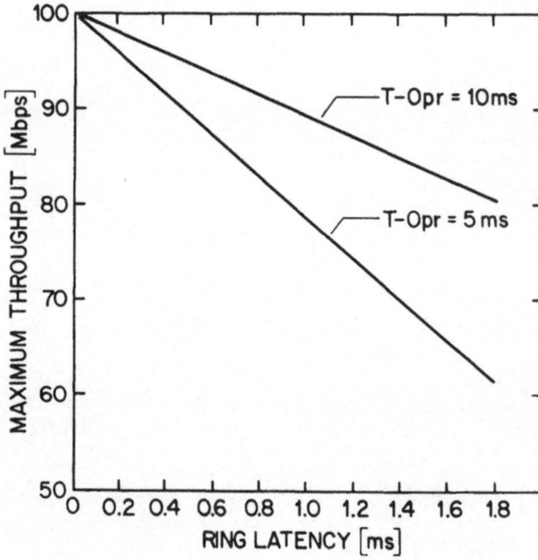

Fig. 13. FDDI maximum throughput versus ring latency for T_Opr = 5 and 10 ms (10 active stations). From [Dykeman 87a]. © 1987 Information Gatekeepers, Inc., Boston, MA.

We can see that the difference between the ring latency and the target token-rotation time has a significant effect on the maximum total throughput. As a rule of thumb, T_Opr should be at least five times the ring latency in order to maintain throughput levels above 80 Mbps. If T_Opr is less than the ring latency, no frame transmissions can occur since all token-rotation timers will always expire owing to the ring latency experienced by the circulating token.

The frame size used by FDDI stations to transmit information will also affect the maximum total throughput. The frame size determines the frame-transmission time, which in turn determines the length of transmissions beyond the expiration of the THT. This time is accounted for in the tot_tx_time expression in Eq. (4.1). Because tot_tx_time is the coefficient of a low-order term, its impact is only significant when the number of transmitting stations is small. Also, the effects of transmissions beyond THT expiration are only significant when a small number of frames is transmitted each time the token is held by a station [i.e., when (T_Opr - r_l)/F is small]. Therefore, under most circumstances the frame size does not significantly affect the maximum total throughput of an FDDI token ring. It should be noted, however, that this would no longer be true if stations were not able to transmit multiple frames per token.

We next address the performance characteristics of the FDDI priority mechanism. Our first result obtained from a detailed simulation model is for an FDDI ring in which there are eight stations, each attempting to transmit frames at one of the eight asynchronous priority levels. The asynchronous-restricted, and synchronous ring access classes are not used in this example. The arrival rate of frames to be transmitted is identical at each station. In total, there are 11 stations connected to the ring, three of which are idle. The target token-rotation time is 0.1 sec, and the eight token-holding time thresholds are, from highest to lowest priority, 0.1, 0.0765, 0.0562, 0.039, 0.025, 0.014, 0.0062, and 0.0015 sec. The total ring latency is 0.0010236 sec, and the total frame length, including MAC framing bits, is 12660 bits.

In Fig. 14, we show the throughput plotted against the arrival rate for each of the eight priority levels, along with the overall ring throughput. At low arrival rates, all classes of traffic receive some bandwidth since all token-holding time thresholds are greater than the total ring latency. However, as the arrival rate of frames increases, the throughput of lower-priority traffic begins to decrease, and is eventually reduced to zero. This characteristic of the asynchronous priority class is potentially very useful. Low-priority traffic (e.g., large file transfers) can be assigned to a priority level for which the throughput is reduced to zero when the total offered load exceeds a certain value. In this way, the low-priority traffic is automatically deferred during busy periods,

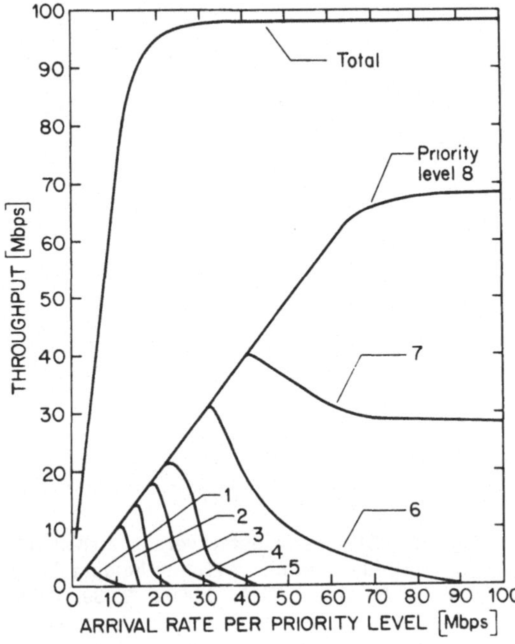

Fig. 14. FDDI throughput versus arrival rate per priority level. (Token-holding time thresholds: 0.1, 0.0765, 0.0562, 0.039, 0.025, 0.014, 0.0062, and 0.0015 sec; ring latency: 0.0010236 sec). From |Dykeman 87a|. © 1987 Information Gatekeepers, Inc., Boston, MA.

until the load on the ring has been reduced. Different levels of low-priority traffic can be defined to be deferred at different loads. Also, multiple levels of higher-priority traffic can be defined to give preferred service to some traffic while not completely shutting off the competing traffic, even when the load offered to the ring exceeds its capacity.

Figure 15 shows the mean delay plotted against the arrival rate for frames of each priority level in our first example. The mean delay of frames in a priority class increases rapidly as the arrival rate approaches the maximum throughput of that class. The delay curve for priority-8 frames levels off in the region where the arrival rate per priority level is between 30 and 60 Mbps. This is due to the fact that in this range the total ring traffic load no longer increases, but the level-8 arrival rate is smaller than level-8 maximum throughput. In other words, level-8 traffic benefits from the decreasing throughput of the lower-priority traffic.

The final results presented in this section address the issue of fairness in the FDDI protocol. If a protocol is completely fair, the transmission delay experienced by a frame will depend on its priority not on the location of the transmitting station.

We consider a system in which ten stations are connected to a 10 kilometer ring. One of these stations transmits priority-8 frames, while the other nine stations transmit priority-7 frames. We shall assume that the arrival rates at each of the stations transmitting priority-7 frames are equal. The "first priority-7 station" refers to the first station transmitting priority-7 frames, sequentially following the station transmitting priority-8 frames on the ring, and so on.

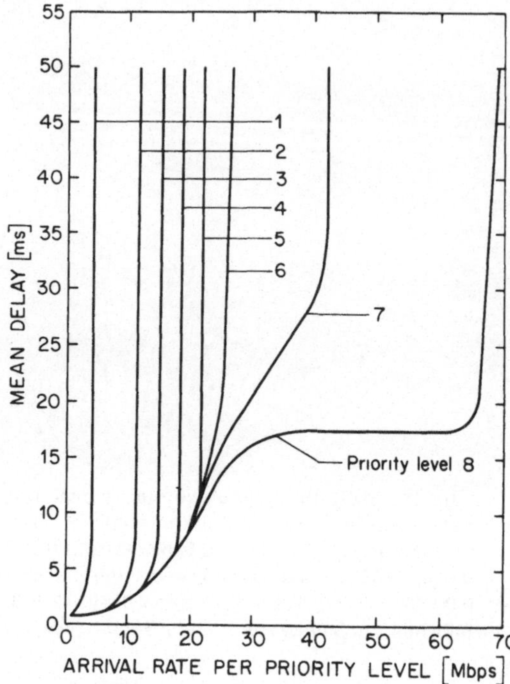

Fig. 15. FDDI mean delay versus arrival rate per priority level. (Token-holding time thresholds: 0.1, 0.0765, 0.0562, 0.039, 0.025, 0.014, 0.0062, and 0.0015 sec; ring latency: 0.0010236 sec). From [Dykeman 87a]. © 1987 Information Gatekeepers, Inc., Boston, MA.

In Fig. 16, we show the throughput for the first and last priority-7 stations, priority-8 station, and for the overall ring, as the arrival rate of high-priority traffic is varied from 12 to 100 Mbps. The arrival rate of priority-7 traffic at each station is fixed at 6.33 Mbps. The token-holding time thresholds for priorities 8 and 7 are 0.01 and 0.002 sec, respectively. In Fig. 16, we see that the throughput of low-priority frames is eventually reduced to zero, but that the low-priority station furthest from the high-priority station sequentially on the ring is cut off first, and the last station to be cut off is the first priority-7 station. When we examine the mean delay curves presented in Fig. 17, we can also observe this unfair behavior. The frames arriving at the last priority-7 station experience higher mean delays than frames at the first priority-7 station when the total throughput exceeds 80 Mbps.

Each time the token is received, a station transmitting priority-i frames normally transmits for a time period equal to T_Pri(i) - T_Pri(low), where low is the lowest active priority level. However, when a high-priority station does not have a full queue of frames to transmit, the unused portion of its bandwidth will be available for use by the lower-priority stations on the ring. This unused bandwidth will circulate sequentially among the stations transmitting lower-priority traffic, beginning with the first low-priority station following the high-priority station, until it is "reclaimed" for use by the high-priority station. Each time the high-priority bandwidth is reclaimed and re-relinquished, it begins circulating with the first station following the high-priority station. This explains the behavior observed in Figs. 16 and 17. When the load offered is low, there will frequently be unused bandwidth, but it will be able to circulate among all low-priority stations, and therefore no unfairness will result. But as the load offered increases, the high-priority station will relinquish portions of its bandwidth for shorter durations, thereby favoring the stations immediately following it on the ring.

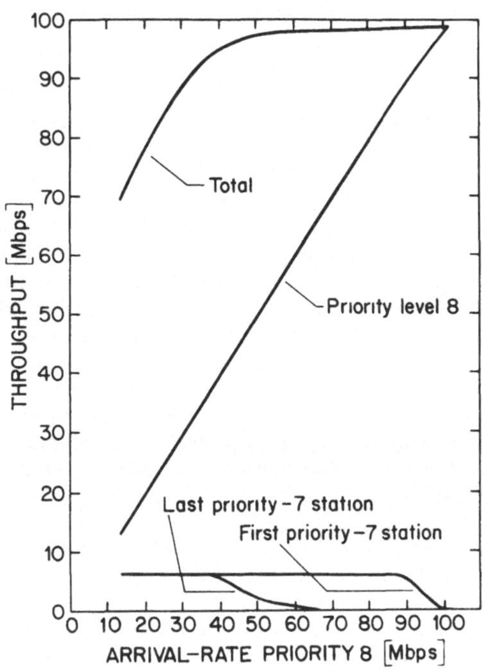

Fig. 16. FDDI throughput versus arrival rate of priority-8 traffic. (One priority-8 station; nine priority-7 stations with a constant arrival rate of 6.33 Mbps; T_Pri(8) = 0.01 sec; T_Pri(7) = 0.002 sec). From [Dykeman 87a]. © 1987 Information Gatekeepers, Inc., Boston, MA.

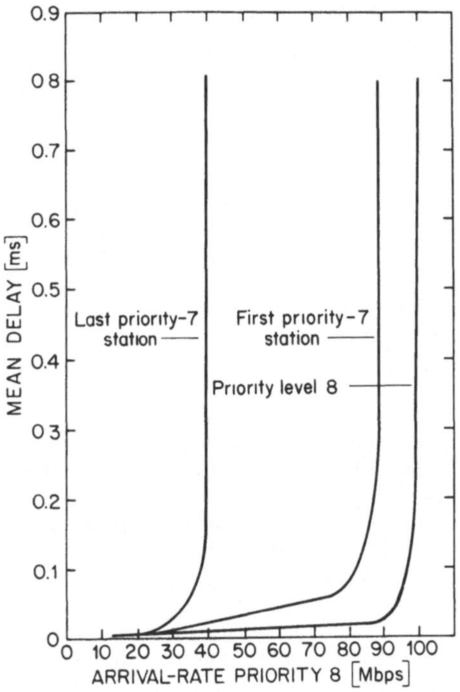

Fig. 17. FDDI mean delay versus arrival rate of priority-8 traffic. (One priority-8 station; nine priority-7 stations with a constant arrival rate of 6.33 Mbps; I_Pri(8) = 0.01 sec; T_Pri(7) = 0.002 sec). From [Dykeman 87a]. © 1987 Information Gatekeepers, Inc., Boston, MA.

We conclude this section by briefly highlighting two interesting effects of the FDDI MAC protocol we have observed:

1) There exists a tradeoff between the efficiency of the FDDI protocol and the effectiveness of its priority mechanisms. We have seen that the maximum total throughput remains high as long as the target token-rotation time is large with respect to ring latency. However, if even one application on the ring has a very short response-time requirement the total throughput may be severely limited, even if that application is rarely active, since the target token-rotation time must be set to one half the minimum response time required. To take advantage of the full effectiveness of the FDDI protocol (e.g., to obtain guaranteed response time), efficiency may have to be sacrificed.
2) A fairness problem may exist within the asynchronous-access class when more than one priority level is used. Stations transmitting low-priority frames following stations transmitting high-priority frames sequentially on the ring have an unfair advantage over other low-priority stations. The problem would become serious if under certain circumstances some stations transmitting frames of a given priority level did not receive any bandwidth, while others were able to continue transmitting. In our experiments, this fairness problem only became evident when the total ring utilization was high (above 80%).

5. INTERCONNECTED TOKEN RINGS

Interconnection Architecture

Local-area networks must be capable of interconnecting a large number of stations over maximum distances of several kilometers. Whenever the limitations of a single ring or bus subnetwork are reached with respect to the maximum number of attachments or maximum distance, means to interconnect subnetworks become necessary. In the system considered here, the functions needed for subnetwork interconnection are provided in specific nodes called bridges. These functions need to be simple in order to achieve the high throughput values required for the interconnection of high-speed subnetworks at reasonable costs. Simplicity of bridges can be achieved through a connectionless interworking of the subnetworks. This implies that bridges do not perform any complex flow or error control; such functions are only provided end-to-end, i.e., between the communicating stations. In case of congestion, bridges simply discard frames they cannot handle momentarily. Discarded frames will be recovered through the end-to-end protocol. Some potential congestion scenarios follow: (i) Traffic from multiple subnetworks is flowing to a single subnetwork; (ii) traffic peaks caused by coinciding demands; (iii) a portion of a network is out of service and traffic is being re-routed through the operational part of the network, or (iv) interconnection of subnetworks running at different speeds.

When network congestion occurs, recovery of frames discarded by the bridges becomes necessary. A potential exposure of the architecture described is that retransmission of lost frames further increases congestion and hereby network performance may be severely degraded. Therefore, a detailed understanding of the network performance characteristics under both normal traffic load and overload conditions is of prime importance. We subsequently report on an investigation in which a hierarchy of interconnected token rings was studied |Bux 85b|. Rings operate as specified in the IEEE 802.5 Standard; as end-to-end protocol, Type 2 of the IEEE 802.2 Logical Link Control Standard is employed |IEEE 84a,b|.

The network structure underlying our study is depicted in Fig. 18. Several token rings are interconnected through bridges and a "backbone" token ring. User stations are only attached to the local rings but not to the backbone; the latter serves to interconnect bridges. This two-level hierarchy provides a high degree of flexibility in wiring a building or a group of buildings.

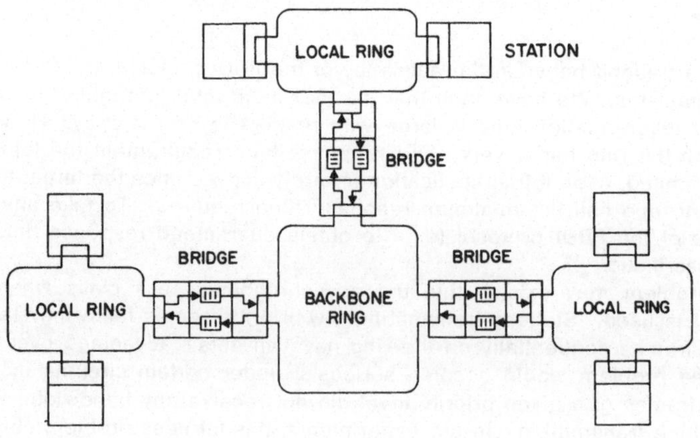

Fig. 18. Multiring network. From |Bux 85b|. © 1985 IEEE.

As described in Section 3, it is feasible to implement access priorities on token rings. In a network of interconnected token rings, such priorities could be used for bridges to reduce their buffer requirements. Stations or bridges may be allowed to transmit more than a single frame per token, namely, up to a maximum number of frames or for a maximum "token-holding time". This possibility may also be used to give better service to bridges as compared to normal ring stations.

We have studied two different kinds of ring operation. The so-called "nonpriority mode" assumes that all stations and the bridges operate at the same ring-access priority level, and are only allowed to transmit a single frame per token. In "priority mode", bridges operate at a higher ring-access priority level than normal stations. In priority mode, bridges are allowed to transmit continuously until their transmit buffer has been emptied. On the backbone, priority access is not employed and each bridge transmits only one frame per token.

Bridges provide a basic routing and store-and-forward function. Based on the destination addresses of frames received on its local ring, the bridge filters out those frames destined to stations on a different ring. Similarly, it also copies all frames from the backbone ring destined to stations attached to the local ring of the bridge. Frames are stored in RAM until they can be transmitted on the local or backbone ring. We assume that the total memory space is partitioned into two separate buffer pools, one for each direction (local ring to backbone; backbone to local ring). The buffer pools are structured in segments of a fixed size. Received frames which do not find a sufficient number of free segments upon their arrival at a bridge are lost. The bridge does not take any action when such a frame loss occurs; responsibility for detecting and recovering from lost frames lies entirely with the attaching stations.

End-to-End Protocol

We consider a system with a connection-oriented end-to-end protocol between the communicating user systems, i.e., a protocol providing connection establishment and termination, frame sequencing, error detection and recovery functions, and means for flow control. Protocols providing the functionality needed for flow- and error-control in LAN's are, for example, Class 4 of the ISO/ECMA Transport Protocol or the IEEE 802.2 Type 2 Logical Link Control Protocol. Depending on this choice, end-to-end flow and error control is performed in layers corresponding to either layer 4 or layer 2 of the OSI reference model. The protocol under consideration in this paper is the IEEE 802.2 Type 2 Logical Link Control Protocol.

Data units provided by a user of the layer-2 services are transmitted in Information (I-) frames. The logical-link control sublayer makes use of so-called Supervisory (S-) frames for control functions, such as connection establishment and termination, positive and negative acknowledgments, flow-control information, etc.

We next give a brief outline of the functions of this protocol. For a detailed specification, the reader is referred to |IEEE 84a|.

Flow Control: Flow control is realized by a window mechanism, i.e., a sender is permitted to transmit up to W (the window size) I-frames without having to wait for an acknowledgment. The receiver uses Receive Ready (RR-) frames to acknowledge I-frames correctly received and to indicate to the sender that more I-frames can be transmitted.

Error Recovery: Any I-frame received with an incorrect Frame Check Sequence (FCS) is discarded. When a received I-frame has a correct FCS, but its send sequence number is not equal to the one expected by the receiver, the receiver returns a Reject (REJ-) frame. It then discards all I-frames until the expected I-frame has been correctly received. The sender, upon receiving a REJ-frame, retransmits I-frames starting with the sequence number received within the REJ-frame.

In addition to REJECT recovery, a time-out mechanism is used. At the instant of transmission of an I-frame, a timer will be started if it is not already running. When the sender receives an RR-frame, it restarts the timer if there are still unacknowledged I-frames outstanding. When the timer expires, the station performs a "checkpointing" function by transmitting an RR-frame with a dedicated bit (the "P-bit") set to one. The receiver, upon receiving this frame, must return an RR-frame with the "F-bit" set to one. When this RR-frame is received by the sender, it either proceeds with transmitting new I-frames or retransmits previous I-frames depending on the sequence number contained in the received RR-frame.

Performance Results

A model of the system described above has to reflect the details of the ring-access mechanism and the bridge functions. Moreover, since frames lost by bridges are recovered through the end-to-end LLC protocol, this also needs to be modeled. Because a model of this complexity is far beyond the type of queueing models which can be analytically treated today, we employed simulation.

Although the simulation model allows us to measure performance attributes at any desired detail, we shall restrict the discussion to two basic performance measures, viz., throughput and end-to-end-delay. Throughput of a connection is defined as the number of bits received at both ends across the higher-layer interface per unit time. In the following examples, we shall show the total throughput of all connections in the network. End-to-end delay is defined as the time elapsed between transferring a data unit across the higher-layer interface at the source node until transferring it across this interface at the sink node. For the subsequent discussion, we need to specify a further quantity called "offered data rate". This is defined as the number of bits generated by an application for transmission per unit time under the condition that the application is not halted because of backpressure.

The transmission rates investigated are 4 Mbps for the local rings and 16 Mbps for the backbone. Bridges are assumed to need 300 μsec to process one frame. The size of each of the two buffer pools is assumed to be 4 kbyte, unless otherwise stated. The maximum I-field length is 0.5 kbyte; the framing overhead and the length of S-frames is 24 bytes, as specified in |IEEE 84a,b|. The time intervals between generation of messages are assumed to be exponentially distributed. The mean message length is 1 kbyte; the coefficient of variation 1.5. Because of message segmentation and supervisory frames, the resulting overall frame-length distribution observed in simulation resembles the bimodal distribution reported in |Shoch 80| with a mean of about 250 bytes.

Figure 19 shows the total throughput (in Mbps) as a function of the total offered data rate (also in Mbps). The traffic pattern assumed is completely symmetrical: Each of the 12 stations attached to a ring generates the same amount of traffic and has a logical link set up to a station on a different ring. I-frame transmission on each logical link is two-way. Bridges operate in nonpriority mode.

We observe that when the offered data rate is increased from zero, throughput initially follows linearly. When the 4 Mbps rings become noticeably loaded, queues of frames waiting to enter them build up in the bridges and eventually overflow of the bridge buffers occurs. The overflow frequency increases with the transmission activity of the stations which, in turn, depends on the LLC window size W. With large window sizes, frame losses occur at lower offered data rates than with small window sizes. Loss of an I-frame leads to the retransmission of one or more I-frames depending on the number of I-frames a station has outstanding when it receives a REJ or has performed checkpointing. Obviously, the window size sets an upper limit to the number of frames to be retransmitted per lost frame. Therefore, the additional traffic created by retransmissions decreases with smaller window sizes. This explains the significant differences between the throughput values pertaining to different window sizes at high offered data rates. For large window sizes, throughput shows a pronounced maximum. When the offered data rate is increased beyond a certain critical value, throughput drops the more rapidly the bigger the window size and approaches an asymptotic value.

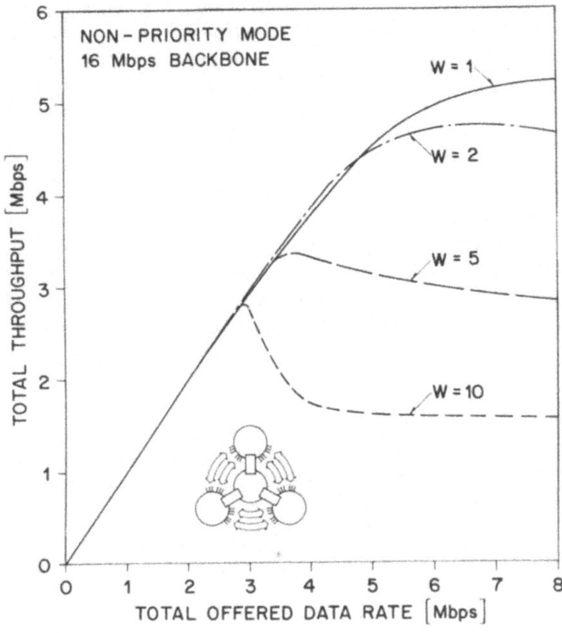

Fig. 19. Total throughput versus total offered data rate for different window sizes W. Non-priority mode. Symmetrical traffic. From |Bux 85b|. © 1985 IEEE.

In Fig. 20, we consider the same scenario, however, bridges employ priority access to the local rings. We observe substantial improvement. In fact, for all the window sizes investigated, no or negligibly few frame losses were observed. The explanation of this rather striking effect is as follows. Priority access for bridges has two basic consequences. First, it shortens the queues of frames waiting to enter the destination rings and hence contributes to a reduction of buffer

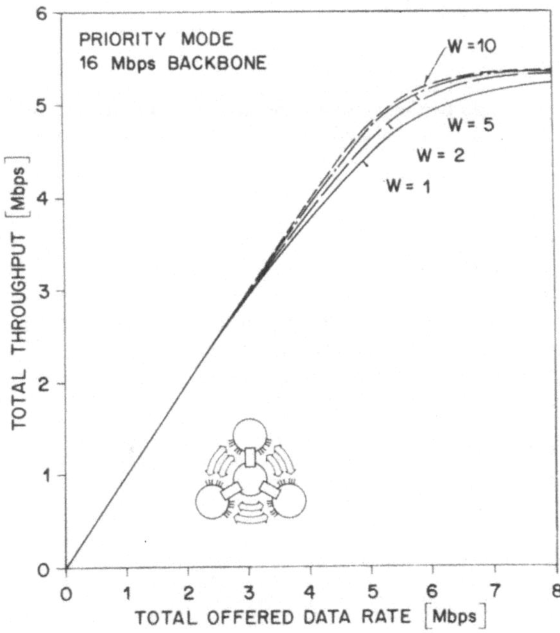

Fig. 20. Total throughput versus total offered data rate for different window sizes W. Priority mode. Symmetrical traffic. From |Bux 85b|. © 1985 IEEE.

overflows. Second, access of stations to their local ring is delayed which slows down both the injection of new I-frames into the network and the returning of acknowledgments. Delayed acknowledgments, on the other hand, further throttle the transmission of I-frames because of the LLC window flow-control mechanism. The overall effect is similar to flow-control schemes suggested for wide-area networks in which traffic entering the network is handled with lowest priority |Gerla 80|, |Giessler 81|, |Lam 79|, |Schwartz 87|.

As will become clear from the next example (Fig. 21), the effectiveness of the priority-mode operation is due to a great extent to the symmetry of the rather unbalanced traffic pattern assumed in this case: Six stations attached to one ring and the same number of stations attached to a second ring transmit I-frames to twelve stations on the third ring. The throughput characteristic observed indicates congestion in the bridge connecting the backbone with the third ring when the offered data rate exceeds a certain value. Again, a small window size is the best choice for this traffic pattern.

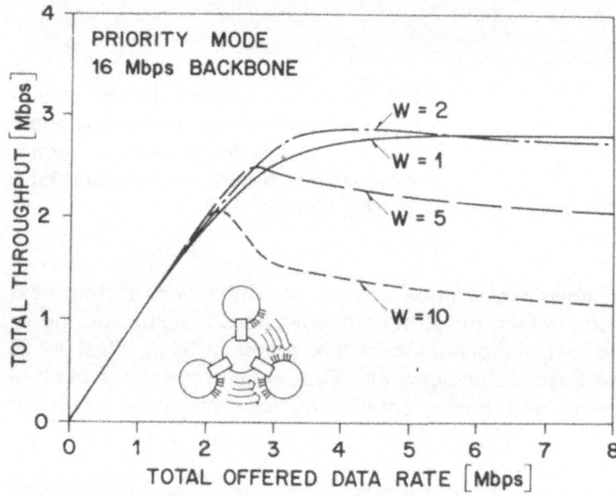

Fig. 21. Total throughput versus total offered data rate for different window sizes W. Priority mode. Asymmetrical traffic. From |Bux 85b|. © 1985 IEEE.

As shown in |Bux 85b|, priority access for bridges has a major advantage with respect to fairness: It avoids the very pronounced differences in the quality of service among inter-ring and intra-ring traffic that exist when no priorities are used.

All the results shown so far have indicated that a small window size is an effective means of minimizing frame loss and congestion owing to retransmissions. However, it is important to understand that a small window size can also be a disadvantage under certain conditions. In Fig. 22, we show the throughput characteristic of a network in which two stations on each ring communicate with two stations attached to a different ring. Bridges employ priority access to the rings. We observe that the total throughput can be substantially improved by increasing the window size. This indicates that at small window sizes, stations are not able to make full use of the available bandwidth, because acknowledgments do not return sufficiently fast.

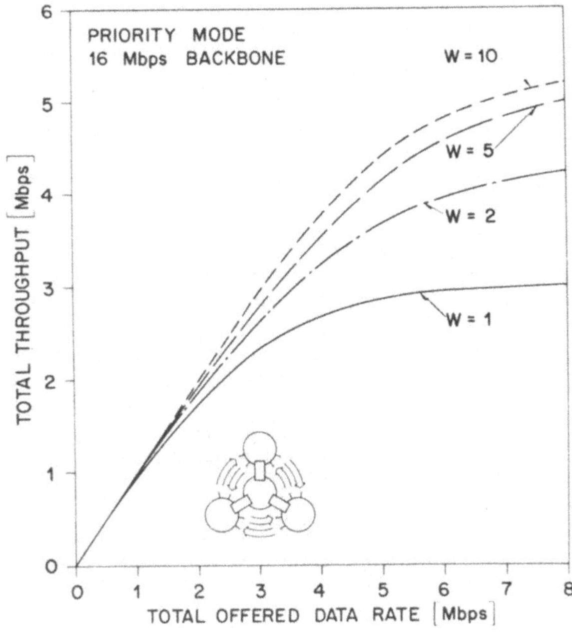

Fig. 22. Total throughput versus total offered data rate for different window sizes W. Priority mode. Symmetrical traffic. From |Bux 85b|. © 1985 IEEE.

Dynamic Flow Control

The results shown so far demonstrate that recovery of frames lost in congested bridges further increases congestion, and hereby network performance can be severely degraded. We have devised a dynamic flow-control mechanism whose goal it is to guarantee robust and efficient network operation under both normal traffic load and overload. Our approach is to enhance the IEEE 802.2 Type-2 Protocol by including the following dynamic flow-control algorithm. Initially, stations use the window size as defined during the set-up of the logical link. Whenever a station needs to retransmit an I-frame (either because of a received REJECT frame or after checkpointing) it sets its window size to one. Afterwards, the window size is increased by one (up to the initial value) for every n-th successfully transmitted (i.e., acknowledged) I-frame.

The rationale behind this algorithm is as follows. Under normal conditions, i.e., no congestion, the actual window size used is the one negotiated between the communicating partners. By setting the window size to one, whenever there is an indication of a possible congestion, we achieve a high responsiveness of the flow-control mechanism in the sense that an immediate and very effective throttling of the network input traffic is performed. Subsequent to reduction, stations again attempt to increase their window sizes. This process is tightly coupled to the reception of acknowledgments. Hereby, we achieve control of the speed by which the window size is increased, by the momentary ability of the network to transport frames successfully.

We next show how the multiring network performs when this enhanced LLC protocol is employed. We first address the question of selecting an appropriate value for the parameter n in the dynamic window-size algorithm. The value of n specifies how many I-frames a station needs to transmit successfully (following a window-size reduction) before it increases its window size by one.

For the scenario previously studied in Fig. 21, in Fig. 23 we show the total network throughput as a function of the total offered data rate for different values of n. The initial window size is ten. We observe substantial improvement in throughput when the window size is dynamically adjusted.

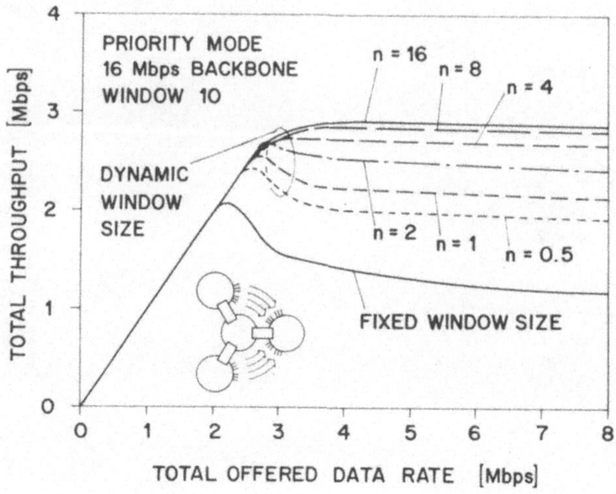

Fig. 23. Total throughput versus total offered data rate for dynamic window-size LLC with initial window size ten and different values of n. Priority mode. (Note: n = 0.5 means that the window size is increased by two per acknowledgment.) From |Bux 85b|. © 1985 IEEE.

The performance gain depends heavily on the choice of parameter n defined above; the larger n the higher the throughput. The incremental gain in throughput decreases, however, as n increases. For example, the throughput difference between the cases n = 8 and n = 16 is rather small. For small values of n, e.g., one or two, the algorithm works too dynamically in the sense that the window is apparently opened very rapidly and hence the throttling effect does not last long enough.

A general observation from numerous simulations has been that, even under extreme overload, the frame-loss frequencies in bridges is never substantially greater than one percent, when the dynamic window-size algorithm with $n \geq 8$ is employed. Under this condition, only a very small fraction of the bandwidth is lost for retransmissions, and hence performance is almost ideal.

Finally, in Fig. 24 we compare the throughput characteristics of the dynamic window size LLC and the fixed window size LLC for three different bridge-buffer sizes. We observe that for larger fixed window sizes, the overload behavior is not improved by bigger bridge buffers. On the other hand, the dynamic window-size algorithm yields a stable throughput behavior even for relatively small bridge-buffer sizes in the sense that throughput never decreases with increasing offered data rate.

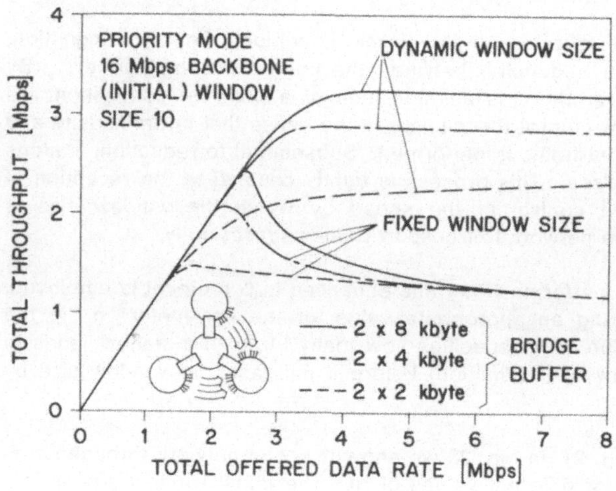

Fig. 24. Total throughput versus total offered data rate for different bridge-buffer sizes. Fixed and dynamic window size LLC with fixed or initial window size ten and n = 8. Priority mode. From |Bux 85b|. © 1985 IEEE.

With the dynamic window, throughput is improved by bigger bridge buffers, however, increasing the buffer size beyond the two times 8 kbyte shown in the figure, does not yield any noticeable improvement.

In all cases where the fixed window-size protocol works without congestion problems, the dynamic flow-control algorithm yields equally good performance since the window size is never or only very rarely reduced. A typical example is the one of Fig. 20 for which the throughput characteristic of the dynamic window LLC with an initial window size of ten is identical to the result for the fixed window of ten, namely, almost ideal.

Concluding this section, we want to point out that the dynamic window mechanism described here has been incorporated into a revision of the IEEE 802.2 Standard.

6. CONCLUDING REMARKS

The objective of this survey paper was to demonstrate what kinds of performance-related problems have arisen in the development of the token-ring technology and architecture. Although we believe to have covered most of the practically relevant issues, the discussion is certainly not complete. We did not address important performance issues arising in the hardware and microcode design of the token-ring adapters, although these may be crucial for system performance |Wong 84|, |Bux 84|. We deliberately excluded considering the impact of higher-layer protocols, i.e., those above LLC from this discussion, since we chose to focus on token-ring specific issues. We want to point out, however, that in many of today's implementations, the nature and implementation of these protocols may have a dominating effect on the performance experienced by the user.

Token-ring technology has reached a level of great maturity with the possible exception of very high speeds. Therefore, one may argue that there is not much potential for further relevant performance work. We do not share this view since there are various important areas in which we are far from having a satisfactory understanding. One is the analytical modeling of cyclic-service systems in which the service given to individual stations (queues) is governed by certain timing rules, e.g., FDDI's timed-token protocol. A second area where only very little theoretical understanding has so far been achieved is the modeling of multiring local-area networks. Any progress towards a better understanding of the complex processes in such systems would be extremely helpful to network architects and designers.

ACKNOWLEDGMENT

The material in Sections 4 and 5 of this paper is based on joint work of the author with Doug Dykeman and Davide Grillo. Their substantial contributions are gratefully acknowledged.

REFERENCES

|ANSI 85a| "FDDI Token Ring Media Access Control (MAC)," Draft Proposed American National Standard, X3T9/84-100, Rev-9, Dec. 13, 1985.

|ANSI 85b| "FDDI Token Ring Station Management (SMT)," Draft Proposed American National Standard, X3T9/85-, Rev-1.3, July 10, 1985.

|ANSI 85c| "FDDI Token Ring Physical Layer Medium Dependent (PMD)," Draft Proposed American National Standard, X3T9/85-, Rev-3.3, Nov. 18, 1985.

|ANSI 85d| "FDDI Physical Layer Protocol (PHY)," Draft Proposed American National Standard, X3T9/85-39, Rev-11, Aug. 10, 1985.

|Arndt 84| K. Arndt and H. Sulanke, "A Queueing System with Relative and Cyclic Priorities," Elektronische Informationsverarbeitung und Kybernetik, Vol. 20, Nos. 7/9, pp. 423-425, Sept. 1984.

|Avi-Itzhak 65| B. Avi-Itzhak, W.L. Maxwell and L.W. Miller, "Queueing with Alternating Priorities," Oper. Res., Vol. 13, pp. 306-318, 1965.

|Boxma 84| O.J. Boxma, "Two Symmetric Queues with Alternating Service and Switching Times," in *Performance '84*, E. Gelenbe (ed.), Elsevier Science Publ. B.V. (North-Holland), Amsterdam, 1984, pp. 409-431.

|Boxma 86a| O.J. Boxma and W.P. Groenendijk, "Pseudo-Conservation Laws in Cyclic-Service Systems," Report OS-R8606, Centre for Mathematics and Computer Science, Amsterdam, June 1986.

|Boxma 86b| O.J. Boxma and B. Meister, "Waiting-Time Approximations for Cyclic-Service Systems with Switch-Over Times," Performance '86 and ACK SIGMETRICS 1986, Perform. Eval. Rev., Vol. 14, No. 1, pp. 254-262, May 1986.

|Brodetskii 73| G.L. Brodetskii and V.P. Vinnitskii, "Some Characteristics of Systems with Cyclic Data Processing," Cybernetics, Vol. 9, No. 3, pp. 407-413, May-June 1973.

|Bux 81a| W. Bux, F. Closs, P. Janson, K. Kümmerle and H.R. Müller, "A Reliable Token-Ring System for Local-Area Communication," Conf. Rec. NTC '81, Piscataway, NJ., IEEE, 1981, pp. A2.2.1-A2.2.6.

|Bux 81b| W. Bux, "Local-Area Subnetworks: A Performance Comparison," IEEE Trans. Commun., Vol. COM-29, No. 10, pp. 1465-1473, Oct. 1981.

|Bux 82| W. Bux, P. Janson, H.R. Müller and D.T.W. Sze, "Method of Transmitting Information between Stations Attached to a Unidirectional Transmission Ring," European Patent Application, Appl. No. 80107706.6, Date of Publ.: 23.06.82, Bulletin 82/25.

|Bux 83a| W. Bux and H.L. Truong, "Mean-Delay Approximation for Cyclic-Service Queueing Systems," Performance Evaluation, Vol. 3, No. 3, pp. 187-196, Aug. 1983.

|Bux 83b| W. Bux, F. Closs, K. Kümmerle, H. Keller and H.R. Müller, "A Reliable Token Ring for Local Communications," IEEE J. Select. Areas Commun., Vol. SAC-1, No. 4, pp. 756-765, Nov. 1983.

|Bux 84| W. Bux, "Performance Issues in Local-Area Networks," IBM Syst. J., Vol. 23, No. 4, pp. 351-374, 1984.

|Bux 85a| W. Bux, "Performance Issues," in *Lecture Notes in Computer Science 184: Local Area Networks: An Advanced Course*, Springer-Verlag, Berlin Heidelberg New York Tokyo, 1985, pp. 108-161.

|Bux 85b| W. Bux and D. Grillo, "Flow Control in Local-Area Networks of Interconnected Token Rings," IEEE Trans. Commun., Vol. COM-33, No. 10, pp. 1058-1066, Oct. 1985.

|Cohen 81| J.W. Cohen and O.J. Boxma, "The M/G/1 Queue with Alternating Service Formulated as a Riemann-Hilbert Problem," in *Performance '81*, F.J. Kylstra (ed.), North-Holland, Amsterdam, 1981, pp. 181-199.

|Cohen 83| J.W. Cohen and O.J. Boxma, "Boundary Value Problems in Queueing System Analysis," in *Mathematics Studies 79*, North-Holland, Amsterdam, 1983.

|Cooper 69| R.B. Cooper and G. Murray, "Queues Served in Cyclic Order," Bell Syst. Tech. J., Vol. 48, pp. 675-689, 1969.

|Cooper 70| R.B. Cooper, "Queues Served in Cyclic Order: Waiting Times," Bell Syst. Tech. J., Vol. 49, pp. 399-413, 1970.

|de Moreas 81| L.F.M. de Moreas, "Message Queueing Delays in Polling Systems with Applications to Data Communication Networks," Dept. of Systems Science, School of Engineering and Applied Science, University of California, Los Angeles, CA, UCLA-ENG-8106, May 1981.

|Dixon 83| R.C. Dixon, N.C. Strole and J.D. Markov, "A Token-Ring Network for Local Data Communications," IBM Syst. J., Vol. 22, Nos. 1/2, pp. 47-62, 1983.

|Dykeman 87a| D. Dykeman and W. Bux, "An Investigation of the FDDI Media Access Control Protocol," in Proc. EFOC/LAN 87, Basel, Switzerland, June, 1987, (IGI Europe), Information Gatekeepers, Inc., Boston, pp. 229 - 236.

|Dykeman 87b| D. Dykeman and W. Bux, "An Investigation of the FDDI Media Access Control Protocol," IBM Zurich Research Laboratory, Research Report, RZ 1591, May 1987.

|Eisenberg 71| M. Eisenberg, "Two Queues with Changeover Times," Oper. Res., Vol. 19, pp. 386-401, 1971.

|Eisenberg 72| M. Eisenberg, "Queues with Periodic Service and Changeover Times," Oper. Res., Vol. 20, pp. 440-451, 1972.

|Eisenberg 79| M. Eisenberg, "Two Queues with Alternating Service," SIAM J. Appl. Math., Vol. 36, No. 2, pp. 287-303, April 1979.

|Farber 73| D.J. Farber, J. Feldman, F.R. Heinrich, M.D. Hopwood, D.C. Loomis and A. Rowe, "The Distributed Computer System," in Proc. 7th IEEE Computer Society Intl. Conf., Piscataway, NJ, IEEE, 1973, pp. 31-34.

|Farmer 63| W.D. Farmer and E.E. Newhall, "An Experimental Distributed Switching System to Handle Bursty Computer Traffic," in Proc. ACM Symposium on Problems in the Optimization of Data Communications, Pine Mountain, GA, 1963, pp. 31-34.

|Ferguson 85| M.J. Ferguson and Y.J. Aminetzah, "Exact Results for Nonsymmetric Token Ring Systems," IEEE Trans. Commun., Vol. COM-33, No. 3, pp. 223-231, March 1985.

|Gerla 80| M. Gerla and L. Kleinrock, "Flow Control: A Comparative Survey," IEEE Trans. Commun., Vol. COM-28, No. 4, pp. 553-574, April 1980.

|Giessler 81| A. Giessler, A. Jaegemann, E. Maeser and J.O. Haenle, "Flow Control Based on Buffer Classes," IEEE Trans. Commun., Vol. COM-29, No. 4, pp. 436-443, April 1981.

|Goyal 87| A. Goyal and D. Dias, "Performance of Priority Protocols on High-Speed Token Ring Networks," submitted to Third International Symposium on the Performance of Data Communication Systems, Rio de Janeiro, Brazil, June 22-25, 1987.

|Grow 82| R.M. Grow, "A Timed Token Protocol for Local Area Networks," Electro/82, Token Access Protocols (17/3), May 1982.

|Hashida 72a| O. Hashida, "Analysis of Multiqueue," Rev. Elect. Commun. Lab., Vol. 20, pp. 189-199, 1972.

|Hashida 72b| O. Hashida and K. Ohara, "Line Accommodation Capacity of a Communication Control Unit," Rev. Elec. Commun. Lab., Vol. 20, Nos. 3-4, pp. 245-253, March-April 1981.

|Humblet 78| P.A. Humblet, "Source Coding for Communication Concentrators," Report ESL-R-798, Electronic Systems Laboratory, MIT, Jan. 1978.

|IEEE 84a| IEEE Standard 802.2, Logical Link Control.

|IEEE 84b| IEEE Standard 802.5, Token Ring Access Method and Physical Layer Specifications.

|Iisaku 81| S. Iisaku, N. Miki, N. Nagai and K. Hatori, "Two Queues with Alternating Service and Walking Time," Trans. Inst. Electron. Commun. Eng. Jpn., Vol. J64-B, No. 4, pp. 342-343, April 1981.

|Johnson 85| M.J. Johnson, "Proof that Timing Requirements of the FDDI Token Ring Protocol are Satisfied," Technical Report 85.8, Research Institute for Advanced Computer Science, NASA Ames Research Center, Aug. 1985.

|Johnson 86| M.J. Johnson, "Fairness of Channel Access for Non-Time-Critical Traffic Using the FDDI Token Ring Protocol," in Proc. Advanced Seminar on Real-Time Local Area Networks, Bandol, France, April 1986.

|Konheim 74| A.G. Konheim and B. Meister, "Waiting Lines and Times in a System with Polling," J. ACM, Vol. 21, No. 3, pp. 470-490, July 1974.

|Kuehn 79| P.J. Kuehn, "Multiqueue Systems with Nonexhaustive Cyclic Service," Bell Syst. Tech. J., Vol. 58, pp. 671-698, 1979.

|Kümmerle 87a| K. Kümmerle, F.A. Tobagi, and J.O. Limb, (eds.) in *Frontiers in Communications: Advances in Local Area Networks,* IEEE Press, New York, 1987.

|Kümmerle 87b| K. Kümmerle and M. Reiser, "Local-Area Networks - Major Technologies and Trends," in *Frontiers in Communications: Advances in Loca Area Networks,* IEEE Press, New York, 1987, pp. 2-26.

|Kurosawa 81| K. Kurosawa and S. Tsujii, "Analysis on Asymmetric Polling Systems with Limiting Service," in Proc. Nat. Telecommunications Conference 1981, New Orleans, Louisiana, Nov. 29 - Dec. 3, 1981, pp. G.4.2.1-G4.2.5.

|Lam 79| S.S. Lam and M. Reiser, "Congestion Control of Store-and-Forward Networks by Input Buffer Limits - An Analysis," IEEE Trans. Commun., Vol. COM-27, No. 1, pp. 127-134, Jan. 1979.

|Neuts 68| M.F. Neuts and M. Yadin, "The Transient Behavior of the Queue with Alternating Priorities with Special Reference to the Waiting Times," Mimeo Series No. 136, Dept. of Statistics, Purdue University, Jan. 1968.

|Nomura 78| M. Nomura and K. Tsukamoto, "Traffic Analysis on Polling Systems," Trans. Inst. Electron. Commun. Eng. Jpn., Vol. J61-B, No. 7, pp. 600-607, July 1978.

|Penney 79| B.K. Penney and A.A. Baghdadi, "Survey of Computer Communications Loop Networks: Parts 1 and 2," Computer Communications, Vol. 2, pp. 165-180 and 224-241, 1979.

|Ross 86| F.E. Ross, "FDDI - A Tutorial," IEEE Commun. Magazine, Vol. 24, No. 5, May 1986.

|Rubin 83| I. Rubin and L.F.M. DeMoreas, "Message Delay Analysis for Polling and Token Multiple-Access Schemes for Local Communication Networks," IEEE J. Select. Areas Commun., Vol. SAC-1, No. 5, pp. 935-947, Nov. 1983.

|Saltzer 80| J.H. Saltzer and K.T. Pogran, "A Star-Shaped Ring Network with High Maintainability," Computer Networks, Vol. 4, pp. 239-244, 1980.

|Schwartz 87| M. Schwartz, "Telecommunications Networks: Protocols, Modeling and Analysis," Addison-Wesley, Reading, MA, 1987.

|Sevcik 87| K.C. Sevcik and M.J. Johnson, "Cycle Time Properties of the FDDI Token Ring Protocol," IEEE Trans. Software Eng., Vol. SE-13, No. 3, pp. 376-385, March 1987.

|Shoch 80| J.F. Shoch and J.A. Hupp, "Measured Performance of an Ethernet Local Network," Comm. ACM, Vol. 23, pp. 711-721, 1980.

|Stallings 84| W. Stallings, "Local Networks - An Introduction," Macmillan, New York, 1984.

|Swartz 80| G.B. Swartz, "Polling in a Loop System," J. ACM, Vol. 27, pp. 42-59, 1980.

|Sykes 70| J.S. Sykes, "Simplified Analysis of an Alternating-Priority Queueing Model with Setup Times," Oper. Res., Vol. 18, pp. 1182-1192, 1970.

|Takacs 68| L. Takacs, "Two Queues Attended by a Single Server," Oper. Res., Vol. 16, pp. 639-650, 1968.

|Takagi 85| H. Takagi, "Mean Message Waiting Times in Symmetric Multi-Queue Systems with Cyclic Service," in Proc. IEEE Int. Conf. on Commun., Chicago, June 23-26, 1985, pp. 1154-1157.

|Takagi 86| H. Takagi, "Analysis of Polling Systems," The MIT Press, Cambridge, MA, 1986.

|Ulm 82| J.M. Ulm, "A Timed Token Ring Local Area Network and Its Performance Characteristics," in Proc. 7th Conference on Local Computer Networks, Minneapolis, IEEE, Feb. 1982, pp. 50-56.

|Watson 84| K.S. Watson, "Performance Evaluation of Cyclic Service Strategies - A Survey," in *Performance '84*, E. Gelenbe (ed.), Elsevier Science Publ. B.V. (North Holland), Amsterdam, 1984, pp. 521-533.

|Wong 84| J. W. Wong and W. Bux, "Analytic Modeling of an Adapter to Local-Area Networks," IEEE Trans. Commun., Vol. COM-32, No. 10, pp. 1111-1117, Oct. 1984.

|Zafiropulo 86| P. Zafiropulo, H.R. Müller and F. Closs, "Data/Voice Integration Based on the IEEE 802.5 Token-Ring LAN," in Proc. EFOC/LAN 86, Amsterdam, The Netherlands, June, 1986, (IGI Europe), Information Gatekeepers Inc., Boston, pp. 67-76.

Vergleich der Kanalzugriffsverfahren CSMA/CD, Token–Bus, Token–Ring und Slotted–Ring für Poisson– und unterbrochene Poisson–Ankunftsprozesse

Ottmar Gihr
Universität Stuttgart
Institut für Nachrichtenvermittlung und Datenverarbeitung
Seidenstr. 36, 7000 Stuttgart 1

Kurzfassung

Die Kanalzugriffsverfahren CSMA/CD(Ethernet), Token–Bus, Token–Ring und Slotted–Ring werden hinsichtlich ihres Verhaltens bei Poisson–Ankunftsprozessen gegenüber unterbrochenen Poisson–Ankunftsprozessen untersucht. Für alle Kanalzugriffsverfahren werden Modelle vorgestellt, die simulativ untersucht werden.

1 Einleitung

In den letzten Jahren fanden Lokale Netze eine weite Verbreitung in unterschiedlichen Einsatzgebieten, wie Büro– und Fabrikautomatisierung, Rechenzentrum oder Ausbildung. Das Anwendungsspektrum ist sehr reichhaltig vom Ersatz der sternförmigen V.24 Verkabelung der Terminals bis zu hochwertiger Dokumenten– und Bildverarbeitung.

Derzeit im Einsatz befindliche Lokale Netze basieren bei dem Übertragungsmedium und dem Kanalzugriffsverfahen zum überwiegenden Teil auf den von IEEE (Institute of Electrical and Electronics Engineers) und ISO (International Organization for Standardization) standardisierten Zugriffsverfahren CSMA/CD [1], Token–Bus [2], Token–Ring [3] und Slotted–Ring [4]. Das Slotted–Ring–Zugriffsverfahren hat bei der Standardisierung noch keinen stabilen Zustand erreicht und wird daher noch weniger eingesetzt. Die Datenübertragungsraten von 1 bis 10 MBit/s genügen für derzeitige Anwendungen, da der Engpaß der Datenübertragung bei Lokalen Netzen nicht in dem Kanalzugriffsverfahren, sondern in den Verarbeitungsinstanzen der höheren Schichten zu suchen ist. Weder die Anschaltung des Rechners über den LAN–Controller noch der Rechner selbst sind in der Lage, ein Lokales Netz auszunützen. Der Anforderungsstrom einer Station an das Lokale Netz wird meist durch einen Poisson–Prozess beschrieben. Die Zugriffsverfahren wurden mit diesem Ankunftsprozeß bereits untersucht und in der Literatur dargestellt [5,6,7,8,9]. Durch die sich jetzt abzeichnende Steigerung der Leistungsfähigkeit der Anschaltungen, sowie des Applikations–Prozessors an sich, werden die Lokalen Netze stärker und in einer oft veränderten Art ausgenützt. Es werden innerhalb einer kurzen aktiven Zeit viele Anforderungen generiert, gefolgt von einer längeren Zeit ohne Nachrichtenübertragungen. Durch einen unterbrochenen Poisson–Prozeß ist diese Verkehrscharakteristik, die oft auch als Burst–Verkehr bezeichnet wird, beschreibbar.

Im nächsten Abschnitt werden die unterschiedlichen Kanalzugriffsverfahren beschrieben und die Simulationsmodelle dafür vorgestellt. In Abschnitt 3 werden die beiden unterschiedlichen Anforderungsströme, Poisson–Prozeß und unterbrochener Poisson–Prozeß (Burst–Prozeß), charakterisiert. Die Ergebnisse, bestehend aus den mittleren Transferzeiten für die 4 Kanalzugriffsverfahren, werden im Abschnitt 4 in einer vergleichenden Studie dargestellt.

2 Kanalzugriffsverfahren

Der Kanalzugriff für eine Paketübertragung wird bei allen Kanalzugriffsverfahren dezentral gesteuert. Alle Stationen beteiligen sich an der Kanalvergabe gleichberechtigt. Beim Token–Bus- und Token–Ring–Verfahren gibt es während des Betriebs eine ausgezeichnete Station, die den Token das erste Mal generiert und überwacht. Es kann jedoch jede Station diese Aufgabe übernehmen, abhängig von den vorhandenen Stationen und ihren Stationsadressen. Bei dem Slotted–Ring–Kanalzugriffsverfahren ist eine Station notwendig, die den Rahmen, der auf dem Ring umläuft, generiert, den Laufzeitausgleich durchführt und Kanäle für synchrone Datenübertragung belegt und den Stationen zuordnet. Die Kanalzugriffsverfahren CSMA/CD (Ethernet) und Token–Bus verwenden als Übertragungsmedium einen gemeinsamen Bus, während die Verfahren Token–Ring und Slotted–Ring physikalisch auf einem unidirektionalen Ring, mit je einem Übertragungsabschnitt zwischen zwei Stationen, aufbauen. Als Übertragungsmedium kommen, neben Glasfaser und verdrillten Kupferleitungen, hauptsächlich Koaxialleitungen zum Einsatz.

2.1 CSMA/CD (Ethernet)

Dem CSMA/CD–Kanalzugriffsverfahren (Carrier Sense Multiple Access with Collision Detection) [1] liegt, bei der Vergabe des Kanals an eine Station, ein stochastischer Algorithmus zugrunde. Jede Station beobachtet passiv den Kanalzustand, bis sie ein Datenpaket zu übertragen hat. Ist der Kanal belegt, so wartet die Station das Ende der laufenden Übertragung ab. Ist der Kanal frei, so beginnt die Station das Datenpaket zu senden. Beginnt keine weitere Station in der Zeit, bis alle anderen Stationen den Kanal als belegt erkannt haben, zu senden, so wird das Datenpaket erfolgreich übertragen. Beginnt mindestens eine weitere Station in dieser Zeit zu senden, so kommt es zu einer Kollision auf dem Übertragungskanal, die durch den Binary–Truncated–Backoff–Algorithmus aufgelöst wird. Jede Station stellt ihre Übertragung um eine zufällige Zeit, die sich aus der Backoff–Zeit ergibt, zurück :

$$BackoffTime = random(0, min(2^n, 2^{10})) * SlotTime.$$

Die Slot–Zeit (*Slot Time*) ist gegeben durch die doppelte Signalverzögerungszeit der entferntesten Stationen auf dem Koaxialkabel (nach Standard : $51.2\mu s$) und n ist die Anzahl der Kollisionen dieses Datenpaketes seit dem ersten Übertragungsversuch.

Alle Stationen sind an einen, eventuell über Repeater gekoppelten, zusammenhängenden Übertragungskanal angeschlossen. Dies führt zu einem logischen Modell nach Bild 1 und einem Modell für die Simulation nach Bild 2. Das Simulationsmodell besteht aus den N aktiv an der Datenkommunikation beteiligten Stationen. Der Anforderungsstrom der Datenpakete wird in eine Warteschlange eingeleitet. Die in der Warteschlange stehenden Datenpakete werden durch die Kanal–Bedieneinheit der

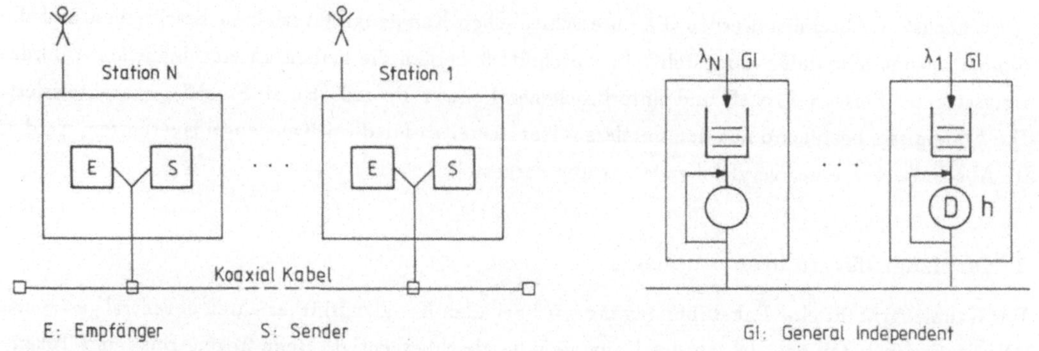

Bild 1 : Logisches Modell der CSMA/CD-Konfiguration

Bild 2 : Verkehrstheoretisches Modell von CSMA/CD

Station nach dem CSMA/CD-Protokoll auf dem Kanal übertragen. Die Übertragungszeit des Datenpaketes wird durch eine Bedienenheit mit konstanter Bedienzeit (deterministisch D, da konstante Paketlänge verwendet wird) der Zeitdauer h repräsentiert.

2.2 Token-Bus

Bei dem Token-Bus-Zugriffsverfahren sind, wie bei dem Ethernet-Verfahren, alle Stationen an ein gemeinsames Übertragungsmedium, den Bus, angeschlossen. Für die Vergabe der Sendeberechtigung bilden die Stationen einen logischen Ring. Stationen können dynamisch in diesen Ring aufgenommen oder entfernt werden. In diesem logischen Ring wird die Sendeberechtigung, die durch ein Token repräsentiert wird, zyklisch weitergereicht. Erhält eine Station die Sendeberechtigung, so darf sie für eine bestimmte Zeit Nachrichten aussenden, bevor sie die Sendeberechtigung an die nachfolgende Station weitergeben muß.

Die Stationen unterscheiden 8 Service-Klassen, die zu 4 Zugriffsklassen (Klasse 6, 4, 2, 0) paarweise zusammengefaßt sind. Die Zugriffsklassen werden in der Reihenfolge 6, 4, 2, 0 bedient. In der Zugriffsklasse 6 (High Priority Class) darf die Station für die Zeitdauer *High Priority Token Hold Time* Nachrichten aussenden. In den Zugriffsklassen 4, 2 und 0 darf die Station Nachrichten senden, wenn die vorgegebene Zykluszeit des Tokens (Token Rotation Time) dieser Zugriffsklasse noch nicht erreicht ist.

Durch dieses deterministische Verhalten kann in der Zugriffsklasse 6, bei geeigneter Wahl der Stationsparameter, eine maximale Zeit angegeben werden, bis die Station auf den Kanal zugreifen kann. Aus diesem Grund wird das Token-Bus-Verfahren bei Realzeitanwendungen bevorzugt.

Bild 3 zeigt das Modell einer Token-Bus-Konfiguration, die aus N Stationen besteht. Station N habe die höchste und Station 1 die niedrigste Stationsadresse. Durch diese Adressen ist der logische Ring, von Station N über Station i zu Station 1 und wieder zu Station N, festgelegt. In dem verkehrstheoretischen Modell für die Simulation (Bild 4), besteht jede Station aus 4 Warteschlangen entsprechend den 4 Zugriffsklassen. Diese Warteschlangen nehmen die zu übertragenden Nachrichten auf, um sie nach der FIFO-Strategie (First In First Out), bei vorhandener Sendeberechtigung

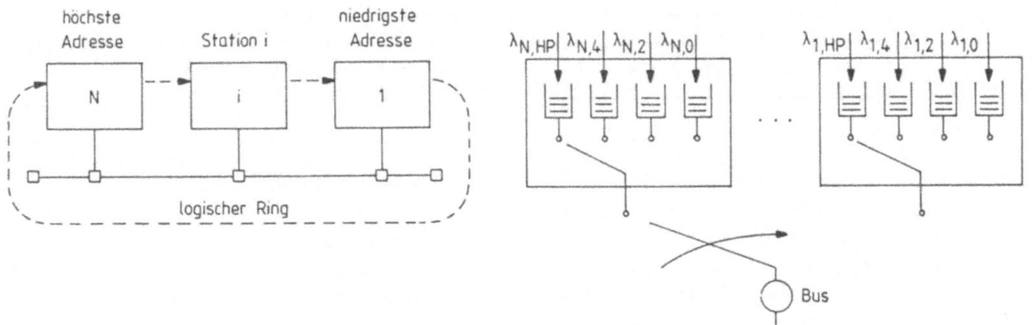

Bild 3 : Modell der Token–Bus–Konfiguration

Bild 4 : Verkehrstheoretisches Modell des Token–Bus

und momentan höchster Priorität, zu übertragen. Die Zirkulation des Tokens wird durch zyklische Abfrage (Polling) dargestellt, wobei die Token–Übertragungszeit die Umschaltzeit zwischen den Stationen ergibt. Die maximale Verweilzeit des Tokens in einer Station ergibt sich aus der *High Priority Token Hold Time* der Zugriffsklasse 6 und den *Token Rotation Times* der restlichen Klassen.

2.3 Token–Ring

In einem Lokalen Netz, das nach dem Token–Ring–Kanalzugriffsverfahren arbeitet, sind alle Stationen durch Kabelsegmente physikalisch so miteinander verbunden, daß sich ein Ring ergibt. Das ankommende Signal wird von der Station empfangen, regeneriert und der nachfolgenden Station weitergesandt. Die Station, die die Sendeberechtigung (Token) besitzt, kann Nachrichten auf dem Ring aussenden. Die von ihr ausgesandte Nachricht läuft über den ganzen Ring und wird von dieser Station wieder vom Ring entfernt. Während die Nachricht auf dem Ring umläuft, prüft jede Station die Empfangsadresse der Nachricht und legt, falls sie angesprochen ist, eine Kopie der Nachricht an.

Bei dem Token–Ring–Zugriffsverfahren werden 8 Prioritätsklassen unterschieden. Erhält eine Station die Sendeberechtigung, so kann sie Nachrichten aussenden bis die *Token Hold Time* aufgebraucht ist oder eine der Stationen über das *Access Control Feld* eine höhere Priorität beantragt. Tritt eine dieser Bedingungen ein, so wird der Token mit der entsprechenden Priorität weitergegeben.

Das *Access Control Feld* ist Teil des Kopfes (Header) jedes Rahmens, der auf dem Ring umläuft (siehe Bild 5). In dieses Feld schreibt eine Station die von ihr beantragte Priorität (Reservierung), falls die bisher dort beantragte Priorität kleiner ist. Sendet die momentan sendeberechtigte Station in einer niedrigeren Priorität, so schickt sie sofort den Token mit der erhöhten Priorität auf den Ring und legt den momentanen Stationszustand auf den Stack. Trifft der Token mit der auf diesen Wert erhöhten Priorität wieder bei der Station ein, so wird der Zustand vor der Unterbrechung wieder hergestellt. Die Station fährt dann mit Übertragungen in der bisherigen Priorität fort, falls im Token keine erneute Reservierung auf eine erhöhte Priorität stattfand. Nach der Übertragung einer Nachricht kann somit die nächste Bedienung einer Priorität zurückgestellt werden, bis Nachrichten höherer Priorität übertragen sind. In der höchsten Prioritätsklasse ist wiederum eine maximale

|<– SFS –>|<——— F C S Coverage ———>|<– EFS –>|

SD	AC	FC	DA	SA	INFO	FCS	ED	FS

SFS = Start-of-Frame Sequence
SD = Starting Delimiter (1 octet)
AC = Access Control (1 octet)
FC = Frame Control (1 octet)
DA = Destination Address (2 or 6 octets)
SA = Source Address (2 or 6 octets)
INFO = Information (0 or more octets)
FCS = Frame-Check Sequence (4 octets)
EFS = End-of-Frame Sequence
ED = Ending Delimiter (1 octet)
FS = Frame Status (1 octet)

Bild 5 : Rahmenformat auf dem Token–Ring

Zugriffszeit der Station angebbar, die direkt von der Anzahl der Stationen im Ring und der *Token Hold Time* abhängt.

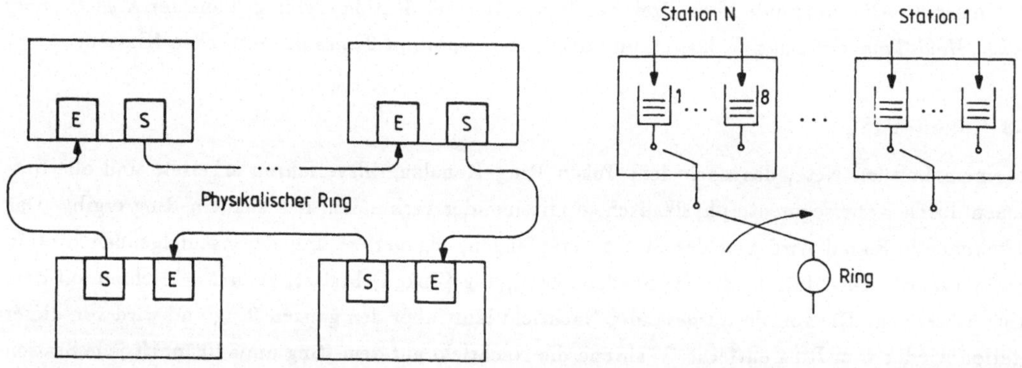

Bild 6 : Modell der Token–Ring–Konfiguration

Bild 7 : Verkehrstheoretisches Modell des Token–Ring

Bild 6 zeigt das Modell, das sich für eine Token–Ring–Konfiguration ergibt. Es besteht aus den über Kabelsegmente vom Sendeteil des Senders zum Empfangsteil der nachfolgenden Station unidirektional verbundenen Stationen. Das sich daraus ergebende verkehrstheoretische Modell (Bild 7) besteht je Station, entsprechend den Prioritätsklassen, aus 8 Warteschlangen. Das Token–Ring–Zugriffsverfahren legt die Reihenfolge der Bedienungen der Warteschlangen fest.

2.4 Slotted–Ring

Das Slotted–Ring–Kanalzugriffsverfahren kann als einziges Verfahren außer dem Paketverkehr (non isochron) auch einen durchschaltevermittelten Verkehr (isochron) übertragen. Eine ausgezeichnete

Station (Monitor) erzeugt synchron Rahmen der Zeitdauer 125 μs (Bild 8), überwacht den Ring, teilt die durchschaltevermittelten Kanäle zu und übernimmt den Laufzeitausgleich. Der Rahmen ist in Zeitschlitze (Slots) der Länge 77 Byte eingeteilt, die entweder für durchschaltevermittelten Verkehr reserviert sind oder den Stationen für paketvermittelten Verkehr zur Verfügung stehen. Die Monitorstation reserviert die Zeitschlitze für durchschaltevermittelten Verkehr und trägt diese Zuordnung in den Rahmenkopf (Frame Header) jedes Rahmens ein.

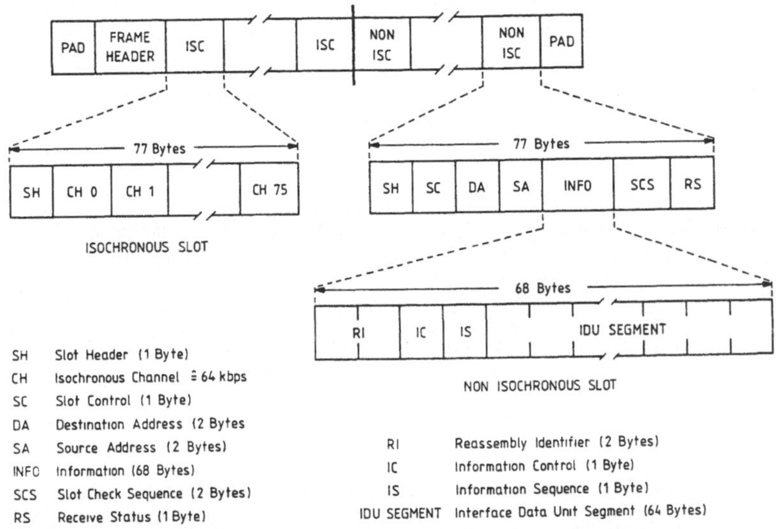

Bild 8 : Rahmenformat beim Slotted-Ring

Ein Zeitschlitz für durchschaltevermittelten Verkehr wird in 76 Kanäle unterteilt. Jeder Kanal überträgt 1 Byte pro Rahmen und hat somit eine Bandbreite von 64 kBit/s. Jeder Kanal stellt eine Voll-Duplex-Verbindung bereit, so daß die Bandbreite doppelt genützt wird. Die ausgesandte Information wird beim Empfänger entnommen, der dafür seine Information in diesem Kanal zum Sender überträgt.

Ein Zeitschlitz für paketvermittelten Verkehr enthält ein Segment (64 Byte) der zu übertragenden Nachricht. Die Nachricht wird beim Sender in Segmente aufgeteilt (Segmenting), beim Empfänger wieder aufgesammelt und zu der ursprünglichen Nachricht zusammengesetzt (Reassembly). Hat eine Station ein Segment zu übertragen, so wird der nächste freie Zeitschlitz mit diesem Segment belegt. Der Empfänger kopiert das Segment und teilt den korrekten Empfang im RS-Feld (Receive Status) dem Sender mit. Der Sender schaltet diesen Zeitschlitz frei und übergibt ihn als freien Zeitschlitz der nachfolgenden Station, die ihn wieder belegen kann. Es kann somit keine Station den Ring monopolisieren.

Bild 9 zeigt das Simulationsmodell und Bild 10 zeigt das verkehrstheoretische Modell [10] für den paketvermittelten Verkehr. Jede Station wird durch eine Warteschlange repräsentiert, in der die Segmente zwischengespeichert werden, bis durch eine der freien Bedieneinheiten die Bedienung erfolgt. Jeder Zeitschlitz für paketvermittelten Verkehr wird durch eine Bedieneinheit repräsentiert, die zyklisch rotiert. Die Zeitschlitze für durchschaltevermittelten Verkehr treten in dem Modell als

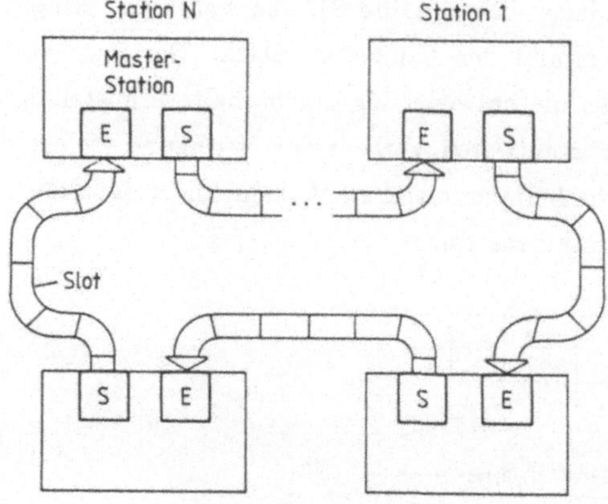

Bild 9 : Modell der Slotted-Ring-Konfiguration

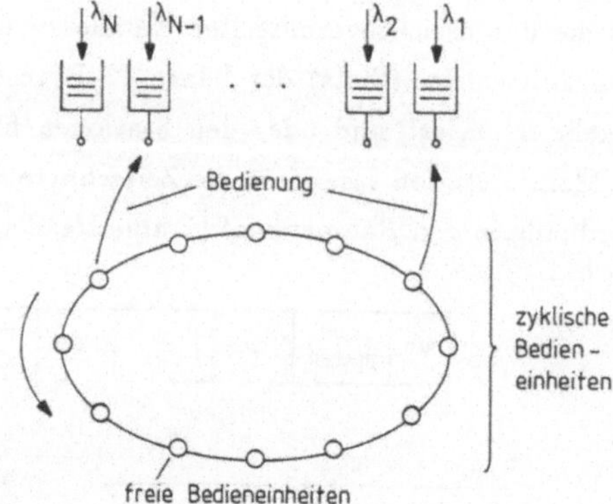

Bild 10 : Verkehrstheoretisches Modell des Slotted-Ring

dauernd belegte Bedieneinheiten auf, da für diese Verkehrsbeziehungen, gegenüber dem sich schnell ändernden paketvermittelten Verkehr, von einem eingefrorenen Zustand ausgegangen wird.

Das Slotted-Ring-Zugriffsverfahren unterscheidet beim Kanalzugriff keine Prioritäten. Es ermöglicht, Sprache und Daten zu integrieren, und es ist vorgesehen, daß mehrere Slotted-Ring-Netze über Brücken (Bridges) auf der Kanalzugriffsschicht (Media Access Control Sublayer) direkt oder über ein "backbone" Slotted-Ring-Netz höherer Datenübertragungsrate gekoppelt werden und so ein größeres zusammenhängendes Netzwerk bilden, das auch als Metropolitan Area Network (MAN) bezeichnet wird.

3 Ankunftsprozesse

Die Nachrichten (Pakete), die bei der Kanalzugriffseinheit zur Übertragung ankommen, werden durch einen Ankunftsprozeß beschrieben. Zwei Arten von Ankunftsprozessen werden für die Untersuchung herangezogen: Der Poisson-Ankunftsprozeß zur Beschreibung eines gleichmäßigen Verhaltens der Station und ein unterbrochener Poisson-Ankunftsprozeß für die Beschreibung von kurzzeitigen Aktivitäten mit längeren Pausen.

3.1 Poisson-Ankunftsprozeß

Die zeitlichen Abstände zweier aufeinanderfolgender Paketankünfte werden als Zufallsvariablen T_{A_j} aufgefaßt. Sind diese Zufallsvariablen unabhängig voneinander und haben dieselbe negativ exponentielle Verteilungsfunktion $F_A(t)$ mit dem Parameter λ_A als Ankunftsrate, so ist der Ankunftsprozeß ein Poisson-Prozess.

$$T_{A_j} : P\{T_{A_j} \leq t\} = F_A(t) = 1 - e^{-\lambda_A t} \qquad f_A(t) = \lambda_A e^{-\lambda_A t} \qquad ; \lambda_A - \textit{Ankunftsrate}$$

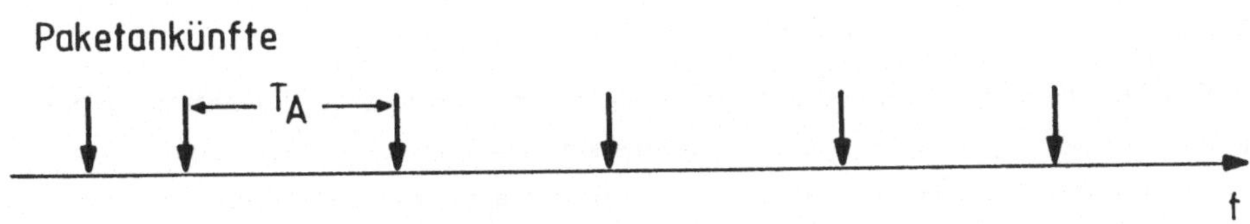

Bild 11 : Poisson–Ankunftsprozeß

3.2 Unterbrochener Poisson–Ankunftsprozeß

Der unterbrochene Poisson–Ankunftsprozeß weist einen aktiven Zustand, in dem Nachrichten mit negativ exponentiell verteiltem Ankunftsabstand ankommen (Poisson–Prozeß), und einen passiven Zustand auf, in dem keine Nachrichten ankommen. Die Verweildauern in den Zuständen werden wiederum mit Zufallsvariablen, die eine negativ exponentielle Verteilungsfunktion haben, beschrieben. Alle Zufallsvariablen seien voneinander unabhängig.

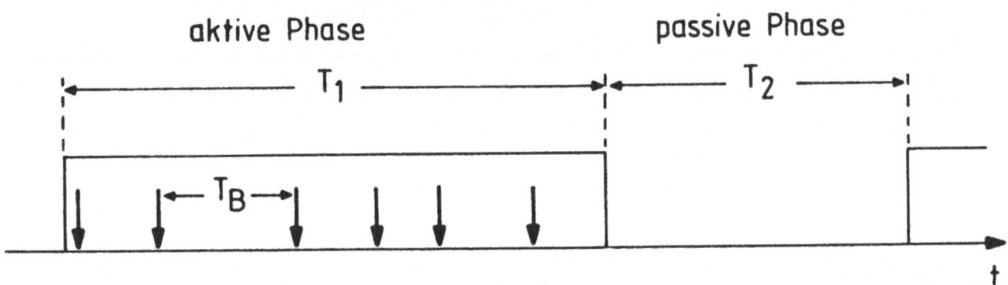

Bild 12 : Unterbrochener Poisson–Ankunftsprozeß

Verweildauer im aktiven Zustand :

$$T_1 : P\{T_1 \leq t\} = F_1(t) = 1 - e^{-\lambda_1 t} \qquad f_1(t) = \lambda_1 e^{-\lambda_1 t}$$

Verweildauer im passiven Zustand :

$$T_2 : P\{T_2 \leq t\} = F_2(t) = 1 - e^{-\lambda_2 t} \qquad f_2(t) = \lambda_2 e^{-\lambda_2 t}$$

Ankunftsabstand im aktiven Zustand :

$$T_B : P\{T_B \leq t\} = F_B(t) = 1 - e^{-\lambda_B t} \qquad f_B(t) = \lambda_B e^{-\lambda_B t}$$

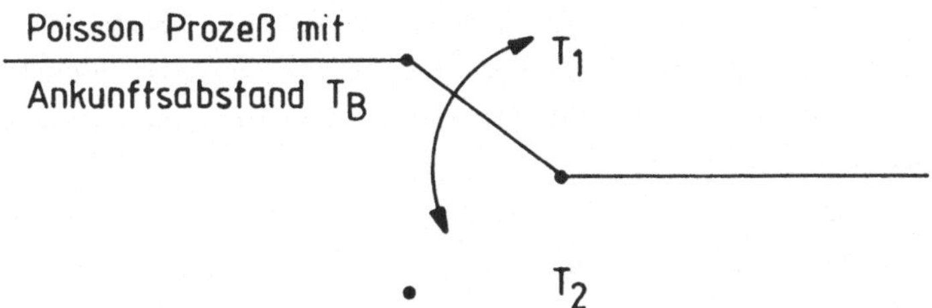

Bild 13 : Unterbrochener Poisson–Prozeß

4 Simulation

Jedes der 4 Kanalzugriffsverfahren wurde anhand seines verkehrstheoretischen Modells in ein Simulationsprogramm umgesetzt. Die Simulationsprogramme sind nach der zeittreuen ereignisgesteuerten (event–by–event) Methode aufgebaut und in der Programmiersprache Pascal auf VAX–Rechnern implementiert. Jeder Simulationslauf umfaßt 10 Teiltests über die der Mittelwert und das Vertrauensintervall berechnet wird.

4.1 Ergebnisse

Die Ergebnisse, die in den nachfolgenden Schaubildern dargestellt sind, beziehen sich auf die mittlere Transferzeit der Nachrichten. Die Transferzeit ist definiert als die Zeit von der Übergabe der Nachricht beim Sender an die Kanalzugriffseinheit bis zum Verlassen der Kanalzugriffseinheit beim Empfänger. Bei der Übergabe der Nachricht an die sendende Kanalzugriffseinheit kann diese bei voller Warteschlange (100 Warteplätze) blockiert sein. Die Nachricht wird in diesem Fall abgewiesen und geht verloren. Der Empfäger kann nicht blockiert sein. Die Transferzeiten für alle 4 Kanalzugriffsverfahren sind über dem Gesamtangebot (A) aufgetragen. Das Gesamtangebot setzt sich aus den Stationsangeboten (A_i) zusammen, die bei allen Stationen gleich groß sind.

$$A = \sum_{i=1}^{N} A_i$$

N – Anzahl der Stationen
A – Gesamtangebot
A_i – Stationsangebot

Das Stationsangebot ist normiert auf die Übertragungsgeschwindigkeit (c) des Lokalen Netzes, die einheitlich zu 10 MBit/s gewählt wurde.

$$A_i = \frac{\lambda_{A_i} d}{c}$$

λ_{A_i} – Paketankunftsrate der Station i
d – Paketgröße
c – Übertragungsgeschwindigkeit = 10 MBit/s

Die Paketgröße wurde für jeden Simulationslauf als konstant angenommen mit d=64 Byte für die Verkehrscharakteristik bei Termialverkehr und d=1024 Byte für die Verkehrscharakteristik bei File Transfer. Der Verkehrsfluß zwischen den n=10 simulierten Stationen sei jeweils gleich groß, so daß ein vollständig symmetrischer Verkehr entsteht. Für den Slotted–Ring ist zu bemerken, daß die Paketgröße d=64 Byte genau in einen Slot paßt.

Für den unterbrochenen Poisson–Ankunftsprozeß wurde $E[T_2] = 10E[T_1] = 100E[T_B]$ gewählt. Dies bedeuted, daß die passive Phase im Mittel 10mal länger als die aktive Phase und diese wiederum im Mittel 10mal länger als der mittlere Ankunftsabstand zweier Nachrichten ist.

In der Tabelle 1 sind die wichtigsten Parameter der einzelnen Kanalzugriffsverfahren, die in der Simulation verwendet wurden, aufgeführt.

Gemeinsam		
Transmission Speed	10	MBit/s
Number of Stations	10	
CSMA/CD (Ethernet)		
Preamble	64	Bit
Address Length	48	Bit
Type Size	16	Bit
Frame Check Sequence	32	Bit
Inter Frame Spacing	96	Bit
Jam Size	48	Bit
Token–Bus (Baseband)		
Address Length	48	Bit
High Priority Token Hold Time	0.1	ms
Minimum Time between Frames	2.4	μs
Token–Ring		
Start of Frame Sequence	16	Bit
Address Length	48	Bit
Frame Check Sequence	32	Bit
End of Frame Sequence	16	Bit
Station Latency	1	Bit
Delay between Stations	500	ns
Slotted–Ring		
Frame Length	125	μs
One Frame on the Ring		
Number of synchronous Slots	0	
Segmentation	60	μs
Reassembly	30	μs

Tabelle 1 : Parameter der Zugriffsverfahren

Für die Kanalzugriffsverfahren Token–Bus und Token–Ring wurden jeweils die höchsten Prioritätsklassen bei der Simulation verwendet. Bei dem Slotted–Ring–Zugriffsverfahren wird für die Untersuchung der synchrone Verkehr (durchschaltevermittelte Verkehr) nicht betrachtet, so daß die gesamte Bandbreite dem Paketverkehr zur Verfügung steht. Für Segmentierung und Reassembly wird pro Segment eine Bearbeitung in der Station vorgesehen, deren Zeitbedarf aber so gewählt wurde, daß dadurch kein Engpaß bei der Übertragung entsteht, da der Schwerpunkt der Untersuchung auf den Kanalzugriffsverfahren liegt.

Bild 14 (Bild 15) zeigt für den Poisson–Ankunftsprozeß und Paketgröße d=64 Byte (d=1024 Byte) die mittleren Transferzeiten über dem Gesamtangebot aller Stationen an den Übertragungskanal. Für die Paketgröße d=64 Byte liegen die Zugriffsverfahren Ethernet und Token–Ring bei kleinem Angebot (A<0.25) sehr nahe beieinander. Ethernet weist im unteren Bereich geringfügig kleinere Transferzeiten auf als der Token–Ring. Ab einem Angebot von A>0.3 beginnt die Transferzeit von Ethernet stark zu steigen, um bei A>0.5 instabil zu werden, während das Token–Ring–Zugriffsverfahren länger bei kleinen Transferzeiten bleibt und bei Angebotswerten A>0.8 instabil wird. Der Slotted–Ring hat

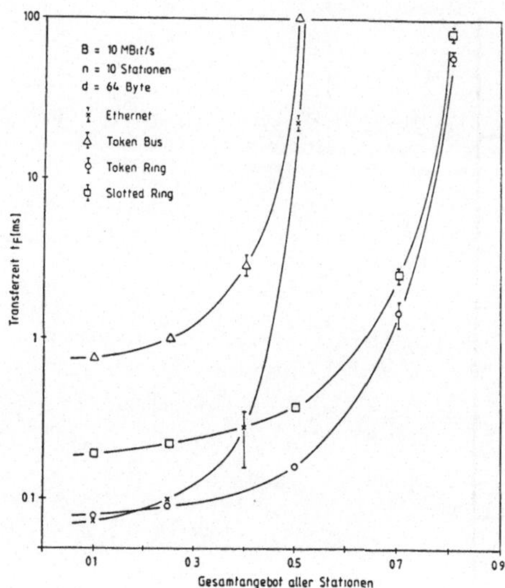

Bild 14 : Poisson-Ankunftsprozeß mit Paketgröße d=64 Byte

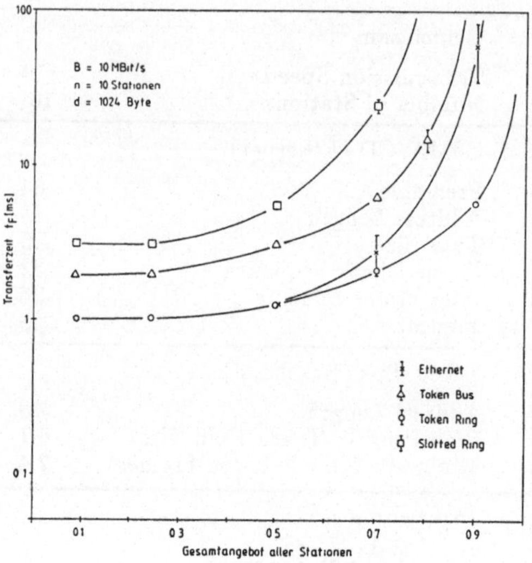

Bild 15 : Poisson-Ankunftsprozeß mit Paketgröße d=1024 Byte

bei kleinen Angebotswerten bereits eine höhere Transferzeit, die in der Bearbeitungszeit für das Aufspalten der Datenpakete in die Segmente begründet ist. Bei höheren Angebotswerten verhält er sich ähnlich dem Token–Ring. Die Stabilitätsgrenze liegt bei symmetrischem Verkehr bei dem Durchsatz D_{max} :

$$D_{max} = \frac{s(((d'-1)DIV64)+1)Byte}{125\mu s}$$

d' = d/Byte

s- Anzahl der Slots/Rahmen (s=2 bei c=10 MBit/s)

Für d=64 Byte und s=2 ergibt sich ein maximaler Durchsatz von D_{max}=8.192 MBit/s. Der Token–Bus hat bei kleinen Angebotswerten bereits längere Transferzeiten und verhält sich bei höherem Angebot A>0.5 instabil, ähnlich dem Ethernet–Verfahren. Bei Ethernet liegt die Instabilität, bei steigendem Angebot, in der steigenden Anzahl der Kollisionen auf dem Übertragungskanal. Beim Token–Bus ist der Overhead durch die Weitergabe des Tokens so groß (der Token wird nach jeder Übertragung weitergegeben), daß die Instabilität ebenfalls bei Angebotswerten A>0.5 eintritt. Durch Änderung der *High Priority Token Hold Time* kann diese Instabilität zu höheren Angebotswerten verschoben werden. Das Vertrauensintervall wurde in den Diagrammen eingezeichnet wenn es nicht zu klein ist und mit den Simulationspunkten zusammenfällt.

Für die Paketgröße d=1024 Byte (Bild 15) sind die Transferzeiten für Ethernet und Token–Ring bei kleinem Angebot identisch. Ethernet wird jedoch schneller instabil, während Token–Ring bis Angebotswerte von A>0.9 stabil ist. Das Token–Bus–Kanalzugriffsverfahren beginnt bei kleinen Angebotswerten bereits mit höheren Transferzeiten und wird auch vor Ethernet instabil. Das Slotted-Ring–Kanalzugriffsverfahren verhält sich ähnlich, mit nochmals erhöhten Transferzeiten bei kleinem Angebot und früherer Instabilität. Die erhöhte Transferzeit bei kleinem Angebot ergibt sich aus der im Mittel von einer Station maximal nutzbaren Bandbreite, die die Hälfte der insgesamt verfügbaren Bandbreite des Paketverkehrs ist. Es muß ein von einer Station belegter Slot als freier Slot weitergegeben werden und kann erst beim nächsten Umlauf des Rahmens von dieser Station wieder verwendet werden.

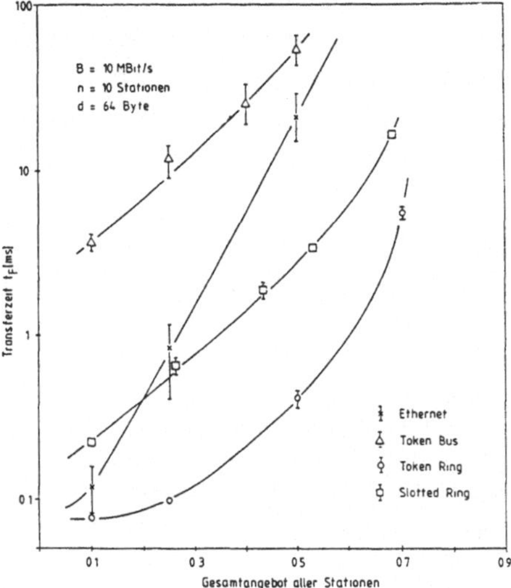

Bild 16 : Unterbrochener-Ankunftsprozeß mit Paketgröße d=64 Byte

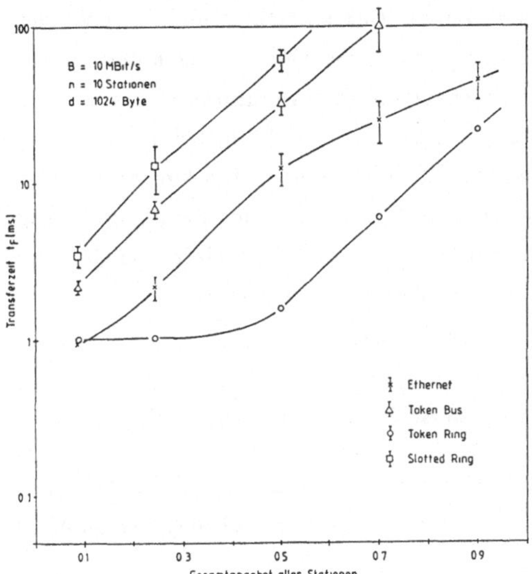

Bild 17 : Unterbrochener-Ankunftsprozeß mit Paketgröße d=1024 Byte

Die Bilder 16 und 17 stellen die Transferzeit für den unterbrochenen Poisson–Ankunftsprozeß, bei den konstanten Paketgrößen d=64 Byte (Bild 16) und d=1024 Byte (Bild 17), dar. Der Token–Ring zeigt für beide Paketgrößen die kleinsten Transferzeiten über den gesamten Angebotsbereich. Ethernet hat bei Paketgröße d=64 Byte und geringem Angebot dieselben Transferzeiten wie Token–Ring, um jedoch bei steigendem Angebot schnell anzusteigen. Das Token–Bus–Kanalzugriffsverfahren liegt in den Transferzeiten wiederum über dem Ethernet–Verfahren. Das Slotted–Ring–Verfahren liegt bei kleinem Angebot und d=64 Byte für die Transferzeiten zwischen Ethernet und Token–Bus, hat jedoch bei Angeboten A>0.2 kleinere Transferzeiten als Ethernet. Bei d=1024 Byte verhält es sich ähnlich wie das Token–Ring–Verfahren. Im oberen Angebotsbereich dieser beiden Schaubilder ist, durch die ungleichen Verluste der Zugriffsverfahren bei vollen Warteschlagen, zu berücksichtigen, daß ungleiche Durchsätze erreicht werden. Vor allem das CSMA/CD–Verfahren neigt durch sein stochastisches Verhalten zu Verlusten. Die verfügbare Bandbreite wird bei hohem Angebot sehr ungleich auf die Stationen aufgeteilt.

Dieser Vergleich der Kanalzugriffsverfahren untereinander und bei verschiedenen Verkehrscharakteristiken zeigt, daß bei den hier gewählten Parametern (Stationszahl, Übertragungsgeschwindigkeit, Paketgrößen, *High Priority Token Hold Time ...*) das Token–Ring–Kanalzugriffsverfahren sich als sehr stabil über den gesamten Angebotsbereich erweist. Ethernet verhält sich bei kleinem Angebot ähnlich dem Token–Ring, um jedoch bei höherem Angebot schneller instabil zu werden. Das Token–Bus Zugriffsverfahren hat, bedingt durch den hohen Overhead, Nachteile über den gesamten Angebotsbereich. Bedingt durch seine Stabilitäts– und Durchsatzgrenzen verhält sich Slotted–Ring unterschiedlich.

Zusammenfassung

In dieser Arbeit wurden die Kanalzugriffsverfahren CSMA/CD (Ethernet), Token–Bus, Token–Ring und Slotted–Ring modelliert und einer vergleichenden simulativen Untersuchung unterworfen. Der Ankunftsprozeß für Anforderungen der höheren Schichten wurde sowohl durch Poisson– als auch unterbrochene Poisson–Ankunftsprozesse modelliert. Die mittlere Transferzeiten zeigen für das Token–Ring–Kanalzugriffsverfahren, bei den hier gewählten Parametern, die kleinsten Werte über den gesamten Angebotsbereich. Bei kleinen Angebotswerten verhält sich CSMA/CD (Ethernet) ähnlich wie Token–Ring, wird jedoch bei höherem Angebot früher instabil.
Es werden derzeit weitere Untersuchungen durchgeführt für :

- veränderte Netzausdehnung,
- schiefe Last,
- veränderte Stationszahl und
- unterschiedliche Prioritäten.

Der Author möchte sich bei E.–H. Göldner, R. Greinert, K. Sauer und M. Schubert für die Unterstützung bei der Durchführung der Simulationen bedanken.

Literatur

[1] IEEE Std 802.3–1985(ISO DIS 8802/3) Carrier Sense Multiple Access with Collision Detection (CSMA/CD); ISBN 0–471–82749–5

[2] IEEE Std 802.4–1985(ISO DIS 8802/4) Token–Passing Bus Access Method and Physical Layer Specifications; ISBN 0–471–82750–9

[3] IEEE Std 802.5–1985(ISO DP 8802/5) Token Ring Access Method and Physical Layer Specifications; ISBN 0–471–82996–X

[4] Draft of Proposed IEEE Standard 802.6 : Metropolitan Area Network (MAN), Media Access Control. (Slotted Ring); P802.6/85–01, Revision E, October 4, 1985

[5] W. Bux; Local–Area Subnetworks: A Performance Comparison; IEEE Transactions on Communications, Vol. COM–29, No. 10, pp. 1465–1473, 1981

[6] S.S. Lam; A Carrier Sense Multiple Access Protocol for Local Networks; Computer Networks, Vol. 4, pp. 21–32, 1980

[7] F.A. Tobagi, V.B. Hunt; Performace Analysis of Carrier Sense Multiple Access with Collision Detection; Computer Networks, Vol. 4, pp. 245–259, 1980

[8] E. Arthurs, B.W. Stuck; A Theoretical Performance Analysis of Polling and Carrier Sense Collision Detection Communication Sytems; Performance of Data Communication Systems, September 14, 1981, Paris, France

[9] J.L. Hammond, P.J.P. O'Reilly; Performance Analysis of Local Computer Networks; 1986, ISBN 0–201–11530–1

[10] M. Zafirovic–Vukotic, I.G. Niemegeers; Analytical Models of the Slotted Ring Protocols in HSLANs; In Proceedings of IFIP WG 6.4 Workshop "High Speed Local Area Networks", A. Danthine, D. Spaniol (ed.), February 16–17, 1987, Aachen, North Holland 1987

ON THROUGHPUT AND DELAY IN S-ALOHA MULTI-HOP NETWORKS

C. Gotthardt V. Brass
Fachbereich Elektrotechnik - Datenverarbeitungstechnik
FernUniversität Hagen, D-5860 Iserlohn

ABSTRACT

This paper presents the study of a multi-hop packet radio network with spatial randomly located stations using the S-ALOHA protocol. We evaluate the one-hop throughput for the case of different transmission probabilities for newly generated and previously collided packets. It is shown that the impact of the retransmission probability on throughput is comparable for fully-connected and multi-hop networks. Our main interest, however, is the evaluation of the mean backlog-time of a packet, i.e. the mean number of slots between the first transmission attempt and the successful transmission of a packet. We show that multi-hop networks have better stability properties than fully-connected networks because the number of users who can interfere with each other is very small. Further, the influence of the station's mobility on the mean backlog-time is treated.

1. INTRODUCTION

Communication of locally distributed mobile stations sometimes is performed by use of radio packet switching. A slotted radio channel is used to carry packets, generated by a cluster of bursty-traffic stations, governed by a multiple access protocol like S-ALOHA or CSMA.

With the S-ALOHA protocol the channel is splitted into equidistant slots of one packet length. On such a packet radio channel under S-ALOHA one station can transmit and all other stations in the transmitter's range can receive the packet. A station transmits its packets independently of other potential channel users. If two or more stations transmit simultaneously a collision occurs. The collided packets are destroyed and must be retransmitted. To avoid further collisions, stations are not allowed to repeat their collided packets in the same future slot. Instead, they must skip a random number of slots before repeating their collided packets. If there is a small traffic volume successful transmissions are highly probable, thus only the transmission-time contributes to the packet-delay. With heavier traffic or increasing number of competing stations collision probability rises. The packet-delay then will comprise in addition a portion resulting from waiting for retransmission.

The number of competing stations in such a mobile local area network can be reduced by reduction of the stations' transmission ranges. In doing so, a spatial separation between stations is gained, because those stations without radio connection cannot disturb each other. A transmitted packet is received only by a subset of all stations. Therefore it is possible that other stations in sufficiently distant locations of the network can transmit successfully at the same time. This matter of fact is called the spatial reuse of the ALOHA-channel.

On the other hand, by reduction of transmission power the packets require more hops through the network to reach their final destination and therefore additional channel capacity is required. To minimize packet-delay between source and destination the number of hops must be minimized. To reach this goal a packet must be routed per hop to a neighbour located as far as possible in the direction of the final destination; loop-ways are not desired. Thus, an appropriate performance measure to characterize such a system is the mean one-hop progress. It is determined by the transmission range and the density of stations, respectively. A small transmission range is favourable since it reduces collision recurrence but a large transmission range reduces the number of neccessary hops, because a successful packet completes a larger part of the whole way.

The literature /LAM73,TOBA75a,TOBA75b,KLEI75/ contains a detailed discussion of mean throughput and delay for S-ALOHA and CSMA random-access protocols. These papers, however, deal with fully-connected networks. In case of partly-connected networks results are known only for the throughput. From /SILV78/ it is known that mean throughput can be more than 1/e in multi-hop ALOHA networks, since two or more packets can be successfully transmitted during the same slot. Of course, this only is possible in sufficiently spatial distant localities of the network. In /TAKA84/ it is shown that the mean progress per hop is optimum, if every station covers on the mean N = 8 neighbour-stations in its transmission range.

Some applications require the system not to work with maximum throughput, but instead to offer a limited mean packet-delay. A solution for the mean packet-delay in not fully-connected S-ALOHA systems is not available today. Requirements for a maximum throughput and a small mean delay, at the same time, are mutually exclusive.

The mean backlog-time of a packet is defined as the time from the first collision of a packet until its successful transmission, measured in slots. This quantity is an appropriate performance measure for packet-delay per hop. In this paper we present a close approximation to compute the mean backlog-time.

In the next paragraph we give a detailed description of our model. In the third paragraph the analytical computations and in the fourth paragraph the numerical computations are presented. Finally, we conclude with a summary and discuss our results.

2. MODEL DESCRIPTION

In this paragraph we give a detailed description of our model. The assumptions introduced here are valid throughout this paper.

Transmission radius R

All stations have the same transmission range, stations inside a circle of radius R around a station P can receive this station's transmissions and are able to transmit packets to P in one hop. All the stations outside this circle cannot transmit (receive) to (from) station P in one hop, because they have no direct connection. A collision occurs if a station P has two or more active stations in its transmission range and one or more of the transmitted packets are adressed to station P. On the other hand, a collision occurs if a station transmits to a neighboured station which itself transmits a packet during the same slot. Collided packets have to be retransmitted in later slots.

Spatial distribution of the stations

The stations are randomly arranged in the plane according to a two-dimensional Poisson-distribution with density λ. This arrangement represents a snapshot of a mobile or semi-mobile network, which can be seen as quasi-stationary. The mean number of stations $N = \lambda\pi R^2$ in a circle of radius R is a measure of the network's connectivity. The system is assumed to comprise a large number of stations M (M>>N) in a homogenous area. Edge effects are neglected.

Channel-access method

The channel-access method is S-ALOHA and the slot-length corresponds to the packet-length plus the maximal signal propagation-delay.

Traffic behavior

The stations are assumed to generate and transmit new packets according to a Poisson-distribution with mean Mp packets/slot. After a collision the station's packet generation process is blocked until the collided packet is successfully retransmitted. Retransmissions of packets result in an additional arrival-process to the channel with mean α (packets/slot). A collided packet will be retransmitted with probability α in each slot following a collision. Thereby two arrival-processes appear which are assumed to be independent but both depend on the number of previously collided packets. The number k of skipped slots before a station decides to retransmit follows a geometrical distribution. The period following a collision until the station's next trial occurs is called mean wait of a packet.

$$\sum_{k=1}^{\infty} \alpha k(1-\alpha)^{k-1} = \frac{1}{\alpha} \tag{1}$$

It is assumed that the successful reception of a packet is confirmed immediately on a separate acknowledgement channel, thus acknowledgement traffic does not contribute to the general network's traffic load.

Distribution of the sources and destinations of packets

The stations communicate at equal probability $1/(M-1)$ with each other in the network.

Routing

The routing strategy MOST-FORWARD within R is used /SILV78/. Each station is assumed to know the position of those terminals within a distance R. If a packet has to be routed over more than one hop, then in each hop it is transmitted to that neighbour being located most forward in the direction of the final destination. If no neighbour exists in the half-circle of the forward direction, the station is forced to send the packet in the reverse direction. If the station actually has no neighbour, then the packet is not transmitted in the current slot.

In the model described the following performance measures are of interest:

1) $S(p,\alpha,N)$ = one-hop throughput

The one-hop throughput is defined as the mean number of successful one-hop transmissions of all stations during one slot. The one-hop throughput is not the effective throughput, because it might require several one-hop transmissions. The real effective throughput is the one-hop throughput divided by the mean number of hops necessary for a source-to-destination transmission. The mean number, say h, of hops per source-to-destination transmission depends on the mean distance between the randomly choosen source-to-destination pairs, established by the actual geometry. A measure of the effective network throughput is, therefore, the mean progress of a packet/slot.

2) $Z(p,\alpha,N)$ = mean progress of a packet in the destination's direction per slot, $Z = S(p,\alpha,N)/h$.

3) $D(p,\alpha,N)$ = mean one-hop delay

The one-hop delay is defined by the backlog-time of a packet plus its transmission-time. The variance of packet-delay mainly results from collisions. If the mean wait of a collided packet is $1/\alpha$ and the packet is collided i times, the mean delay is determined as i/α. The parameter controlling the distribution of a packet's waiting-time before its next transmission trial has considerable influence on the mean backlog-time and therefore is an important parameter in system design. /KLEI75/ has shown, that mean throughput and backlog-time in S-ALOHA systems mainly depend on the mean of that distribution. For this reason our model can be applied quite generally.

It is trivial to mention that the one-hop delay can only be evaluated if a neighboured station exists; otherwise the one-hop delay is infinite. The stations are assumed to have at least one neighbour; an insulated station occurs in our model with probability exp(-N).

3. ANALYSIS

3.1 A common binomial arrival-process

The most simple measure to evaluate is the one-hop throughput, the mean number of successful transmissions per slot. Since all stations are assumed to have an identical traffic behavior, we can take a randomly picked up station and evaluate its mean number of packets/slot received successfully. The total network-throughput results from summing over the one-hop throughputs of all stations.

First we evaluate mean throughput using the simplifying assumption that new and collided packets constitute an united arrival-process with transmission probability p. After a collision the affected packets are restored to contribute to the same arrival-stream like new packets and therefore are retransmitted with probability $\alpha = p$ in each subsequent slot. It follows that a station transmits a packet with probability p in the next slot regardless of former collisions. We study a randomly selected receiver Q with a sender P and i other stations in its transmission range R and evaluate the mean throughput:

$$S_i = p(1-p)(1-p)^i \tag{2}$$

where

p = probability that station P sends

(1-p) = probability that receiver Q is not active

$(1-p)^i$ = probability that the i other stations are not active.

This throughput applies under the condition that there are P plus i other stations in the transmission range of the receiver Q. Due to the Markovian-character of the Poisson-distribution we easily can determine the probability that there exist i other stations inside the circle with radius R around Q, conditioned on the fact that there is at least one transmitter in a distance R of Q /SILV78/:

$$p\{A_i\} = \frac{(\lambda\pi R^2)^i}{i!} e^{-\lambda\pi R^2} = \frac{N^i}{i!} e^{-N} \tag{3}$$

The conditioned one-hop throughput of a multi-hop S-ALOHA system with an infinite number of users therefore is:

$$E[S|i+2] = (1-e^{-N})\, p(1-p)(1-p)^i \tag{4}$$

From (4) the unconditioned one-hop throughput can be computed:

$$E[S] = S = \sum_{i=0}^{\infty} (1-e^{-N})\ p(1-p)(1-p)^i \frac{N^i}{i!} e^{-N}$$

$$E[S] = (1-e^{-N})\ p(1-p)\ e^{-pN} \qquad (5)$$

In a finite-user system, for example M = 20, the number of stations B_i in a subarea Ax of A is binomial-distributed.

$$p\{B_i\} = \binom{M}{i} \left(\frac{\lambda Ax}{\lambda A}\right)^i \left(1- \frac{\lambda Ax}{\lambda A}\right)^{M-i} = \binom{M}{i}\left(\frac{\lambda Ax}{M}\right)^i \left(1- \frac{\lambda Ax}{M}\right)^{M-i} \qquad (6)$$

The mean throughput of multi-hop S-ALOHA system comprising M stations is therefore:

$$E[S] = \sum_{i=0}^{M-2} \{1-(1- \frac{\lambda\bar{\pi}R^2}{M})^{M-1}\}\ p(1-p)(1-p)^i \binom{M-2}{i}\left(\frac{\lambda\pi R^2}{M}\right)^i \left(1- \frac{\lambda\pi R^2}{M}\right)^{M-i-2}$$

$$= \{1 -(1- \frac{N}{M})^{M-1}\}\ p(1-p)(1-\frac{pN}{M})^{M-2} \qquad (7)$$

In the limiting case when M increases to infinity we again get formula (5), because:

$$\lim_{M\to\infty}(1- \frac{pN}{M})^{M-2} = e^{-pN} \qquad (8)$$

3.2 Two separate arrival-processes

The assumption, that new and collided packets are transmitted with equal probability is now revoked, instead of this we investigate the general case $\alpha \neq p$. In this context we again consider the above established throughput equation (5).

$$E[S] = (1-e^{-N}) \sum_{i=0}^{\infty} \underbrace{p(1-p)^{i+1}}\frac{N^i}{i!} e^{-N} \qquad (9)$$

This equation might be interpretated to contain in the marked part the throughput contribution of one station in an (i+2)-user S-ALOHA system. In doing so, we get:

$$E[S] = (1-e^{-N}) \sum_{i=0}^{\infty} E[S|i+2] \frac{N^i}{i!} e^{-N} \qquad (10)$$

Taking into account the number j of backlogged packets and the resulting arrival-processes of new and backlogged packets, we are able to determine the mean throughput of an I-user S-ALOHA system. Both arrival-processes are completely defined by the number j of backlogged packets. Both processes assume a finite number of sources, their interarrival times are geometrically distributed, and therefore both are binomial-distributed. The arrival-processes of newly generated and collided packets X and Y, respectively, are:

Arrival-distribution of new packets:
$$P\{X=x\} = \binom{I-j}{x} p^x (1-p)^{I-j-x} \tag{11}$$

Arrival-distribution of collided packets:
$$P\{Y=y\} = \binom{j}{y} \alpha^y (1-\alpha)^{j-y} \tag{12}$$

Using these processes we can determine the mean throughput as:

$$\begin{aligned} E[S|I,j] &= \frac{1}{I}\cdot P\{X=1\}\cdot P\{Y=0\} + \frac{1}{I}\cdot P\{X=0\}\cdot P\{Y=1\} \\ &= \frac{I-j}{I} p(1-p)^{I-j-1} (1-\alpha)^j + \frac{i}{I} (1-p)^{I-j} \alpha(1-\alpha)^{j-1} \end{aligned} \tag{13}$$

Furthermore:
$$E[S,j|I] = E[S|I,j]\cdot P\{j|I\} \tag{14}$$

Applying a wellknown equation for conditioned probabilities and inserting equations (13) and (14) into equation (10) we get after summing over all j:

$$E[S] = (1-e^{-N}) \sum_{i=0}^{\infty} \sum_{j=0}^{i+2} E[S|i+2,j]\cdot P\{J=j|i+2\}\cdot\frac{N^i}{i!} e^{-N} \tag{15}$$

This equation contains one unknown, namely $P(J=j|i+2)$. $P(J=j|i+2)$ is the probability that j packets are backlogged in an (i+2)-user S-ALOHA system. To evaluate this distribution the (i+2)-user S-ALOHA system can be specified as a time-discrete Markov-chain with (i+2) states /KLEI75/. This Markov-chain is completely defined by the number of backlogged packets. State k of this chain is defined by k packets being backlogged. We recall here that stations having a collided packet cannot generate a new packet before their collided packet is transmitted successfully. The transition-probabilities of the Markov-chain according to /KLEI75/ are:

$$p(i,j) = \begin{cases} 0 & j \leq i-2 \\ i\alpha(1-\alpha)^{i-1}(1-p)^{I-i} & j = i-1 \\ (1-\alpha)^i (I-i)p(1-p)^{I-i-1} + [1-i\alpha(1-\alpha)^{i-1}]\,(1-p)^{I-i} & j = i \\ (I-i)p(1-p)^{I-i-1}\,[1-(1-\alpha)^i] & j = i+1 \\ \binom{I-i}{j-i} p^{j-i} (1-p)^{I-j} & j \geq i+2 \end{cases} \tag{16}$$

The stationary state-probabilities $P(J=j|i+2)$ can be determined by the system of linear equations:

$$P\{J=j|I\} = \sum_{k=0}^{I} P\{J=k|I\}\cdot p(k,j) \tag{17}$$

where:

$$\sum_{k=0}^{I} P\{J=k|I\} = 1 \tag{18}$$

The Markovian model provides the stationary probabilities for the number of backlogged packets in an (i+2)-user S-ALOHA system. From this probabilities it is possible to determine the mean value as usual:

$$E[J|i+2] = \sum_{k=0}^{i+2} k \cdot P\{J=k|i+2\} \tag{19}$$

By dividing this mean by the number of stations (i+2), the mean number of backlogged packets per station is gained:

$$E[B|i+2] = \sum_{k=0}^{i+2} \frac{k}{i+2} \cdot P\{J=k|i+2\} \tag{20}$$

This mean backlog again can be interpreted as a conditioned mean of blocked stations in an (i+2)-user S-ALOHA system, comparibly to what we have done to get the mean throughput of equation (10). By weighting the number of backlogged stations with their corresponding probabilities taken from the Poisson-distribution and summing up, the condition is resolved whereby we get the mean backlog of a station in a multi-hop S-ALOHA system.

$$\begin{aligned} E[B] &= (1-e^{-N}) \sum_{i=0}^{\infty} E[B|i+2] \cdot \frac{N^i}{i!} e^{-N} \\ &= (1-e^{-N}) \sum_{i=0}^{\infty} \sum_{k=0}^{i+2} \frac{k}{i+2} P\{J=k|i+2\} \frac{N^i}{i!} e^{-N} \end{aligned} \tag{21}$$

Of course, the most interesting and most important performance measure is the mean backlog-time. The packet-delay possibly is influenced by the mobility of stations. This is taken into account by assuming a new geographical configuration for any new packet, whilest a retransmission is assumed to take place in an unchanged situation. The packet-delay is substantially influenced by the mobility of stations. This is taken into account by the fact that by determining the mean backlog-time we have to observe the channel for several slots in which a packet will be repeated. In the meantime the number of neighbours could have changed since all stations are mobile. As the mobility of the stations is relatively slow compared to the random waiting-time of backlogged packets we argue that this effect can be neglected. For the above computation it is assumed that the number of neighbours is constant until the packet is transmitted successfully.

Now we demonstrate for the case $\alpha = p$ that the mean backlog-time decreases, if the topology is assumed to change during each slot. It can be shown that the above made assumption considering the system as quasi stationary results in an upper bound for the mean backlog-time. By this assumption we get the mean backlog-time of a multi-hop system by the same way of computation. We sum up the backlog-times of a fully-connected system with (i+2) users (i = 0,1,2,...), weight these times with their probability of occurrence, and have:

$$E[D] = \sum_{i=0}^{\infty} \frac{\sum_{k=0}^{i+2} \frac{k}{i+2} \cdot P\{J=k|i+2\}}{\sum_{j=0}^{i+2} E[S|i+2,j] \cdot P\{J=j|i+2\}} \frac{N^i}{i!} e^{-N} \tag{22}$$

Using the simplifying assumption that new and collided packets form a combined arrival-process with transmission-probability p (α = p), we easily calculate the mean backlog-time of an I-user S-ALOHA system by means of a simple equation. Then is:

$$T_b = E[W]\cdot E[Z] \tag{23}$$

where

E(W) = mean number of repetitions until successful transmission of a packet

E(Z) = mean wait between two transmission attempts of a packet

E(Z) is known already, equation (1).

$$E[Z] = \sum_{k=1}^{\infty} k\alpha(1-\alpha)^{k-1} = \frac{1}{\alpha} = \frac{1}{p} \quad \text{für } p = \alpha \tag{24}$$

As mentioned above, the mobility of stations has an impact on the number of repetitions neccessary. Two limiting cases can be distinguished:

a) The system is quasi stationary, i.e. the topology remains constant from the first to the successful transmission.

The probability P(W=j) that a packet has exactly j retransmissions until successful transmission is:

$$P\{W=j\} = [1-P\{A\}][1-P\{B\}]^{j-1}\, P\{B\} \tag{25}$$

P(A) = P(new packet is successfully transmitted)

P(B) = P(previously collided packet is successfully transmitted)

The mean number of retransmissions is:

$$E[W] = \sum_{j=1}^{\infty} j\cdot P\{W=j\} = \frac{1-P\{A\}}{P\{B\}} \tag{26}$$

Since new and collided packets have the same transmission-probability and constitute a combined arrival-process, the probability for a successful transmission is the same for both:

$$P\{A\} = P\{B\} = (1-p)^{i-1}$$

$$E[W] = \frac{1-(1-p)^{i-1}}{(1-p)^{i-1}} \qquad T_b = \frac{1-(1-p)^{i-1}}{p(1-p)^{i-1}} \tag{27}$$

Using the backlog-time of an fully-connected I-user S-ALOHA system we can now establish the mean backlog-time in a multi-hop S-ALOHA system, to be:

$$E[W|i] = \frac{1-(1-p)^{i-1}}{(1-p)^{i-1}} \qquad E[D|i] = \frac{1}{p}\, E[W|\, i] = \frac{1-(1-p)^{i-1}}{p(1-p)^{i-1}}$$

$$E[D] = \sum_{i=0}^{\infty} E[D|i+2]\cdot P\{i\} = \sum_{i=0}^{\infty} \frac{1-(1-p)^{i+1}}{p(1-p)^{i+1}} \frac{N^i}{i!} e^{-N} = \frac{1}{p}\{\frac{1}{1-p} e^{\frac{pN}{1-p}} -1\} \tag{28}$$

b) The stations are assumed to move so quickly that for each retransmission attempt a new topology must be considered. Then a new number of neighbours must be choosen randomly for each transmission attempt, independently of the number of neighbours during the last transmission. In this case we have for the number of retransmissions:

$$P\{W=j\} = \sum_{n_0=0}^{\infty} \sum_{n_1=0}^{\infty} \cdots\cdots \sum_{n_j=0}^{\infty} (1-p)^{n_j+1} \left(1-(1-p)^{n_{j-1}+1}\right)\cdots\left(1-(1-p)^{n_1+1}\right) \cdot\left(1-(1-p)^{n_0+1}\right) \frac{N^{n_0}}{n_0!} e^{-N} \frac{N^{n_1}}{n_1!} e^{-N} \cdots \frac{N^{n_j}}{n_j!} e^{-N} \quad (29)$$

$$P\{W=j\} = (1-p)\, e^{-pN} \left[1-(1-p)\, e^{-pN}\right]^j$$

$$E[W] = \frac{1}{1-p} e^{pN} - 1 \qquad E[D] = \frac{1}{p}\left[\frac{1}{1-p} e^{pN} - 1\right] \quad (30)$$

The differences in delay between cases a) and b) depend on the traffic intensity of the stations. Under a low traffic load (small transmission-probabilities p, small density of stations N) packet-delay in general is small, because only few packets collide there. Case a) and b) therefore differ only unsignificantly. If the traffic intensity increases, then packets collide more often and have to be retransmitted. If for example in case a) a station with a collided packet happens to have 10 neighbours instead of the mean, N = 5, then the local traffic can be expected to be much higher than elsewhere in the net, and therefore this station is expected to require more retransmission attempts than other stations, having a smaller number of neighbours. As a consequence the packet-delay will be substantially increased in this case and the mean delay must be expected to be higher than in case b). If the network topology is assumed to change during each slot, then the above mentioned abnormal situation with 10 neighbours quickly changes to a more normal situation. Therefore a packet can much easier be retransmitted in case b), and mean packet-delay can be expected to be smaller than in case a). From figure 1 we see that station's mobility has negligible influence on the backlog-time when the transmission probability is less or equal to 1/N.

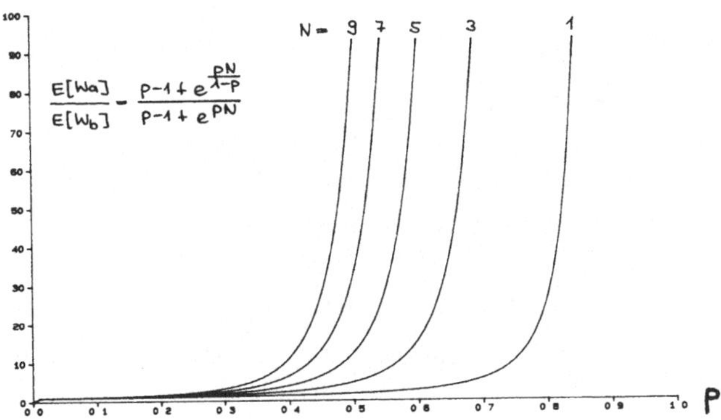

Figure 1: Relation of fixed- (a) to changing- (b) topology delay over p

4. NUMERICAL RESULTS

In this section we describe the model considered more precisely which is required to perform a simulation study. Furthermore, we describe the traffic model used for packet generation and the packet scheduling algorithm used.

We have simulated a model of the network, where the structure, traffic behavior and operational algorithms are such that all stations perform in a statistically identical manner and all links are able to carry the same traffic load. Associated with each station of the network, there is a random process which defines the points in time (the transmission slot) where the next transmission is attempt occurs. For the traffic behavior we assumed a heavy traffic situation, where each station has always at least one packet available for transmission. We simulated the S-ALOHA protocol for a network structure, where the stations are positioned by a random number generator in the plane, cf. figure 2, resulting in network-structures with randomly generated topologies, where the number of neighbourstations to a selected station obeys the Poisson-distribution. The network simulated comprises M = 50 stations distributed over a circle with radius r = 5km. We assumed throughout that all stations have a circlic transmission range of constant radius R = 2km (antennas with omnidirectional characteristic). Regular structures were not considered during this simulation.

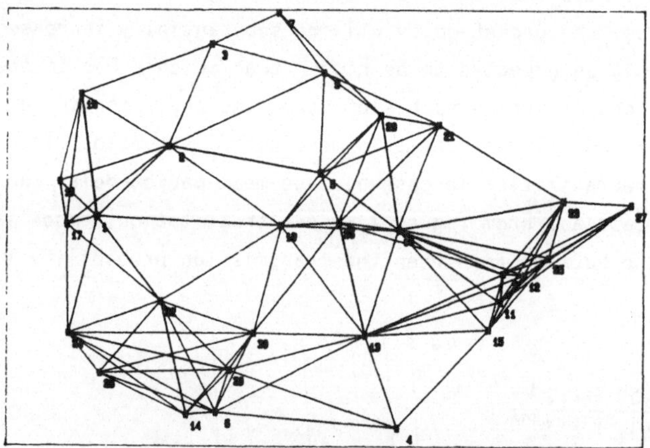

Figure 2: Example of a randomly generated network with a mean number of N=7 neighbours. The lines indicate which stations are in radio-contact.

Further, it is assumed that unlimited buffer resources are available at each station with a common queue for all outgoing links. Each station usesits own connectivity-matrix which includes information about which stations are in radio-contact and which not.

This matrix CM is defined by a boolean function as follows:

$$CM: S \times S \rightarrow (0,1) \text{ with}$$

$$CM(s1,s2) = \left\{ \begin{array}{l} 1, \text{ iff s1 and s2 are in radio-contact} \\ 0, \text{ otherwise} \\ S \text{ is the set of stations, with card}(S) = M \end{array} \right\}$$

Therefore the number of hops neccessary to reach a destination station is assumed to be known in advance and is recorded in a so called <u>distance-matrix</u>. Based on that matrix each station is able to select one neighbour which has a distance in hops to each final destination which is one less than its own. Therefore, a <u>static</u> shortest path scheme is used as a routing algorithm. The path length there is defined by the number of a route's hops. A single transmitter per station is assumed, whose ervice discipline for transmitting packets is First-Come - First-Served.

<u>Figure 3 (a)</u>
shows computational results for the one-hop throughput for α = 0.8p; 1.0p; 1.2p

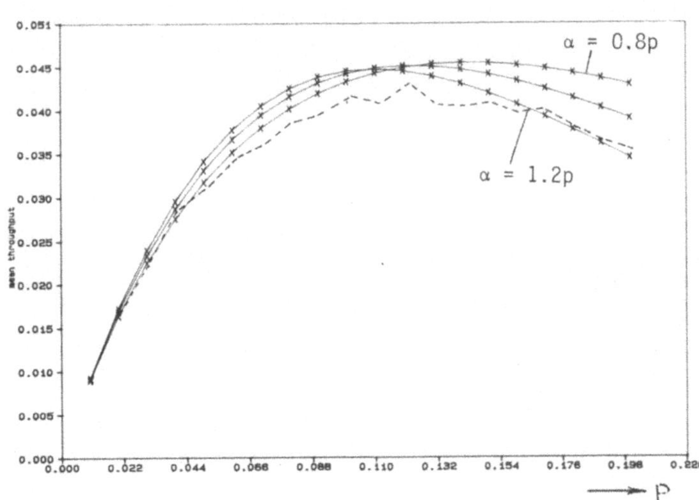

<u>Figure 3 (b)</u>
shows computational results for the one-hop throughput for α = 0.1 and 0.2, respectively

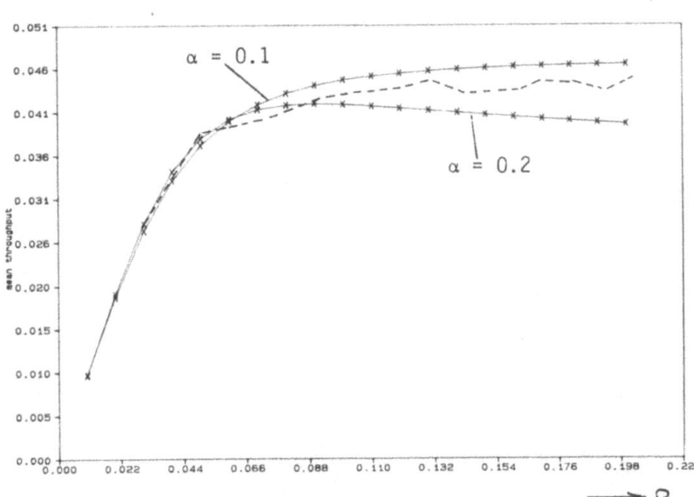

The dotted lines in both figures show the simulation results.

The three curves in figure 3 (a) differ in the factor b ($\alpha = bp$), which relates the transmission probability of collided packets to the transmission probability of newly generated packets. In all three cases we have the same maximum one-hop throughput where a higher factor b shifts the maximum value to a smaller transmission probability p. In figure 3 (b), the one-hop throughput for a constant retransmission probability α is shown, where the mean throughput for $\alpha = 0.1$ is better than for $\alpha = 0.2$.

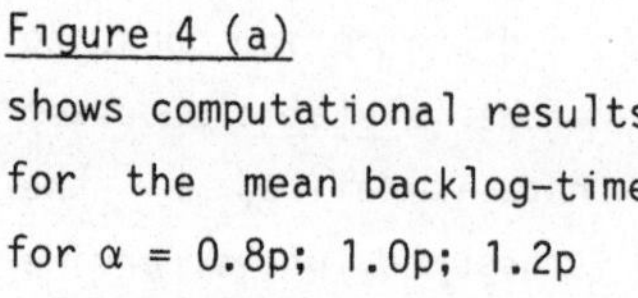

Figure 4 (a)
shows computational results for the mean backlog-time for $\alpha = 0.8p;\ 1.0p;\ 1.2p$

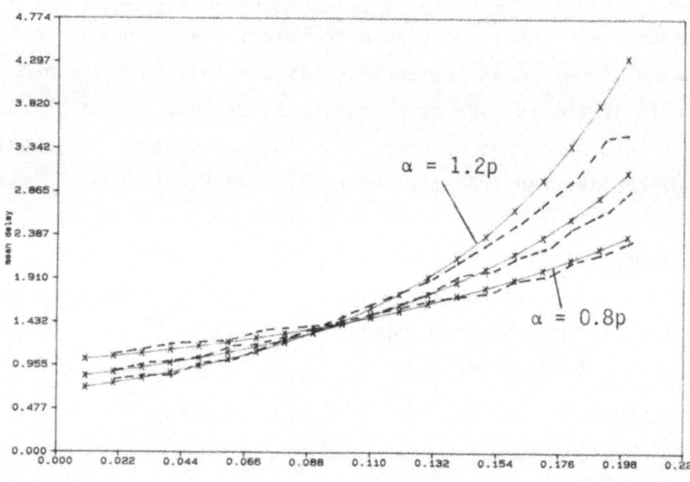

Figure 4 (b)
shows the analytic mean backlog-time for $\alpha = 0.1$ and 0.2, respectively

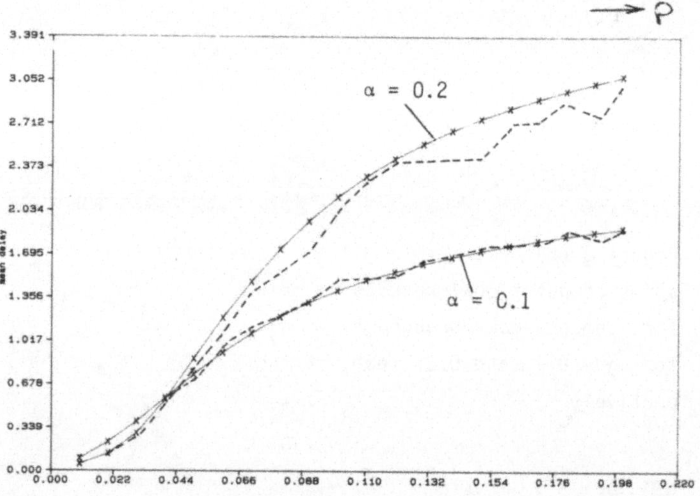

From figure 4 we see that the retransmission probability α has an serious infuence on the mean backlog-time. In figure 4 (a), $\alpha = bp$, the mean backlog-time always increases to infinity. However, in case of a constant transmission probability for collided packets, cf. 4 (b), the mean backlog-time does not increase to infinity and therefore the system never becomes instable. In all figures the mean number of neighbours to a station is seven.

5. SUMMARY

We have investigated a S-ALOHA multi-hop network with randomly distributed mobile stations. Using a Markovian Model for describing an i-user S-ALOHA system we have solved for the mean one-hop throughput and the backlog-time in a multi-hop network. The one-hop throughput is evaluated for the case where collided packets have a differing transmission probability from newly generated packets. It is shown that this backlog transmission probability has an important influence on mean throughput and on the stability of the system. Our main interest was to evaluate the mean backlog-time. Further, we used a praticable routing algorithm, which approximates the MOST-FORWARD routing rule. The system was observed with respect to a number of different parameters. The numerical results are validated by simulation and give a short impression of the system's behavior. It was shown that like in other systems high throughput results in a high backlog-time, but that there exists an optimal parameter set to either optimize the mean throughput or the mean backlog-time.
The continuous support of the paper by B. Walke is gratefully acknowledged.

REFERENCES

/KLEI75/ L. Kleinrock, S.S. Lam: Packet Switching in a Multiaccess Broadcast Channel: Performance Evaluation; IEEE Transactions on Communications, Vol. COM-23, No. 4, pp. 410-423, April 1975.

/LAM73/ S.S. Lam, L. Kleinrock: Packet-switching in an slotted satellite channel; National Computer Conference (NCC 1973), pp. 703-710

/SILV78/ J. Silvester, L. Kleinrock: Optimum Transmission radii for Packet Radio Networks or Why six is a magic number; in Conf. Rec., Nat. Télecommun. Conf., pp 4.3.1-4.3.5, Dec 1978.

/TAKA84/ H. Takagi, L. Kleinrock: Optimal Transmission Ranges for Randomly Distributed Packet Radio Terminals; IEEE Transactions on Communications, Vol. COM-32, No. 3, pp. 246-257, March 1984.

/TOBA75a/ F.A. Tobagi, L. Kleinrock: Packet Switching in Radio Channels: Part I - Carrier Sense Multiple-Access Modes and Their Throughput-Delay Characteristics; IEEE Transactions on Communications, Vol. COM-23, No. 12, pp. 1400-1416, Dec 1975.

/TOBA75b/ F.A. Tobagi, L. Kleinrock: Packet Switching in Radio Channels: Part II - The Hidden Terminal Problem in Carrier Sense Multiple-Access and the Busy-Tone Solution; IEEE Transactions on Communications, Vol. COM-23, No. 12, pp. 1417-1433, Dec 1975.

Munich Simulation Computer: Leistungsbewertung und Modellierung

W. Hahn, H. Anger, A. Hagerer
Fachbereich Informatik
Universität Passau

Leistungsfähige Spezialsysteme für die Logik-Simulation werden mehr und mehr benötigt, um den Zeitaufwand für die funktionale Evaluierung von VLSI-Entwürfen gering zu halten. Der Bericht skizziert die Leistungsbewertung des Munich Simulation Computer (MuSiC), eines hochparallel organisierten Ereignisflußrechners für die schnelle Simulation digitaler Systeme. Zwecks Überprüfung der getroffenen Leistungsaussagen ist ein in OCCAM formuliertes und auf einem Multi-Transputernetz implementiertes Modell in Entwicklung. Konzepte dieser Modellierung werden vorgestellt.

1. Einleitung

Steigende Ansprüche an Kompaktheit, Zuverlässigkeit und Leistung führten zur weitreichenden Formalisierung und damit möglichen Rechnerunterstützung des Entwurfsverfahrens für integrierte Schaltungen. Im Vordergrund stehen dabei Werkzeuge für die Entwurfsschritte "Konstruktion" und "Verifikation", die, dader praktischen Anwendbarkeit formaler Verifikationsmethoden zur Zeit noch enge Grenzen gesetzt sind, vorerst eine Überprüfung eines Entwurfes nur mittels Simulation erlauben. Derzeit noch weitgehend eingesetzte, programmierte Simulatoren auf Universalrechnern haben für die funktionale und logische Evaluierung der immer komplexer werdenden Schaltungen jedoch einen prohibitiv hohen Rechenzeitbedarf. Trotz ihrer prinzipiellen Eignung ist daher darüberhinaus an einen Einsatz in weiterführenden Anwendungsgebieten, wie

- Verifikation von Mikroprogrammen
- Entwicklung neuer Systemarchitekturen
- Evaluierung von Systemprogrammen auf simulierten Prozessoren

nicht zu denken.

Hauptursache für die geringe Simulationsgeschwindigkeit programmierter Simulatoren ist die Notwendigkeit, die Auswertung der durch parallele Signalflüsse simultan aktivierten Schaltungselemente zu sequentialisieren. Die Strategie einer Hardware-Implementierung des Simulationsalgorithmus in Form eines Spezialsystems (/1/,/2/) ist ein begrenzter Ausweg. Bei massiver Ausnutung der Parallelität bei der Komponentenauswertung wird eine Leistungssteigerung gegenüber programmierten Simulatoren in der Größenordnung um 10^3 erreicht.

Demgegenüber ist das Arbeitsprinzip des bisher schnellsten Simulationsrechners, der Yorktown Simulation Engine (YSE) von IBM (/3/), die zustandsorientierte Simulationsstrategie auf der Basis einer Übersetzung einer Entwurfsbeschreibung in ausführbaren Code, bei der zu jedem Simulationszeitpunkt alle Komponenten ausgewertet werden. Obwohl mittels Pipelinetechnik effizient in Hardware umsetzbar, besitzt diese Simulationsstrategie den Nachteil, daß ein großer Teil der Rechenleistung auf nutzlose, weil im Ergebnis unveränderte Komponentenauswer-

tungen verwendet wird.
Ein weiterführender Vorschlag ist der *Munich Simulation Computer* (*MuSiC*) (/4/) mit ereignisgesteuerter Simulationsstrategie und ebenfalls der Fähigkeit zu hochgradig paralleler Komponentenauswertung. Im Anwendungsgebiet Logiksimulation, insbesondere bei geringem Ereignisaufkommen, wird dadurch die Leistung der YSE übertroffen. Eine 256-Prozessor-Version von MuSiC kann mehr als 8 Millionen Gatter und Flipflops mit einer Geschwindigkeit im Bereich von 10^9 bis 10^{10} Gatterauswertungen pro Sekunde simulieren.
Operationsprinzipien und Organisation von MuSiC werden kurz erläutert. Darauf aufbauend erfolgt eine Skizzierung der Leistungsvorhersage unter Beachtung leistungsbeeinflußender Entwurfsmerkmale. Anschließend werden die Konzepte einer Modellierung des MuSiC-Entwurfes mittels OCCAM auf einem Multi-Transputernetzwerk vorgestellt.

2. MuSiC - Entwurf

Der zu simulierende Entwurf einer digitalen Schaltung ist ein vernetztes System von Grundkomponenten, die durch Leitungen miteinander verbunden sind. Die formale Beschreibung des vernetzten Systems ist ein Graph. Bei der Simulation wird dieser Graph als ein Programm im Sinne von Datenflußprogrammen angesehen. Die Auswertung von Komponenten ist damit gleichwertig zur Ausführung von Befehlsknoten. Ausnutzbare Parallelität ist direkt ablesbar.
Dem realen Signalfluß entspricht im Programmgraphen ein Datenfluß. Die Nutzung der Tatsache jedoch, daß zu einem Zeitpunkt nur ein Teil der Elemente aufgrund von Signaländerungen an den Eingängen aktiviert wird und nur dieser Teil auszuwerten ist, bedeutet für die Simulation, daß die Modellierung des realen Signalflusses mittels Ereignisfluß (Änderungsdatenfluß) (/5/) und nicht mittel Datenfluß erfolgt. Prinzipien datengetriebener Rechnerarchitekturen sind daher die Grundlage für die Implementierung des Ereignisflußkonzeptes in MuSiC:

- ein MuSiC-Programm ist eine Menge von Knoten (Templates).
  Jedes Template enthält die Operationsbeschreibung des modellierten Elementes, Speicherplätze für die Operanden und Verweise auf nachfolgende Knoten.
- Programmabarbeitung bedeutet wiederholende Anwendung von:
  . Aktivierung eines Templates erfolgt durch die Ankunft einer Operandenänderung (Feuerungsregel).
    Es werden nur Templates zur Ausführung freigegeben, bei denen keine Operandenänderung mehr eintreffen kann (Selektionsregel). Diese Templates werden einem Operationswerk zwecks Auswertung der Operation mit den veränderten Operanden zugeführt.
    Die Implementierung berücksichtigt, daß bei ereignisgesteuerter Templateaktivierung und vollständig asynchroner Ausführung der Operationen der aktivierten Templates sich das Problem der Mehrfachselektion eines Templates stellt: es wird ein Template durch

die Ankunft eines neuen Operandenwertes schon zur Ausführung ausgewählt, obwohl diesem Template später noch weitere Operandewerte zugestellt werden. In MuSiC wird daher, indem auf dem Programmgraphen bzw. auf der Menge der Templates eine Rangfunktion definiert wird:

- Templates für speichernde Elemente besitzen den Rang 0,
- eine Template besitzt den Rang i, falls
  mindestens ein direktes Vorgängertemplate den Rang i-l besitzt,
  jedes Vorgängertemplate einen Rang j besitzt, mit j<i,
  das Template kein ein speicherndes Element beschreibendes Template ist,

bei der Programmabarbeitung sichergestellt, daß nur Templates zur Ausführung freigegeben werden, bei denen keine Operandenänderung mehr eintreffen kann, denn sobald alle Templates der Ränge i<j ausgewertet sind, liegen alle Operanden für Templates des Ranges j vor. Die Menge der Templates mit gleichem Rang sind voneinander datenunabhängig und können nebenläufig bzw. in beliebiger Reihenfolge ausgeführt werden.

Ein Ergebnis einer Auswertung wird nur dann nachfolgenden Knoten zugestellt, wenn es von dem zuvor zugestelltem Ergebnis abweicht (Verteilungsregel).

Die Entscheidung, ob ein Ergebnis weiterzuleiten ist, kann auf dem Vergleich berechnetes Ergebnis mit zuvor berechnetem, gespeichertem Ergebnis basieren. Dieses Verfahren bedeutet jedoch eine Leistungsminderung, z.B. durch erhöhtes Transportaufkommen von Ergebnissen. Die Ermittlung der Ergebniswertveränderung basiert in MuSiC hingegen auf der simultanen Auswertung der Templates mit den neuen, veränderten und den alten Operanden.

MuSiC kann aus bis zu 2048 Verarbeitungseinheiten (PU, processing unit) bestehen. In jeder PU können 32K Templates gespeichert werden, so daß Entwürfe in der Größenordnung um 60 Millionen Gatter simuliert werden können.

Die PU's sind in fünf Kommunikationsnetze eingebettet:

- Program Load Communication Net
  Die durch Übersetzung eines Entwurfes auf einem Host-System generierte Menge von Templates wird über dieses Netz auf die PU verteilt.
- Result Packet Communication Net (RPCN)
  Die in der PU i ermittelten Ergebnisse werden über ein Delta-Netz /6/ zur PU j (i≠j) transportiert. Das Netz besteht aus ($\lceil \log_2 N+1 \rceil$) 2x2-Routern (N: Zahl der PU's), so daß I/O-Kommunikation mit dem Host ebenfalls über dieses Netz abgewickelt werden kann.
- Operation Packet Communication Net
  Die Ausführung eines aktivierten Templates ist nicht an die das Template speichernde PU gebunden. Zur Ausführung aufbereitete Templates (Operationspakete) können im Rahmen einer Lastverteilung über dieses Netz zu einer anderen, nicht ausgelasteten PU transportiert werden.

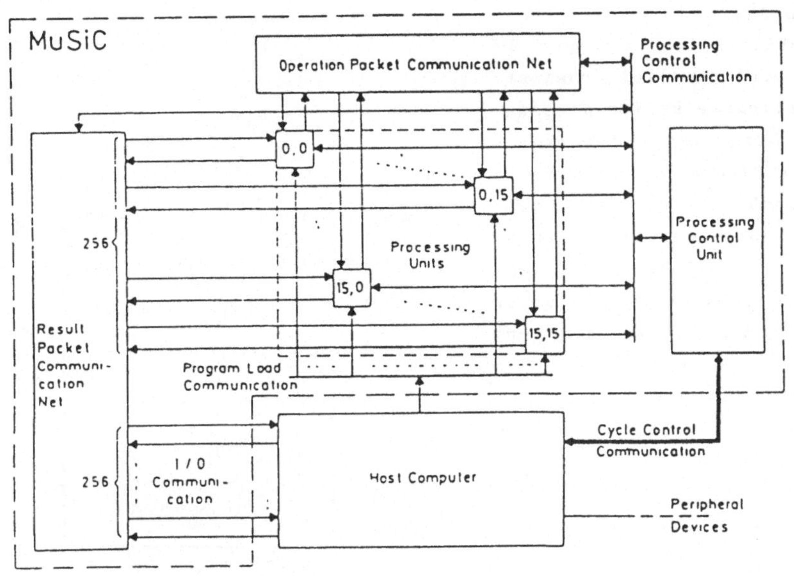

Abb. 1: MuSiC - Organisation

- Processing Control Communication Net
  Signale zur Steuerung und Uberwachung des Fortschreitens der Programmarbeitung von Rang zu Rang werden über dieses Netz übertragen.
- Cycle Control Communication Net
  Die Steuerung der Simulationszyklen wird vom Host-System aus vorgenommen.

Eine Varbeitungseinheit besteht aus drei Hauptkomponenten:

- Cell Block
  Ein Zellblock enthält acht unabhängige Speichereinheiten zur Aufnahme der Templates und führt die Aktualisierung der Templates bei Ankunft neuer Operandenwerte mittels der Pipelinestufen Update Preparation Unit (UPU) und Template Memory Unit (TMU, im Update-Modus arbeitend) sowie die Bereitstellung zuvor aktivierter Templates als Operationspakete mittels der Pipelinestufe TMU (nun im Fetch-Modus arbeitend) durch. In der Stufe External Input Unit (EIU) werden vom RPCN Ergebnispaketteile seriell entgegengenommen und zu einem Paket zusammengesetzt. Die Stufe Input Distribution Unit (IDU) führt die Zustellung des Ergebnispakets zu der UPU/TMU-Einheit aus, inder das vom Paket addressierte Template gespeichert ist.
- Operation Packet Distribution Unit
  Aus den von den TMU's und zwei Puffern, in denen Operationspakete anderer PU's gespeichert werden können, angebotenen Operationspaketen wird eines ausgewählt und der Funktionseinheit zugestellt.

- Function Unit
  Die Funktionseinheit einer PU enthält drei parallel arbeitende Operationswerke für die Ausführung bitweiser und wortweiser Operationen und algorithmisch beschriebener Funktionen. Weitere Pipelinestufen übernehmen die Verteilung der Ergebnispakete (Result Distribution Unit, RDU) und die Aufsplittung eines Ergebnispaketes in sequentiell zu versendende Ergebnispaketteile (External Output Unit, EOU).

Jede Pipelinestufe benötigt für die Ausführung eines Arbeitsschrittes vier Takte. Ausgenommen davon sind das wortweise arbeitende Operationswerk und das Operationswerk für algorithmisch beschriebene Funktionen.

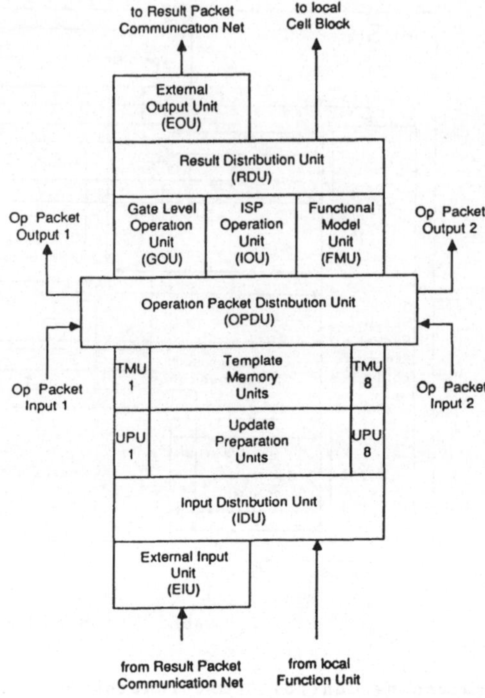

Abb. 2: Pipelinestufen einer PU

3. Leistungsvorhersage

Die nachfolgend vorgestellte Prognose der Leistung von MuSiC verwendet folgende Einstellung der Basisgrößen:

- N = 256 Verarbeitungseinheiten
- Operationsausführungszeit einer Pipelinestufe von $\delta = 120$ ns. (Taktabstand $\tau = 30$ ns.)
- Vektorisierbarkeit von Operanden bis zum Grad v = 8, d.h. Zusammenfassung von bis zu acht Signalen in einem Operanden,
- Fan Out ≤ 2

und beschränkt sich auf die Simulation auf Gatterebene mittels bitweiser Operationen.
Prinzipiell werden Gatter niedrigen Funktionsniveaus (AND, OR, EXOR, AND/OR, MULTIPLEXER, ...) mit bis zu vier Eingängen durch eine Maschineninstruktion modelliert, wobei die Vektorisierbarkeit von Operanden die Abbildung mehrerer, gleichartiger Gatter auf eine Instruktion erlaubt, was die Zahl der von MuSiC auszuführenden Instruktionen u.U. deutlich reduziert.

Bezeichnet f(v) den Anteil der Schaltungselemente an der Gesamtanzahl n_{rk} von

Elementen auf einem Rang, die mittels eines Templates mit Operanden des Vektorisierungsgrades v nachgebildet werden können, so sind maximal

$$n_f = \sum_{v=1}^{8} \frac{f(v) \cdot n_{rk}}{v}$$

Templates zu speichern.

Das Ereignisflußprinzip reduziert die Zahl der aktivierten Templates. Bezeichnet p_e die Wahrscheinlichkeit für die Änderung eines Einzelsignales, so erfolgt eine Templateaktivierung mit der Wahrscheinlichkeit

$$p_t = \sum_{v=1}^{8} [\ f(v) \cdot (1 - (1 - p_e)^v)\ ] \quad ,$$

woraus sich eine Anzahl von $p_t \cdot n_f$ auszuwertenden Templates ableitet. Die Auswertung kann bei N PUs minimal in der Zeit

$$t_E = \frac{p_t \cdot n_f}{N} \cdot \delta \qquad [\ s\ ]$$

erfolgen.

Bedingt durch die ranggeordnete Simulation ist eine Latenzzeit für die Pipeline TMU_i - OPDU - GOU - RDU - EOU - [(1 + $\lceil \log_2 N \rceil$) Stufen des RPCN] - EIU - IDU - UPU_j - TMU_j + [7 für die Processing Control Unit] einzubeziehen:

$$t_{rk}^- = t_E + t_\lambda \qquad [\ s\ ]$$

$$t_\lambda = (\ 43 + 4 \cdot (1 + \lceil \log_2 N \rceil)\ ) \cdot \tau \qquad [\ s\ ].$$

Das natürliche Leistungsmaß für den Ereignisflußrechner MuSiC wäre an sich die Anzahl der je Sekunde berechneten Ereignisse. Der besseren Vergleichbarkeit wegen wird im folgenden Text jedoch jeweils als Quotient aus Gesamtzahl der Gatter und der zur Auswertung der daraus jeweils aktivierten Gatter benötigten Zeit die Anzahl der Gatterauswertungen je Sekunde [G/s] bestimmt, die ein System wie z.B. die YSE haben müßte, um einen Entwurf gleicher Gatterzahl in der gleichen Zeit simulieren zu können. In diesem Sinne ergibt sich als folgende obere Leistungsgrenze

$$G_{rk}^+ = \frac{n_{rk}}{t_{rk}} \cdot 10^9 \qquad [\ G/s\ ].$$

Für n → ∞ und obigen Basiswerten ergibt sich daraus ein Grenzwert von

$$G_{\infty,p_e}^+ = \frac{8 \cdot 256}{(1 - (1 - p_e)^8\ ) \cdot 120} \cdot 10^9 \qquad [\ G/s\ ].$$

Für z.B. $p_e = 0.4$ bedeutet dies eine Simulationsleistung von

$$G_{\infty,0.4}^+ = 1.74 \cdot 10^{10} \qquad [\ G/s\ ].$$

Eine Betrachtung der Grenzfälle

$$p_e \sim 0.1 : \qquad G_{rk}^+ \rightarrow \frac{1}{p_e} \cdot \frac{n_{rk} \cdot N}{n_f \cdot \delta} =: g_\sim$$

$$p_e \rightarrow 0 : \qquad G_{rk}^+ \rightarrow \frac{n_{rk}}{t_\lambda} =: g_0$$

ergibt, daß

(1) die Leistung von MuSiC umgekehrt proportional zur Signaländerungswahrscheinlichkeit p_e ist,

(2) die Leistung für kleine Werte von p_e durch die Kosten der Rangeinteilung begrenzt wird,

während sich bei hoher Signaländerungswahrscheinlichkeit, also bei häufigem Auswerten nahezu aller Templates, die Operandenvektorisierung, bedingt durch die damit verbundene Reduzierung der Zahl der Templates, leistungssteigernd auswirkt.

In einer Abschätzung der von MuSiC erbrachten, mittleren Leistung muß allerdings der Einfluß von

- Fan Out > 2,
- ungleichmäßiger Lastverteilung und
- Blockaden beim Transport der Ergebnispakete durch das Result Packet Communication Net

Berücksichtigung finden.

(1) 2 < Fan Out $\leq$ 32

Ein Funktions-Template enthält Verweise auf bis zu zwei Nachfolge-Templates. Die Nachbildung eines Elementes mit höherem Fan Out geschieht mittels Einschubs von Copy-Templates, die jeweils bis zu vier Nachfolge-Templates besitzen können. Mit z.B. zehn Copy-Templates, in zwei Stufen angeordnet, kann so Fan Out auf 32 erhöht werden. Bezahlt werden muß dies mit bis zu zehn, bezogen auf den Simulationsfortschritt, unproduktiven Instruktionen. Die Erhöhung der Zahl aktivierter Templates pro Rang erhöht die Auswertungszeit und verringert gemessen in Gatterauswertungen pro Sekunde die Leistung von MuSiC.

(2) dynamische Lastverteilung

Die gleichmäßige Verteilung der Funktions- und Copy-Templates eines Ranges auf die N Verarbeitungseinheiten, die N als statische Last bezeichnete Mengen von l_{st} Templates ergibt, führt während der Programmabarbieung aufgrund der ereignisgesteuerten Aktivierung dennoch für jede Verarbeitungseinheit zu einer unterschiedlichen Zahl aktivierter Templates. Die Auswertung aller aktivierten Templates eines Ranges ist erst mit der Ankunft des letzten Ergebnispaketes abgeschlossen. Bei gleichartigem Verhalten aller Verarbeitungseinheiten (p_e unabhängig von der PU) wird die Zeitdauer für die Simulation eines Ranges von der Tätigkeitszeit der Verarbeitungseinheit mit der maximalen Anzahl aktivierter Templates beeinflußt. Unter der Annahme der Gleichwahrscheinlichkeit und der Unabhängigkeit der Aktivierung eines in einer Verarbeitungseinheit gespeicherten Templates ist die Wahrscheinlichkeit für die Aktivierung von k Templates binomialverteilt:

$$P\{X = k\} = \begin{bmatrix} l_{st} \\ k \end{bmatrix} \cdot p_t^k \cdot (1-p_t)^{l_{st}-k}$$

Der Preis, der für die Synchronisation mittels Ränge zu bezahlen ist, beträgt demnach

$$E\,[\,\max\,\{X_1,\ldots,X_N\}\,]\; - \;E\,[X]$$
$$= \sum_{k=1}^{l_{st}} k\cdot[\,F^N(k) - F^N(k-1)\,] \; - \; p_t\cdot l_{st}$$

mit $X, X_1, \ldots, X_N$ unabhängig identisch gemäß einer Binomialverteilung mit Verteilungsfunktion F verteilten Zufallszahlen. Die Schwierigkeiten bei der Berechnung von $E\,[\,\max\,\{X_1,\ldots,X_N\}\,] - E\,[X]$ sind in /7/ erörtert worden. Die ebenfalls dort ermittelten numerischen Ergebnisse zeigen, daß

- die relativen Synchronisationskosten

  $$c_r = \frac{E\,[\,\max\,\{X_1,\ldots,X_N\}\,] - E\,[X]}{E\,[\,X\,]}$$

  zwischen 0.5 und 0.04 liegen, für Aktivierungswahrscheinlichkeiten p_t zwischen 0.04 und 0.4,
- die Sychronisationskosten mit steigender Anzahl an Verarbeitungseinheiten nur sehr langsam ansteigen, d.h. eine Verdoppelung der Zahl der Prozessoren bedeutet nahezu eine Verdoppelung der Leistung.

(3) Ergebnispakettransport

Die Zeit t_r für den Transport aller auf einem Rang erzeugten Ergebnispakete zu ihren Ziel-TMU's, wo sie die Templates weiterer Ränge aktivieren, muß in die Abschätzung der mittleren Simulationszeit für einen Rang einbezogen werden. Dazu wird folgendes Modell des Ergebnispaketflußes verwendet:

- der Strom von Ergebnispaketen durch das RPCN wird aufgrund von Konflikten um den konstanten Faktor $1/c_{th}$ zeitlich gedehnt (c_{th} normalisierter Durchsatz: Verhältnis von Durchsatz für eine spezifische Umgebung zu maximalem Durchsatz (/6/)),
- mit konstanter Wahrscheinlichkeit p_{ex} wird ein Ergebnispaket zu einer anderen Verarbeitungseinheit, als der, in der es erzeugt wurde, transportiert (externes Ergebnispaket),
- die Zieladdressen externer Ergebnispakete genügen einer Gleichverteilung, so daß ein Flußgleichgewicht besteht: die Zahl von einer Verarbeitungseinheit an das RPCN abgegebenen Pakete ist gleich der vom RPCN empfangenen Pakete,
- externe und lokale Ergebnispakete, die gleichzeitig in der IDU ankommen, können parallel weitergeleitet werden, wenn sie verschiedene der n_{UPU} im Update-Modus arbeitenden UPU/TMU-Einheiten addressieren, was mit Wahrscheinlichkeit $(1 - (1/n_{UPU})^2)$ erfolgt. Im Konfliktfall wird das externe Paket mit Priorität weitergeleitet und das lokale um eine Operationsausführungszeiteinheit δ verzögert,
- ein Template besitzt im Mittel $E[T_{FO}]$ Nachfolgetemplates (mittlerer Fan Out).

Aufgrund der änderungsgesteuerten Ergebnispaketerzeugung fallen in einer Verarbeitungseinheit im Mittel bei $n_t = \Sigma\, n_t(v)$ aktivierten Templates

$$n_{ta} = \sum_{v=1}^{8} (1 - (1 - p_e)^v) \cdot n_t(v)$$

Operationspakete an, die Ergebnispakete erzeugen. Es werden im Mittel

$$n_{ex} = p_{ex} \cdot E[T_{FO}] \cdot n_{ta}$$

externe und

$$n_{lc} = (1-p_{ex}) \cdot E[T_{FO}] \cdot n_{ta}$$

lokale Ergebnispakete erzeugt.

Bei konfliktfreiem Transport betragen die Transportzeiten inklusive der Latenzzeiten für die jeweiligen Pipelinestufen

$$t_{ex} = \left(\frac{n_{ex}}{c_{th}} + 5 + \lceil \log_2 N \rceil \right) \cdot \delta \qquad [s]$$

$$t'_{lc} = (n_{lc} + 4) \cdot \delta \qquad [s].$$

Unter Einbeziehung der Bearbeitungszeit für die aktivierten Templates

$$t_{ta} = (n_{ta} + 2) \cdot \delta \qquad [s]$$

ergibt sich eine Rangbearbeitungszeit von

$$t'_{rk} = \max \{ t_{ta}, t'_{lc}, t_{ex} \} \qquad [s].$$

Mittels der Wahrscheinlichkeit, mit der ein lokales Ergebnispaket ein externes innerhalb der Rangbearbeitungszeit in der IDU vorfindet, und der Wahrscheinlichkeit der UPU-Addressengleichheit, berechnet sich die mittlere Anzahl von Addresskonflikten:

$$n_{cl} = n_{lc} \cdot \left[\left(\frac{n_{ex}}{t'_{rk}}\right) \cdot \left(\frac{1}{n_{UPU}}\right)^2 \right]$$

Die Transportzeit für die lokalen Ergebnispakete verzögert sich auf

$$t_{lc} = (n_{lc} + 4 + n_{cl}) \cdot \delta \qquad [s].$$

Die Rangbearbeitungszeit beträgt dann

$$t^-_{rk} = \max \{ t_{ta}, t_{lc}, t_{ex} \} \qquad [s].$$

Als eine Abschätzung der unteren Leistungsgrenze von MuSiC erhält man folglich

$$G^-_{rk} = \frac{n_{rk}}{t^-_{rk}} \qquad [G/s].$$

Die Grafik veranschaulicht die Leistungsgrenzen im Falle der Simulation eines in "two-level"-Logik entworfenen Pipelineprozessors mit 10^6 Gattern und Flipflops. Die Leistungsangaben zur YSE sind aus /3/ entnommen.

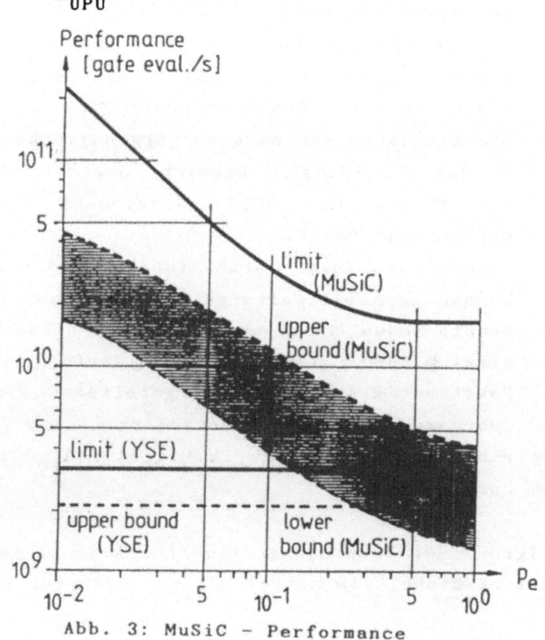

Abb. 3: MuSiC - Performance

4. Verifikation der Leistungsvorhersage

Der Entwurf von MuSiC wird mittels eines Modelles hinsichtlich Leistung und Funktionalität überprüft. Das MuSiC-Modell, in OCCAM formuliert und auf einem Multi-Transputernetz ausführbar, ist zugleich ein funktionsfähiger Simulator, so daß mit der Simulation verschiedener, charakteristischer Entwürfe auch Abhängigkeiten der Leistung von Eigenschaften zu simulierender Schaltungen, z.B. die die Leistung wesentlich beeinflußende Signaländerungswahrscheinlichkeit, quantifiziert werden können.

Die prozeßorientierte Programmiersprache OCCAM offeriert dem Anwender gleichzeitig die Vorteile einer Hochsprache und die effiziente Nutzung der Kooperationsfähigkeit von Transputern (/9/) in einem Netz. Mittels OCCAM kan ein System in Form einer Menge miteinander verbundener Prozesse entworfen werden. Jeder Prozeß, für sich gesehen eine unabhängige Entwurfseinheit, führt eine Reihe von Aktionen repetierend aus. Aktionen sind elementare Prozesse, Sequenzen von Prozessen oder Mengen parallel ablaufender Prozesse. Elementare Prozesse, Prozesse mit einer einzigen Aktion, sind die Zuweisung, Eingabe und Ausgabe. Ein- und Ausgabe werden für die Kommunikation zweier paralleler Prozesse über unidirektionale Kanäle benötigt. Der Nachrichtenaustausch ist synchronisiert und ungepuffert. Die Terminierung des Ausgabeprozesses verzögert sich solange, bis ein den Kanal benutzender Eingabeprozeß gestartet wird. Im Anschluß an den Nachrichtenaustausch können die kommunizierenden Prozesse unabhängig voneinander die Abarbeitung ihrer Aktionenfolge fortsetzen.

Der Transputer implementiert das OCCAM-Konzept der Nebenläufigkeit von Prozessen und der Kommunikation mittels Zeitscheibentechnik und Datenaustausch via Speicher, falls das Prozeßsystem nur auf einem Transputer abläuft, oder, im Falle der Ausführung auf mehreren Transputern, mittels echter Parallelausführung und Datenaustausch über parallel zum Prozessor arbeitende physikalische Kanäle (Links). Struktur und logisches Verhalten der OCCAM-Programme bleiben unbeeinflußt von der Art und Weise, wie ein Prozeßsystem auf ein Netz von Transputer mittels Plazierungsanweisungen abgebildet wird.

Die Programmierung aller MuSiC-Algorithmen in OCCAM und die Strukturierung der OCCAM-Module gemäß den Registertransferstufen von MuSiC bedeuten zunächst eine Modellierung des MuSiC-Systems unter dem Aspekt höchster Modelltreue, darüberhinaus aber auch die Erstellung einer formalen Spezifikation eines Ereignisflußrechners für die Simulation digitaler Systeme.

Das in OCCAM formulierte Simulationsmodell stellt sich als ein System hierarchisch gegliederter, paralleler Prozesse dar, deren Tätigkeiten eine zyklisch repetierende Folge von ereignisgesteuerter Operationsausführung und Änderungsgesteuerter Signalwertweitergabe ist.

Grundbausteine des Prozeßsystems sind Registertransferstufen modellierende Basisprozesse. Die Bereitstellung standardisierter Basisprozesse für einfache, aus einem Flipflop mit vorgelagerten Set/Reset-Netzen bestehende Registertransferstufen bzw. für datenverarbeitende Stufen (Abb. 4) sowie die Funktionswertermittlung kombinatorischer Netze mittels automatisch generierter Tabellen erleichtert die fehlerfreie Modellbildung.

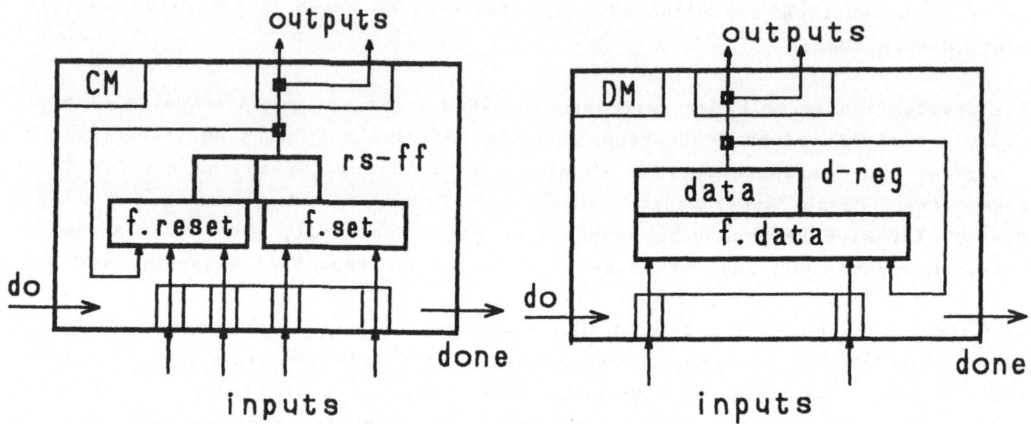

Abb. 4: Basisprozesse für Kontroll- und Datenmodule

Problematisch bezüglich OCCAM, jedoch lösbar, ist die angestrebte Ereignissteuerung der Operationen und der Kommunikation, denn im Gegensatz zur Kommunikation zwischen Hardware-Registertransferstufen, bei der ein Registerzustand einer Nachfolgestufe zur Abfrage angeboten wird, ohne daß die Abfrage zwingend ist, läßt das Kanalkonzept von OCCAM nur synchronisierte ("handshaking") Kommunikation zu. Eine im Rumpf eines Basisprozesses enthaltene Funktion operiert deshalb statt auf Kanälen auf "shared variables", die der Aufnahme der Eingangssignale der Registertransferstufe dienen, und zwecks Funktionswertweiterleitung auf den "shared variables" der Eingangssignale der folgenden Registertransferstufen. Zugriffskonflikte werden mittels eines über einen speziellen Kanal (s. Abb. 4: DO - DONE) versendetes, synchronisierendes Kontrolltoken vermieden. Die Semantik der Kontrolltoken entspricht der taktgesteuerten "Master-Slave"-Funktionsweise realer Flipflops: zum einen Verarbeitung der Eingangsituation bei konstant bleibendem Ausgang und zum anderen Umsetzen des Ausganges bei verändertem Ausgangswert. Dabei gilt vorteilhaft, daß Rechenleistung für die registertransferspezifische Funktion nur in Anspruch genommen wird, wenn eine Änderung der Eingangssituation dies notwendig macht. Die in das Modell übertragene Synchronität des Entwurfes rechtfertigt diese gegenüber der Kommunikation zwischen auf einem Transputer ablaufender Prozesse via Kanäle im übrigen etwa zehnmal schnellere Form des Datenaustausches.

Objekte der nächsten Hierarchiestufe des Prozeßsystems, mit denen z.B. eine

Pipelinestufe modelliert werden kann, setzen sich, selber wiederum als Prozesse ausgeprägt, aus mehreren, derartigen Basisprozessen zusammen. Die Aktionen aller Basisprozesse eines derartiger Funktionsprozesses werden durch das über die die Basisprozesse verbindenden Fädelungskanäle gesendete Kontrolltoken sequentialisiert. Da ein Funktionsprozeß jeweils nur auf einem Transputer abläuft, bedeutet dies keinen Laufzeitverlust, sondern vermeidet unnötige Prozeßwechsel.

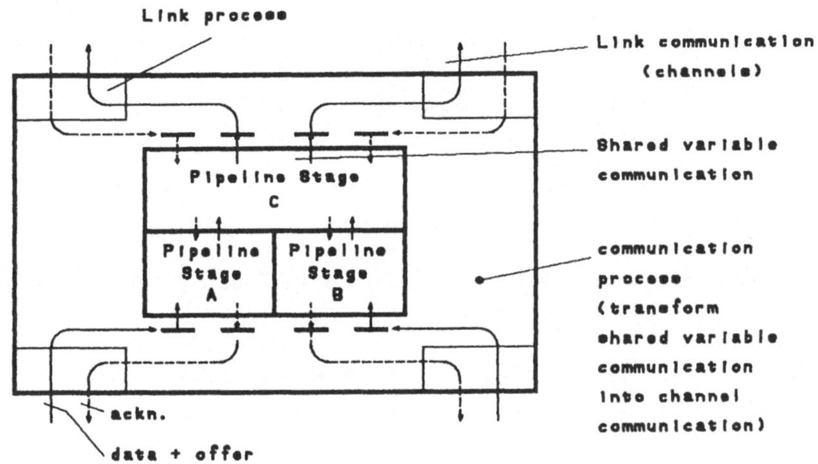

Abb. 5: Ein-Transputer-Prozeßsystem

Ein oder mehrere Funktionsprozesse, die auf einem Transputer ablaufen sollen, werden in einen Kommunikationsprozeßrahmen (Abb. 5) eingebettet, der folgende Aufgaben erfüllt:

- Broadcast-Kommunikation bezüglich der Kontrolltoken,
- Wechsel der Kommunikationsmethode: von "shared variables" für On-Chip-Kommunikation zu Kanalkommunikation für Inter-Transputer-Kommunikation,
- Abbildung von Signalpfaden zwischen auf unterschiedlichen Transputern plazierten Pipeline-Stufen auf den physikalischen Kanal zwischen den beteiligten Transputern und Koordinierung der Kommunikation.

Im in der Entwicklung befindlichen Modell wird jeweils konkret eine Verarbeitungseinheit (PU) auf ein Netz aus fünf Transputern abgebildet. Dabei werden die acht TMU/UPU-Einheiten zwar auf nur zwei Transputer verteilt. Dennoch läßt eine Abschätzung des Code-Umfanges für die Modell-Pipelinestufen TMU und UPU erwarten, daß pro TMU/UPU Speicherplatz für 2 kTemplate zur Verfügung steht, so daß Entwürfe bis zu einer Größe von 112 kTemplate (rund 300000 Gatter) für Leistungsmessungen herangezogen werden können.
Freie Links der TMU/UPU-Transputer bilden einen Ring, über den das Kontrolltoken in die Modell-Verarbeitungseinheiten eingespeist und parallel zu den anderen Transputern zur nächsten Modell-Verarbeitungseinheit weitergeleitet werden kann.

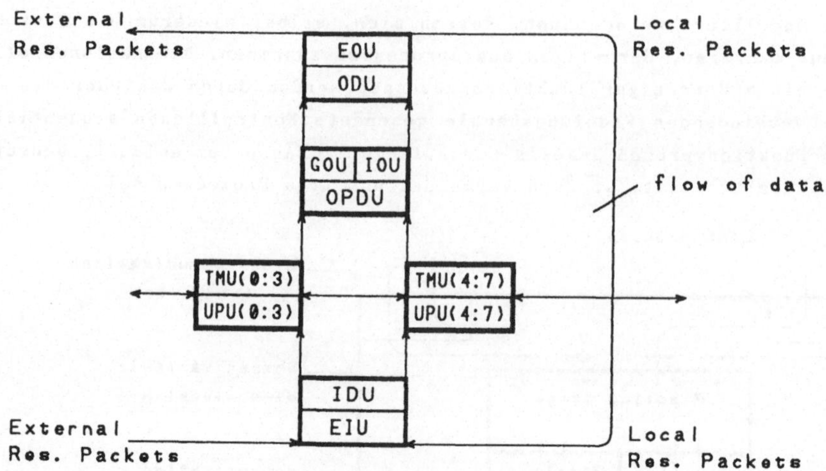

Abb. 6: Verteilung der PU-Pipelinestufen auf ein Netz aus 5 Transputern

Das dreistufige Result Packet Communication Net des Modelles verbindet sieben Modell-Verarbeitungseinheiten. Über den freien, achten Ein- bzw. Ausgang wird die Kommunikation mit dem Host-System abgewickelt. Bedingt durch die Beschränkung auf vier Links pro Transputer, wird ein Modell-2x2-Router mit seinen zwei Eingängen und zwei Ausgängen auf einem Transputer ausgeführt.

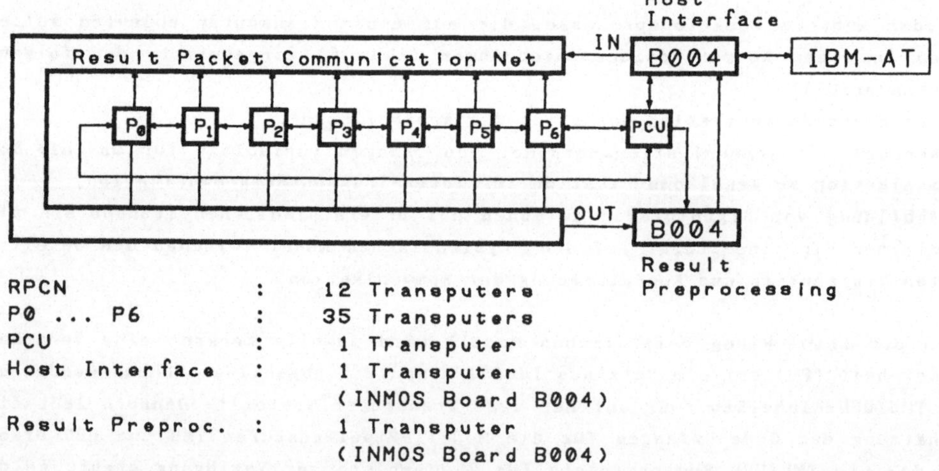

RPCN	:	12 Transputers
P0 ... P6	:	35 Transputers
PCU	:	1 Transputer
Host Interface	:	1 Transputer (INMOS Board B004)
Result Preproc.	:	1 Transputer (INMOS Board B004)

Abb. 7: Abbildung der MuSic-Simulationsmodell auf ein Multi-Transputer-Netz

Auf einem weiteren Transputer werden die Tätigkeiten der Processing Control Unit nachgebildet. Steuersignale, im Entwurf über das Processing Control Communication Net und über das Program Load Communication Net übertragen, erreichen die Verarbeitungseinheiten über Busleitungen, die auf dem Kontrolltoken-Ring nachgebildet werden. Auf eine Modellierung des Operation Packet Communication Net wird zunächst verzichtet. Die Funktionen des Host-Systems von MuSiC werden von Prozessen auf einem Entwicklungsboard (IMS B004, 2 MB RAM) erfüllt. Komplettiert wird das Modell durch ein Prozeßsystem, auf einem weiteren Entwicklungsboard ablaufend, das eine Zwischenpufferung und Blockung von Simulationsergebnissen und Meßwerten als Vorbereitung für die Ausgabe auf ein residentes Speichermedium vornimmt.

5. Schlußbemerkung

Mit dem Munich Simulation Computer ist ein Rechner entwickelt worden, der auf einer für derartige Systeme neuartigen Architektur und auf einem den realen Abläufen in den simulierten Objekten entsprechenden Operationsprinzip (Änderungsdatenfluß) basiert. Die Leistung einer 256-Prozessor-Version von MuSiC liegt im Bereich zwischen 10^9 und 10^{10} Gatterauswertungen pro Sekunde und kann weiter gesteigert werden. Der Überprüfung von Entwurf und Leistung dient ein Modell, das mittels OCCAM als Modellbeschreibungssprache und mittels eines Multi-Transputernetzes für die Modellausführung entwickelt wird. Die Ergebnisse der hier skizzierten Arbeiten werden darüber hinaus als Grundlagen für Weiterentwicklungen in Richtung auf einen hardware-beschleunigten Simulator und einen universeller einsetzbaren Rechnertyp verwendet.

Literatur:

/1/ KITAMURA, Y., et al.:
Hardware Engines for Logic Simulation
in: E. Hörbst (ed.): Advances in CAD for VLSI - Vol. 2:
Logic Design and Simulation, edited by E. Hörbst
North Holland, 1986, pp. 165 - 192

/2/ ABRAMOVICI, M., LEVENDEL, Y.H., MENON, P.R.:
A Logic Simulation Machine.
19th ACM/IEEE Design Automation Conference 1982
June 1982, pp. 65 - 73

/3/ PFISTER, G.F.:
The Yorktown Simulation Engine: Introduction
Proc. ACM/IEEE 19th Design Automation Conference
June 1982, pp. 55 - 59

/4/ HAHN, W.:
Event-Flow Computation as Key to fast Digital Design Simulation
Microprocessing and Microprogramming
Vol. 18, Dec. 1986, pp. 27 - 38

/5/ FISCHER, K.:
Ereignisfluß-Modelle für die effiziente Simulation digitaler Systeme.
Dissertation, Fakultät für Mathematik und Informatik, Universität Passau, Juli 1986

/6/ DIAS, D.M., JUMP, J.R.:
Analysis and Simulation of Buffered Delta Networks.
IEEE Trans. on Comp., Vol. C-30, April 1981, pp. 273 - 282

/7/ FISCHER, K. , KÄMPKE, T.:
A Contribution to the Performance Evaluation of the MuSiC Eventflow Computer Architecture.
MIP - 8615, August 1986

/8/ DOWSING, R.D.:
Simulating Hardware Structures in OCCAM.
Software & Microsystems, Vol. 4, No. 4,
August 1985, pp. 77 - 84

/9/ MAY, D., SHEPARD, R.:
OCCAM and the Transputer.
in: Reijns, G.L., Dagless, E.L. (Eds.):
Concurrent Languages in Distributed Systems.
North Holland, 1985, pp. 19 - 33

Modellierung und Analyse des Catalog Managements des BS 2000

Ralf Mahnkopf

Edwin Schicker

SIEMENS AG

Otto-Hahn-Ring 6

8000 München 83

1. Einleitung

Das BS 2000 ist das Betriebssystem der Großrechner der Firma Siemens. Eine wichtige Funktion des BS 2000, das Catalog Management System (CMS) war Gegenstand einer mehrstufigen Simulationsuntersuchung von Dezember 1985 bis Mai 1987. Das CMS verwaltet die Charakteristika aller gültigen BS 2000-Dateien in einem Systemkatalog. Jeder Datei ist ein Eintrag im Systemkatalog zugeordnet, der alle wesentlichen Informationen über die Datei (Dateiname, Schutzeigenschaften, Zuordnung von logischen und physikalischen Seiten der Datei etc.) enthält. Der Zugriff auf Einträge im Systemkatalog ist nur über das CMS möglich. Da Benutzer-Tasks während ihres Zugriffs auf den Systemkatalog unterbrechbar sind, existieren Sperrmechanismen, um Inkonsistenzen in Katalogeinträgen und Verklemmungen beim Zugriff zu vermeiden. Aufgrund vorhersehbarer Mehrbelastung der Betriebssystemkomponente CMS durch eine geplante Vervielfachung der zulässigen Taskanzahl und um den Anforderungen künftiger schnellerer Anlagen gerecht zu werden, wurde eine weiterentwickelte CMS-Version mit neuem Sperrmechanismus entworfen. Die Sperren sollen möglichst kurz gehalten werden bzw. möglichst wenig Einträge blockieren.

Ein Leistungsvergleich der beiden CMS-Versionen sollte bereits vor der Implementierung Aussagen über die Leistungsverbesserungen liefern, die durch die Einführung neuer Sperrmechanismen erzielbar sind und evtl. Ansätze für die Optimierung dieser Mechanismen aufzeigen.

Zu diesem Zweck wurde in einer ersten Phase je ein Simulationsmodell der aktuell eingesetzten (alten) und der weiterentwickelten (neuen) CMS Version erstellt.

Als Simulationsziel war zu überprüfen, ob das neue CMS tatsächlich die erhofften Vorteile hinsichtlich des Durchsatzes bringt. Insbesondere sollten die Auswirkungen bei verschiedenen Laststärken untersucht werden.
In einer zweiten Phase der Simulationsuntersuchung wurde ein Vergleich zwischen dem "neuen" CMS und einer sogenannten Task-Lösung durchgeführt. Bei dieser werden Betriebsmittelanforderungen über eine zentrale Servertask abgewickelt und nun nicht mehr über Betriebssystemaufrufe synchronisiert.

Im Bild stellt sich das wie folgt dar :

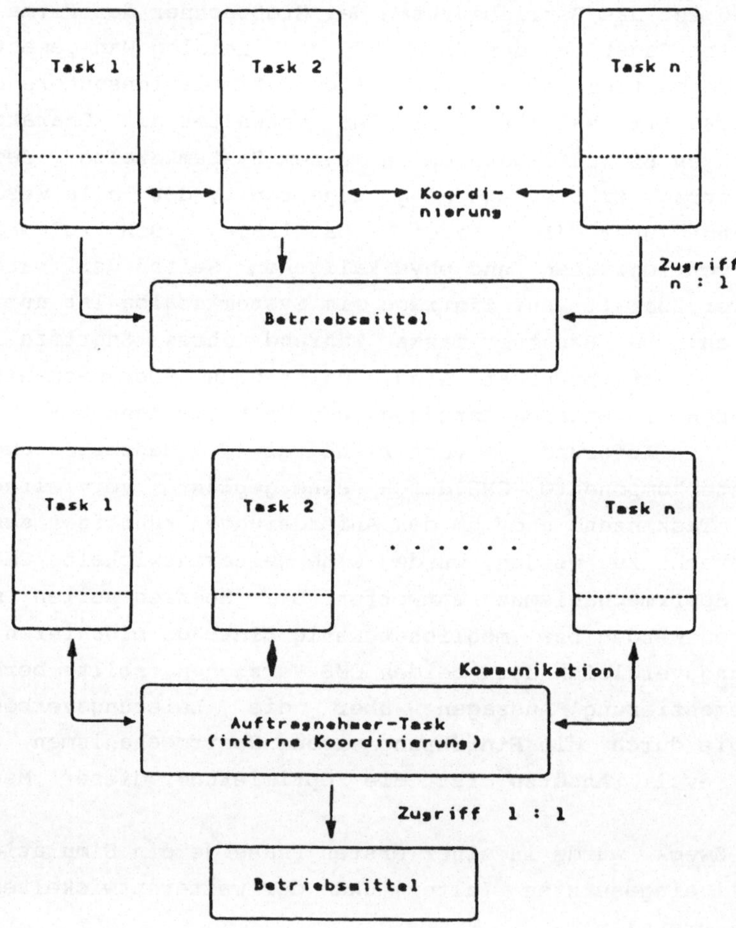

Bild 1 : Prozedur- versus Prozeßlösung

Dem Wegfall der Synchronisierung zwischen den Benutzer-Tasks stehen zusätzliche Task-Wechsel gegenüber. Anhand eines dritten Modells wurde untersucht, ob bei der Prozeßlösung mit Performance-Einbußen zu rechnen ist. Die wichtigsten Meßgrößen waren Durchsatz und mittlere Verweildauer eines Auftrags auf Anlagen verschiedener Leistungsfähigkeit .

2. Systembeschreibung

2.1 Das Catalog-Management-System (CMS)

Das priviligierte Katalogverwaltungssystem (CMS) ist Teil des Datenverwaltungssystems (Data Management System, DMS) /1/ des BS2000. Zur Verwaltung der Charakteristika aller BS2000-Dateien benötigt das CMS einen Systemkatalog. Dieser ist aufgeteilt in Blöcke (Seiten) fester Länge. Jedem Benutzer wird beim Eintrag der Benutzerkennung ein Primärblock zugeteilt. Zusätzlich benötigte Blöcke werden dynamisch zugewiesen und mit dem Primärblock verkettet. Die Charakteristika einer Datei werden in einem Katalogeintrag abgelegt. Ein Block kann mehrere Einträge aufnehmen.
Der Systemkatalog gliedert sich also logisch in die Informationseinheiten Katalogeintrag, Block (Menge von Katalogeinträgen) und Benutzerkennungen (verkettete Liste von Blöcken). Dieses sind genau die Einheiten, die ein Task beim Zugriff für sich sperren kann. Wir sprechen von einer Sperre (bzw. einem Lock) auf einen Eintrag, einen Block oder eine Benutzerkennung.
Zum Lesen oder Schreiben eines Katalogeintrages wird der ihn enthaltende Block in einen CMS-Puffer gespeichert. Informationen über die Puffer (z.B. Alter, Nummer des enthaltenen Blocks etc.) werden in einer Tabelle verwaltet. Beim Zugriff auf diese Tabelle muß diese exklusiv gesperrt werden. Ebenso müssen die Tabellen der Lock-Verwaltung beim Zugriff exklusiv gesperrt werden. Das Synchronisieren der Benutzer-Tasks bei diesen Tabellenzugriffen bezeichnet man als Serialisierung.
Die verschiedenen CMS-Versionen unterscheiden sich im Mechanismus der Sperrung der oben aufgeführten Informationseinheiten des Systemkatalogs. Als Lock-Strategie wird das Verfahren bezeichnet, für wie lange Zeit und welche Katalogeinheit (Eintrag, Block oder Kennung) gesperrt wird. Eine Sperrung der großen Katalogeinheit

Kennung behindert natürlich mehr andere Tasks als die Sperrung einer Datei.

Das CMS stellt mehrere Prozeduren zur Verfügung, durch deren Aufruf ein Benutzer-Task auf den Katalog zugreift. Exemplarisch seien an der Prozedur OPEN die Unterschiede der Lock-Strategien aufgezeigt.

2.2 Lock-Strategie der Prozedur OPEN beim alten CMS

Bild 2 schematisiert die Lock-Strategie der Prozedur OPEN beim alten CMS. Dabei stehen die Muster

▒▒ für exklusive Locks

░ ░ für share Locks (mehrere Tasks können gleichzeitig lesen)

Die Breite der Symbole (▒▒ , ▒ , ░) soll die Größe der gesperrten Katalogeinheiten verdeutlichen.

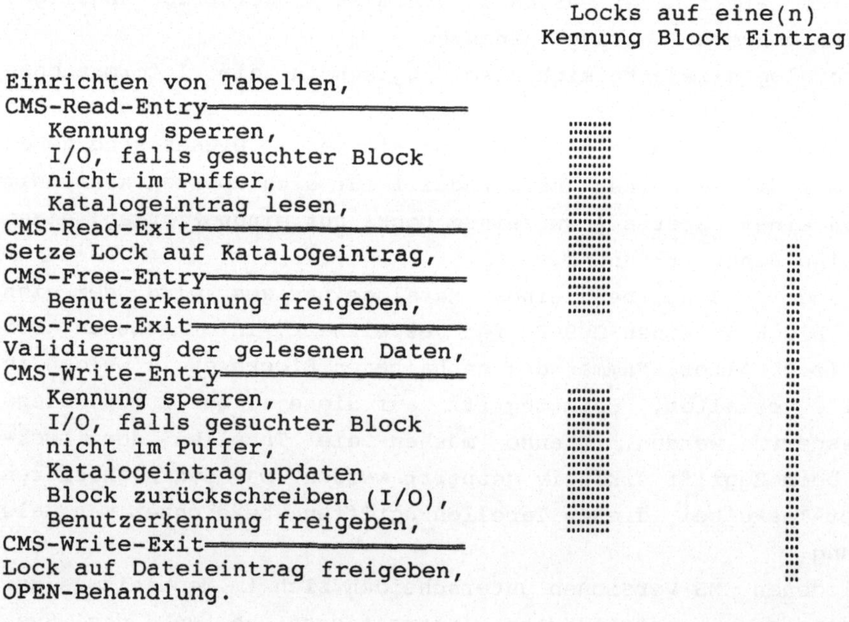

Locks auf eine(n)
Kennung Block Eintrag

Einrichten von Tabellen,
CMS-Read-Entry
    Kennung sperren,
    I/O, falls gesuchter Block
    nicht im Puffer,
    Katalogeintrag lesen,
CMS-Read-Exit
Setze Lock auf Katalogeintrag,
CMS-Free-Entry
    Benutzerkennung freigeben,
CMS-Free-Exit
Validierung der gelesenen Daten,
CMS-Write-Entry
    Kennung sperren,
    I/O, falls gesuchter Block
    nicht im Puffer,
    Katalogeintrag updaten
    Block zurückschreiben (I/O),
    Benutzerkennung freigeben,
CMS-Write-Exit
Lock auf Dateieintrag freigeben,
OPEN-Behandlung.

Bild 2 : Prozedur OPEN mit Lock-Strategie beim alten CMS

Das alte CMS kennt nur die Katalogeinheiten Kennung und Eintrag. Es zeichnet sich durch relativ einfache Algorithmen aus, sperrt aber große Katalogeinheiten für lange Zeit. Dies führt bei vielen Benutzer-Tasks zu hoher Konflikthäufigkeit.

2.3 Lock-Strategie der Prozedur OPEN beim neuen CMS

Das Schema der Lock-Strategie der Prozedur OPEN beim neuen CMS verdeutlicht Bild 3 (Verwendung der Muster analog zu 2.2).

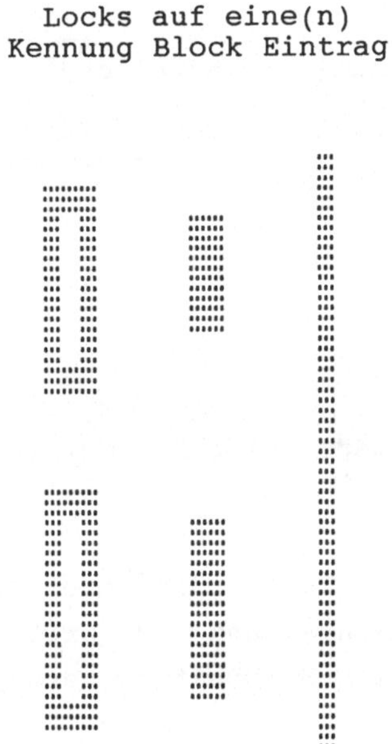

```
                                          Locks auf eine(n)
                                        Kennung Block Eintrag

Einrichten von Tabellen,
CMS-Read-Entry
   Setze Lock auf Katalogeintrag,
   Kennung nur lesend sperren,
   Gesuchten Block sperren,
   I/O, falls gesuchter Block
   nicht im Puffer,
   Blocksperre freigeben,
   Katalogeintrag lesen,
   Benutzerkennung freigeben,
CMS-Read-Exit
Validierung der gelesenen Daten,
CMS-Write-Entry
   Kennung nur lesend sperren,
   Gesuchten Block sperren,
   I/O, falls gesuchter Block
   nicht im Puffer,
   Katalogeintrag updaten
   Block zurückschreiben (I/O),
   Blocksperre freigeben,
   Benutzerkennung freigeben,
   Lock auf Dateieintrag freigeben,
CMS-Write-Exit
OPEN-Behandlung.
```

Bild 3 : Prozedur OPEN mit Lock-Strategie beim neuen CMS

Das neue CMS kennt außer den Katalogeinheiten Kennung und Eintrag auch den Block. Es unterscheidet außerdem zwischen lesendem und schreibendem Zugriff. Dadurch werden die Sperralgorithmen im Vergleich zum alten CMS komplizierter. Der Vorteil liegt jedoch darin, daß große Katalogeinheiten meist nur lesend und relativ kurz gesperrt werden. Die Konflikthäufigkeit der Benutzer-Tasks verringert sich.

2.4 Prozedur OPEN bei der Task-Lösung

Die Task-Wechsel zwischen Benutzer-Task und CMS-Servertask bei der Prozedur OPEN verdeutlicht Bild 4.

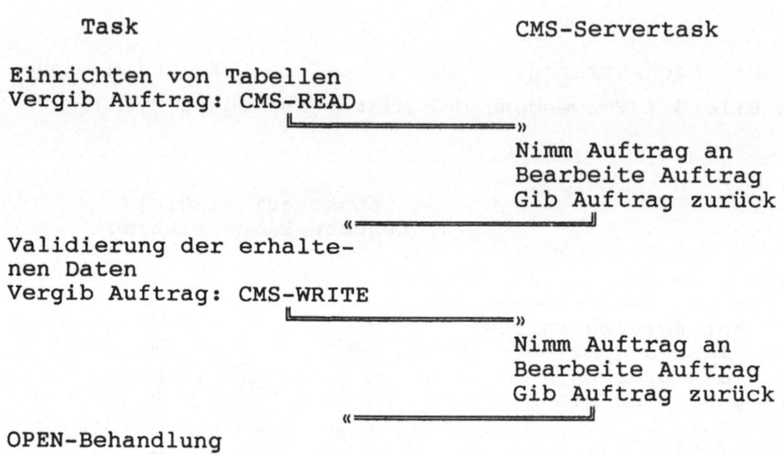

Bild 4 : Prozedur OPEN bei der Task-Lösung

Die Benutzer-Task, die die Prozedur OPEN durchführt, gibt die Aufträge CMS-Read und CMS-Write an eine CMS-Servertask weiter. Die READ und WRITE-Bearbeitung entspricht Bild 3. Der Lock-Mechanismus wurde zwar eins zu eins vom neuen CMS übernommen, die Locks müssen jedoch nicht mehr über Betriebssystemaufrufe abgewickelt werden. Außerdem ist die Serialisation nicht mehr nötig, da die Tabellen nun innerhalb der Servertask verwaltet werden und damit nur noch sie selbst auf die Tabellen zugreift.

3. Modellierung

3.1. Das Simulationssytem BORIS

Zur Modellierung der CMS-Versionen wurde das Simulationssystem BORIS /2/ verwendet. BORIS steht für Blockorientiertes Interaktives Simulationssystem und ist ein Werkzeug zur interaktiven Modellierung und Simulation zeitdiskreter, ereignisorientierter Systeme. Ein BORIS-Modell besteht aus Komponenten (Blöcken), die untereinander Informationen austauschen.
Komponenten entsprechen den kleinsten Funktionseinheiten des zu modellierenden Systems, die nachgebildet werden sollen. Komponenten empfangen Daten, analysieren und verarbeiten diese und geben Informationen an andere Komponenten weiter. Sie werden als PASCAL-Prozeduren unter Verwendung der BORIS-Sprachelemente realisiert. Der Prozedurkopf legt die Anschlüsse und Zustände der Komponente fest, der Prozedurrumpf beschreibt die Funktion der Komponente. Die auszutauschenden Informationen werden als PASCAL-Datentypen beschrieben.
Komponenten, die eine logische Einheit bilden und miteinander kommunizieren, können zu einem **Netz** zusammengefaβt werden. Elemente eines Netzes können Komponenten oder schon definierte Netze sein. Komponenten können in (Sub-) Komponenten zerlegt werden. Dadurch ergibt sich die Möglichkeit, hierarchische Strukturen aufzubauen, top-down zu modellieren und das Modell nach und nach zu verfeinern. Dies entspricht dem Prinzip des Stepwise Refinement /3/.

3.2 Das CMS Modell

Für jedes der drei CMS-Systeme wurde ein entsprechendes Modell entworfen. Exemplarisch werden am Modell des neuen CMS (Bild 5) die Funktionen der Komponenten erklärt .
Dieses Modell /4/ spiegelt die statische Struktur der wichtigsten Betriebsmittel wider. Es enthält auch eine Komponente Lastgenerator, welche die Last erzeugt und ins System einschleust. Sie generiert Benutzer-Tasks, die an das System zur Bearbeitung übergeben werden. Die Komponente Queuel ist eine Warteschlange, welche die rechenbereiten Tasks aufnimmt. Die Komponente CPU bearbeitet die Tasks und verwaltet die Tabellen für die Locks. Die Komponente IO stellt einen Verzögerungsbaustein dar für das Lesen oder Schreiben

von Blöcken von Platte. Diese Komponenten ähneln sich in allen drei Modellen. Unterschiede gibt es naturgemäß bei den Komponenten, welche die Börsen nachbilden. Börsen sind Warteschlangen, welche zusammen mit den Tabellen der Komponente CPU die Lock-Strategien realisieren. Entsprechend den verschiedenen Lock-Strategien finden sich verschiedene Börsenkomponenten. Im Modell des neuen CMS (Bild 5) kann man drei Blöcke von Börsen sehen, welche von links nach rechts die Sperrung eines Eintrags, eines Blocks und einer Benutzerkennung repräsentieren.

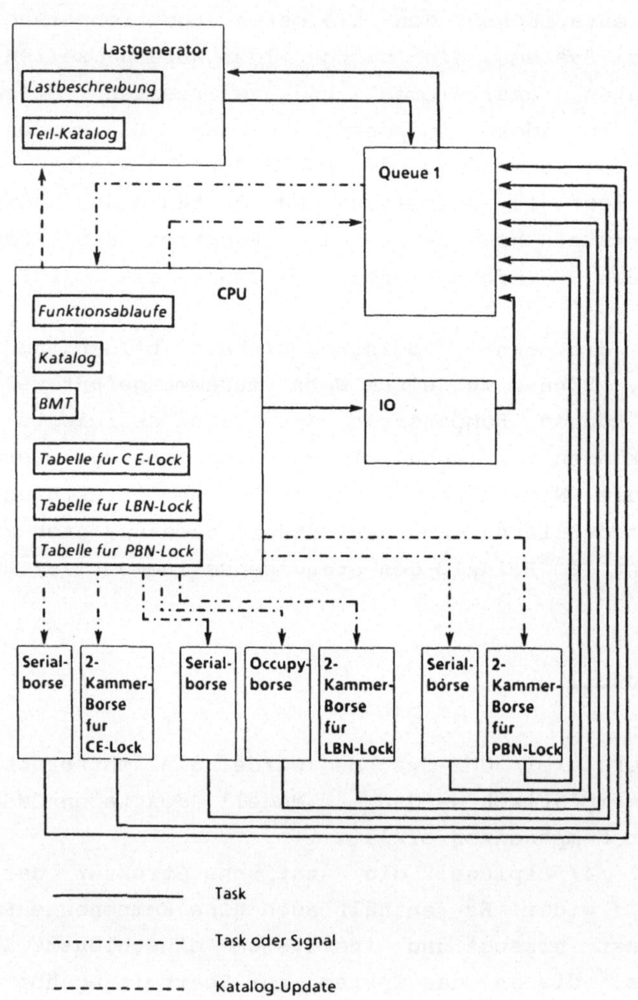

Bild 5 : Das Modell des neuen CMS

Ein typischer Ablauf eines Prozesses durch das System stellt sich wie folgt dar :

(1) Erzeugung der Task durch den Lastgenerator,

(2) Einreihung der Task in Queuel nach Priorität geordnet,

(3) Warten der Task bis zum Vorrücken an den Kopf der Warteschlange,

(4) Task-Wechsel, falls CPU frei oder höhere Priorität als der rechnende Task,

(5) Bearbeitung der Task in der CPU je nach Task-Typ bis zur Anforderung eines Betriebsmittels, dem Ende der CPU-Zeitscheibe, dem Ende der Abarbeitung des Task oder bis zu einer Verdrängung,

(6) Falls Betriebsmittel IO angefordert --> Einreihen der Task in den IO Wartepool,

(7) Falls Betriebsmittel Börse angefordert

Börse frei --> weiter bei (5)

Börse belegt --> Einreihen in die entsprechende Börse,

(8) Falls Ende der CPU-Zeitscheibe erreicht oder Verdrängung --> Einreihung der Task in Queuel und weiter bei (3),

(9) Falls Ende der Abarbeitung erreicht --> Rückkehr der Task zum Lastgenerator

3.3 Modelleinschränkungen

Aufgrund des Untersuchungsziels "Laufzeitvergleich unter verschiedenen Lockstrategien" wurden der Katalog und vor allem die Algorithmen der Sperrmechanismen exakt nachgebildet.
Abstraktionen wurden an folgenden Punkten vorgenommen :

(1) Task-Management

Das komplizierte Task-Management des BS 2000 wurde stark vereinfacht. Für einen Vergleich der Lock-Strategien (altes/neues CMS) genügte eine FIFO-Warteschlange. Bei der Modellierung der Task-Lösung mußte natürlich mit Prioritäten gearbeitet werden, da die CMS Servertask höher priorisiert als die Benutzer-Tasks sein muß.

(2) Lastbeschreibung

Bei den zu verarbeitenden Prozeduren wurden nur die wichtigsten Fälle simuliert. Denn das Simulationsziel war nicht die Untersuchung von Sonder- und Fehlerfällen, sondern der Vergleich des Verhaltens der Systeme bei der Verarbeitung von normalen Prozeduren.

(3) Memory Management

Die komplexen Algorithmen des Memory Managements wurden nicht nachgebildet. Auftretendes Paging wurde über Wahrscheinlichkeiten gesteuert.
Zusammenfassend kann gesagt werden, daß alle Merkmale, in denen sich die Sperrmechanismen unterscheiden, nachgebildet wurden. Komponenten, welche bei den 3 Systemen (altes, neues CMS und Tasklösung) in gleicher Weise auftreten, wurden soweit möglich abstrahiert.

4. Ergebnisse

Im folgenden wurden aus der Vielzahl von Messungen drei repräsentative herausgegriffen (siehe auch /5/) :

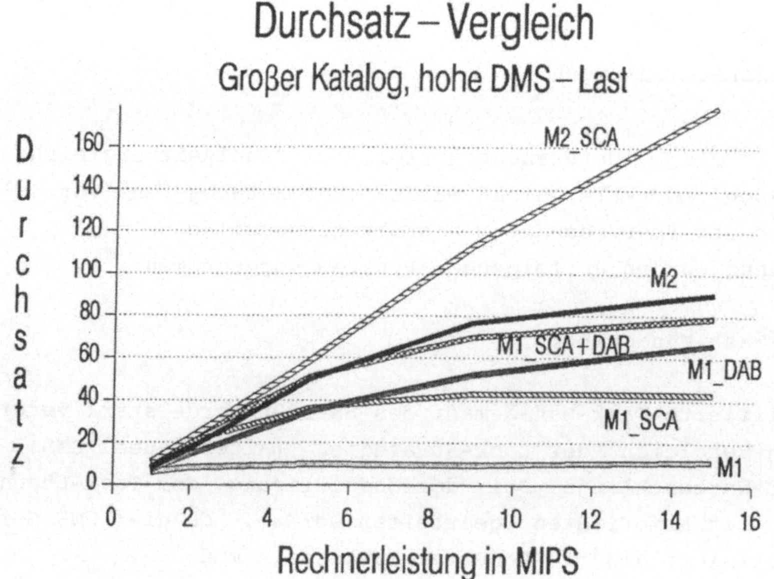

Bild 6 : Vergleich von altem und neuem CMS

Das Modell des alten CMS (M1) wurde mit bzw. ohne Speedcat (SCA) und mit bzw. ohne Data-Access-Buffer (DAB) gemessen. Das Tool SCA erspart das sequentielles Suchen nach dem Block, in dem ein gewünschter Katalogeintrag steht. Dadurch werden I/O's gespart, allerdings auf Kosten eines höheren CPU-Bedarfs. DAB unterdrückt alle Read-I/O's, indem der Katalog im Arbeitsspeicher resident gehalten wird. Auch dies erhöht den CPU-Verbrauch.
Das Modell des neuen CMS (M2) wurde zum Vergleich mit und ohne SCA gemessen. Zugrunde liegen Rechnerleistungen von 1,2,5,9 und 15 MIPS. Der Durchsatz gibt die Anzahl der abgearbeiteten DMS-Aufrufe pro Sekunde an.
Bei einer Rechnerleistung von 1 MIPS schneiden beide Modelle fast gleich gut ab. Bei höheren Leistungen wird der Unterschied sehr deutlich : Während Modell M1 aufgrund der Vielzahl an I/O-Zugriffen die Sättigung bereits erreicht, liegt der Durchsatz bei Modell M2 erheblich höher.

Das folgende Bild 7 zeigt einen Vergleich des Durchsatzes beim neuen CMS und der Task-Lösung.

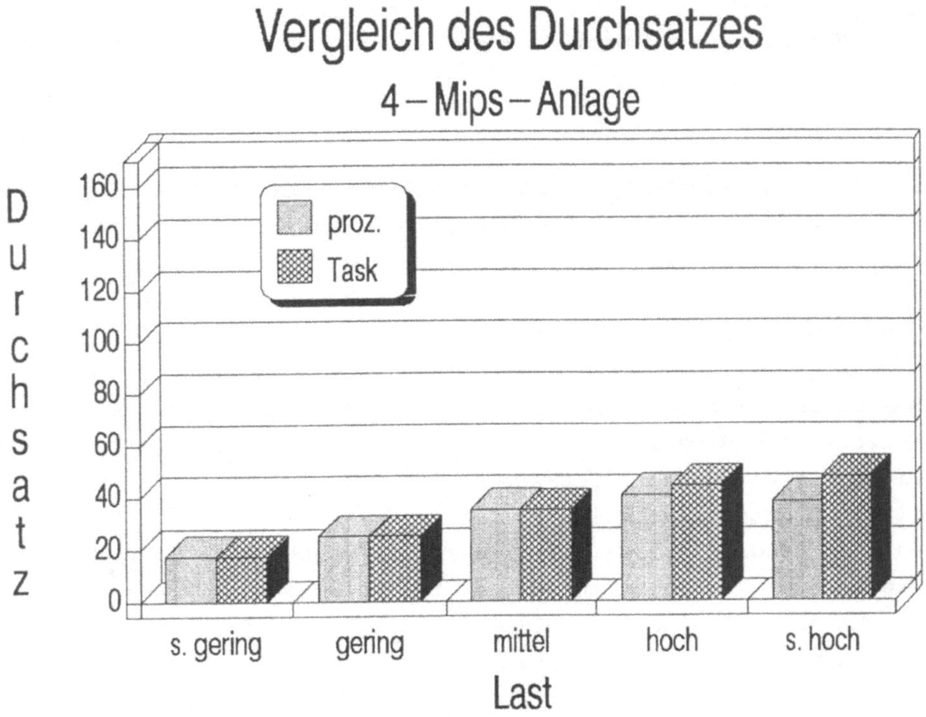

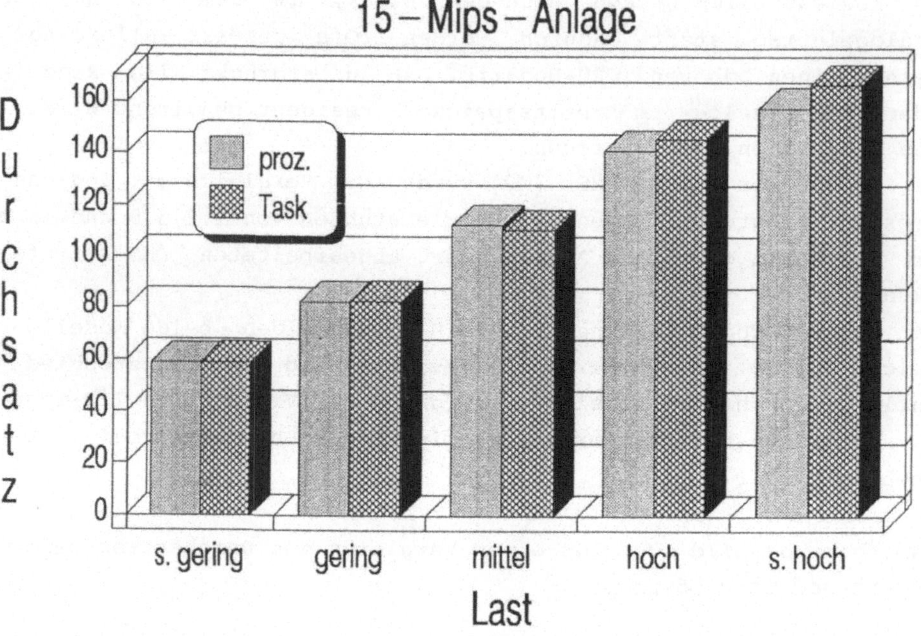

Bild 7 : Vergleich von prozeduraler Lösung (neues CMS) und Task-Lösung

Die Task-Lösung erlaubt eine bessere Struktur des Systems und braucht weniger Systemaufrufe. Die Serialisation innerhalb der Servertask entfällt. Die Task-Lösung erfordert jedoch pro Auftrag zwei zusätzliche Taskwechsel. An der häufig frequentierten Funktion CMS wurde nun überprüft, ob dadurch mit Performance-Einbußen zu rechnen ist.

Die Messungen wurde bei unterschiedlich hoher Auslastung des Systems für verschiedene Rechnerleistungen durchgeführt. Sehr ge-ringe Last entspricht etwa einer CPU-Auslastung von 50%. Sehr hohe Last entspricht praktisch Sättigung der CPU (Auslastung > 97%). Der Durchsatz gibt die Anzahl der abgearbeiteten DMS-Aufrufe pro Sekunde an. Die beiden Lösungen schneiden bei diversen Lasten praktisch gleich gut ab. Bei einer 15-MIPS-Anlage im Vergleich zu einer 4-MIPS-Anlage ist der gemessene Durchsatz für beide Lösungen erwartungsgemäß höher.

5. Probleme

Die Prozeduren, welche das CMS aufrufen, waren in Form von Ablaufdiagrammen beschrieben. Die größte Schwierigkeit bei der Erstellung der CMS-Modelle lag darin, diese Ablaufdiagramme in eine blockorientierte Sichtweise umzusetzen. Die Modellierung einer entsprechenden statischen Struktur (vergleichbar etwa mit der von GPSS Blöcken) hätte bei der Vielzahl von Abläufen zu einem schwer überschaubaren Geflecht von Modellkomponenten geführt.
Als Alternative bot sich die Modellierung der in Bild 5 dargestellten statischen Struktur an. Die Ablaufdiagramme werden als Informationseinheiten betrachtet, die sich durch das Modell bewegen. Diese Lösung bedeutet jedoch den Fluß großer Datenmengen durch das System. Auf die Simulation mit BORIS bezogen heißt das umfangreiche Parameterübergaben bei Prozeduren und somit lange Laufzeiten des Simulators.
Aus diesem Grunde werden die Ablaufdiagramme schließlich als "abstrakter" Code aufgefaßt und in einem modellierten Speicher des Systems abgelegt. Ein Benutzer-Task besteht dann nur noch aus einem Befehlzeiger, der den momentanen Abarbeitungsstand markiert sowie zusätzlichen Attributen. Die Komponente CPU dient als Interpreter, welche die Benutzer-Task nun entlang des abstrakten Codes "Ablaufdiagramm" führt. Der Datenfluß durch das System ist deutlich geringer. Ein weiterer Vorteil ist die Unterprogrammfähigkeit. Gleiche Teile der Ablaufdiagramme können als abstrakter Unterprogrammcode abgelegt werden. Bei der Abarbeitung speichert sich die Benutzer-Task den (die) alten Befehlszeiger in einem Stack.
Günstig ist bei dieser Lösung die hohe Flexibilität. Da die Ablaufdiagramme des neuen CMS sich im Modellierstadium noch mehrfach änderten, konnten diese Änderungen durch Eingabe des entsprechenden neuen Codes schnell nachvollzogen werden.

6. Zusammenfassung

Die Modellierung einer Betriebssytemkomponente wurde am Beispiel CMS aufgezeigt. Die Simulation hat sich dabei als nützliche Entscheidungshilfe bei der Bewertung von Software-Komponenten erwiesen. Problematisch war die Umsetzung einer prozeßorientierten Sichtweise in eine blockorientierte. Verschiedene Lösungen dieser Umsetzung wurden diskutiert und die Vor- und Nachteile kurz skizziert.

Literaturverzeichnis

/1/ Koch, R.
Datenverwaltungssytem BS 2000
Technische Beschreibung
Siemens AG, München, 1987

/2/ Benutzerhandbuch BORIS Version 2.0
Ausgabe Februar 1987
Siemens AG, München, ZTI SOF 3

/3/ Wirth, N.
Program Development by Stepwise Refinement
Communications of the ACM 14, No. 4
1971 221-227

/4/ Kramer, H., Mahnkopf, R., Rosenbaum, U.
Simulation des Katalogverwaltungssytems des BS 2000
Siemens AG, München, ZTI SOF 313, 1986

/5/ Schicker, E.
Erster Ergebnisbericht zur CMS-Simulation
Software-Entwicklungsprojekt-Band 502.33
Siemens AG, München, K D ST, 1986

LEISTUNGSANALYSE MIT INT^3: EINER INTERAKTIVEN, INTELLIGENTEN UND INTEGRIERTEN PC-MODELLIERUNGSUMGEBUNG

Axel Lehmann, Helena Szczerbicka
Universität Karlsruhe
Institut für Rechnerentwurf und Fehlertoleranz
(Prof.Dr.D.Schmid)

Kurzfassung

Dieser Beitrag beschreibt ein Konzept und die Realisierung eines Prototyps zur interaktiven, wissensbasierten Unterstützung von Anwendern in den verschiedenen Phasen eines Modellierungsprozesses. Das vorliegende Konzept geht von der Annahme aus, daß in zunehmendem Maße Systemspezialisten aus verschiedensten Anwendungsbereichen detaillierte Systemanalysen als Grundlage für Entscheidungen benötigen, die den praktischen Einsatz dieser Systeme, deren Entwurf, Auswahl und Modifikation betreffen. Es wird davon ausgegangen, daß die meisten dieser Systemspezialisten nur über Grundkenntnisse im Bereich der Systemanalyse verfügen, einen mangelnden Uberblick über Modellierungsmethoden und -werkzeuge besitzen, aber andererseits mit zunehmender Verbreitung von PC's einen direkten Zugang zu leistungsfähigen Verarbeitungssystemen haben.
Ausgehend von diesen Annahmen berichtet dieser Beitrag über die Zielsetzung, den aktuellen Realisierungsstand und eine typische Anwendung unseres INT^3- Konzepts: über eine interaktive, intelligente und integrierte Modellierungsumgebung, die z. Zt. in einer Standard-PC-Umgebung (IBM-XT/AT) schrittweise realisiert wird. Neben einer Beschreibung des konzeptionellen Ansatzes konzentriert sich dieser Beitrag auf die Demonstration des Einsatzes des INT^3-Prototyps zur schrittweisen Lösung eines typischen Leistungs- und Kapazitätsplanungsproblems für ein Rechensystem im Dialogbetrieb.

1.Motivation

Die rasche Innovation und Verbreitung von Rechen- und Kommunikationssystemen erfordert in zunehmendem Maße die Bereitstellung benutzerfreundlicher, wirtschaftlicher Methoden und Werkzeuge für Analysen, die u.a. zur Bewertung der Leistung, Zuverlässigkeit, Testbarkeit und Wirtschaftlichkeit dieser Systeme verwendet werden können. Diese Bewertungen bilden häufig die Grundlage für weitreichende Entscheidungen, die beispielsweise die Synthese neu zu entwerfender Systeme oder auch detallierte Analysen bereits existierende Systeme betreffen.

Als Hilfsmittel werden zunehmend Modellierungsverfahren angewendet, wobei aufgrund der Vielzahl und der Komplexität existierender Modellierungsmethoden und -werkzeuge das Fachwissen und die Erfahrung von Experten erforderlich ist, um eine zielgerichtete und effiziente Anwendung zu erzielen. Um dieses zu gewährleisten, ist eine problem- und kostenorientierte Auswahl von Modellierungstechnik, -werkzeug und Maschinenumgebung vorzunehmen, bei der neben den Analysezielen des Benutzers zu berücksichtigen sind:

* Notwendigkeit statistischer oder deterministischer Analysen,
* Nachbildung stabiler bzw. instabiler Systemzustände,
* Nachbildung des dynamischen Systemverhaltens auf unterschiedlichen Detaillierungsebenen, um beispielsweise 'top down' eine schrittweise Verfeinerung der Modellbildung zu erreichen,

* Dekomposition eines Modells in Submodelle, die eine separate Analyse der Submodelle ggf. unter Verwendung verschiedener Lösungstechniken ermöglicht,
* Wunsch nach qualitativer oder quantitativer Analyse,
* Notwendigkeit eines 'rapid prototyping' aus Zeitgründen und
* Eingangs-, Ausgangs- und Steuerparameter eines ausgewählten Modellierungswerkzeugs.

Somit müssen für die verschiedenen Phasen einer Systemanalyse - d.h. zur Problemanalyse, Modellerstellung, -anwendung und -auswertung - außer Kenntnissen über zahlreiche Modellierungsmethoden und -werkzeuge auch Wissen über numerische und statistische Analyseverfahren und über Simulationstechniken verfügbar sein.

Als besonders schwierig erweist sich dabei meist der Umstand, daß die mit der eigentlichen Problemstellung vertrauten Systemspezialisten als Anwender von Modellierungstechniken
* meist keinen umfassenden Uberblick uber Alternativen und Anwendungskriterien von Modellierungs-und Bewertungsmethoden besitzen,
* nicht mit der zielgerichteten Auswahl von Modellierungsverfahren und der Erstellung konzeptioneller, werkzeugunabhängiger Modelle vertraut sind,
* oft nicht die Vielfalt der zur Verfügung stehenden Modellierungswerkzeuge, deren Effizienz und Anwendbarkeit fur einen bestimmten Problembereich kennen,
* selten die notwendige Erfahrung und Ubersicht über Lösungstechniken, wie numerische oder Simulationsanalysen besitzen und
* vielfach nicht auf genügend praktische Erfahrungen mit der Validierung und Kalibrierung von Modellen sowie der Planung, Durchführung und Interpretation von Experimenten verfugen.

Aus diesen Gründen sind Systemspezialisten meist auf das Fachwissen von Modellierungsexperten und deren praktische Erfahrungen angewiesen. Vor dem Hintergrund dieser Situation, in der sich neben einer Vielzahl von Domänenspezialisten u.a. auch Auszubildende und Studenten befinden, ist das Ziel der vorliegenden Arbeiten darin zu sehen, diesen Anwendern auf ihrem Standard-PC eine INT^3-Modellierungsumgebung zur Verfügung zu stellen: interaktiv, intelligent und integriert. Dieser Beitrag faßt Zielsetzung, Konzept und Realisierungsstand unserer Arbeiten an dem INT^3-Projekt zusammen.

2. Realisierung einer INT^3-Umgebung

Das Forschungsprojekt HECTOR ('HEterogeneous Computers TOgether') ist Gegenstand einer Kooperation zwischen IBM Deutschland und der Universität Karlsruhe mit dem Ziel, ein heterogenes, schnelles Universitätsrechnernetz zu realisieren. Im Zusammenhang mit diesem Projekt werden auch zahlreiche Forschungs- und Lehrprojekte durchgeführt, mit dem Ziel, Einsatzbereiche für 'personal computers' in Lehre und Ausbildung an einer Universität zu erschließen. Das Projekt INT^3 ist eines dieser Lehr- und Forschungsprojekte, das mit Unterstützung der IBM seit 1985 am Institut für Rechnerentwurf und Fehlertoleranz durchgeführt wird [Hec86].

2.1 Allgemeine Randbedingungen

Bei der Konzeption einer INT^3-Modellierungsumgebung stand das Bestreben im Vordergrund, soweit als möglich bereits verfügbare, weit verbreitete Hardware und Software einsetzen zu können. Der konzeptionelle Ansatz geht dabei von folgenden Voraussetzungen aus:
* Hardware-Grundausstattung: Standard-PC vernetzt mit leistungsfähigen 'host'-Systemen;
* Gewährung von Graphikunterstützung zur Modellkonstruktion und Ergebnisauswertung;

* Einsatz von Expertensystemen als Klassifikations- und Beratungssysteme;
* Bereitstellung von Modellierungswerkzeugen für unterschiedliche Anwendungsbereiche;
* Wissensstand der Anwender entsprechend dem eines Informatikstudenten im Hauptstudium ohne Kenntnisse im Bereich der Modellbildung;
* geplanter Einsatz des Prototyps in einem Praktikum 'Leistungsanalyse von Rechensystemen'.

Wenn man die verschiedenen Schritte einer Modellierung von der Problemspezifikation bis zur Experimenteauswertung analysiert, mit denen sich ein Anwender auseinandersetzen muß, so können die folgenden getrennten Modellierungsphasen unterschieden werden, [Leh87]. Diese Phasen unterscheiden sich voneinander sowohl in bezug auf die erforderlichen Kenntnisse des Benutzers als auch hinsichtlich des Angebots an Hilfsmitteln, die ihm dabei zu Verfügung stehen:

1. Spezifikation des aktuellen Problems bezüglich der Analyseziele in einer formalisierten Beschreibung des betrachteten Systems, seiner Last und der zu analysierenden Kenngrößen;
2. Auswahl einer Modellierungsmethode, die problemangepaßt und effizient anwendbar ist (z. B. einfache Warteschlangennetze, erweiterte Warteschlangenetze, Markov Ketten, Petri-Netze etc.);
3. Konstruktion eines konzeptionellen, werkzeugunabhängigen Modells;
4. zielgerichtete Auswahl von Modellierungswerkzeug und Lösungstechnik, einschließlich einer werkzeugabhängigen Implementierung des konzeptionellen Modells;
5. Validierung und Kalibrierung des ausführbaren Modells, einschließlich der Experimenteplanung;
6. Experimentelle Anwendung des validierten Modells;
7. Statistische Analyse und graphische Interpretation der Modellierungsergebnissen.

Unserer Konzept geht davon aus, Hilfsmittel so bereitzustellen, daß auch Anwendern mit unterschiedlichen Modellierungskenntnissen in jeder der oben beschriebenen Modellierungsphasen Unterstützung geboten werden kann.

2.2 Konzept

In bezug auf die zuvorgenannte Zielsetzung haben wir unter den genannten Randbedingungen einen konzeptionellen Entwurf einer INT^3-Umgebung für die Anwendung auf Standard-PC's entsprechend der Darstellung in Bild 1 festgelegt. Dieses Konzept, das in [LKK86a] ausführlich erläutert ist, kann durch folgende Kennzeichen zusammenfassend charakterisiert werden:

* der Benutzer erhält Zugang zu den unterstützenden Werkzeugen der Modellierungsphasen über eine beratende Dialogkomponente (Dialog-Ebene); diese Komponente hat außerdem die Aufgabe, als zentrale Datenverwaltungskomponente die in den einzelnen Phasen gesammelten Problem- und Modellinformationen zu verwalten, an andere Werkzeuge weiterzugeben und Zugriffsrechte und Datenkonsistenz zu überwachen;
* in der Werkzeug-Ebene werden dem Anwender verschiedene Hilfsmittel zur Modellkonstruktion, -ausführung und auswertung angeboten, insbesondere in Form von:
  - -> Grafik-Software zur Problemspezifikation (Phase 1), zur Modellkonstruktion (Phase 3) und zur Ergebnisdarstellung (Phase 7);
  - -> Expertensysteme als Klassifikationssystem zur Auswahl einer problemgerechten Modellierungsmethode (Phase 2) sowie als Beratungssysteme zur Auswahl einer effizienten Lösungstechnik (Phase 4), zur Validierung und zur Experimenteplanung(Phase 5);
  - -> Modellierungswerkzeuge und Modellbibliotheken für den Aufbau und die Ausführung von Modellen (Phasen 4 und 6);

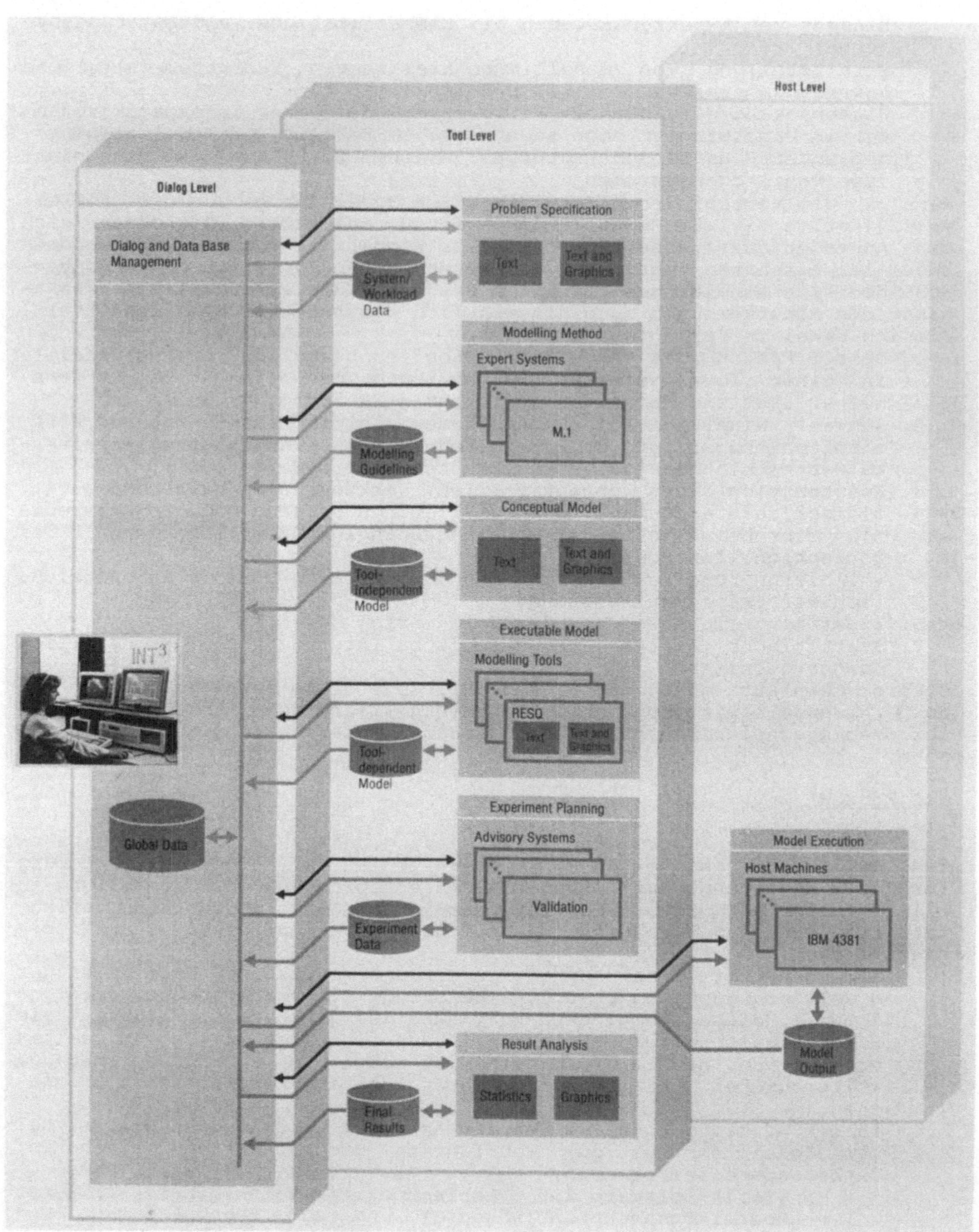

Bild 1: Allgemeines Konzept von INT[3]

-> Interpretationshilfen zur Bewertung der Ergebnisse,

* in der 'host'-Ebene können die Modelle ausgeführt werden, deren Software-Realisierung oder Komplexität ein leistungsfähigeres System als einen INT^3-PC erfordert. Die Modellierungsergebnisse werden jedoch zur Interpretation wieder auf den PC transferiert.

2.3 Stand der Realisierung

Zur Implementierung des zuvor erläuterten INT^3-Konzeptes waren neben den in Kapitel 2.1 genannten allgemeinen Randbedingungen noch einige andere Voraussetzungen zu beachten, die sich u.a. aus der zur Verfügung stehenden Hardware und Software sowie dem vorgesehenen Zeitplan zur Realisierung eines Prototyps ergaben. Zu diesen projektbedingten Randbedingungen zählen :

* Verwendung des IBM-XT/AT als Standard-PC mit folgender Grundausstattung:
  - -> 512 KB Hauptspeicher
  - -> 10 MB Winchester-Plattenspeicher
  - -> 1 Monochrombildschirm (mit Hercules-Karte)
  - -> 1 Farbgraphikbildschirm (mit EGA-Karte)
  - -> 2 asynchrone Schnittstellen für:
    Maus-Anschluß und
    Dateitransfer zu einer IBM 4381 ('host')
* Einbeziehung bereits verfügbarer Modellierungswerkzeuge, wie u.a.:
  - -> RESQ ('Research Queueing Package' [SMK82])
  - -> SIMLAB (mit SIMANIMATION und SIMSCRIPT II.5)[Joh86]
* Nutzung bereits vorhandener Graphik- und Treiber-Software, soweit als möglich;
* Verwendung (auf IBM-PCs) verfügbarer 'expert system shells' zur Realisierung der Klassifikations- und Beratungssysteme; aufgrund einer Voranalyse im Rahmen einer Diplomarbeit [Rol86] kam dazu nur das Rahmensystem M.1 [Fra84] in Frage;
* Test und Einsatz einzelner Unterstützungswerkzeuge durch fachkundige Studenten und anschließender Einsatz im Rahmen des Praktikums 'Leistungsanalyse von Rechensystemen'.

Der gegenwärtige Realisierungsstand des INT^3-Prototyps auf IBM-XT/AT und IBM 4361/4381 als 'host'-System umfaßt bislang die folgenden selbstentwickelten bzw. angepaßten Software-Werkzeuge:

* eine Dialog- und Datenverwaltungskomponente für die Dialog-Ebene [Fre87];
* ein Expertensystem zur problemangepaßten Auswahl einer Modellierungsmethode (in Phase 2), implementiert unter Verwendung des regelbasierenden Rahmensystems M.1 [Fra84], [Rol86];
* ein Softwarepaket für die graphische Struktur- oder Zustandsbeschreibung des zu modellierenden Systems (in Phase 1) und für den graphischen, werkzeugunabhängigen Entwurf konzeptioneller Warteschlangen Modelle (in Phase 3) [Fre85],[Rec87];
* ein Expertensystem zur zielgerichteten Auswahl einer Lösungstechnik (in Phase 4) [Rol86];
* das Modellierungswerkzeug RESQ, das sowohl zur numerischen Analyse als auch zur Simulation einfacher und erweiterter Warteschlangeetzen, [SMK82];
* eine Erweiterung und Anpassung des Simulationspakets SIMLAB [Joh86] zur interaktiven, animierten Simulation von einfachen und erweiterten Warteschlangennetzen (Phasen 4 und 6), [Her87];
* ein Softwarepaket für die graphische Auswertung und die Interpretation der RESQ-Modellierungsergebnisse (in Phase 7);
* ein Übersetzer für konzeptionelle Modelle in RESQ-Syntax [Rec87].

3. Experimentelles Anwendungsbeispiel

Als Beispiel für die Anwendung unseres INT^3-Konzepts und seines aktuellen Realisierungsstandes sei ein Rechnerkonfigurations- und Kapazitätsplanungsproblem aus der Sicht eines Planungsbeauftragten betrachtet. Dessen Aufgabe bestehe global gesehen darin, aufgrund einiger Prognosen hinsichtlich der aktuellen und künftig zu erwartenden interaktiven Last, Vorschläge für einen leistungsgerechten Systemausbau zur Entscheidung vorzubereiten. Genauer gesagt besteht die Aufgabe darin, zu analysieren, wieviele Benutzer an Bildschirmarbeitsplätzen im Mehrprogrammbetrieb bei unterschiedlichen Mehrprogrammfaktoren noch effizient arbeiten können, so daß auch das System günstig ausgelastet aber nicht überdimensioniert ist. Die Konfiguration des Rechners sei gekennzeichnet durch eine bekannte Verarbeitungseinheit (CPU), einen bestimmten Arbeitsspeicherausbau (mem) und zwei periphere Plattenspeicher (disks).

Bei Anwendung der gegenwärtig verfugbaren INT^3-Umgebung durch den Planungsbeauftragten, könnte die Analyse dieses zunächst nur vage umrissenen Problems folgendermaßen ablaufen. Der Planungsbeauftragte tritt als INT^3-Benutzer zunächst in die Dialog-Ebene ein (siehe Bild 1), in der er u.a. nach einer Bezeichnung seines Problems, nach möglicherweise schon zurückliegenden Analysen dieser Problemart etc. befragt wird. Danach offeriert die Dialog- und Datenverwaltungskomponente dem Benutzer den Eintritt in die einzelnen Modellierungsphasen unter Angabe der dort jeweils verfügbaren Werkzeuge und unter Bereitstellung der in der jeweiligen Phase benötigten Modell- und Problemdaten. Es sei angenommen, daß der Planungsbeauftragte im folgenden die Unterstützung von Werkzeugen in jeder Modellierungsphase in Anspruch nimmt, was nicht notwendigerweise der Fall sein muß.

• Phase 1: Problemspezifikation

In der Phase der Problemspezifikation soll ein Benutzer sein Wissen um Ziel und Gegenstand seiner Analyse möglichst genau beschreiben. Die Aufgabe unterstützender Werkzeuge besteht u.a. darin, dem Benutzer vor Eintritt in den eigentlichen Modellbildungsprozeß sein Problem selbst zu verdeutlichen, vorhandenes Problem- und Systemwissen zu sammeln, um es in den späteren Modellierungsphasen - auch zur Nutzung durch andere Werkzeuge - direkt einbringen und auf Konsistenz prüfen zu können.
In diesem Bemühen erhält der Benutzer von INT^3 gegenwärtig Unterstützung von zwei Werkzeugen zur graphischen und alphanumerischen Problembeschreibung. Mittels der Graphik können Struktur- oder Zustandsübergangsbeschreibungen eines Systems auf dem Farbgraphikschirm vorgenommen werden. Mit dem anderen Werkzeug kann auf dem Monochromschirm im Dialog mit dem Benutzer Wissen über das Analyseobjekt gesammelt werden, beispielsweise im vorliegenden Beispiel [Sta84]:

- Objekte des Systems und deren Namen ?
  -> TERMINAL,HOST,CPU,DISK,MEMORY
- Attribute der Objekte ?
  Zahl ?
  Objekttyp (Subsystem/ Element) ?
  -> (im vorliegenden Beispiel wäre das gesamte Rechensystem HOST als Subsystem anzusehen, das bei detaillierterer Betrachtung aus Elementen CPU, MEMORY und DISK besteht);
  etc.
- Verbindungen zwischen den Objekten ?
  -> (diese können auf unterschiedlichen Detaillierungsebenen beschrieben werden, im Beispiel:
  abstrakt: TERMINAL1 -> HOST; HOST -> TERMINAL1; etc.
  detaill.: TERMINAL1 -> CPU ; CPU -> DISK1; etc.)
- Ablauf von Transaktionen ('routing)'?
  -> (entsprechend der beschriebenen Struktur auf zwei Ebenen möglich, z.B.:
  TRANSAKTION1 := TERMINAL1 -> HOST -> TERMINAL1; etc.)
- Ziele der Modellierung ?

Bild 2: Beratungssystem zur Auswahl einer Modellierungsmethode

Fällt das zu analysierende Problem in die Klasse der diskreten oder in die Klasse der kontinuierlichen Probleme ?
>> **d(iskret).**

Welche globale Zielsetzung verfolgen Sie mit der Modellierung?
Leistungsanalyse, Zuverlässigkeitsanalyse, Zustandsanalyse, Funktionstest
>> **L(eistungsanalyse).**

Sie wollen eine Leistungsanalyse vornehmen. Welche der folgenden Detailziele sollen dabei untersucht werden?
Engpassanalyse, Auslastung, Durchsatz, Antwortzeit, Wartezeit, Warteschlangenlänge, Prüfung auf Blockierungen.
>> **Au(slastung), Wartez(eit).**

Die Beschreibung eines Systems oder seines dynamischen Verhaltens kann struktur-, ablauf- oder zustands- orientiert erfolgen. In welcher Form können Sie Ihr System, bzw. dessen Verhalten modellieren?
>> **s(truktur), a(blauf).**

Ist es notwendig, im Modell den Wettbewerb mehrerer Aufträge um die Zuteilung von Betriebsmitteln des modellierten Systems erfassen zu können?
>> **ja.**

Ist es notwendig, im Modell das Entstehen und den Abbau von Warteschlangen nachbilden zu können?
>> **ja.**

Ist es im Hinblick auf das Analyseziel wichtig, die Auftragsreihenfolge deterministisch oder probabilistisch beschreiben zu können?
>> **p(robabilistisch).**

Ist es notwendig, im Modell nachbilden zu können, daß der Ablauf eines Prozesses an bestimmten Stellen mit der Erfüllung einer Bedingung (z.B. mit der Synchronisation mehrerer Prozesse/Aufträge bzw. Funktionen) verknüpft ist?
>> **ja.**

Für die geschilderte Problemstellung erscheint die Verwendung des folgenden Modellierungsverfahren am geeignetsten:

modell = erweitertes Warteschlangennetz (100%),
da m-4-1-3 und m-4-1-2

(2a): Ausschnitt des Benutzerdialogs

Globalziel ? -> LEISTUNGSANALYSE ;
Detailziele dieser Leistungsanalyse ? -> CPU-AUSLASTUNG, MEMORY-AUSLASTUNG, HOST-ANTWORTZEIT, etc.;

Diese Spezifikation kann der Anwender in textueller Form im Dialog eingeben oder mit graphischer Unterstützung (ähnlich der in Phase 3 beschriebenen graphischen Konstruktion der Modellstruktur). Die Objekte und Attribute der Elemente und Subsysteme werden auf dem Monochromschirm spezifiziert. Die Plazierung von Elementen und Subsystemen wird ebenso wie deren Verbindungen mit Hilfe der Maus eingegeben. Der textuelle Teil vervollständigt die graphische Spezifikation. Diese Probleminformationen werden in einer Datei gesammelt, die ebenfalls Teil der globalen Datenbasis ist (Bild 1) und die in den weiteren Phasen der Modellierung bei Bedarf anderen Werkzeugen übergeben wird.

• Phase 2: Auswahl einer problemgerechten Modellierungsmethode (durch Anwendung eines Expertensystems)

Bild 2a zeigt den Dialog des Planungsbeauftragten mit einem Prototypen eines Klassifikationssystems, realisiert mit dem regelbasierenden Expertenrahmensystems M.1 [Rol86]. Der Benutzer wird durch dieses Werkzeug in der Auswahl einer ziel- und wissensorientierten Modellierungsmethode unterstützt und kann erkennen, welche Merkmale des Systems und der Problemstellung die Auswahl einer angemessenen Modellierungstechnik beeinflussen. Er kann seine Angaben durch Gewißheitsfaktoren unterschiedlich gewichten und dadurch auch vages Wissen, d.h. Vermutungen, äußern. Wie das vorliegende Dialogbeispiel in Bild 2a zeigt, wird das Ergebnis der Beratung - in unserem Beispiel ein erweitertes Warteschlangennetz - als Modellierungsmethode mit einem Gewißheitsfaktor gewichtet vorgeschlagen.

Auf Nachfrage kann der Benutzer von der Erklärungskomponente eine Erklärung bekommen, warum diese Schlußfolgerung gezogen wurde, d.h. aufgrund welcher Angaben und Regeln diese Modellierungsmethode vorgeschlagen wurde (siehe Bild 2b). Der Benutzer wird dann gefragt, ob er eine einzelne der gegebenen Antworten revidieren möchte. Auf diese Art ist es möglich, den Einfluß einzelner Antworten auf den Schlußfolgerungsprozeß beurteilen zu können.

Soll die vom System ausgesprochene Empfehlung erläutert werden?
>> **ja.**

Sie wollen eine Leistungsanalyse durchführen; dafür kommen als Modellierungsverfahren in Frage:
Warteschlangennetze, Petri-Netze, Zustandsübergangsdiagramm

Als Modellierungsverfahren wurde ein erweitertes Warteschlangennetz empfohlen, da :
- Leistungsanalysen typischerweise mit Warteschlangennetzen lösbar sind,
- Sie die Modellierung von Betriebsmittelwettbewerb für notwendig erachten,
- Sie die Modellierung von Wartesituationen für notwendig erachten,
- Sie die Modellierung von Bedingungen für den Prozessablauf für notwendig erachten.

Die Analyse Ihres Problems mit Hilfe eines erweiterten Warteschlangennetzes ist möglich, da:
- Sie eine struktur- und ablauforientierte Beschreibung Ihres Systems geben können.

(2b): Erläuterung der vorgeschlagenen Entscheidung

Es ist geplant, um redundante Fragen an den Benutzer zu vermeiden, soviel Informationen als möglich direkt aus der globalen Datenbasis in das Expertensystem zu übernehmen.

● **Phase 3: Erstellung eines konzeptionellen Modells**

In dieser Phase bietet die INT³-Umgebung dem Benutzer die Möglichkeit, die Struktur eines konzeptionellen Modells graphisch mit Hilfe spezieller Symbolmenus und einer Maus auf dem Farbgraphikschirm zu konstruieren und Namensgebung und sonstige textuelle Angaben parallel dazu auf dem Monochromschirm vorzunehmen. In diesem Beispiel wird aufgrund der Empfehlung aus Phase 2 ein Menu mit der Symbolmenge erweiterter Warteschlangennetze [SMK82] auf beiden Schirmen angeboten (siehe Bilder 3a und 3b). Für eine hierarchische Modellkonstruktion steht dem Benutzer zusätzlich auch ein Subsystem-Symbol zur Verfügung (z.B. für HOST).
Das Ergebnis der Konstruktion eines werkzeugunabhängigen, konzeptionellen Modells, das der Planungsbeauftragte für das vorliegende Beispiel entwickelt haben könnte, ist auf unterschiedlichen Detaillierungsebenen in den Bildern 4a und 4b dargestellt. Die Terminals werden als IS-Warteschlangen ('infinite server') modelliert und der Rechner HOST ist durch ein Subsystem-Symbol repräsentiert. Das Subsystem HOST, das in Bild 4b detailliert dargestellt ist, besteht aus einem passiven Warteschlangenelement (MEMORY), das den Hauptspeicher darstellt, und aktiven Elementen, wie CPU und die beiden Plattenspeicher DISK. Eine mehrstufigere Modellhierarchie als im vorliegenden Fall ist ebenfalls möglich. Es ist ferner vorgesehen, daß detaillierte Beschreibungen von Subsystemen Modellbibliotheken entnommen und in das Modell direkt eingefügt werden können,[Rec87].
Das Graphikwerkzeug bietet auch typische Editierfunktionen.

Bild 3: Werkzeug zur graphischen Konstruktion erweiterter Warteschlangennetze

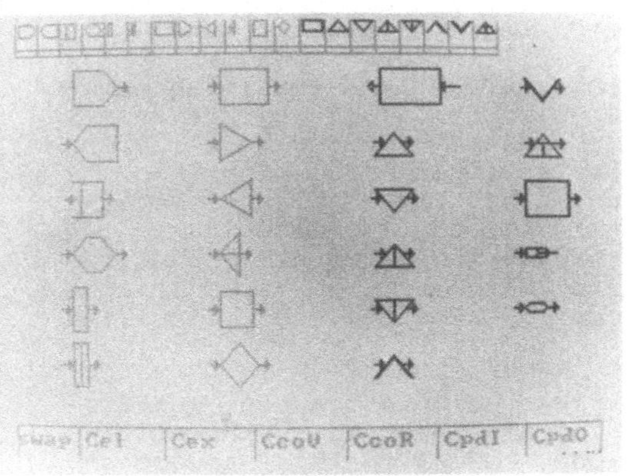

(3a): Symbolmenge

Dies sind alle verfügbaren Symbole. Während der Modellierung können Sie über die Funktion HELP Informationenen über die verfügbaren Symbole und die Kommandos erhalten.

source	submodel	token pool	or allocate
sink	fusion	allocate	transfer
queue	fission	release	subsystem
single server	split	create	inputpad
multiple server	set node	destroy	outputpad
infinite server	dummy node	and allocate	

Zur Fortsetzung drücken bitte Sie einen Knopf der Maus.

(3b): Bedeutung der Symbolmenge

Zur vollständigen Modellstrukturbeschreibung wird der monochrome Bildschirm benutzt, auf dem durch Ausfüllen symbolspezifischer Masken die Knoten bzw. Symbole des konzeptionellen Modells vom Benutzer beschrieben werden sollten (siehe Bild 5). Dabei erscheint auf dem Monochromschirm immer genau die Maske des Symbols, das gerade mittels Maus auf dem Farbgraphikschirm selektiert wird (siehe Pfeil in Bild 5a).
Zur Spezifikation von Abläufen ('routings' von Transaktionen) erhält der Benutzer auch graphische und textuelle Unterstützung (siehe Bild 6). Die 'routing'-Spezifikation fängt mit der Auswahl des Ausgangknotens eines Ablaufs mit der Maus an. Die Liste der Nachfolgeknoten im Ablauf ist automatisch in der Maske angezeigt. Der aktuelle Ablauf ist in der Graphik des Modells mit Farbe markiert. Das System läßt bedingte wie auch stochastische Verzweigungen zu und führt soweit als möglich Konsistenzprüfungen durch.

Bild 4: Graphische 'top-down' Konstruktion der Modellstruktur (Anwendungsbeispiel)

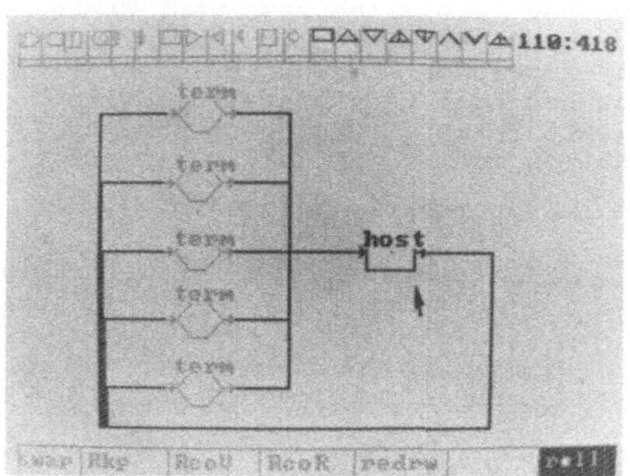

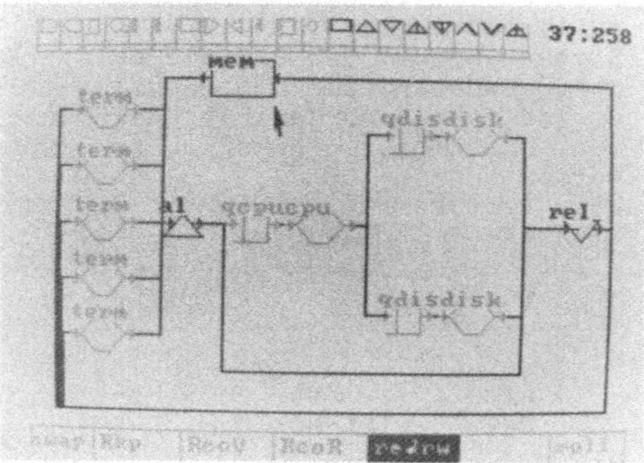

(4a): Globale Darstellung (4b): Detallierte Darstellung

• Phase 4: Erstellung eines werkzeugabhängigen, ausführbaren Modells

Wegen der Verwendung des passiven Elements MEMORY im Warteschlangenmodell wurde im vorliegenden Fall die Simulation als Lösungstechnik ausgewählt. Beratung kann der Benutzer in dieser Phase der Modellierung von einem Expertensystem erhalten, das entsprechend dem in Phase 2 realisiert ist [Rol86]. Sollte der Planungsbeauftragte an einer detaillierten, quantitativen Simulation interessiert sein, so müßte in der vorliegenden INT^3-Realisierung das ausführbare Modell in der Modellierungssprache RESQ [SMK82] implementiert werden. Eine automatische Generierung eines ausführbaren RESQ-Modells aus einem konzeptionellen Modell wird gerade realisiert.

Ist der Benutzer jedoch an einer interaktiven, animierten und qualitativen Simulation interessiert, so würde ihm als Werkzeug zur Modellrealisierung SIMLAB mit der Simulationssprache SIMSCRIPT II.5 angeboten. Hiermit kann der Benutzer das Ablaufgeschehen im System direkt mitverfolgen, bediente und wartende Transaktionen - je nach Zugehörigkeit in unterschiedlichen Farben - in ihrem Ablauf mitverfolgen. Außerdem kann der Benutzer interaktiv in die Simulation eingreifen, Parameterwerte zu definierten Zeitpunkten verändern, deren Auswirkungen beobachten, ggf. sogar in einem Ein-Schritt-Modus.

Bild 5: Beschreibung eines Knotens ('single server')

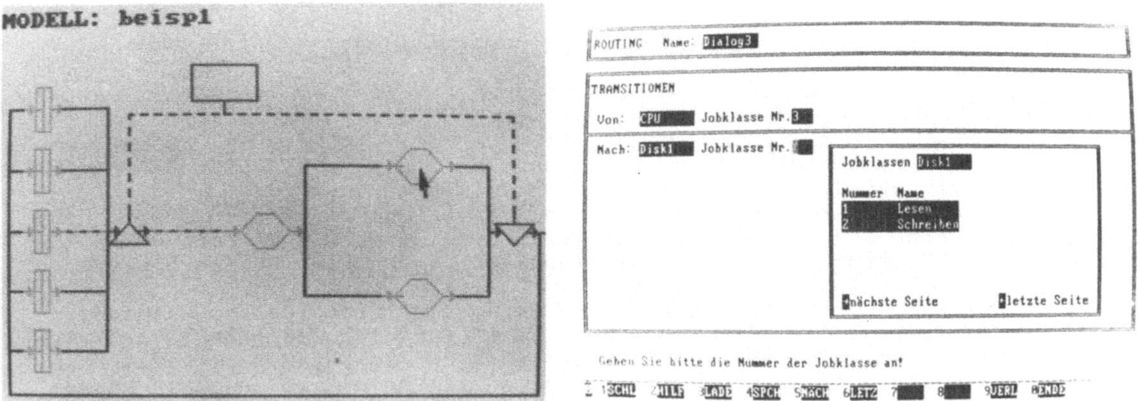

(5a): Selektion des Knotens (5b): Beschreibungsmaske

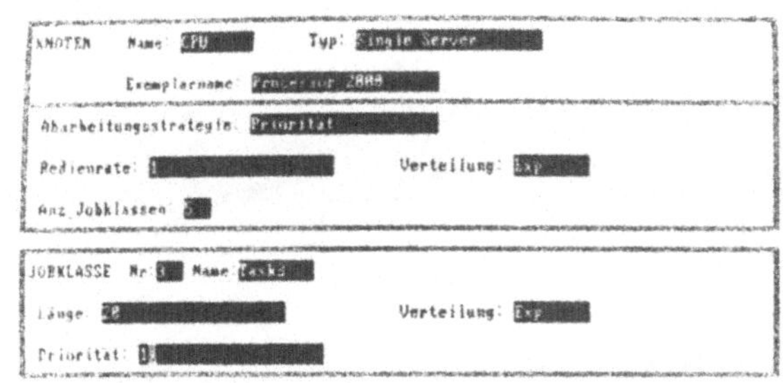

Bild 6: Ablaufspezifikation ('routing')

• Phase 5: Modellvalidierung und Experimenteplanung

In dieser Phase ist eine Unterstützung durch ein Beratungssystem geplant. Im aktuellen Beispiel hat der Benutzer alle Parameter des Modells sowie die Länge eines Simulationslaufes und die Anzahl der Experimente selbst festzulegen. Unterstützende Werkzeuge hierzu werden zur Zeit entwickelt.

• Phase 6: Experimentelle Anwendung

Im Falle einer Lösung des konzeptionellen Modells mit RESQ wurde die RESQ-Eingabedatei zur Modellausfuhrung an den 'Host'-Rechner IBM 4381 übertragen, und die Ergebnisdateien der Simulationsexperimente wurden an den PC zurück übergeben.
Das SIMSCRIPT-Modell wird interaktiv auf dem PC ausgeführt. Die Animation der Simulation wurde durch eine Bewegung der Punkte, welche ablaufende Transaktionen im Modell darstellen, erreicht (Bild 7).

• Phase 7: Auswertung von Modellierungsergebnissen

In dieser Phase stellt INT^3 ein Werkzeug zur Verfugung, das die quantitativen Modellierungsergebnisse entsprechend der in Phase 1 vom Benutzer genannten Zielsetzung darstellt. Da die interaktiv steuerbare Simulation weitestgehend nur zur anschaulichen, qualitativen Analyse des dynamischen Systemverhaltens dient, werden nur die statistisch relevanten RESQ-Modellierungsergebnisse graphisch ausgewertet. Der Benutzer kann die darzustellenden Variablen, deren Wertebereiche sowie Art und Bereich der Skalierung festlegen. Auch eine Parametrisierung der Kurven ist moglich.
Als Beispiel einer Ergebnisanalyse des gestellten Problems zeigt Bild 8 die Auslastung des Hauptspeichers als Funktion der Zahl aktiver Terminals dar, mit zwei Mehrprogrammfaktoren als Parameter. Damit kann im vorliegenden Beispiel der Planungsbeauftragte die gewünschten Ergebnisgrößen und die interessierenden Abhangigkeiten ubersichtlich interpretieren. Die Ergebnisse von Bild 8 zeigen in diesem Fall, daß bei Anschluß von 12 Terminals und einem Mehrprogrammfaktor von 5 eine maximale Auslastung bei minimaler Antwortzeit (nicht abgebildet) erzielt werden kann.

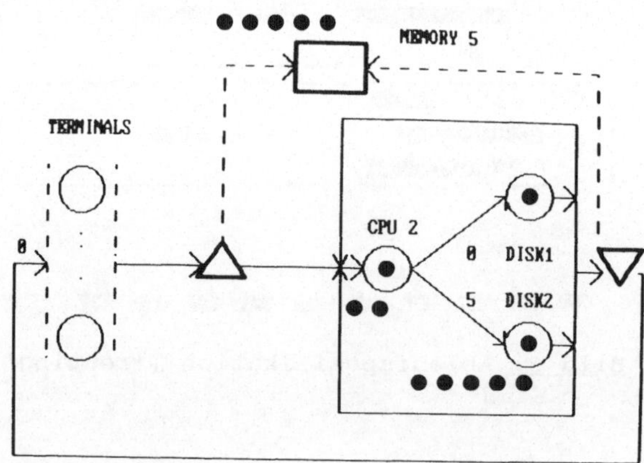

Bild 7: Interaktive, animierte Simulation (Anwendungsbeispiel) Momentaufnahme während der Simulation

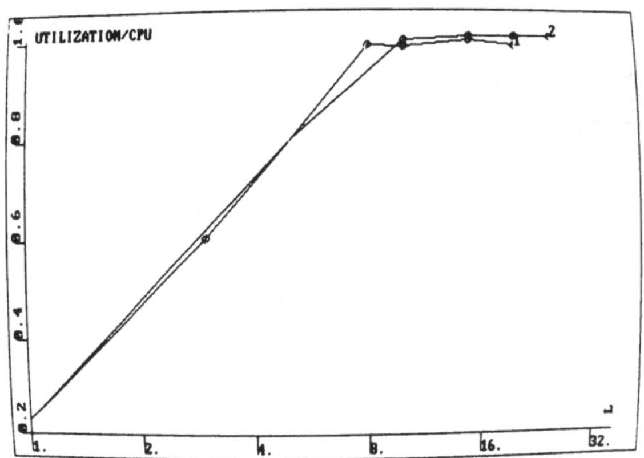

Bild 8: Graphische Auswertung der RESQ-Simulationsergebnisse
L: Zahl der Benutzer

	Kurve 1:	Kurve 2:
Mehrprogrammfaktor:	5	10
MEMORY - Kapazität:	K	2K

4. Zusammenfassung und Ausblick

Dieser Beitrag gibt einen Uberblick über Zielsetzung, Konzept und aktuellen Realisierungsstand einer INT³-Modellierungsumgebung: einer **interaktiven, intelligenten** und **integrierten** Unterstützung von Benutzern in den verschiedenen Phasen eines Modellierungsprozesses an ihrem PC-Arbeitsplatz. Begonnen als Lehr- und Forschungsprojekt im Rahmen der Kooperation zwischen der Universität Karlsruhe und IBM Deutschland wird zunächst versucht, Teilnehmern eines Fortgeschrittenen-Praktikums 'Leistungsanalyse von Rechensystemen' einen Prototyp einer INT³- Umgebung zur Verfügung zu stellen, auch um so erste Erfahrungen von einem eingegrenzten Anwenderkreis hinsichtlich Akzeptanz, Effizienz und Verbesserungmöglichkeiten zu erhalten. Diese Erfahrungen sollten Hinweise enthalten u.a. auf:

* die Eignung als Lehr- und Ausbildungssystem,
* den Komfort und die Handhabbarkeit der Unterstützungswerkzeuge,
* die Adaptionsfähigkeit der INT³-Umgebung an den Wissensstand der Benutzer.

Aufgrund der bisherigen Arbeiten an INT³, daraus resultierenden Anregungen und Überlegungen werden folgende Themenschwerpunkte im Mittelpunkt künftiger Arbeiten unseres Instituts im Bereich der Leistungsanalyse von Rechensystemen stehen:

-> Unterstützung einer semi-automatischen Modellsynthese u.a. unter Einsatz graphischer und wissensbasierter 'front-ends' zu Modellierungswerkzeugen;

-> Kombination von quantitativer und qualitativer Modellbildung, [Bür87];

-> Methoden zur zuverlässigkeitsbezogenen Leistungsmodellierung von Systemen mit stufenweise reduzierbarer Leistung ('gracefully degrading systems').

Schlußwort

Einige dieser Arbeiten sind auch Gegenstand des von der Deutschen Forschungsgemeinschaft geförderten Forschungsvorhabens 'Wissensbasierte, zuverlässigkeitsbezogene Leistungsmodellierung von Kommunikations- und Rechensystemen',(Az.:Schm 623/1-1).
Abschließend möchten wir uns besonders bedanken bei Herrn Prof.Dr.D. Schmid für sein Interesse an unseren Arbeiten, IBM Deutschland für die maschinelle Ausstattung und Förderung dieses Projektes und unseren Studenten A.Freiberg, C.Freiseis, G.Roll, G.Recktenwald und B.Hergeth für engagierte Mitarbeit und Diskussionen sowie für die Implementierungen des INT^3-Prototyps.

Literatur

[Bür87] Bürle,G.:
The Role of Qualitative Reasoning in Modelling.
Proceedings of IMACS, Int. Sympos. on AI, Expert Systems and Languages in Modelling and Simulation, Barcelona, 2-4 June 1987.

[Fra84] Framentec (Ed.):
M.1 Product Description.
Framentec, Monte Carlo, 1984.

[Fre85] Freiseis, C.:
Interaktive graphische Generierung von Modellen zur Verkehrsanalyse.
Diplomarbeit, Universität Karlsruhe, Institut für Informatik IV, 1985.

[Fre87] Freiberg,A.:
Realisierung einer Dialog- und Datenverwaltungskomponente für INT3.
Diplomarbeit, Universität Karlsruhe, 1987.

[Hec86] Hector: Heterogeneous Computers Together.
Proceedings, Wissenschaftliches Symposium der IBM, Bad Neuenahr, Juni 1986.

[Her87] Hergeth,B.:
Bereitstellung von Werkzeugen zur interaktiven animierten Simulation im Rahmen von INT3.
Diplomarbeit, Universität Karlsruhe, 1987.

[Joh86] Johnson. Glen D.:
PC SIMANIMATION, User's Guide and Casebook.
CACI- Europe, Los Angeles, June 1986.

[Leh87] Lehmann, A.:
Taxonomy and Application of Expert Systems in Simulation.
Proc. of IMACS, Int. Sympos. on AI, Expert Systems and Languages in Modelling and Simulation, Barcelona, June 1987.

[LKK86a] Lehmann,A.; Knödler,B.; Kwee,E.; Szczerbicka,H.: Rechnergestützte Modellierung in einer PC-Umgebung. März, 1986.

[LKK86] Lehmann,A.; Knödler,B.; Kwee,E.; Szczerbicka,H.: Dialog-Oriented and Knowledge-Based Modeling in a typical PC-Environment. Proc. of the SCS-86-Multiconference, San Diego, USA, Jan. 1986.

[MaS85] MacNair,E.A.; Sauer,C.H.: Elements of Practical Performance Modeling. Prentice Hall, 1985.

[Rec87] Recktenwald,G.: Dialogorientierte Generierung von RESQ-Verkehrsmodellen ausgehend von einer graphischen Modellbeschreibung. Diplomarbeit, Universität Karlsruhe, 1987.

[Rol86] Roll, G.: Expertensysteme als Beratungssysteme zur Konzeption, Losung und Anwendung von Verkehrsmodellen. Diplomarbeit, Universität Karlsruhe Inst. für Informatik IV, 1986.

[SaC81] Sauer,C.H.;Chandy,K.M.: Computer Systems Performance Modelling. Prentice-Hall Inc.,Englewood Cliffs,1981.

[Sta84] Staaden, M.: Dialogorientierte Problemspezifikation zur Verkehrsanalyse in Rechensystemen. Diplomarbeit, Universität Karlsruhe, Institut fur Informatik IV, 1984.

[SMK82] Sauer,C.H.; MacNair,E.A.; Kurose,J.F.: The Research Queueing Package: Past,Present, and Future. in: AFIPS,Vol. 51,Proceed. of the National Computer Conference,1982.

On Selected Performance Issues of Database Systems

Theo Härder
University of Kaiserslautern*
West Germany

Abstract
Performance, data integrity, and user-friendly access to data are considered to be cardinal properties of database management systems (DBMSs). But performance will steadily receive more attention as more interactive application are designed and implemented for almost all domains of our life. (According to some rumors, DBMS is sometimes said to be another name for performance problems). In this paper, we discuss specific performance problems of centralized 'conventional' DBMSs, DB/DC systems, and DBMSs for 'non-standard' applications (e.g. engineering, office, etc.). Then, a short survey attempts to sketch the solutions achieved and the problems remaining for the most important performance-critical functions/components in DBMSs. Finally, some aspects and mechanisms are discussed how database management system performance could be controlled and improved by measurement/monitoring techniques and subsequent adaption of DB-schema design.

1. Introduction

25 years ago, database management systems began to emerge in selected areas of computer applications, primarily for administrative and business tasks. From their early beginning, they rapidly more and more entered fields which now depend heavily on their proper function and performance behavior. The denotation 'DBMS' is often used as a buzzword for systems performing some data management jobs including simple file systems. To obtain a more accurate separation, we postulate a few mandatory concepts for DBMSs. Data description should be done 'outside the application program' which has to refer to the DB-schema when specifying an operation (DDL/DML). Record-type crossing DB-operations are also one of its important characteristics. Moreover, the notion of transactions [Gr80, HR83] became a fundamental DBMS concept within the last 10 years. Atomic transactions were originally introduced to relieve the application programmer from all aspects of concurrency and failure [EGLT76]. A transaction is defined to be a unit of work meaningful in the application environment; it must be executed completely or leave the database unaffected ('all or nothing' approach) despite the presence of failures and concurrent activities. These essential properties should be embodied in a DBMS deserving its name.

Today, the variety of DBMS is so diverse, and the spectrum of database (DB) applications is so broad and multi-facetted that it appears to be impossible to pay attention to all their uses and their specific problems in an equally balanced manner. For example, we don't want to focus on high data volume DBMSs tailored to specific applications (image processing and analys , document retrieval, etc.) where prime operational demands are data entry, archive management and fast output of 'long fields', statistical DBMSs where access and retrieval of unidentified data plays the dominant role, or to database machines (DBM) where various kinds of hardware support, specialized multi-processor architectures, and customized machine instructions are key issues to optimize DBMS processing for an anticipated workload of a (narrow) class of applications. Furthermore, DBMSs in (heterogeneous) networks [Ef87] are excluded, since they are in an early stage of research and not yet subject to serious performance evaluations. This incomplete list is meant to name only a few of the approaches we cannot deal with. Numerous further architectures and proposals for DBMSs were developed - in many cases paper designs: so-called PhD-machines. Such approaches are neither well understood nor have gained wide agreement on their importance yet.

Instead, we concentrate on DBMSs and their applications whose performance problems are at least well known, 'heavily attacked' by researchers and system designers during recent years, or even solved to a satisfactory extent. Moreover, we are able to report practical experience and consequences of performance behavior in most of these cases. Hence, we mainly evaluate three particular classes of DBMS architectures and applications:

- **Centralized DBMS**: The database is allocated to a single system granting a single user view of the centralized data to all its applications. DBMSs provide navigational or relational interfaces [As76] coupled with a host language as well as stand-alone query languages which allow for non-procedural specification of DB operations. Typical workloads consist of mixes of long batch transactions as well as short or medium

---

* This work was performed while the author was visiting at IBM Almaden Research Center, San Jose, CA 95120

sized transactions including complex ad hoc queries and report generation. The execution of some job types is usually planned according to resource availability to avoid worst contention problems (e.g. batch). Nevertheless, the DBMS is expected to cope with moderate to high degrees of concurrency.
The underlying computer system may consist of only one processor or a tightly coupled multi-processor architecture. To effectively exploit the 'horse power' of the system, the DBMS can be either assigned to one process (single server) making some kind of multi-tasking mandatory for performance reasons or to several processes (multiple servers) each running in single- or multi-tasking mode [Hä79, St81].

- **DB/DC system** as kernel part of a transaction system: Many on-line business applications are highly repetitive, perform a short sequence of simple operations, and require fast interaction with the database thereby reading and updating only a few records. Because of the repetitive nature of such business procedures, they can be designed and implemented in advance; hence, an entire application (say a banking application comprising deposit, withdrawal, transfer of money and inquiry of account information) may be preplanned and accomplished by a few transaction types ('canned' transactions) which are invoked with actual parameters (e.g. amount of money to be withdrawn) from an automatic teller machine (ATM)) or terminal (parametric user).
  In these applications, the type of data model is less important, because DB operations are so simple. Much more important are their tough operational needs in highly concurrent environments for obvious reasons Although achievable with full-function centralized DBMSs, demanding performance requirements imply either highly specialized systems (e.g. IMS Fast Path) or solutions based on a group of individual computer systems which are either losely or closely coupled. Since availability and modular growth are other key issues in high performance transaction systems (HPTS), recent developments pursue approaches based on a couple of local computer systems sharing or managing the same database. Although such a database may be partitioned or (partially) replicated, the cooperating (independent) DBMS have to derive the view of a centralized database allowing for serializable transactions.

- **DBMS for non-standard applications**: While 'classical' DBMSs - primarily designed for business usage - have satisfactory success in these areas, they have so far failed to be a help for other important types of applications suffering from urgent data management problems. Non-standard applications in engineering and office systems or for processing high volumes of scientific data show substantially different sizes and usage patterns of data; thus, their effective management needs adjusted data structures and operations as well as more appropriate mechanisms to specify integrity constraints. Furthermore, operational requirements call for DBMS coupling to graphical workstations with considerable local processing capacity. Design and construction support implies time-critical operations, since the system's responsiveness determines the pace and the non-disruptive flow of workstation interaction. In such systems, DBMS processing has to anticipate sequences of powerful operations tied together by a comparably small number of long transactions [KLMP84], rather than extremely high concurrency of very short transactions typical for DB/DC systems.

To motivate the challenges of the future that DBMS researchers and designers are currently faced with, we will give a number of particularly high performance requirements for various DB applications serving as the focal points of architectural and algorithmic ability to overcome such 'giant' workloads.

- **Complex queries**: The first example focuses on very large data sets, complex ad hoc queries, and fast retrieval ability. A particular scenario assumes an average throughput of 15 transactions per second in a management information system. The workload consists of complex and unpredictible queries to be processed in a 40 GB database (Bank of America). Since only part of the query predicates may be evaluated via existing access paths, a sequential scan of the entire DB is necessary in the worst case to select the qualified records. Assuming that direct access is provided for 2/3 of the requests (short and medium size transactions), the remaining fraction would burden the system with a tremendous sequential load. As a consequence, it is very demanding to obtain an average response time of < 30 sec for such 'exhaustive search' transactions.

- **High transaction rates**: Banking has always been the fore-runner in DBMS applications as far as extraordinary transaction rates and data volumes to be updated are concerned [Bu85]. While current systems achieve transaction throughput up to 400 transactions per second (TPS) of the DEBIT-CREDIT type [Anon85], future HPTS are expected to master throughputs of > 1KTPS [GGGHS85] with an average response time of < 1 sec. A further challenge is the concurrent processing of batch and on-line transaction mixes operating on the same database which adds an extra burden to concurrency control.

  The evolutionary trend towards DB-driven public communication networks provokes additional increase of transaction processing capacity. Future telephone networks consult a (potentially distributed) DB during call processing to establish caller permission, perform name-address translations, select personal profiles, establish routes, etc. [HGLW87]. Because DB services are currently utilized only for a small fraction of calls (e.g. to validate a credit card number or to derive routing instructions specific to a given call ('800' toll-free calls in USA)), processing demands are moderate, that is, about 100 simple queries per DB-node carrying a copy of the database of > 10^5 records. Future call processing, however, is anticipated to rely entirely on timely and individual DB information requiring short read transactions with a throughput demand of > 10^4 TPS accessing > 10^7 records without any 'observable' delay. Moreover, some of these transactions might execute more complex ad hoc queries and random updates to maintain up-to-date DB information.

- **Object-oriented access and modification**: No specific requirements in terms of transaction throughput are known - primarily, because it is hard to characterize the sophisticated and varying workloads of engineering applications by means of some 'standard' transaction. On the other hand, focus is on reponse time of powerful operations to be useful for interactive construction. Such operations include 'vertical access' to an entire complex object where up to 10^3 - 10^5 heterogeneous records (linked to each other) have to be fetched or written instantaneously, or 'horizontal access' where spatial or similarity search may stimulate sequential scans through large inventories or object libraries. This kind of complex object handling, however, is supposed to take place in low concurrency environments.

These examples emphasize future performance demands in a wide variety of application areas. They avoid the impression that 'all problems are already solved' in the DBMS field. Instead, they should stimulate research activities to design and evaluate new system structures and algorithms. Such performance goals imposed on current DBMSs would really stress and saturate them long before they are reached. It is questionable whether they can be met by systems evolving from existing ones or whether they necessarily imply entire new concepts and approaches.

On the other hand, current DBMSs have their strengths and weaknesses. Sometimes they don't cope well even with strongly reduced workloads. In the following, we will first investigate some specific performance problems in the selected application areas before we evaluate the ability of functions and algorithms in current DBMS to overcome bottlenecks and performance-critical issues.

2. Some important performance problems observed

After having recognized the DB-performance challenges we have to meet in the near future, we are going to discuss three major performance problems, each one chosen from a different area of our three types of DBMS applications. Note, that most of the difficult problems are provoked by data sharing which cannot easily be circumvented. *Hardware can be shared for economical reasons, but data sharing is intrinsic in the applications (J. Saltzer).*

At first, a short remark about the measure of DBMS performance. When we analyze a complex computing system and its internal behavior, a large number of particular measures are needed to characterize and evaluate its performance behavior: path lengths, # of I/O's, storage usage, CPU utilization, # of process switches, # of messages and many more. Here, we use application-oriented measures to describe and compare DBMS performance, since 'MIPS numbers' are not well related to 'what a user expects and gets'. Hence, the externally observable performance is usually expressed in 'throughput' and 'response time'. DEBIT-CREDIT [Anon85] gained common acknowledgment as **the** transaction used as a unit measure in DB/DC systems because of its simplicity. Throughput figures expressed in **TPS** (transactions per second of the DEBIT-CREDIT type) are used as a measure for comparing transaction system performance (response time requirements are explicitly defined by limiting 95% of the transactions to < 1 sec.). TPS is at the most reasonable for homogeneous sets of short transactions to reduce the system performance to a 'golden' number that allows easy comparison (which sometimes is questionable); however, it entirely fails to capture workloads consisting of mixes of short, medium and long transactions with varying degrees of read and write accesses. Since there are no agreed-upon measures available, we use general measures like elapsed time for throughput when appropriate.

2.1 Concurrency control

DBMSs are multi-user systems where data sharing is inherent. Thus, synchronization of accesses to shared data plays a dominant role in such systems, and concurrency control algorithms are designed to assure consistent data references. That is, they have to provide serializability of transactions (single user view of data) which implies that all read operations within a transaction are repeatable [BHG87]. As a consequence, all data referenced have to be protected until a transaction commits (CL3 = consistency level 3 [GPLT76]). A less rigorous form of synchronisations is isolating only modified data in a transaction, but not guaranteeing the same view of the referenced data in the course of transaction execution (CL2). Such a method - typically adhered to in current 'navigational' DBMSs [Gr78] - requires some discipline and responsibility of the application programmer to derive consistent results and DB changes. However, much of the synchronization burden is removed from the DBMS to utilize available resources more efficiently.

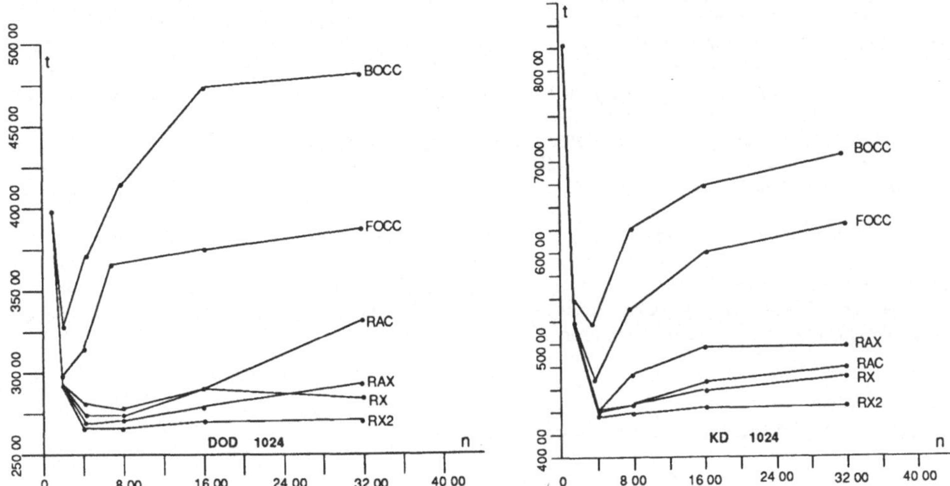

Figure 1: Simulation of various CC algorithms - elapsed time (t)

To illustrate the effect of concurrency control, we borrow some of the results gained in the most accurate and extensive empirical performance evaluation obtained for concurrency control (CC) algorithms for centralized DBMS so far [Pe86]. A detailed simulation model (40000 PL/I statements), that was driven by object reference strings (ORSs) derived from real-life DB applications, was used to evaluate the ability of various CC methods to cope with high degrees of parallelism and related issues (deadlocks, transactions to be aborted). Pessimistic methods prevalent in existing DBMS, e.g. standard locking (RX) [Gr78] and some refinements (RAX, RAC), were compared to various optimistic algorithms (BOCC, FOCC [KR81, Hä84]) to decide the general question of superiority. The effect of a 'relaxed' locking protocol RX2 (RX for CL2) on performance was also explored for comparison purposes. 6 ORSs with up to more than 10^5 references and more than 2000 transactions each were available as 'workload models'. An accurate characterization and description of the workloads may be found in [Pe86]. Here, we only want to emphasize that the ORSs reflected transaction mixes consisting of short, medium, and sometimes long transactions with varying percentages of read and update operations, that is, CC was not confined to 'easy or delightful' situations. Fig. 1 depicts selected results for various degrees of parallelism (n) for two different ORSs (executed with a buffer size of 1024 frames (2 MBytes)). Since all system and environment parameters remained unchanged during the throughput tests shown, CC methods and their behavior under increasing concurrency may be compared to each other. Apparently, the following properties are revealed:

- Very low parallelism does not effectively exploit the system resources (I/O waits causing idle times).
- Low parallelism yields the best throughput results, not necessarily the shortest response times, since arriving messages may be delayed in the input queue until they are served - subject to availability of processing capacity (n).
- Pessimistic methods are clearly superior to optimistic methods when concurrency increases. They block a transaction issuing a conflicting request. Such conflicts are often resolved when a transaction keeping such a blocking lock commits; hence, the blocked transaction may proceed with its normal execution after some waiting time. This kind of blocking frequently avoids the occurrence of a deadlock; therefore, extra overhead is often saved for deadlock resolution, that is, selecting, aborting and reprocessing a victim transaction.
- Given only moderate data contention (conflict potential) as in the chosen ORSs, optimistic methods begin to exhibit 'thrashing behavior', since many transactions fail to pass validation. Hence, they are aborted and

reprocessed. Without additional measure (load control, declare a victim as 'golden'), they may repeatedly fail (livelocks, etc.).

- High contention is effectively reduced by RX2, because only short read locks are applied which removes many obstacles for concurrent transactions.

These results only allow for a crude evaluation of CC methods. Obviously, optimistic methods don't seem to perform well in highly concurrent situations with unpredictable usage patterns; here, pessimistic methods are the clear winners. Unfortunately, there does not exist a consistent performance ordering among them. Various applications and different patterns of data contention don't deliver conclusive results. It comes as no surprise that CL2 protocols always yield the best results. However, it should be emphasized that it is desirable to delegate all CC issues to the DBMS (CL3 protocols) in order to remove them from the programmer's responsibility and not to potentially compromise integrity of shared data or consistency of results. This is especially true in relational DBMSs where the user has a more abstract view of the data than in navigational ones [As76].

It must be anticipated that CC remains a performance-critical DBMS problem - for complex transactions as well as for specific reference patterns, e.g. hot spot data [Ga85, Re82]. Note, the inherent properties of the performance behavior expressed in Fig. 1 are related to data contention and not machine speed or other characteristics. Therefore, they don't vanish with higher MIPS rates, in contrary, they tend to get much worse with substantially increased concurrency.

2.2 Linear growth of transaction processing power

The problems reported so far occur at 'very' low degrees of parallelism compared to transaction systems with hundreds or thousands of TPS. However, they are primarily related to transaction duration and data set sizes referenced (complex queries, etc.). Transaction systems execute short transactions, and it should be easy to overcome such CC-induced problems - at least in ranges of n < 50.

In transaction systems, the focus is on 'high performance', availability and modular growth (including horse power (MIPS) and data storage (disks)). The challenging goal is to provide **linear growth** of transaction processing power measured in TPS. Availability and system growth beyond certain limits imply the use of multi-processor architectures for catering the necessary MIPS power. (Currently, uniprocessor speed is limited to < 30 MIPS (Fujitsu) and typically ranges from 16 to 21 MIPS 'for the high end'). To satisfy high availability requirements, loose coupling of processors seems to be mandatory to have better failure isolation and a stronger barrier against propagation of errors to working subsystems [Ra78]. Tight coupling is much more susceptible to failures of various kinds. An empirical field study fully supports this evaluation. It has revealed that NonStop systems based on a loosely coupled architecture have a MTBF of 10 years [Gr85] whereas tightly coupled architectures currently achieve MTBF of 2 months (after having installed a number of improved robustness mechanisms [Gr87]).

For this reason, the major proposals for HTPS architectures - called DB-sharing and DB-distribution - are based on loose processor coupling. Fig. 2 illustrates the characteristic properties of these system architectures. It is obvious that load balancing and load control play a performance-critical role in such architectures. In our context, load balancing provides a policy for sharing central resources (data, processors, etc.) among competing processes (DBMS servers) whereas load control tries to adjust and schedule these processes dynamically [Re86].

In so-called DB-sharing systems [Sh85], multiple DBMSs share the DB at the disk level. Hence, there is no need for data partitioning among the DBMSs. Accordingly, a DBMS is able to execute an entire transaction locally. To do so, it possibly has to request resources such as data pages (currently in a 'foreign' system buffer (SB)) or locks (managed by a 'remote' lock manager) from another DBMS. Since inter-system cooperation is assumed to be considerably more expensive than local requests, the efficiency of the approach strongly depends on how effectively locality of reference can be maintained in each of the participating systems. Appropriate load partitioning is mandatory to help balance the usage of system resources and to minimize the share of remote data and lock requests [Re86].

As opposed to DB sharing, each processor owns a partition of the disks in DB-distribution systems [HR86]; the corresponding DBMS is handling all data requests for a disk partition which is statically allocated (reachability in failure situations requires two paths to the disks). Furthermore, the processor is reponsible for all consistency issues of that partition, that is, concurrency control, logging and recovery are performed locally. A transaction accessing data in several partitions is executed by means of subtransactions that are agents of the transaction in every DBMS executing some operations for it. Atomicity of transactions is assured by 2-phase commit protocols. Since transaction loads do not match the static data partitions perfectly, we also have to cope with minimizing external activities for a transaction, e.g. calls for function shipping or invocations of remote subtransactions. Again, load balancing and load control are of paramount importance for system performance.

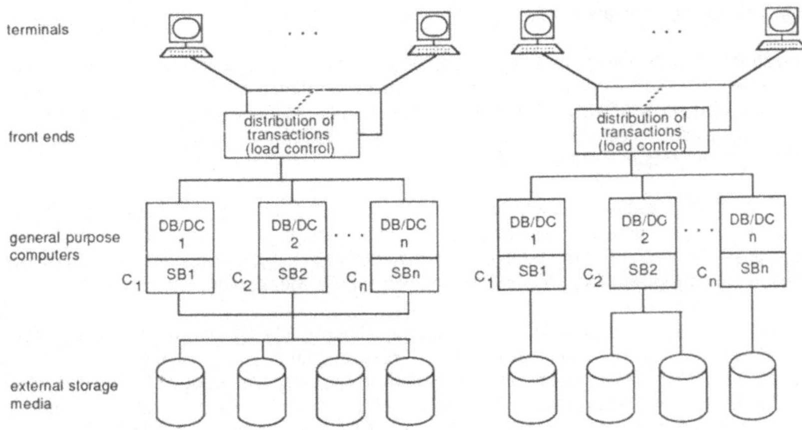

Figure 2: DB-sharing and DB-distribution architectures

Experience in todays multi-processor systems show that it is difficult to accomplish even only 'almost' linear speed up at the application level. Performance degradation is due to resource management overhead, unbalanced resource utilization, and coordination effort for both tightly as well as loosely coupled architectures (II in Fig. 3a). In particular, memory contention, cache invalidation, and expensive context changes (process switches) cause early saturation and asymptotic system behavior in tightly coupled systems. Message costs and delays, unevenly distributed system load, etc. are major concerns in loosely coupled systems. The 'ideal' behavior should be achievable - at least in close approximation - for 'non-related' applications when cheap processes, efficient messages ('the two myths of computer science'), etc. are available (I in Fig. 3a). A good example is Tandem's NonStop system which incorporates linearity of performance and modular extensibility which is transparent to the application, since it provides add-on performance across processors connected by a fast bus and by a moderate bandwidth LAN [HC85]. This desirable behavior, however, requires message-based, location-transparent process cooperation which is more expensive than memory-based data synchronization and message exchange. Hence, one might argue that 'lack of optimization of the simple case' is the price for linearity in application-level growth. Rumor: Some systems (e.g. ELXSI) claim *more than linear growth of performance* by avoiding main memory contention and reducing message delays with large caches and extremely fast buses, respectively.

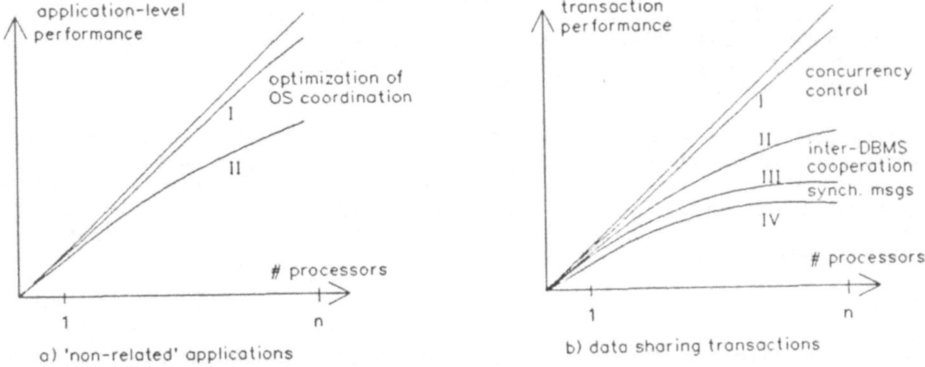

Figure 3: Linear growth of transaction processing performance

Even more challenging is the **linearity goal** in data-related applications such as HPTS. Inherent to such DB/DC applications is the need of synchronization across the multi-processor architecture, since consistency of data accesses among all activities must be preserved. The specific problem occurring in all types of HPTS may be coined as a 'concurrency/throughput cycle' which may be characterized as follows [RS85]:

- The higher the number of transactions, the higher data contention has to be anticipated (given a constant size or a less than proportional growth of the DB). Locality of references frequently observed in data sharing applications increases the conflict probability; some data elements that show normal traffic under low degrees of concurrency gradually become hot spots when exposed to higher traffic. This situation then serializes transactions accessing them.

- Synchronization must be performed across processor boundaries as soon as the DB/DC system spans more than one processor. External messages for synchronization protocols imply long delays for a transaction, since it synchronously waits for the answer (activation/deactivation of participating processes on sender/receiver sides, delay of message transfer by e.g. buffered transfers, message handling/queuing and transmission for request and reply).

- The aggregated time delays for synchronization requests extend the allocation of the transaction's resources thereby obstructing or blocking other transactions, e.g. lock wait. Increased lock contention, however, cause an increase of the multi-programming level to keep the desired throughput level. Hence, an increased multi-programming level, in turn, implies even higher contention.

An additional problem occurs in DB-sharing systems. Since every DBMS has its own system buffer, a particular piece of data may reside in more than one buffer at a time. Thus, such systems have to cope with fully replicated data; the shared disk pool must be considered as the hardcopy of some recent state of the replicated DB. Because pages are units of data transfer between buffer and disk and, more important, among the buffers of the participating DBMSs, an additional problem burdens the cooperation of independent DBMSs:

- Larger lock granules are necessary to prevent conflicting updates of a page residing in multiple buffers (e.g. two different records are inserted into the same or an overlapping physical location). The choice of pages as opposed to records as smallest locking units is overly restrictive and may frequently cause fictitious conflicts when traffic to a page becomes higher.

- A page update in one buffer makes all copies of the page in the remaining buffers invalid. This buffer invalidation problem costs extra message and/or write overhead to prevent other DBMSs from accessing 'old' data.

Fig. 3b illustrates how performance of data-related applications might be affected by multi-processor DBMSs (as a personal view). Let us assume that we have already an 'optimized' multi-processor system at the OS-level which behaves nearly linear when executing 'non-related' applications (I). Performance is assumed to be substantially influenced by concurrency control requirements amplified by the need of providing synchronization across processor boundaries. Of course, the chosen CC algorithm and the degree of contention play an extraordinary role. This is especially true when transactions become more and more complex (10 - 100 data references and/or DML operations vs. less than 10 in DEBIT-CREDIT) - but yet of moderate size - and when multi-step transactions (several inputs and outputs from/to terminal) are used. Then, transaction duration and resource allocation get much longer which then amplifies the negative feedback in the 'concurrency/throughput cycle'. Note, such resource demands are expected in future on-line transactions - batch transactions involving (very) long readers and writers embody much larger resource consumptions.

In special cases, the expected overhead indicated by (II) in Fig. 3b may be greatly reduced by effective *load and data partitioning* such that each individual system performs like a 'quasi' centralized DBMS. To narrow the performance gap (I-II) for moderate size on-line transactions, however, new ideas and new synchronization concepts are urgently needed when special partitioning options are not applicable. As a consequence, effective CC is very important in centralized DBMSs and even more important in multi-processor DBMSs where much higher degrees of concurrency have to be dealt with.

A further problem influencing the overall performance is sketched by the degradation indicated by (II-III) in Fig. 3b. Here, we want to summarize all inter-system related aspects of cooperation, control, and coordination which may be very expensive depending on frequency of use and efficiency of implementation. Load balancing and load control is considered to be extremely important in such systems to assign transactions to the (current) location of (most) data to be accessed. Since the data set referenced is not (entirely) known in advance and since data is concurrently requested by multiple DBMSs, transaction assignment is always 'fuzzy' due to incomplete and imprecise in-advance knowledge. Flexible transaction assignment to DBMSs keeping the required data currently in their system buffers is a key property of DB-sharing systems (affinity problem [Tr83]). To effectively use this optimization potential, however, appears to be a tough and difficult scheduling and balancing problem which is very sensitive to load compatibilities and changes. Load balancing does not vanish in DB-distribution systems; but it is facilitated, since knowledge about data location is 'perfect' due to the static partitioning. On the other hand, load changes and static data allocation may lead to an unbalanced overall use of resources. Hence, load control helps to enhance locality, to reduce message delays, and minimize 'external' lock requests which are all of prime importance to system performance. Another aspect in DB-sharing systems is the integrated solution of CC and buffer invalidation [RS85, Ra87] which optimizes various aspects of lock requests and exchange of the valid copy of a data page. To summarize the discussion, effective load balancing as well as 'external message saving' solutions for all cooperation services are necessary to avoid broadening the gap (II-III).

Another issue influencing inter-DBMS cooperation is emphasized as a separate aspect in Fig. 3b, since some approaches try to support and optimize data and message exchange by special efforts, e.g. by providing specialized hardware. Hence, the availability of synchronous messages for requesting locks or synchronous access for fetching data from a 'remote' DBMS resp. memory could close the gap (III-IV). Such hardware-based solutions for efficient DBMS cooperation - sometimes called close coupling [HR86] - help to improve the overall system performance. However, it should be clear that they alone will not solve the basic performance problems of HPTS, as illustrated in Fig 3b

We already mentioned that the Tandem NonStop system provides linear performance growth for 'non-related' applications at the extra expense of exclusively using location-transparent messages. Hence, these costs are part of all corresponding performance figures, no matter whether the system consists of one or of n processors; in other words, a comparison should be only made with systems of the same type, since the simple case of local transaction processing is not treated differently. Nevertheless, the transaction system ENCOMPASS [Bo81] proved to be extensible and linear in performance for a large DEBIT-CREDIT application. In [TSR86], measurements are reported where a transaction system was stepwise extended from 2 to 32 processors which were connected within a node by a fast processor bus (DYNABUS) and across nodes via a LAN (FOX). The 32 processor system reached 95% of the linearly extrapolated throughput of the 2 processor system; it should be mentioned that the system growth was transparent to the application programs.

This example does not mean that the linearity goal in data-related applications is already achieved. A closer look into the application reveals that in this case partitioning can be performed perfectly, since value ranges of account numbers of the most important record type ACCOUNT may be assigned to various processors. Such a distribution supports overall I/O balancing which has a strong influence on throughput. (Furthermore, HISTORY, BRANCH, and TELLER record types can be partitioned such that remote references are not required; record locks reduce the hot spot problem). Moreover, the transaction load was very simple and permitted local DBMS processing without interference. Hence, the challenging performance problem of Fig. 3b remains for more general cases.

Again, since data sharing is intrinsic in applications, hardware and sheer processing power will not fix the most critical problems in high volume transaction processing. But 'problem-aware' DB-schema design as well as organizational support (preplanning of transaction processing to avoid data contention) will greatly reduce the burden of such transaction systems.

2.3 Overhead of complex object mapping

The performance problems described in the following are only marginally related to parallelism; they occur in low concurrency as well as single user environments. In contrast, they are caused by complex retrieval and update operations on large heterogeneous data structures - representing complex objects consisting of a potentially large number of subobjects which are linked to each other by network relationships.

We have accomplished DB-based prototype implementations of various engineering applications (e.g. VLSI design, solid modeling, map handling in geographic information systems [Hä87]) to explore the DBMS load of powerful mapping functions applied to complex objects thereby using an existing CODASYL DBMS. As a result, we are able to present some empirical data to make the performance problem more specific. 3D geometric modeling, which is considered to be one of the most demanding areas for DBMS use, may serve as our example. Since detailed description is too space-consuming, we need some intuition to 'communicate and reveal the inherent characteristics of modeling and managing engineering objects'. Hence, our goal is primarily to illustrate the order of magnitude of DBMS work involved.

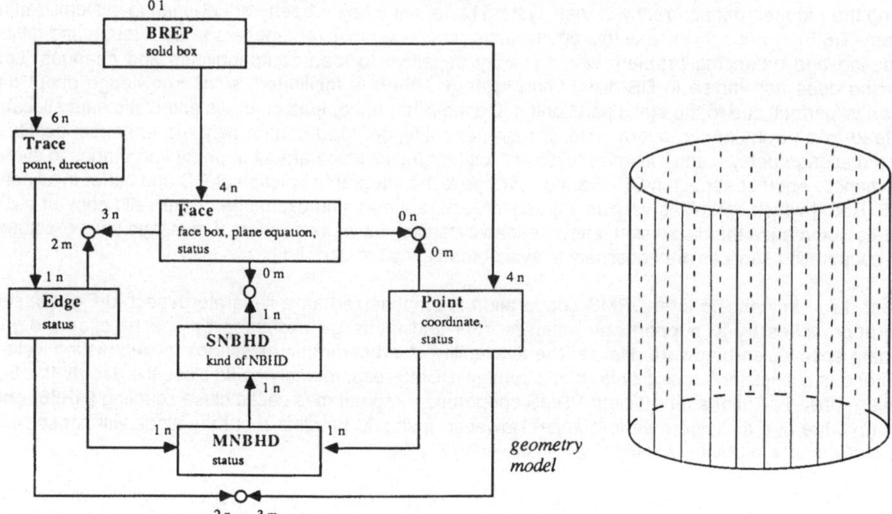

Figure 4: Geometric model of solids - boundary representation

The schema of the boundary representation (BREP) of 3D solids (geometrical part) is given by Fig. 4a and a sample object by Fig. 4b. A complex object is represented by sets of records of the various types interrelated as shown by the schema (arrows describe 1:n-relationships and circles inter-connection records (n:m)). Creating a new complex object, e.g. a cylinder approximated by n faces (POLYCYL(50)), requires the insertion of a single BREP record together with records describing all faces, edges, points, etc. and their inter-relationships. Depending on the accuracy of approximation, one quickly obtains hundreds or thousands of records to be inserted or modified. We provided powerful ADT-like operations (e.g. CREATE POLYCYL(n)) for the applications by enhancing the DBMS by an application layer which translates the ADT operations into sequences of operations (DML operations) passed to the DBMS interface (data model). Fig. 5a indicates the 'growth of costs' in terms of DML operations caused by refining the DB representation to better approximate the real object. In a second step, the straightforward DBMS use was improved by using locality in the application layer. Further optimization included context knowledge of the application which, however, made the approach less general. It did not reduce the overhead to a feasible size. None of these adaptions improved the mapping costs to a limit acceptable in interactive applications. On the other hand, it is questionable whether an overly accentuated application dependency is a step in the right direction. Fig. 5b depicts the distribution of DML operations over the various object types (record and set types of a CODASYL schema). Apparently, this distribution is far from being uniform. Furthermore, locality of reference depends heavily on the type of operation performed. As a consequence, selective access to subobjects is desirable.

The bad performance of 'conventional' DBMS for handling complex objects has triggered broad research efforts to overcome such deficiencies. Main areas of research are

- the design of data models embodying some degree of object orientation [HMMS87, SS86]; as illustrated in Fig. 4a, a substantial share of the relationships are non-hierarchical (non-disjoint and sometimes recursive) which should be honored by providing non-redundant representations and support of symmetrical traversal.

- the development of extensible DBMS architectures, multi-dimensional access path structures, efficient storage clusters and other implementation techniques for complex objects [PSSWD87, Sc86].

- object handling 'performed near by the application'; this implies fetching complex objects into an object buffer possibly located in a workstation (Checkout), accumulation of updates in the object buffer, and propagation of sets of updates after finishing some construction work (Checkin [BKK85, KLMP84]).

- the exploitation of parallelism in single-user operations on large heterogeneous data structures [HHM86].

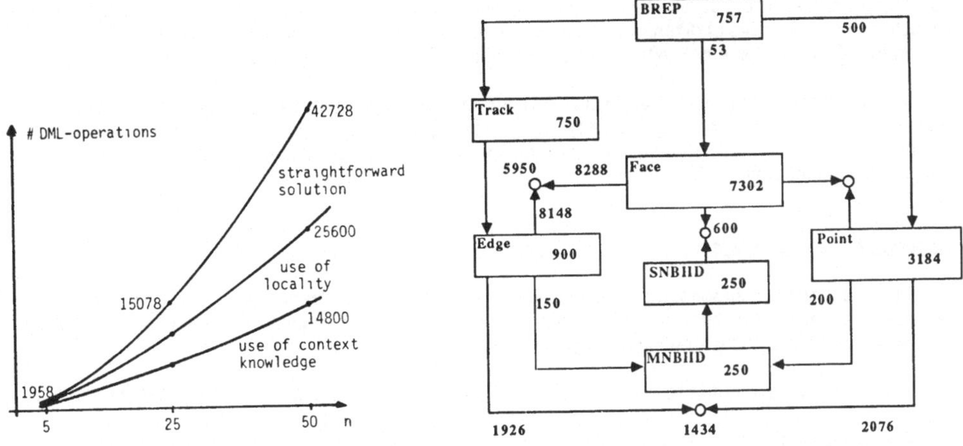

Figure 5: Costs of generating the DB representation of simple solids

All these research efforts are addressing the performance problem expressed in Fig. 5. We can hope to get an answer soon by prototype systems for non-standard DBMS whether or not the performance gap can be bridged in a satisfactory manner.

3. Performance-critical functions and algorithms in DBMS

Thus far, we have described three characteristic performance problems of different areas of DBMS use. Due to limited space these examples can only outline the wide spectrum of performance problems to be solved. For this reason, it is helpful to identify the most important concepts, issues, and algorithms in performance-critical DBMS areas and to characterize their current approaches and solutions. After more than 15 years of intensive research and development, an immense amount of ideas, concepts and results are available which may serve as a base for future system design. Where are we now?

Fastl I/O

Most DBMS are CPU-bound (long execution paths, process switching overhead). Nevertheless, special functions like writing log records or fetching large sets of data records or long fields (document container) require fast I/O. Currently, chained or queued I/O for magnetic disks is often proposed [We86] to relieve the I/O bottleneck for those functions. Chained I/O allows for n consecutive block transfers with only one seek (positioning of read/write head) and one rotational delay thereby reducing access time for a large object (1 MByte) by a factor of 5-7 (4 Kbyte blocks) compared to n consecutive single block transfers.

During the last few years so-called solid state disks entered the market place. They offer non-volatility of data, block-oriented access via channels (3-4.5 Mbytes/sec) and capacities of about 500 Mbytes at a price per byte ratio currently 50 times higher than that of a magnetic disk. Since seek and rotational delay don't apply, an access time of 1-2 ms for an arbitrary block (4 Kbytes) may be achieved. Benchmark tests have revealed that about 2500 block I/O's per sec. are possible when four 4.5 Mbyte channels are available [Ga87]. That is, using a dedicated channel to the log file, about 600 log writes per sec. can be performed which is an increase of a factor of 10 compared to conventional disk technology.

Due to the enormous price difference of solid state disks, only selective use of them may be economic. However, a characteristic observation of DBMS's data usage makes such an application quite realistic. Locality of reference (to blocks on disks) is extremely distinct; there are many almost dead files and on the other hand, high reference rates are directed against comparably small subsets of the DB (dictionaries, indices, special data structures). For example, more than 50% of the disk references were observed to a

200 Mbyte subset of a 500 Gbyte DB [Ga87] which is 0.4‰ of the whole DB size. Hence, appropriate allocation of data to disks of different types may drastically improve I/O behavior at reasonable costs. How big should the electronic store be? What are the most effective organizational schemes for it?

It was stated elsewhere [RK84] that physical clustering of records within a page and consecutive pages is a prime performance-determining property. Since divergent processing requirements of records don't coincide with the one static allocation (to be chosen without redundancy), suitable clustering of records is a major challenge of physical DB design.

Support of locality of reference

Empirical observations suggest that the 'famous' 80-20 rule of data reference is conservative. Indeed, it may be repeatedly applied to approximate the above stated reference behavior. The static clustering property on disk is complemented by the dynamic data requests of transactions.

In a DBMS, a database buffer has to be maintained for purposes of interfacing main memory and disk. Since a physical access to a database page on disk (block) is much more expensive than an access to a database page in the buffer, the main goal of the database buffer manager is the minimization of I/O for a given buffer size. Locality of reference observed in all DBMS applications is exploited to optimize page replacement in the buffer (sequentiality of reference would make all conceivable efforts obsolete).

Replacement algorithms often used for buffer management are LRU, CLOCK or GCLOCK [EH84]. With navigational DBMS interfaces (one record at a time), there does not seem to be any potential for further improvement. Relational interfaces, however, offer more knowledge about 'future' accesses due to their non-procedural, set-oriented request specification. Hence, an interesting research problem is how such knowledge can be made available to the buffer manager for the prediction of future reference behavior. Some preliminary investigations show some remarkable optimization potential [SS82]. Which means should be introduced to allow the buffer manager to accept 'advice' from the query interpretation or compilation process in order to use the context information of high-level DB languages? How could kind of prefetching or preplanning algorithms be integrated into DB buffer management to further reduce synchronous waiting times for the requesting transaction?

Large buffer sizes will relieve the replacement problem; with 50-100 Mbytes main memory or buffer sizes available in the near future, locality of reference should be facilitated. "The 5 Minute Rule for Trading Memory for Disc Accesses and the 10 Byte Rule for Trading Memory for CPU Time" were postulated in [GP87]. In short, pages referenced within 5 minute intervals should be kept main memory resident. It seems reasonable that this '5 minute rule' should be reflected in good DBMS architectures - given fast growing and cheap memory capacities. Again, what are the consequences for physical DB design?

Access path structures

There is one unanimously agree on access path structure for key and sequential access - the ubiquitous B-tree [Co79]. Hashing is often important; the usefulness of extendible hashing is sometimes discussed in a controversial manner [LC86]. Structures supporting multiple attribute access seem to have a spectrum of usage in 2D and 3D applications. Currently, there are no such access path structures approved in practical applications. Simulating this type of access by concatenated keys and a B-tree is only useful if the full search key or a sizeable prefix of it is specified; furthermore, it does not guarantee the clustering property.

A relatively new proposal for a storage structure preserving the cluster property is the grid file [NHS84]; moreover, it allows for symmetric access via all key attributes and observes the '2 disk access principle' for fetching the first record to be qualified. Research for suitable multi-dimensional access path and storage structures may proceed along the guidelines given by the grid file.

Query optimization

Relational DB languages prohibit any decisions by the application program or user to influence the efficiency of query execution. As a consequence, the DBMS is solely responsible for all performance issues. Since access paths available are transparent to the user, appropriate access path selection has to be done by the DBMS; furthermore, since access paths may be created and dropped dynamically, binding of application program requests to access paths has to be kept dynamically, i.e. repeated execution of the same query may be accomplished via different access paths.

For the purpose of query optimization, the DBMS has to collect and maintain a selection of statistical data [Se79] describing essential quantitative properties of index structures, relations, clustering, etc. While effective algorithms for query translation and optimization are developed and precompilation avoids much of the runtime overhead of interpretive approaches, accuracy of optimization is compromised due to an inherent fundamental problem. Query optimization rests on two 'fatal' assumptions which are not easy to circumvent:
1. All data elements and all attribute values are uniformly distributed.
2. Search predicates used in queries are independent.

In distributed DBMS, even more simplifying assumptions (message traffic, locations of query processing) are necessary; they jeopardize the quality of selection decisions.

Empirical investigations show that only few attributes have uniform value distributions; even worse, some attributes have highly skew distributions. The optimizer problem may be illustrated by the following qualification clause where linear interpolation is applied (although questionable) due to lack of more accurate information:

SALARY $\geq$ 100K AND AGE BETWEEN 20 AND 30.

Assume, the salary range is from 10K to 1M; another obvious problem is the independence assumption among SALARY and AGE. A typical optimizer approach includes the derivation of all possible access plans and their evaluation based on cost formulae (# of I/O's, CPU time).

Hence, the minimum cost plan is chosen for execution. Often, the optimizer calculates uniformly wrong such that its relative decision is usually good (or at least useful). However, sometimes it is badly wrong (by a factor of 100-1000). Our considerations will convince the reader that query optimization is a 'hard' and partly elusive problem. What are the best heuristics to improve its decision accuracy? What are suitable approaches that allow self-checking and self-correcting measures during query execution?

The use of abstract access specification languages and rules for query translation and optimization facilitates the maintenance of the optimizer component and supports exchange of cost criteria or revision of heuristics. First steps in this direction are described in [Fr87].

Integrity control

Semantic integrity control which should be enforced by the DBMS is currently a neglected problem area. Designers and implementors, however, become aware of the importance of such a DBMS function. As a consequence, many efforts are directed towards enhancing current systems with more powerful system-enforced integrity control. Preservation of referential integrity [Da81] is a primary goal which requires efficient access path support to become feasible. Application-defined integrity constraints are even more complex and, thus, more expensive to control [Ya82]. At the moment, little experience exists about the extra burden of this function for DBMS performance.

Concurrency control

As indicated in section 2.1, concurrent activities must be synchronized in order to guarantee serializable schedules for transactions. It is safe to say that the method of choice for CC is locking [CS84, Pe86]. This empirically derived observation does not mean that no performance problems have to be anticipated under locking protocols. On the contrary, 'unfavorable transaction mixes' may lead to severe lock contention which is characterized by long and frequent lock waits or even deadlock situations. CC will remain a major performance problem. However, refined discussion of this issue is in order. Optimiziation of CC implies the solution of two important aspects
- granularity of locks
- handling of hot spot data elements.

Currently, many DBMSs use page locking for reasons of simplicity (implementation, lock management). While such a granularity seems to be efficient for most kinds of data, it represents a 'huge' obstacle on certain kinds of objects like catalogs, addressing tables, index structures (B*-trees), where usually many transactions are competing for access Hence, selective use of smaller granularities (e.g. tuple or entry locks) on such objects may greatly improve concurrent activities. As a consequence, lock management should provide a hierarchy of lock granules depending on the traffic characteristics of the object to be locked, e.g. relation, tuple, and field locks.

To optimize access to very active data items sometimes called 'high traffic' or 'hot spot' data element, strict two-phase locking protocols [Gr78] are inappropriate. They would cause serious drawbacks, at least for special resource usage patterns or high traffic situations, because they serialize transaction processing at such points of contention.

Although small lock granules may relieve the problem to a certain degree, very high transaction rates are dependent on tailored mechanisms to avoid jam-like situations Solutions to the problem of hot spot synchronization were proposed by [Ga85] and [Re82]. The escrow paradigm as an extension of these approaches is a special synchronization method on aggregate field values [ON87]. We feel, however, that we need suitable generalizations of these methods to be integrated into lock management. Demanding high performance requirements of transaction systems will dictate 'the use of a variety of methods of concurrency control by a single transaction' [GK85]. Note, a performance problem of an otherwise 'bottleneck-free' system will be visible as soon as its transaction load is increased by a factor of 2. Then, some data elements will turn into hot spots. Under the assumption of growing transaction loads, concurrency control is <u>the permanent</u> performance problem.

Another aspect of the concurrency problem should be mentioned. Transaction systems running 'canned' transactions may better control resource usage since some preplanning is possible due to DB design and organizational measures. However, if such on-line transactions are scheduled together with ad hoc queries, concurrency control has to solve 'some tough nuts'. The situation becomes even worse when business requirements imply mixed operation of batch and on-line transactions as well as ad hoc queries (batch transactions are characterized by long read or write sequences).

Logging and recovery

Minimization of I/O is a good design principle. This applies for logging as well [Gr81]. Therefore, page logging (often used in DBMSs) is not a good approach to collect the necessary redundancy for failure recovery. Page logging produces an enormous (giantic) amount of log data; chained I/O would only reduce the output time, not the space requirements. Note, page logging for 1 KTPS would definitely cause some problems with tape handling (for archive logging).

Page logging implies at least page locking [HR83], that is, the lock granule must cover the log granule. Our discussion on concurrency control, however, has identified the need for small lockable units. Therefore, an appropriate technique based on entry logging is the given choice.

Implementation techniques for DB recovery are based on four key concepts: propagation of updates (indirect vs. update-in-place), buffer replacement, commit processing (EOT), and checkpointing. A classification tree derived in [HR83] summarizes all conceivable logging and recovery schemes. A detailed performance discussion can be found elsewhere [BHG87, Re84]. Here we restrict our considerations to a few remarks about optimal mechanisms.

Checkpoints should be supported as a means for limiting the costs for partial REDO during crash recovery [HR83]. Fuzzy checkpoints seem to avoid synchronous I/O and reduce the number of page writes as far as possible [Gr78, Li79]. Therefore, such algorithms are superior to all other techniques.

Minimization of I/O applies to modified data pages, too. Therefore, buffer replacement and EOT processing of data pages should observe this principle. A brief guideline may be sketched as follows: Avoid replacement of pages while the referencing transaction is active (NOSTEAL) and don't propagate modified pages at EOT (NOFORCE), since synchronous writes burden the response time [HR83]. It is sufficient to write enough REDO information to the log at EOT. If time-critical applications cannot afford one synchronous I/O at EOT to write its log data to a safe place, a technique called 'group commit' may be applied. Committing transactions are delayed for a few ms until a couple of committing transactions has filled a log page. Roughly speaking, the group of transactions is committed as soon as the log page has reached the disk. Note entry or record logging is a prerequisite for group commit and I/O minimization. Appropriate logging and recvovery techniques sketched so far relax the I/O problem of DBMSs if any. Hence, current I/O architectures (supporting fast I/O) are considered to be no critical bottleneck.

Load balancing and control

As already explained in section 2.2, load balancing and load control are very important for many performance criteria (utilization, response time, etc.); nevertheless, their role in a DBMS is not well understood because complex dependencies exist among data, memory use, blocking times, synchronization needs, recovery issues, etc.

As a first attempt to approach part of the problem we consider an example from [Pe87]. Fig. 6 corresponds to Fig. 1 where the results of CC protocols are compared. With growing multi-programming level (n = parallelism) some protocols produce an increasing number of rollbacks. Even worse, after immediate restart the same transaction may be aborted over and over again. Without load control, e.g. dynamic adaption of the multi-programming level, such conflicting (long) transactions would have no chance to complete because of an infinite number of rollbacks. Hence, reducing the multi-programming level temporarily until the trouble-makers have finished is a much more reasonable policy than trying to keep parallelism (and probably processor utilization) high.

Fig. 6 illustrates that increased parallelism beyond a certain n is useless in some situations, since the extra work consists of executing transactions repeatedly. Hence, load balancing and control may drastically improve performance 'by reducing system utilization'.

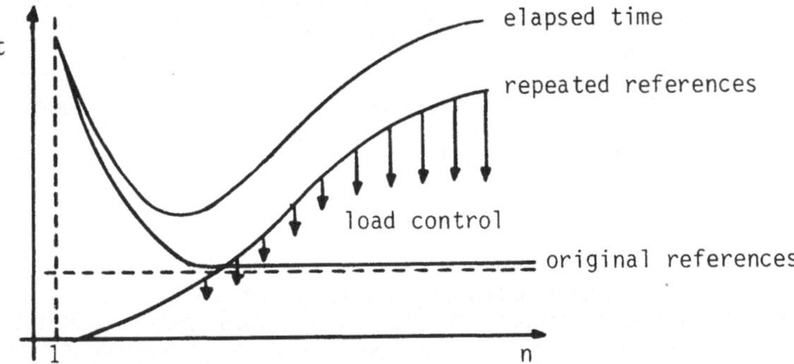

Fig. 6: Influence of load control on transaction throughput

Other areas, e.g. multi-processor DBMSs such as DB-sharing or DB-distribution architectures, have an urgent need for effective balancing and control schemes. Distributed systems, in general, are characterized by a lack of global knowledge. Therefore, every kind of decision or agreement is much more difficult to obtain. As a consequence, hierarchical workload management schemes seem to be appropriate where relatively coarse decisions are made at the global level, but refinement and adjustment is performed locally.

Further important issues which are critically influencing the DBMS performance are the embedding and cooperation of DBMS and OS [Hä79, St81]. Here, we don't want to go into detail but simply sketch some problems by means of the following questions. Does the OS support the transaction concept [We86]? How many times has a data element to be copied until it gets to the user? How many processes (servers) should be allocated to the execution of the DBMS code [HP84]? What are good DB/DC system structures which efficiently and effectively support the cooperation among TP monitor and DBMS [HM86]?

4. Performance monitoring and control

After having illustrated the most important performance-critical DBMS functions, we give a brief overview about approaches, techniques, and tools which aim at analyzing and controling the DBMS performance. These issues don't address an add-on problem, since performance analysis and control are important tasks in every phase of the DBMS's lifetime. Design, implementation and operation of a DBMS require accurate performance data to avoid bad design decisions, to optimize key components based on empirical data and to identify and remove system bottlenecks. Hence, DBMS implementors as well as DB administrators and application programmers [EHRS81] benefit from appropriate performance analysis support.

Monitoring the DBMS performance

For these reasons, it seems to be a good software engineering principle to provide integrated performance measurement and monitoring tools in DBMS [KKMP84]. Such tools enable simple collection and examination of characteristic and prognostic statistics and trace data. For purposes of system implementation/optimization these data are used

- to detect weaknesses of functions and algorithms
- to demonstrate and validate quantitative system properties
- to understand internal control flows and dependencies among algorithms.

Apparently, various tasks of daily system operation are greatly supported by such tools:

- Different aspects should be monitored to facilitate every day maintenance of the system. Important characteristics include disk space, storage structures (balanced trees, overflow areas, cluster properties), communication lines, activities of terminals, transaction programs, DBMS servers, etc.
- System tuning serves to continuously balance and improve resource utilization. Redistribution of system workload or adjustment of acess paths as well as adaptation of runtime options might be necessary to make full use of system resources.
- Capacity planning determines when the system workload will overload system resources. It provides decision support for either rescheduling the workload or allocating new resources to prevent the overload.
- Procurement is a long-term planning task. Which hardware is required based on predicted workload to achieve the desired performance?

Experiences in several DBMS performance analysis projects have shown that a number of different measurement techniques are needed to collect the required data. Among the most useful techniques are recording of internal events (logical and physical page references, process switches, lock waits, etc.), queue length evaluation, dynamic path length determination, measurement of various elapsed or execution times, analysis of storage structures [EHRS81]. Obviously, such techniques must be integral part of the DBMS; they must be designed to use minimal system resources while providing a wide range of performance metrics. For example, MEASURE [TRS86] assigns 169 different standard counters to 13 entity types such as processors or disks. In addition, three types of user-defined counters are available to measure application-specific performance parameters such as transaction types, transaction response time, and queuing delays. Control measurements have shown that such a performance measurement tool can be integrated causing very low overhead (0.02% to 1.14% of the system's CPU time, depending on the number of entities measured and the number of concurrent measurements).

The usefulness of built-in monitoring and explanation facilities is underlined by the fact that more and more DBMSs are offering such services. Examples are UDSMON [SIE], MEASURE [TAN] and IMS/VS Monitor Facility [Mc77]. Due to enhanced system complexities, increased difficulty of performance observation and control, and larger performance parameter space, such services are even more important and convenient in the distributed case (distributed DBMS, DB-Sharing, etc.) [TRS86].

Simulation of DBMS algorithms

We want to emphasize another use of specific measurement data. Measurement identifies bottlenecks and performance problems. However, reimplementation of algorithms and their iterative optimization under control of DBMS measurement is a tiresome and very expensive procedure. For purposes of algorithm optimization (in development and research), simulation is a much more appropriate way. Trace driven simulation as opposed to random number driven simulation helps to accomplish the required accuracy of workload conditions. The trace input for a wide spectrum of DBMS algorithm simulations is an 'object reference string' (ORS) obtained by appopriate measurement tools from real DB applications of different types and sizes. Appropriately chosen ORSs define the workload model for which behavior and optimization of a specific algorithm should be investigated. Such ORSs may be conveniently converted to reflect varying levels of parallelism [HPR85].

We believe that trace driven simulation allows for a much higher degree of realism than standard simulation techniques (or even analytic models) because it does not require any idealizations or assumptions that cannot be validated. With those techniques, important phenomena and events like hot spot data reference, buffer replacement and convoy patterns cannot be modelled adequately. The drawback is that results based on traces do not yield 'general' results, as e.g. analytic models do; they are confined to the given type of application. But since they are very close to a real system's behavior, they are extremely helpful to select or optimize an algorithm for a given system (and type of workload).

For research problems, they offer valuable insights and practical solutions as well, if the workload model is appropriately chosen. As already illustrated in section 2.1, an extensive empirical study on locking algorithms' performance was successfully performed. Other research areas, where trace driven simulation was beneficially used, are buffer management [EH84], logging and recovery [Re84], and synchronization in multi-processor DBMSs [Ra87].

DB design

Logical as well as physical DB design may have an enormous influence on the overall system performance. It is impossible to discuss here the manifold design approaches and philosophies in detail [Sc78, TF80, Ya82, RK84, ST85]. Logical or conceptual schema design involves aggregation of entity and relationship information to relations or record/set types according to the chosen data model. Application-dependent knowledge, frequency of operations, cardinalities of relations, etc. determine a specific DB schema (which may possibly deviate from third normal form relations). We want to point out only a single design problem dictated by processing requirement. A carelessly designed DB schema may contain some hot spot data (e.g. a counter to be updated by every transaction) which will become a permanent insurmountable obstacle for all conceivable synchronization efforts and limit the system's throughput. Therefore, hot spots should be avoided by DB schema design as far as possible [GGGHS85].

Physical DB design is responsible for mapping logical to physical storage structures and for providing efficient processing support by access paths. Since access paths are redundant structures which speed up retrieval, but burden maintenance operations, each of them has to be appropriately selected from the available set of access path types and tailored to frequently requested operations. Due to the inherent design complexity, optimal solutions require stable and effective heuristics; particular approaches adhered to different optimization philosophies are described in [RK84, Ya82]. Here we briefly want to emphasize the important role of physical clustering for DBMS performance; typically, there are always conflicting processing requirements which prevent the perfect cluster property (physical contiguity of records/tuples in processing order) for all traversals (without redundancy). For example, sort order may be interpreted as a special cluster criterion. When sorted sequences of tuples are represented by physical clusters, particular operations involving joins or set-theoretic operators are extremely fast; otherwise, they have to be explicitly sorted for gaining satisfactory algorithmic performance. A sort run, however, stresses the utilization of system resources.

Restructuring of storage objects

Another essential aspect should be mentioned in this context. It is mandatory for a high degree of data independence [NF76] to isolate logical from physical DB structures as far as possible (The application programmer should not be bothered by the complexities of the physical (internal) DB schema). Clean separation of logical and physical issues and a flexible DBMS mapping mechanism between them are prerequisites to adjust the internal schema to the changing processing requirements during the lifetime of a DB installation [Ma83]. This applies not only to restructuring of storage areas, reorganization of overflow tables, or rebalancing of trees, but also to dynamically creating or dropping access paths (B-trees, hash tables), redefining physical clusters, etc. Perfect isolation of logical objects (as in the relational model) guarantees continued availability of all functions; only response time may be affected. The usefulness of such a system property was empirically investigated in [EHR80]. More than 50 different internal schemes were designed and generated for the same logical schema; performance improvements of a factor of 6 and more could be gained without changing any bit in the COBOL application programs.

Finally, we should mention that several researchers have proposed to equip the DBMS itself with self-tuning and self-optimization abilities. So far, little progress has been made to approach this goal in a realistic environment.

5. Conclusions

We have discussed performance problems of centralized DBMSs, DB/DC systems, and DBMSs for non-standard applications; for these areas, different types of 'extreme' performance requirements were described to emphasize the desired performance properties of future DBMS generations.

Some important performance problems - concurrency control, linear growth of transaction throughput, handling and manipulation of complex objects - have been investigated and analyzed in detail. These considerations have revealed a number of performance-critical DBMS functions. DB research and development experiences have issued many concepts and algorithms to the various problem areas. To facilitate the evaluation of this work and to give some guidelines and directions for future investigations, we have discussed selected results for the most important performance-critical DBMS functions (OS support not considered here must be regarded as well).

It was stated that an integrated approach to DBMS monitoring and control is needed to cope with the manifold issues of performance analysis and interpretation. Such tools and mechanisms serve to improve the system behavior - either by optimization of algorithms or by DB-schema adjustment. Since performance-related dependencies among various functions are complex, the built-in performance analysis tools should be equipped with powerful 'explanation' facilities.

Acknowledgements

D. Gawlick, J. Gray, B. Lindsay, C. Mohan, J. Palmer, and K. Shoens shared their great knowledge and experience on all issues of DBMSs with me. They provided helpful and timely information in enlightening discussions ("what is your favorite DBMS problem?"). J. Ch. Freytag and K. Meyer-Wegener have read an earlier version of this paper and have contributed to clarify and improve the presentation of important issues.

Bibliography

[Anon85] Anon et al.: A Measure of Transaction Processing Power, Datamation, April issue, (1985).

[As76] Astrahan, M.M., Blasgen, M.W., Chamberlin, D.D., Eswaran, K.P., Gray, J.N., Griffith, P.P., King, W.F., Lorie, R.A., McJones, P.R., Mehl, J.W., Putzolu, G.R., Traiger, I.L., Wade, B., Watson, V.: System R: Relational Approach to Database Management, ACM TODS **1:2**, (1976), pp. 97-137.

[BHG87] Bernstein, P.A., Hadzilacos, V., Goodman, N.: Concurrency Control and Recovery in Database Systems, Addison Wesley, Reading, Ma., (1987).

[BKK85] Bancilhon, F., Kim, W., Korth, H.F.: A Model of CAD Transactions, Proc. 11th Conf. on VLDB, Stockholm, (1985), pp. 25-33.

[Bo81] Borr, A.: Transaction Monitoring in ENCOMPASS, Proc. 7th Conf. on VLDB, Cannes (1981), pp. 155-165.

[Bu85] Burman, M.: Aspects of a High-Volume Production Online Banking System, Proc. IEEE Spring CompCon, San Francisco, (1985), pp. 244-248.

[Co79] Comer, D.: The Ubiquitous B-tree, ACM Computing Surveys **11:2**, (1979), pp.397-434.

[CS84] Carey, M., Stonebraker, M.: The Performance of Concurrency Control Algorithms for DBMS, Proc. 10th Conf. on VLDB, Singapore, (1984), pp. 107-118.

[Da81] Date, C.J.: Referential integrity, Proc. 7th Conf. on VLDB, Cannes, (1981), pp. 2-12.

[Ef87] Effelsberg, W.: Datenbankzugriff in Rechnernetzen, Informationstechnik it **29:3**, (1987), pp. 140-153.

[EGLT76] Eswaran, K.P., Gray, J.N., Lorie, R.A., Traiger, I.L.: The Notions of Consistency and Predicate Locks in a Database System, CACM **19:11**, (1976), pp. 624-633.

[EH84] Effelsberg, W., Härder, T.: Principles on Database Buffer Management, ACM TODS, **9:4**, (1984), pp. 560-595.

[EHR80] Effelsberg, W., Härder, T., Reuter, A.: An experiment in learning DBTG database administration, Information Systems, **5:2**, (1980), pp. 137-147.

[EHRS81] Effelsberg, W., Härder, T., Reuter, A., Schulze-Bohl, J.: Leistungsmessung von Datenbanksystemen - Meßmethoden und Meßumgebung, Informatik-Fachberichte, **41**, Springer, (1981), pp. 87-102.

[Fr87] Freytag, J.C.: A Rule-Based View of Query Optimization, Proc. SIGMOD '87, San Francisco, CA., (1987), pp. 173-180.

[Ga85] Gawlick, D.: Processing "Hot Spots" in High Performance Systems, Proc. IEEE Spring CompCon, San Francisco, (1985), pp. 249-251.

[Ga87] Gawlick, D.: Personal Communication, (February 1987).

[GGGHS85] Gray, J., Good, B., Gawlick, D., Homan, P., Sammer, H.: One Thousand Transactions per Second, Proc. IEEE Spring CompCon, San Francisco, (1985), pp. 96-101.

[GK85] Gawlick, D., Kinkade, D.: Varieties of Concurrency Control in IMS-VS, Tandem Research Report TR85.6, (1985).

[GLPT76] Gray, J., Lorie, R.A., Putzolu, F., Traiger, I.L.: Granularity of Locks and Degrees of Consistency in a Shared Data Base, Proc. IFIP Working Conference on Modelling of Database Management Systems, Freudenstadt, Germany, (1976), pp. 365-394.

[GP87] Gray, J., Putzolu, F.: The 5 Minute Rule for Trading Memory for Disc Accesses and The 10 Byte Rule for Trading Memory for CPU Time, Proc. ACM SIGMOD Conf. 1987, San Francisco, (1987), pp. 395-399.

[Gr78] Gray, J.N.: Notes on Database Operating Systems, Operating Sytems - An Advanced Course, Lecture Notes in Computer Science 60, Bayer, R., Graham, R.M., Seegmueller, G. (eds.), Springer-Verlag, (1978), pp. 393-481.

[Gr80] Gray, J.N.: A Transaction Model, Research Report RJ 2895, IBM Research Laboratory, San Jose, CA., (1980).

[Gr81] Gray, J., McJones, P., Blasgen, M., Lindsay, B., Lorie, R., Price, T., Putzolu, F., Traiger, I.: The Recovery Manager of the System R Database Manager, ACM Computing Surveys **13:2**, (1981), pp. 223-242.

[Gr85] Gray, J.N.: Why do Computers Stop and What Can Be Done About It, Proc. 'Büroautomation '85', Berichte des German Chapter of the ACM 25, Teubner-Verlag, (1985), pp. 128-145.

[Gr87] Gray, J.N.: Personal Communication, (1987).

[Hä79] Härder, T.: Die Einbettung eines Datenbanksystems in eine Betriebssystemumgebung, Datenbanktechnologie, Teubner Verlag, (1979), pp. 9-24.

[Hä84] Härder, T.: Observations on Optimistic Concurrency Control Schemes, Information Systems **9:2**, (1984), pp. 111-120.

[Hä87] Härder, T.: Database Support for Engineering Applications, Proc. Int. Workshop on Information in Manufacturing Automation, Dresden, GDR, (1987).

[HC85] Horst, R., Chou, T.: The Hardware Architecture and Linear Expansion of Tandem NonStop Systems, Tandem Technical Report 85.3, Cupertino, CA., (1985).

[HGLW87] Herman, G., Gopal, G., Lee, K.C., Weinrib, A.: A Datacycle Architecture for Very High Throughput Database Systems, Proc. SIGMOD'87 Conf., San Francisco, CA., (1987), pp. 97-103.

[HHM86] Härder, T., Hübel, Ch., Mitschang, B.: Use of Inherent Parallelism in Database Operations, Proc. Conf. on Algorithms and Hardware for Parallel Processing, CONPAR'86, Aachen, LNCS 237, Springer-Verlag, (1986), pp. 385-392.

[HMMS87] Härder, T., Meyer-Wegener, K., Mitschang, B., Sikeler, A.: PRIMA - a DBMS Prototype Supporting Engineering Applications, Proc. VLDB '87, Brighton, U.K., (1987).

[HM86] Härder, T., Meyer-Wegener, K.: Die Zusammenarbeit von TP-Monitoren und Datenbanksystemen in DB/DC-Systemen: Existierende Systeme und zukünftige Entwicklungen, Informatik - Forschung und Entwicklung **1:3**, (1986), pp. 101-122.

[HP84] Härder, T., Peinl, P.: Evaluating Multiple Server DBMS in General Purpose Operating System Environments, Proc. Conf. on 10th VLDB, Singapore, (1984), pp. 129-140.

[HPR85] Härder, T., Peinl, P., Reuter, A.: Performance Analysis of Synchronization and Recovery Schemes, IEEE Database Engineering, **8:2**, (1985), pp. 50-57.

[HR83] Härder, T., Reuter, A.: Principles of Transaction-Oriented Database Recovery, ACM Computing Surveys **15:4**, (1983), pp. 287-318.

[HR86] Härder, T., Rahm, E.: Mehrrechner-Datenbanksysteme für Transaktionssysteme hoher Leistungsfähigkeit, Informationstechnik it, **28:4**, (1986), pp. 214-225.

[KKMP84] Kinzinger, H., Küspert, K., Meyer-Wegener, K., Peinl, P.: Integrated Environment for Performance Measurement and Evaluation in a DB/DC System, Computer Performance, **5:4**, (1984), pp. 207-221.

[KLMP84] Kim, W., Lorie, R., McNabb, D., Plouffe, W.: Nested Transactions for Engineering Design Databases, Proc. 10th Conf. on VLDB, Singapore, (1984), pp. 355-362.

[KR81] Kung, H.T., Robinson, J.T.: On Optimistic Methods for Concurrency Control, ACM TODS, **6:2**, (1981), pp. 213-226.

[LC86] Lehman, T., Carey, M.: A Study of Index Structures for Main Memory Database Management Systems, Proc. 12th Conf. on VLDB, Kyoto, (1986).

[Li79] Lindsay, B.G. et al.: Notes on distributed databases, IBM Research Report RJ 2571, San Jose, CA., (1979).

[Ma83] March, S.T.: Techniques for structuring database records, ACM Computing Surveys, **15:1**, (1983), pp. 45-79.

[Mc77 McGee, W.C.: The information management system IMS/VS, IBM Systems Journal, **16:2**, (1977), pp. 84-168.

[NF76] Navathe, S.B., Fry, J.P.: Restructuring for large databases: three levels of abstraction, ACM TODS, **1:2**, (1976), pp. 136-158.

[NHS84] Nievergelt, J., Hinterberger, H., Sevcik, K C.: The grid file: an adaptable, symmetric multikey file structure, ACM TODS **9:1**, (1984), pp. 38-71.

[ON87] O'Neil, P.E.: The Escrow Transactional Method, Proc. Int. Workshop on High Performance Transaction Systems, Asilomar, CA., (1985).

[Pe86] Peinl, P.: Synchronisation in zentralisierten Datenbanksystemen - Algorithmen, Realisierungsmöglichkeiten und quantitative Bewertung, Universität Kaiserslautern, (1986), Dissertation.

[Pe87] Peinl, P.: Load Balancing Policies vs. Concurrency Control - an Empirical Comparison of DBMS Performance Criteria, Univ. of Stuttgart, submitted for publication, (1987).

[PSSWD87] Paul, H.-B., Schek, H.-J., Scholl, M.H., Weikum, G., Deppisch, U.: Architecture and Implementation of the Darmstadt Database Kernel System, Proc. Conf. SIGMOD'87, San Francisco, CA., (1987), pp. 196-207.

[Ra78] Randell, B. et al.: Reliability Issues in Computing System Design, ACM Computing Surveys **10:2**, (1978), pp. 123-166.

[Ra87] Rahm, E.: Performance Analysis of Primary Copy Synchronization in Database Sharing Systems, Interner Bericht 165/87, FB Informatik, Univ. Kaiserslautern, (1987).

[Re82] Reuter, A.: Concurrency on High-Traffic Data Elements, Proc. Conf. on Principles of Database Systems, Los Angeles, CA., (1982), pp. 83-93

[Re84] Reuter, A.: Performance analysis of recovery techniques, ACM TODS, **9:4**, (1984), pp. 526-559.

[Re86] Reuter, A.: Load Control and Load Balancing in a Shared Database Management System, Proc. Conf. on Data Engineering, Los Angeles, CA., (1986).

[RK84] Reuter, A., Kinzinger, H.. Automatic Design of the Internal Schema for a CODASYL-Database System, IEEE Transactions on Software Engineering, Vol. SE-10, No 4, (1984), pp. 358-375.

[RS85] Reuter, A., Shoens, K : Synchronization in a Data Sharing Environment, Research Report, IBM Research Laboratory, San Jose, CA., (in preparation), (1985).

[Sc78] Schkolnick, M.: A survey of physical database design methodology and techniques, Proc. 4th Conf. on VLDB, (1978), pp. 479-487.

[Sc86] Schwarz, P., Chang, W., Freytag, J.C., Lohman, G., PcPherson, J., Mohan, C., Pirahesh, H.: Extensibility in the Starburst Database System, Proc. Int. Workshop on Object-Oriented Database Systems, Asilomar, CA., (1986).

[Se79] Selinger, P.C. et al.: Access path selection in a relational database management system, IBM Research Report RJ 2429, San Jose, CA., (1979).

[Sh85] Shoens, K : The AMOEBA Project, Proc. IEEE Spring CompCon, San Francisco, (1985), pp. 102-105.

[SIE] UDS (BS2000) - Verwalten und Bedienen, Siemens AG, München, Bestell-Nr.: U932-J-Z55-4, (1986).

[SS82] Sacco, G., Schkolnick, M.: A Mechanism for Managing the Buffer Pool in a Relational Database System using the Hot Set Model, Proc. 8th Conf. on VLDB, Mexico City, (1982), pp. 257-262.

[SS86] Schek, H.-J., Scholl, M.H.: The Relational Model with Relation-Valued Attributes, Information Systems, **11:2**, (1986).

[St81] Stonebraker, M.: Operating System Support for Database Management, CACM, **24:7**, (1981), pp. 412-418.

[ST85] Schkolnick, M., Tiberio, P.: Estimating the cost of updates in a relational database, ACM TODS, **10:2**, (1985), pp.163-179.

[TAN] MEASURE Users's Guide, Part no. 82440, Tandem Computers Inc., Cupertino, CA.

[TF80] Teorey, T J., Fry, J.P.: The logical record access approach to database design, ACM Computing Surveys, **12:2**, (1980), pp. 179-211.

[Tr83] Traiger, I.: Trends in System Aspects of Database Management, Research Report RJ3845, IBM Research Laboratory, San Jose, CA., (April 1983).

[TSR86] Some papers about performance aspects of transaction systems, Tandem Systems Review, Cupertino, CA., (1986).

[We86] Weikum, G.: Pros and Cons of Operating System Transactions for Database Systems, Proc. ACM/IEEE Fall Joint Computer Conference, Dallas, (1986).

[Ya82] Yao, B., et al.: Database design techniques I+II, Lecture Notes in Computer Science 132+133, Springer, (1982).

Simulation and Evaluation of Distributed Database Systems

Jian Cai
Institut für Informatik
Universität Stuttgart
Azenbergstr. 12
D-7000 Stuttgart 1

ABSTRACT

We present a detailed simulation model for a distributed database system, which includes the following components: communication system, concurrency control, buffer management, recovery, and deadlock detection. The simulation is carried out using GPSS-Fortran and is driven by a reference-string recorded from a real database application. Reference-strings reflect the real reference patterns more accurately than random references. Three concurrency control schemes and two buffer replacement policies are compared. Due to the very detailed level of investigation in our approach, we yield a thorough understanding of the performance aspects of both buffer management and concurrency control in a distributed database environment.

1. Introduction

Using simulation techniques, much work in performance analysis has been done for centralized database systems /EfHä84//CaSt84//PeRe84/ /KiLa83//ACL85/, while little has been done for distributed databases. Besides, each of those studies was concentrated on only one aspect, either concurrency control or buffer management. But it is not reasonable to estimate system performance without considering other components, such as recovery mechanism and, for distributed systems, network communication, since they also consume system resources and interact in various ways. Analyzing and designing such a complicated system containing all components mentioned above requires a reasonable application of different simulation techniques.

In this paper we describe a detailed simulation model for a locally distributed database system. By means of this model we simulate three different concurrency control schemes and two buffer replacement policies. The work load und the underlying database are characterized by traces recorded in real-life database applications.

2. The problem of driving a simulation system

The mostly used simulation technique is Monte-Carlo-simulation, which generates system work load randomly.

In a database system, system work load is reflected in its object references. These references present a variety of patterns which cannot be generated using random numbers without great difficulties. In some simulation approaches, references, especially reference localities, are generated according to some probabilistic distribution functions, such as Gauss and exponential distribution. This is right or to some extent right, if it is viewed from the final reference statistic, such as reference frequences at some pages. But if we observe a real system, we can find that its reference behavior varies from time to time, and, if we pick out references registered in a time interval, we can find it is often that these references do not agree with any distribution functions. If we do want to use random numbers to generate references analogous to those in real environment, we should use complex random number generator combining different probability distribution functions, so that the generated references have various behaviors in different time. In fact, we could not do this until we have registered references from a real system and thoroughly analysed them. Instead of doing this analysis and re-creation of references, it is reasonable that we directly use those reference-strings.

In the strict sense, a reference-string does not present a stochastic process. Therefore, our simulation using reference-string seems to be not a probabilistic but a deterministic simulation. However, we still think our simulation is a Monte-Carlo-simulation. First, a reference-string has by nature, to some degree, stochastic property. Second, we can use a large number of reference-strings to drive the simulation many times. Third, we still use random number generator for simulating time consumptions at CPU, I/O-channel and so forth. It is in this sense that we think simulation driven by reference-strings, compared to simulation driven by randomly generated work load, is more natural, and the results are more meaningful.

3. System architecture and simulation model

The distributed database system simulated is a partitioned system with respect to data distribution. The communication network is assumed to be Ethernet-like. Each data object is assigned exactly to one site called home site. For some site which is not the home site of a data object, that data object is called a remote one. Each transaction has also a home site, where it is excuted. A transaction runs only at its home site; it requests a remote data object from other sites by means of request-message, and processes the granted data locally. No sub-transaction is assumed.

Fig.1 shows the architecture of the simulated system at one site. Every site has the same structure.

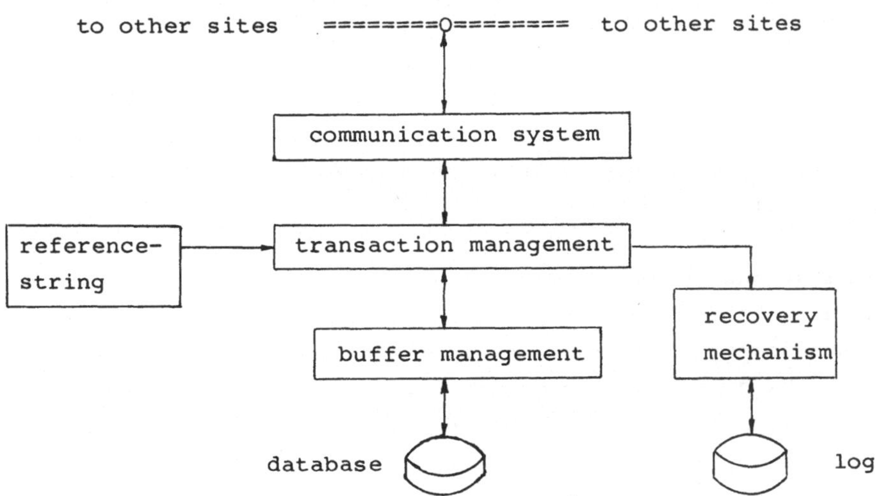

Figure 1: Architecture of the distributed DB-system

The communication system here is responsible for sending and receiving messages. Messages received are assumed to be error-free and in the same order in which they were sent.

Transaction-management (TM) manages the concurrency control as well as transaction start, abort and commit. All three concurrency control schemes considered here use locks for synchronisation. Locking granules are pages. Since transactions set their locks during run-time and locks are nonpreemptive, deadlock can occur. A transaction-wait-for-graph (TWFG) is used to detect deadlocks, and this TWFG is centrally managed by one site (not shown in Fig.1).

Data objects read or written by transactions are addressed in the buffer. The buffer management excutes a routine for buffer replacement.

In the following we give some details for each component and describe the schemes we simulated. Readers with detailed interest are referred to the literature.

3.1. Transaction management

During system initialization, the multiprogramming level is set for each site, which means the number of concurrent transactions at that site. When one transaction has committed, the TM starts a new transaction. When a transaction is aborted because of a deadlock, this transaction will be delayed for some time and then restarted. Concurrency control is the most important task of the TM. We have simulated three concurrency control schemes, all of which use locks for synchronisation.

The first one is the primary copy scheme/Ston78/. Two types of locks are used: shared locks for reading and exclusive locks for writing. Each data object is assigned to one site called primary site (home site). Locks concerning an object are directed to its primary copy and managed by its primary site. Hence, for each lock request for a remote data object, a request message must be sent to the primary site.

The second one is called true-copy-token scheme, introduced in /MiWi82/ and in several related papers. This scheme uses the same lock types as the primary copy scheme. Each valid copy, either the primary copy or copies currently existing at other sites, is associated with either a shared true-copy-token (STCT) or an exclusive true-copy-token (ETCT). To read an object, a transaction can get any one of the shared copies of that object. To write an object, a transaction must make all copies of that object "shrink" so that this transaction possesses a copy with an ETCT and the other copies lose their TCT's. This scheme makes it sometimes possible to read a remote data object immediately without inquiring its home site.

The third scheme achieves high transaction parallelism by using highly compatible locks: R-lock, A-lock and C-lock /BEHR80/. The idea is to prepare a new version for one object in parallel to activities of other transactions which are still reading the old version of the same object. The R-lock stands for read-lock. The A-lock indicates that a new version of the object is being prepared, whereas the C-lock indicates that two committed versions of the object are available: the new one just created (with C-lock) and the old one.

3.2. The communication system

We consider Ethernet-like structures only. Experiments show that message collision in an Ethernet occurs very seldom, if the work load is not heavy. Since our simulated system has only six sites, we think the probability of message collision is so small that it can be ignored. Therefore, we can model the Ethernet simply as a one-queue-one-server.

At each time only one message can be transfered by the network which is represented by the server. The service-time is just the message delay caused by the signal transmission and propagation. The queue simulates the sensor of the Ethernet which detects if the communication line is currently occupied or free. If it is occupied, the message to be sent must wait, until the line is free. At each site there are two separate message buffers, one for messages to be sent out, and one for messages received. Both of them are managed on the first-in-first-out principle.

3.3. Buffer management

At each site we use only one global buffer for all transactions. The buffer size is determined at system initialization and will not be changed during run time. The buffer management, therefore, is nothing but a routine for buffer replacement. We have simulated two buffer replacement policies.

The first replacement policy is the so-called LRU, which is simple and efficient. A stack is used to implement this policy. The page just referenced currently is placed at the top of the stack. The page at the bottom of the stack is hence the "least recently used". If a page in the buffer must be replaced due to a page reference fault, the page at the bottom in the stack will be removed.

The second policy is just a modified LRU. We call it delta-LRU. In database systems, some buffer pages must be written into the database or into a log on the disk, before they can be replaced. An example are pages which have been updated in the buffer but have not yet been written to nonvolatile storage. It is clear that a replacement of these buffer pages costs more than others which can be replaced immediatly without writing them first.

Delta-LRU is different only in the following way: If the page at the bottom of the stack cannot be replaced before it has been written into the secondary storage, then another buffer page will be selected for replacement. This selection is done in an order of bottom-up in the stack. Naturally, if no such page exists at all, the one at the bottom of the stack is to be replaced, after it has been written out.

3.4. Recovery mechanism

We use the recovery mechanism described in /Elha82/. A special log file called "safe" is organized on a disk so that it consists of physically adjecent disk slots. At commit time the pages updated by the transaction will be written sequentially into the safe using chained I/O, which is much faster per block than random I/O. Fig.2 shows this sheme.

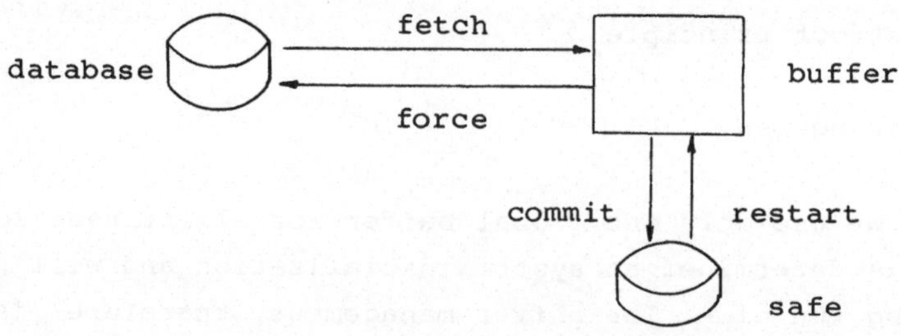

Figure 2: recovery mechanism

All pages updated by a transaction must stay in the buffer until the transaction's commitment. If a data page which was updated by a transaction earlier is to be replaced, it must be forced to the database. If, on reference, a data page fault occurs, it will be fetched from the database. Because we do not consider system crashes, we ignore the restart processing.

3.5. The database reference string

We use a reference-string recorded from an IMS-database application to drive the simulation. The reference-string used contains 578 transactions, with totally 12332 references, so that the average transaction length is about 21 references. Each reference is considered to be either read or write. Among the 12332 references are 6072 of write-type. There are 242 read-only transactions, while there are also several transactions with over one hundred references of write-type. 62 percent of the 576 transactions have less than 10 references. These statistics show that this reference-string includes almost all types of transactions: both very short and very long transactions, both read-only and update-intensive transactions. Because no reference-string from a real distributed environment is available, we use random numbers to distribute these references to every site. Typically, each site has 70% local references, whereas the other 30% remote data pages are treated as from other sites with equal probability.

3.6. Simulation model

The system resources considered are CPU, I/O-channel and communication system. Since we want to simulate the system in detail, all resource consumptions must be taken into account. For each message sending and receiving and for each I/O, CPU time is required; for each transaction reference, i.e. reading or writing a page, CPU time is required; even for process state chaging and lock management CPU time is considered, but for simplicity, we integrate these into the first two kinds of CPU consumption.

In Fig.3 we show our simulation model which includes also the simulation flow chart. For clarity, CPU and I/O are shown for each usage. Here we take the primary copy scheme as an example. Because of some differences in the concurrency control, the simulation flow charts for the true-copy-token scheme and for the RAC-locking scheme are a bit different.

This model is self-explanatory and easy to understand. Perhaps only the re-scheduling needs some explaining: transactions waiting for locks must be re-examined, after the corresponding locks have been

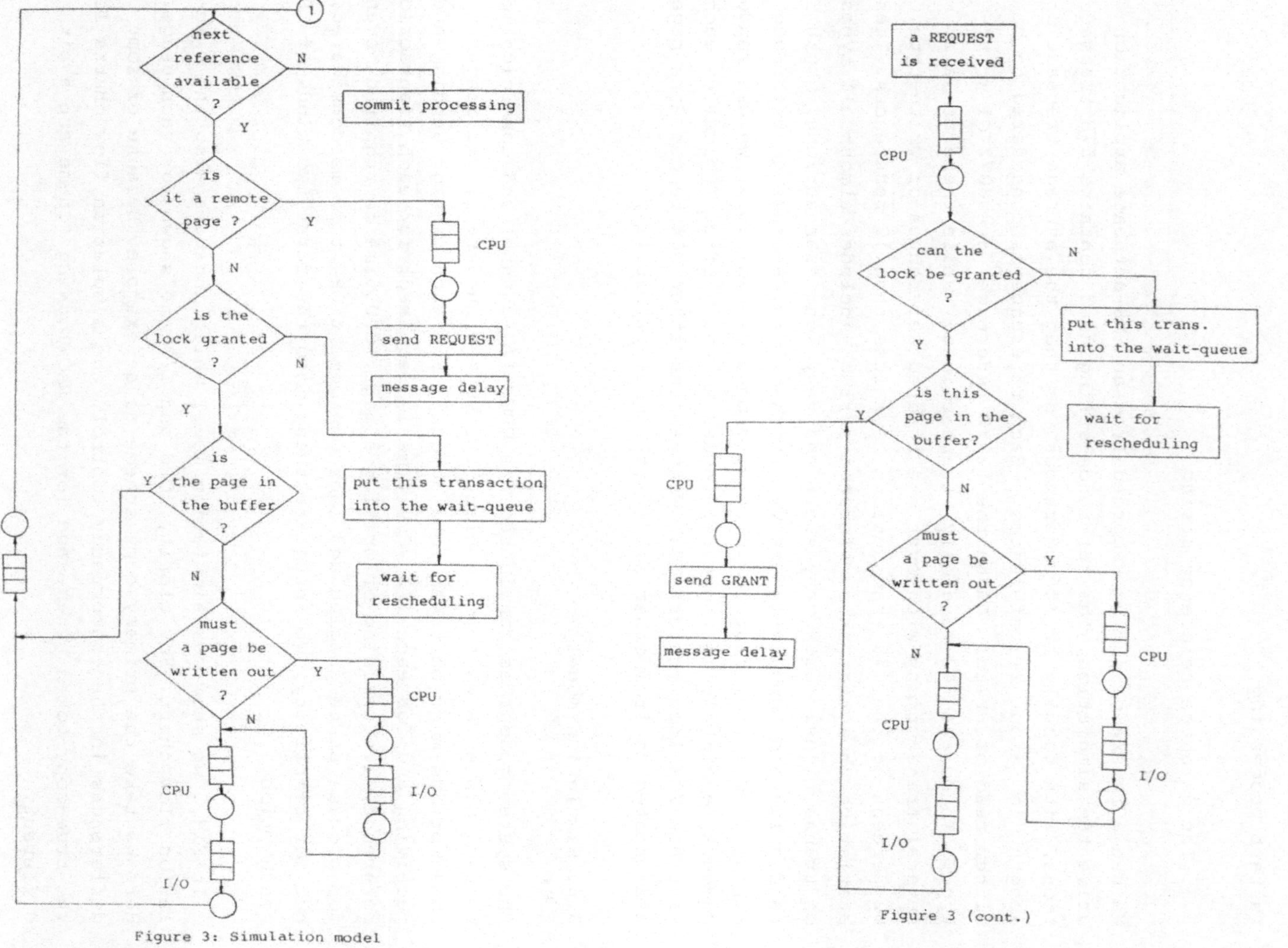

Figure 3: Simulation model

Figure 3 (cont.)

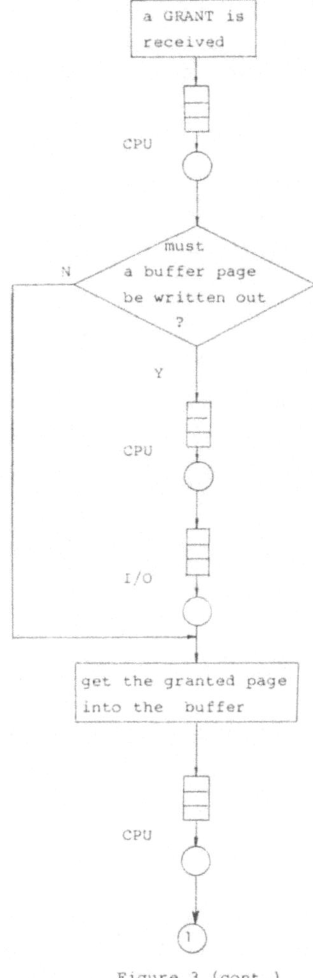

Figure 3 (cont.)

released, in order to check if the locks can now be granted to them. The lock-management schedules transactions's lock-requests on the principle of first-come-first-served.

The commit processing uses distributed two-phase commitment protocol. In each phase, CPU, I/O and messages are simulated. For simplicity, these are omitted in Fig.3.

We use GPSS-Fortran /Schm84/ as the simulation language. CPU, I/O and communicaton system are represented each by a facility in GPSS. We find that the functions GPSS-Fortran offers are quite enough for programming our simulation.

3.7. Simulation parameters

The parameters for the simulation are listed below:

network data transfer rate	10 mbps
number of sites in the network	6
multiprogramming level	4 (per site)
CPU speed	1 mips
buffer size	400 - 800 pages
local reference percentage	70%
DB-disk I/O speed (for one page)	7 - 60ms, mean 30ms
safe-disk I/O speed (for one page)	7ms
CPU for sending message	2000 - 5000 instructions
CPU for receiving message	1000 - 4000 instructions
CPU for one I/O	2500 instructions
CPU for one reference	15000 - 30000, mean 20000 instructions

The work load of the system is determined by the number of sites in the network, the multiprogramming level and the reference-string. The parameters we set here are typical for a small distributed database system.

4. Results

Since our simulation program is very flexible, many parameters can be varied and a lot of experiments can be done. Because of the limit of the scope and size of this paper, we can present only a few results.

Since our simulation is half-probabilistic and half-deterministic, we do not use confidence interval or variance to qualify the precision of the results. The simulation reaches its end, when all references in the string have been processed. In the example given, the simulation time (the time which would be spent in a real database system) is approximately 120 seconds.

Fig.4 shows that the true-copy-token scheme and the primary copy scheme have better throughput than the RAC-locking. This is mainly because the RAC-locking leads to more buffer I/O. The RAC-locking needs more buffer frames, since updating a page in the buffer requires that a new version be generated so that for each write operation

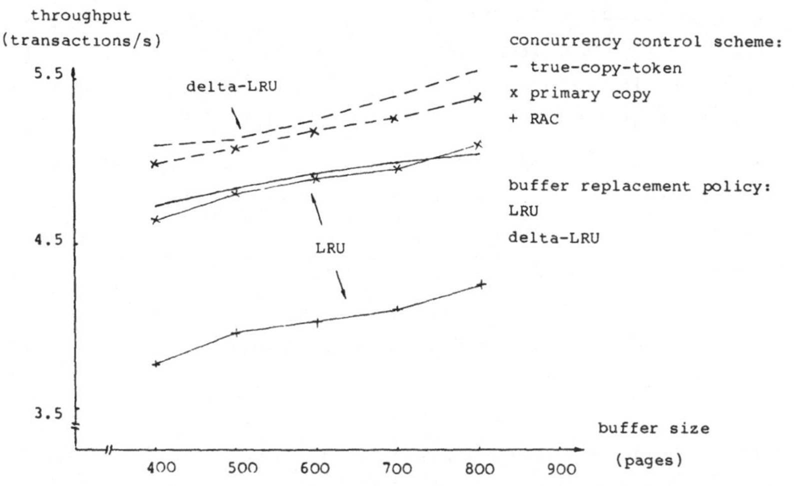

Figure 4: throughput versus buffer size

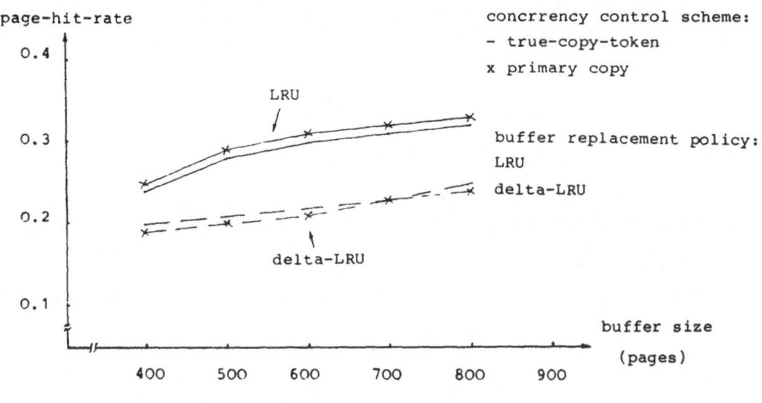

Figure 5: page-hit-rate versus buffer size

ofen a couple of pages must be kept in the buffer. Generally speaking, the RAC-locking is not appropriate for distributed database systems. Besides the need for more buffer it suffers from the problem of frequently checking the dependency-graph and this costs CPU time as well as messages. The true-copy-token scheme and the primary copy scheme achieve comparable performance. It is expected that the true-copy-token scheme will overtake the primary copy scheme, when fewer remote pages are updated so that remote pages can be mostly kept in the buffer and immediately read by transactions locally. As an extreme case, when all transactions are of read-only type, the true-copy-token scheme will require almost no messages for concurrency control, since all copies will not lose their true-copy-token and can be granted to transactions immediately, whereas with the primary copy scheme a transaction must still make requests to the home site each time it wants to read a copy of a remote object. In this case (all transactions are of read-only type), the true-copy-token scheme behaves as the same as the primary copy scheme except that the former saves request (and hence also grant) messages which the latter must pay for.

Surprisingly, all three locking schemes have very low throughput. The reasons are: First, the transactions are quite long, some of them are even very long; second, there are a great number of updates, this does not favour a system with partitioned data distribution; third, those very long transactions should be treaded as batch-transactions.

In Fig.5 we see that delta-LRU for both true-copy-token and primary copy schemes results in a lower page-hit-rate than LRU. However, in Fig.4 we see that delta-LRU gains higher throughput than LRU. We conclude that the performance of buffer replacement policies should be measured not by page-hit-rate, but by their total I/O costs.

Other results, which are not shown here, state that our system is rather CPU-bound, although I/O is also critical. If query interpreter and optimization are considered, CPU should be more critical. Because of the high data transfer rate of the network, messages cost rather CPU than transmission time. An increase of message number loads the CPU more than the network. Beside CPU, there is also a jam at I/O channels. When the buffer size is 400 pages, the I/O utilization is as high as the CPU utilization (about 60%). As a result, I/O operations must spend much time in waiting for I/O channel usage. For reducing I/O amount, the buffer management, which involves both buffer allocation and replacement policy, is of great importance. The physical database design, especially the access methods of databases, and the design of recovery mechanism have also great influence on I/O requirement.

5. Conclusion

The simulation model described above is quite detailed and is in accordance with the reality. Reference-strings reflect real work load and make the simulation accurate. We did not make many assumptions or simplifications which we think could result in less accuracy. By changing parameters and using different reference-strings, a large quantity of experiments can be carried out. In this sense, our simulation system can be used as a test bed for various researches. For example, we can adjust the simulation program with a few efforts for simulating all kinds of algorithms for deadlock detection and recovery in distributed systems. A very important point of view is that more than one components should be simulated together so that their relationship, compatibility and influence on one another can be studied.

Acknowledgement

The author is greatly indebted to Prof. Andreas Reuter for his guidance in this research; his comments and suggestions helped to correct this paper. Prof. Bernd Walter advised me to compare the three distributed concurrency control schemes.

REFERENCES

/ACL85/ R. Agrawal, M.J. Carey, M. Livny: "Models for Studing Concurrency Control Performance: Alternatives and Implementations", Proc. of ACM-SIGMOD, Inter. Conf. on Management of Data, Austin, Texas, 1985.

/BEHR80/ R. Bayer, K. Elhardt, H.Heller, A.Reiser: "Distributed Concurrency Control in Database Systems", Proc. of VLDB Conf. 1980, Canada.

/Cai87/ Jian Cai: "Pufferverwaltung in verteilten Datenbanksystemen" Dissertation, Institut für Informatik der Universität Stuttgart, Mai 1987.

/CaSt84/ M. J. Carey, M. Stonebraker: "The Performance of Concurrency Control Algorithms for Database Management Systems", Proc. of the 10th VLDB Conf., Aug.1984, Singapore.

/EfHä84/ Wolfgang Effelsberg, Theo Härder: "Principles of Database Buffer Management", ACM Trans. on Database Systems, Vol.9, No.4, Dec.1984.

/Elha82/ K. Elhardt: "Das Datenbank-Cache: Entwurfsprizipien, Algorithmen, Eigenschften", Dissertation, Institut für Mathematik und Informatik, TU München, 1982.

/KiLa83/ W. Kiessling, G. Landherr: "A Quantitative Comparision of

Lockprotocols for Centralized Databases", Proc. of VLDB 1983, Florenz.

/KüSc83/ P. J. Kühn und K. M. Schulz als Herausgeber: "Messung, Modellierung und Bewertung von Rechensystemen", 2.GI/NTG-Fachtagung, Stuttgart, Feb. 1983.

/MiWi82/ T.Minoura, G.Wiederhold: "Resillent Extended True-Copy Token Scheme for a Distributed Database System", IEEE Trans. on Software Eng., Vol.SE-8, No.3, May 1982.

/PeRe84/ P. Peinl, A. Reuter: "Empirical Comparision of Database Concurrency Control Schemes", Proc. of the 10th VLDB Conf., Aug.1984, Singapore.

/Schm84/ B. Schmidt: "Der Simulator GPSS-FORTRAN Version 3", Fachberichte Simulation, Band 2, Springer-Verlag,1984.

/Ston78/ M. Stonebraker: "Concurrency Control and Consistency of Multiple Copies of Data in Distributed INGRES", Proc. of 3rd Berkley Workshop on DDBM and Communication Networks, 1978.

Ein funktionales Konzept zur Analyse von Warteschlangennetzen und Optimierung von Leistungsgrößen

G. Bolch, G. Fleischmann und
R. Schreppel

Universität Erlangen-Nürnberg
Institut für Mathematische Maschinen
und Datenverarbeitung (IV)

Zusammenfassung

Die Analyse von Warteschlangennetzen konnte in den letzten Jahren einige Fortschritte verzeichnen, jedoch erweist sich die Behandlung allgemeiner geschlossener Netze noch immer als schwieriges Problem. Da Produktformlösungen nur für spezielle Netztypen angegeben werden können und numerische Verfahren sehr aufwendig sind , gewinnen approximative Methoden stärker an Bedeutung. In der vorliegenden Arbeit wird ein funktionales Konzept zur Charakterisierung von Wartesystemen beschrieben , welches nicht nur eine relativ einfache approximative Analyse allgemeiner geschlossener Warteschlangennetze ermöglicht , sondern auch bei der Optimierung von Leistungsgrößen erfolgreich einsetzbar ist. Die einzelnen Knoten des Systems werden hier durch den Zusammenhang zwischen der mittleren Anzahl der Aufträge in der Station und dem Durchsatz charakterisiert. Ein Vergleich mit Ergebnissen aus der Mittelwertanalyse, beziehungsweise mit Simulationsergebnissen zeigt die Brauchbarkeit der gefundenen Lösungen.

1. Einleitung

Die Komplexität moderner Rechenanlagen machen Messung , Modellierung und Bewertung des Systemleistungsvermögens dringend erforderlich. Leistungsuntersuchungen, die den Entwurf einer Rechenanlage begleiten haben oft entscheidenden Einfluß auf die Entwicklung des Systems und damit dem Gelingen des gesamten Projektes. Zur Modellierung von Rechenanlagen werden oft Warteschlangenmodelle eingesetzt, die zur Herleitung und Optimierung von Leistungsgrößen wie Durchsatz, Wartezeiten, Auslastungen u.s.w. benutzt werden. Wir behandeln im Folgenden ein neues analytisches Verfahren zur approximativen Analyse allgemeiner geschlossener Warteschlangennetze mit einer Auftragsklasse. Es wird gezeigt, daß sich das vorgestellte Konzept leicht auf den Fall mehrerer Auftragsklassen erweitern läßt.

Jeder Knoten i des Netzes besteht aus einer Warteschlange, in die ankommende Aufträge eingereiht werden und m_i identischen Bedieneinheiten mit Bedienrate μ_i zur Abfertigung der Aufträge.
Für die Analyse werden folgende Größen als bekannt vorausgesetzt.

N Anzahl der Knoten
K Anzahl der Aufträge

Für alle Knoten $i=1,\ldots,N$

e_i Besuchshäufigkeiten
μ_i Bedienraten
m_i Anzahl der Bedieneinheiten

Für die sogenannten BCMP- oder Produktformnetze gibt es exakte Analyseverfahren /BCMP 75/.

Produktformwarteschlangennetze können vier verschiedene Knotentypen besitzen:

Typ-1 $(M/M/m)-FCFS$
Typ-2 $(M/G/1)-PS[RR]$
Typ-3 $(M/G/\infty)$ = Infinite Server (IS)
Typ-4 $(M/G/1)-LCFSPR$

Eine weit verbreitete Methode zur Berechnung von Leistungsgrößen für BCMP-Netze ist die Mittelwertanalyse /RELA 80/, bei der sich die Behandlung von Knoten mit mehreren Bedieneinheiten jedoch als speicherplatz- und zeitintensiv erweist. Deshalb sind schon sehr frühzeitig approximative Verfahren entwickelt worden, die sehr gute Ergebnisse liefern, aber auch nur für BCMP-Netze anwendbar sind /NECH 81, CHNE 82/. Für allgemeine Netze gibt es zum Beispiel die RTP-Methode /ABS 84/ und die Methode von Marie /MARI 79/, wobei nach unseren Erfahrungen beide Methoden sehr aufwendig sind und bei der RTP-Methode erhebliche Abweichungen auftreten können.
Das hier behandelte Verfahren ist im Gegensatz dazu sehr einfach, leicht durchschaubar, und auch für allgemeine Netze anwendbar, wobei für alle untersuchten Beispiele zufriedenstellende Genauigkeit erreicht wurde. Außerdem läßt sich das zugrundeliegende Konzept sehr einfach dazu verwenden geschlossene Formeln für optimale Leistungsgrößen zu ermitteln.

2. Analyse von Warteschlangennetzen

2.1 Ein funktionales Konzept zur Charakterisierung von Wartesystemen

Betrachtet man den funktionalen Zusammenhang zwischen dem Durchsatz λ_i und der mittleren Auftragszahl k_i eines Knotens mit lastunabhängigen Bedienraten, so ergibt sich ein Kurvenverlauf wie in Abbildung 1. dargestellt.

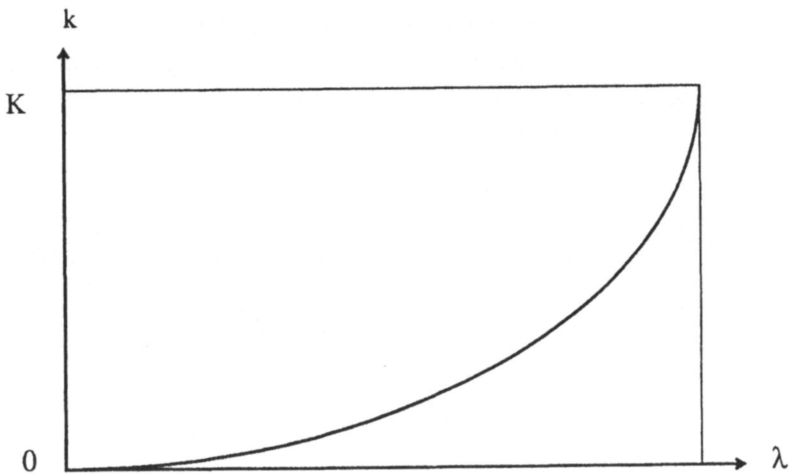

Abb.1:

Dieses Bild stimmt mit der intuitiven Vorstellung und der Erfahrung überein. Bei niedrigem Durchsatz ist zu erwarten, daß die mittlere Auftragszahl klein ist, da die Abstände zwischen zwei ankommenden Aufträgen sehr groß sind. Bei hohem Durchsatz sind die Ankunftsabstände klein, folglich befinden sich viele Aufträge im System, wobei eine weitere geringfügige Erhöhung der Ankunftsrate die Zahl der Benutzer stark erhöht. Für Knoten, bei denen die Zahl der Bedieneinheiten größer oder gleich der Zahl der Aufträge ist (*M/G/∞–Knoten*), ergibt sich für den Zusammenhang zwischen λ_i und k_i eine Gerade $k_i = \lambda_i/\mu_i$, da ankommende Aufträge sofort bedient werden. Allgemein kann festgestellt werden, daß sich die Funktion umso mehr einer Geraden nähert, je größer die Zahl der Bedieneinheiten ist. Wir wollen den Zusammenhang zwischen dem Durchsatz λ_i und der Zahl der Aufträge im Knoten i darstellen als

$$k_i = f_i(\lambda_i)$$

Allgemein können folgende Eigenschaften für die Funktionen f_i festgehalten werden.

- $f_i(0) = 0$
- Monotonie : $f_i(\lambda_i) < f_i(\lambda_i + \Delta\lambda_i)$ für $\Delta\lambda_i > 0$
  f_i ist streng monoton steigend mit λ_i
- Definitionsbereich : $0 \leq \lambda_i \leq \mu_i * m_i$
  Dies folgt aus der Bedingung $0 \leq \rho_i \leq 1$

für IS Knoten gilt: $0 < \lambda_i \leq K * \mu_i$
Dies folgt aus der Bedingung $k_i = \lambda_i / \mu_i$ *und* $k_i \leq K$

Die Monotonie ist bei lastabhängigen Bedienraten im Allgemeinen nicht gegeben und muß deshalb im Einzelfall überprüft werden. Bei "Multiple Server Knoten" bleibt die Monotonie jedoch erhalten.

Zur Analyse von BCMP-Netzen können folgende Funktionen verwendet werden, wobei nur die Letzte exakte Ergebnisse liefert und die anderen geeignete Approximationen sind. Die Auslastung ρ_i berechnet sich aus der bekannten Beziehung $\rho_i = \lambda_i / (m_i * \mu_i)$.

Typ-1(M/M/1),Typ-2,4 $\quad k_i = \dfrac{\rho_i}{(1 - \frac{K-1}{K} * \rho_i)}$

Typ-1(M/M/m) $\quad k_i = m_i * \rho_i + \dfrac{\rho_i}{1 - \frac{K - m_i - 1}{K - m_i} * \rho_i} * P_m(\rho_i)$

Typ-3 $\quad k_i = \dfrac{\lambda_i}{\mu_i}$

Fur die Wartewahrscheinlichkeit $P_m(\rho_i)$ gilt /KLEI 75/ :

$$P_m(\rho) = \frac{\dfrac{(m^* \rho)^m}{m!*(1-\rho)}}{\sum\limits_{k=0}^{m-1} \dfrac{(m^* \rho)^m}{k!} + \dfrac{(m^* \rho)^m}{m!*(1-\rho)}}$$

Sie kann nach /BOLC 83/ folgendermaßen geschätzt werden.

$$P_m(\rho_i) = \frac{\rho_i^{m_i} + \rho_i}{2} \; f\ddot{u}r \; \rho_i > 0.7, \quad P_m(\rho_i) = \rho_i^{\frac{m_i+1}{2}} \; f\ddot{u}r \; \rho_i < 0.7$$

Die approximative Formel für Typ-1(M/M/1),Typ-2 und Typ-4 Knoten kann auf zwei verschiedene Arten hergeleitet werden. Zum einen erhält man diesen Zusammenhang durch Umformung der Formel für die mittlere Verweilzeit der approximierten Mittelwertanalyse nach Bard/Schweitzer /BARD 79/,/SCHW 79/ $(t_i = \frac{1}{\mu_i}(1 + \frac{K-1}{K} k_i))$, zum anderen durch Verwendung der Formel für offene Netze $k_i = \frac{\rho_i}{1 - \rho_i}$ und der Einführung eines Korrekturfaktors $(K - 1)/K$. Der Korrekturfaktor ist motiviert durch die Tatsache , daß bei einer Auslastung von $\rho_i = 1$ sich an diesem Knoten i alle K Benutzer aufhalten ,d.h. für $\rho_i = 1$ gilt $k_i = K$. Die Idee des Korrekturfaktors wurde auch bei Typ-1 (M/M/m) Knoten verwendet.

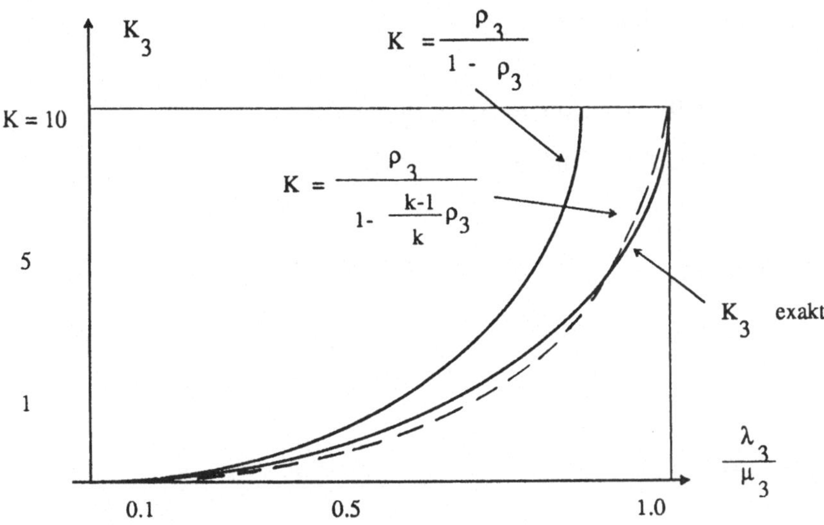

Abb.2:

Abbildung 2 zeigt für ein Beispiel mit 5 Knoten , wie gut der approximative Verlauf von $f_3(\lambda_3)$ (Knoten 3) mit dem exakten übereinstimmt. Zum Vergleich ist auch der approximative Verlauf ohne Korrekturfaktor eingezeichnet.

2.2 Analyse mit Hilfe die Systemgleichung

Wenn wir davon ausgehen, daß die Funktionen f_i für alle Knoten i gegeben sind , so läßt sich folgende Systemgleichung für geschlossene Netze aufstellen.

$$\sum_{i=1}^{N} k_i = \sum_{i=1}^{N} f_i(\lambda_i) = K$$

Mit der Beziehung $\lambda_i = \lambda * e_i$ erhalten wir eine Gleichung zur Bestimmung des Gesamtdurchsatzes λ.

$$\sum_{i=1}^{N} f_i(\lambda * e_i) = g(\lambda) = K$$

Ein Algorithmus zur Bestimmung des Gesamtdurchsatzes λ mit Hilfe der Systemgleichung kann wegen der Monotonie von $f_i(\lambda_i)$ wie folgt aussehen .

Intervallschachtelung zur Bestimmung von λ .

untere Grenze $\lambda_u = 0$, obere Grenze $\lambda_o = \min\limits_i \frac{m_i \mu_i}{e_i}$
(bei M/G/ ∞ -Knoten ist m_i durch K zu ersetzen).

while $g(\lambda) \neq K$ *do*

$$\lambda = \frac{\lambda_u + \lambda_o}{2}$$

$$g(\lambda) = \sum_{i=1}^{N} f_i(\lambda\, e_i)$$

if $g(\lambda) > K$ *then* $\lambda_o = \lambda$

if $g(\lambda) < K$ *then* $\lambda_u = \lambda$

end

Die übrigen Leistungsgrößen ergeben sich dann wie folgt

Für $i = 1,\ldots,N$

Durchsätze $\lambda_i = \lambda * e_i$

Auslastungen $\rho_i = \frac{\lambda_i}{m_i \mu_i}$

mittlere Auftragszahlen $k_i = f_i(\lambda_i)$

Antwortzeiten $t_i = \frac{k_i}{\lambda_i}$

2.3 Bewertung

Die Vorteile des Verfahrens liegen zum einen in der einfachen Berechnung der Leistungsgrößen und zum anderen in der Tatsache , daß die Anzahl der Aufträge K und vorallem die Anzahl der Bedieneinheiten m_i den Rechenaufwand nicht beeinflußen.
Wir betrachten nun die Ergebnisse für ein repräsentatives Beispiel, ein Rechensystem mit $N = 4$ Stationen und K = 10 Aufträgen. Alle übrigen Eingabegrößen sind in der

nachfolgenden Tabelle zusammengefaßt.

i	e_i	μ_i	m_i
1	1	1.9	4
2	0.4	0.9	∞
3	0.4	5.0	1
4	0.2	1.5	1

Tab.1:

Wir verwenden die Mittelwertanalyse zur Berechnung der Leistungsgrößen und erhalten dabei folgendes Ergebnis.

i	λ_i	ρ_i	k_i	t_i
1	6.021	0.792	3.990	0.662
2	2.408	0.267	2.676	1.111
3	2.408	0.481	0.856	0.355
4	1.204	0.802	2.476	2.056

Tab.2:

Für das gleiche Beispiel ergibt sich mit der hier vorgestellten "Summationsmethode" folgendes Ergebnis.

i	λ_i	ρ_i	k_i	t_i
1	5.764	0.758	4.156	0.721
2	2.305	0.256	2.561	1.111
3	2.305	0.461	0.788	0.314
4	1.152	0.768	2.492	2.162

Tab.3:

Es wurden zahlreiche BCMP-Netze mit der Summationsmethode untersucht wobei sich herausstellte, daß dieses Verfahren für die Praxis ausreichend genaue Ergebnisse liefert. In allen untersuchten Fällen ergaben sich für den Durchsatz und die Auslastung sehr gute Werte. Für die Anzahl der Aufträge und die Verweilzeit wurde eine Abweichung von 15% nicht überschritten, wobei im Mittel die Fehler bei 5% lagen. Es ist zu erwarten, daß sich die

Ergebnisse noch weiter verbessern lassen, wenn man geeignetere Funktionen f_i findet.

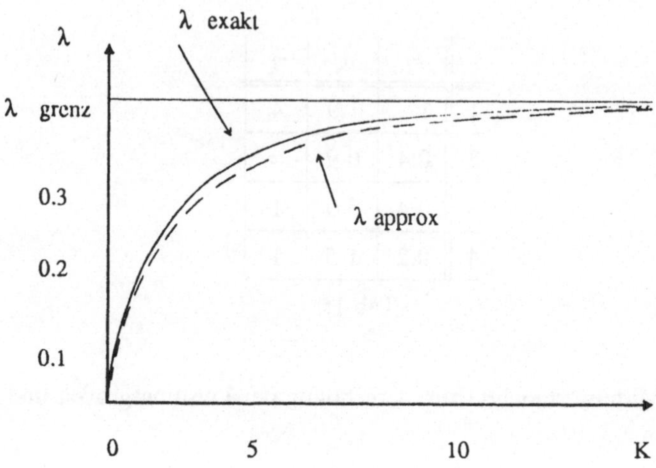

Abb.3:

In Abbildung 3 wird für ein Beispiel mit 5 Knoten der Durchsatz λ in Abhängigkeit von der Zahl der Aufträge im System dargestellt . Die obere Kurve zeigt die mit der Mittelwertanalyse errechneten Durchsätze , die untere die Ergebnisse der Summationsmethode. Das Bild macht deutlich, daß sich für den Gesamtdurchsatz λ, und damit für die Einzeldurchsätze λ_i und die Auslastungen ρ_i, sehr gute Werte ergeben.

2.4 Verallgemeinerung der Summationsmethode

Im vorigen Kapitel wird die Summationsmethode zur iterativen Berechnung des Durchsatzes λ für BCMP-Netze mit einer Auftragsklasse eingeführt. Wir wollen nun zeigen, daß das Verfahren auch für andere Knotentypen und mehrere Klassen anwendbar ist.
Bei der Einführung zusätzlicher Knotentypen verfolgen wir dieselbe Vorgehensweise , die wir bereits für die M/M/m-Knoten in BCMP-Netzen verwendet haben, d.h. es wird die Funktion $f_i(\lambda_i) = k_i$ für Einzelbedienstationen des entsprechenden Typs herangezogen und mit geeigneten Korrekturfaktoren versehen. Die Korrekturfaktoren werden notwendig , da sonst für $\rho_i = 1$ die Zahl der Benutzer k_i unendlich groß wird .
Für M/G/m-FCFS Wartesysteme ist folgende Approximationsformel bekannt /BOLC 83/ .

$$k = m * \rho + \frac{\rho}{1-\rho} * a * P_m \quad mit \quad a = \frac{c_b^2 + 1}{2}$$

c_b bezeichnet den Variationskoeffizienten des Bedienprozesses. Die Approximationsformel für den Knoten i vom Typ M/G/m-FCFS kann beispielsweise lauten (mit der Bedingung $\rho_i = 1 \rightarrow k_i = K$) :

$$k_i = m_i * \rho_i + \frac{\rho_i}{1 - \frac{K - m_i - a}{K - m_i} * \rho_i} * a * P_m$$

Für die Wartewahrscheinlichkeiten P_m können auch die Näherungen aus Abschnitt 2.1 verwendet werden. Im Folgenden werden, anhand eines konkreten Beispiels, die Ergebnisse aus der Summationsmethode mit Simulationsergebnissen verglichen. Die Konfidenzintervalle bei den Simulationsergebnissen liegen unter 5 Prozent. Die Untersuchung weiterer Modelle ergab für den Durchsatz Abweichungen zwischen 5 bis 10 %.

i	e_i	μ_i	m_i	c_b
1	1.00	0.20	4	0.30
2	1.13	0.08	7	2.40
3	0.33	0.80	1	1.00
4	3.33	0.12	10	3.90
5	0.67	0.05	∞	1.00

Tab.4: Eingabegrößen (K = 30)

i	λ_i	ρ_i	k_i	t_i
1	0.339	0.424	2.045	6.038
2	0.382	0.682	5.848	15.328
3	0.112	0.140	0.184	1.643
4	1.127	0.939	17.465	15.502
5	0.228	0.152	4.399	19.330

Tab.5: Simulationsergebnisse

i	λ_i	ρ_i	k_i	t_i
1	0.322	0.402	1.649	5.115
2	0.364	0.650	5.437	14.927
3	0.106	0.132	0.152	1.434
4	1.073	0.894	18.415	17.155
5	0.215	0.143	4.319	20.000

Tab.6: Summationsmethode

Liegen Warteschlangenmodelle mit mehreren Auftragsklassen vor, so müssen die für die Analyse wichtigen Beziehungen erweitert werden:

$$\lambda_{ir} = \lambda_r * e_{ir} \qquad \lambda_i = \sum_{r=1}^{R} \lambda_{ir} \qquad \rho_{ir} = \frac{\lambda_{ir}}{\mu_{ir}}$$

Es gibt jetzt r Systemgleichungen zur Bestimmung der λ_r

$$\sum_{i=1}^{N} k_{ir} = k_r$$

Die Funktionen $f_{ir}(\lambda_{ir}) = k_{ir}$ können wie folgt aussehen /KLEI 75/.

$$k_{ir} = \frac{\lambda_{ir}}{\mu_{ir}} \qquad \text{für } M/G/\infty\text{-Knoten.}$$

$$k_{ir} = \frac{\rho_{ir}}{1 - \frac{K_r - 1}{K_r}\rho_i} \qquad \text{für } M/G/1\text{-}PS \text{ und } M/M/1\text{-}FCFS \text{ Knoten.}$$

Zur Bestimmung des Durchsatzes λ_r wird wieder Intervallschachtelung durchgeführt.

3. Optimierung von Leistungsgrößen

Bei der Optimierung von Leistungsgrößen wie Antwortzeiten, Durchsätze , Auslastungen u.s.w., werden als Entscheidungsgrößen häufig die optimalen Bedienraten gesucht, welche unter bestimmten Nebenbedingungen eine vorgegebene Leistungsgröße optimieren /HETO 85/. Als häufigste Nebenbedingungen treten Kostenbeschränkungen auf, die im linearen Fall folgendermaßen aussehen.

$$\sum_{i=1}^{N} c_i \mu_i = C$$

N ist die Anzahl der Stationen im Warteschlangennetz. Die Parameter c_i geben die Einzelpreise der Bedienraten μ_i in der Station i an. Die Gesamtkosten C des Warteschlangennetzes sind ebenfalls vorgegeben. Realistischer ist eine nichtlineare Kostenfunktion , bei der die Gesamtkosten wie folgt berechnet werden.

$$\sum_{i=1}^{N} c_i \mu_i^{\alpha_i} = C, \quad \alpha_i > 1$$

Bei der Optimierung von Leistungsgrößen gehen wir aus von einem Warteschlangennetz, bestehend aus N Stationen bekannten Typs (Funktionen $f_i(\lambda_i) = k_i$), mit Besuchshäufigkeiten e_i und m_i identischen Bedieneinheiten. Die Gesamtzahl der Aufträge K im Netz ist ebenfalls gegeben.

Gesucht sind die optimalen Bedienraten μ_i , die eine bestimmte Leistungsgröße des vorgegebenen Warteschlangennetzes optimieren. Im folgenden sind einige Optimierungsprobleme zusammengestellt.

A. *Kostenminimierung bei festem Durchsatz.*
Hier wird der Gesamtdurchsatz λ fest vorgegeben und es sollen die Bedienraten μ_i^* gefunden werden, die diesen Durchsatz λ möglichst billig erreichen, bei verschiedenen Kostenfunktionen.

B. *Durchsatzmaximierung unter Kostenbeschränkung.*
Bei einem vorgegebenen Warteschlangennetz soll der Gesamtdurchsatz λ maximiert werden , wobei die Bedienraten μ_i einer Kostenbeschränkung unterliegen. Die Gesamtkosten werden vorgegeben und können als Geldmittel interpretiert werden, die zur Verfügung stehen um die einzelnen Stationen mit den entsprechenden Bedienraten zu kaufen.

C. *Antwortzeitminimierung bestimmter Stationen.*
Hier soll die Antwortzeit einer ganz bestimmten Station im Warteschlangennetz minimiert werden, wobei jedoch ein fest vorgegebener Durchsatz λ eingestellt wird . Die Bedienraten μ_i unterliegen dabei wieder einer Kostenbeschränkung , da sonst die Antwortzeit beliebig gesenkt werden kann.

Wir wollen hier an einem Beispiel die prinzipielle Vorgehensweise zur Optimierung mit Hilfe der Systemgleichung aus Abschnitt 2.2 deutlich machen.

3.1 Durchsatzmaximierung bei linearer Kostenbeschränkung

Wir gehen aus von einem Warteschlangennetz mit N Stationen und K Aufträgen einer Klasse. Gegeben sind außerdem die relativen Besuchshäufigkeiten e_i und Funktionen f_i zur Charakterisierung der Wartesysteme . Wir vernachlässigen hier den Korrekturfaktor $(K-1)/K$ um möglichst einfache Formeln zu erhalten. Dies ist für große Auftragszahlen gerechtfertig, da er dann nahe bei 1 liegt. Aber auch wenn man den Korrekturfaktor berücksichtigt erhält man geschlossene Formeln, mit denen man, vor allem bei kleinen Werten von K , noch bessere Werte für die Optima erhält.

Mit $\lambda_i = \lambda * e_i$, $m_i = 1$ und $\rho_i = \dfrac{\lambda_i}{\mu_i}$ erhalten wir

$$f_i(\lambda_i) = \begin{cases} \dfrac{\lambda e_i}{\mu_i - \lambda e_i} & f\ddot{u}r\ Typ\,1,2,4\ und\ m_i = 1 \\ \dfrac{\lambda e_i}{\mu_i} & f\ddot{u}r\ Typ\,3(IS) \end{cases}$$

Gesucht sind nun die Bedienraten μ_i , welche den Gesamtdurchsatz maximieren und eine lineare Kostenbeschränkung erfüllen

$$\sum_{i=1}^{N} c_i \mu_i = C$$

Die Parameter c_i (Einzelpreise) und die Gesamtkosten C sind ebenfalls vorgegeben . Neben der Kostenbeschränkung müssen die Bedienraten auch noch die Systemgleichung erfüllen , welche hier wie folgt aussieht.

$$\sum_{i=1}^{N} f_i(\lambda) = \sum_{\neq IS} \frac{\lambda e_i}{\mu_i - \lambda e_i} + \sum_{IS} \frac{\lambda e_i}{\mu_i} = K$$

Eine Lösung dieser Optimierungsaufgabe läßt sich nun mit Hilfe der sogenannten Lagrangemultiplikatoren /BRSE 80/ angeben. Dazu stellen wir die Lagrangefunktion $L(\lambda,\mu_1,\ldots,\mu_N,y_1,y_2)$ mit Zielgröße λ und zwei Lagrangemultiplikatoren y_1 und y_2 auf.

$$L(\lambda,\mu_1,\ldots,\mu_N,y_1,y_2) = \lambda + y_1\Big(\sum_{i=1}^{N} c_i\mu_i - C\Big) + y_2\Big(\sum_{\neq IS} \frac{\lambda e_i}{\mu_i - \lambda e_i} + \sum_{IS} \frac{\lambda e_i}{\mu_i} - K\Big)$$

Durch Differenzieren nach λ,μ_i,y_1,y_2 erhält man daraus eine notwendige Bedingung für die optimalen Bedienraten μ_i und den maximalen Durchsatz λ^* .

$$\frac{\partial L}{\partial \lambda} = 0 \qquad \frac{\partial L}{\partial y_1} = 0$$

$$\frac{\partial L}{\partial \mu_i} = 0,\ i = 1,\ldots,N \qquad \frac{\partial L}{\partial y_2} = 0$$

Durch Auflösen des Gleichungssystems erhalten wir folgendes Ergebnis.

$$\lambda^* = \frac{C^*K}{\left(\sum_{i=1}^{N}\sqrt{c_i e_i}\right)^2 + K\sum_{\neq IS} c_i e_i}$$

$$\mu_i^* = \begin{cases} \lambda^* e_i \left(\dfrac{\sum_{j=1}^{N}\sqrt{e_j c_j}}{K\sqrt{e_i c_i}} + 1 \right) & \text{für} \neq IS \\[2ex] \lambda^* e_i \left(\dfrac{\sum_{j=1}^{N}\sqrt{e_j c_j}}{K\sqrt{e_j c_j}} \right) & \text{für } IS \end{cases}$$

Wie wir aus obiger Gleichung sehen , ist der maximale Gesamtdurchsatz λ^* direkt proportional zu den Gesamtkosten C , wenn man von einer linearen Kostenfunktion ausgeht. Bei komplizierteren Optimierungsaufgaben mit nichtlinearer Kostenfunktion , mehreren Stationstypen und weiteren Nebenbedingungen lassen sich die optimalen Bedienraten μ_i^* nicht mehr explizit angeben. Sie können aber mit Hilfe von Iterationsverfahren aus dem

Gleichungssystem ermittelt werden.
Im Folgenden sind weitere Ergebnisse für einige der oben angesprochenen Optimierungsaufgaben dargestellt.

Kostenminimierung bei linearer Kostenfunktion

$$\mu_i^* = \begin{cases} \lambda e_i \left(\frac{\sum_{j=1}^{N} \sqrt{e_j c_j}}{K \sqrt{e_{i_i}}} + 1 \right) & f\ddot{u}r \neq IS \\ \lambda e_i \left(\frac{\sum_{j=1}^{N} \sqrt{e_j c_j}}{K \sqrt{e_{i_i}}} \right) & f\ddot{u}r = IS \end{cases}$$

Die minimalen Kosten berechnen sich nach der Formel:

$$C^*(\mu) = \sum_{i=1}^{N} \mu_i^* c_i$$

Im Falle einer *nichtlinearen Kostenfunktion* kann das folgende Iterationsverfahren angewendet werden.

Initialisiere : $\mu_i = 1 \quad f\ddot{u}r \quad i = 1,\ldots,N$

Iteriere, bis die Abweichungen hinreichend klein sind:

$$y = \lambda * \left(\frac{1}{K} \sum_{i=1}^{N} \sqrt{\alpha_i c_i e_i \mu_i^{\alpha_i - 1}} \right)^2$$

$$\mu_i = \begin{cases} \sqrt{\frac{\lambda y e_i}{\alpha_i c_i \mu_i^{\alpha_i - 1}}} + \lambda e_i & f\ddot{u}r\ Typ(i) \neq IS \\ \left(\frac{\lambda y e_i}{\alpha_i c_i} \right)^{\frac{1}{\alpha_i + 1}} & f\ddot{u}r\ Typ(i) = IS \end{cases}$$

Minimierung der Antwortzeit:

Die Optimierungsaufgabe, die Gesamtverweilzeit T zu minimieren, ist äquivalent zu der Aufgabe den Gesamtdurchsatz zu maximieren. Dies folgt aus dem Gesetz von Little $T = K/\lambda$ und der Tatsache, daß die Zahl der Aufträge K im geschlossenen Netz konstant ist. Bei der Minimierung der Antwortzeit interessieren deshalb nur einzelne Knoten im

Warteschlangennetz. Für lineare Kostenfunktionen können wieder geschlossene Formeln angegeben werden.

3.2 Beispiel zur Durchsatzmaximierung

Wir wollen nun an einem konkreten Beispiel den Durchsatz bezüglich der Bedienraten optimieren , wie es in Abschnitt 3.1 hergeleitet worden ist, und das Ergebnis mit Hilfe der Mittelwertanalyse bewerten.

Es soll ein Warteschlangennetz mit N=5 Stationen vorliegen, wobei 3 Stationen vom Typ (M/M/1) mit jeweils einer Bedieneinheit sind und zwei vom Typ $(M/G/\infty)$. Im Warteschlangennetz zirkulieren $K=20$ Aufträge und es seien die folgenden Besuchshäufigkeiten gegeben.

$e_1 = 1,\ e_2 = 0.2,\ e_3 = e_4 = 0.5,\ e_5 = 0.3$

Die Kosten für die Bedienraten μ_i sollen linear ansteigen , wobei folgende Einzelpreise zugrunde liegen.

$c_1 = 10,\ c_2 = c_3 = 5,\ c_4 = 2,\ c_5 = 1$

Die Gesamtkosten werden mit C = 100 festgelegt. Nach Abschnitt 3.1 ergeben sich die optimalen Bedienraten μ_i^* , welche den Durchsatz λ^* innerhalb der vorgegebenen Kostenbeschränkung erzielen zu

$\mu_1^* = 6.189,\ \mu_2^* = 1.689\ ,\mu_3^* = 3.812,\ \mu_4^* = 1.096\ ,\mu_5^* = 1.236$

mit einem Durchsatz $\lambda^* = 6.189$.

Für die optimalen Bedienraten μ_i^* erhalten wir mit der Mittelwertanalyse einen exakten Durchsatz von $\lambda = 6.569$. Wir können nun zur Kontrolle die als optimal ermittelten Bedienraten etwas variieren , wobei jedoch die Nebenbedingung

$$\sum_{i=1}^{N} c_i \mu_i = 100$$

erhalten bleibt, und den damit erzielten Durchsatz mit den optimalen vergleichen. Wählen wir etwa

$\mu_1 = 7\ ,\mu_2 = 1.7\ ,\mu_3 = 3.8\ ,\mu_4 = 1\ ,\mu_5 = 0.5$

so ergibt die Mittelwertanalyse einen Durchsatz von $\lambda = 6.473$ (< 6.569) , welcher unter dem Durchsatz liegt , den die optimalen Bedienraten μ_i^* erzielen . Ähnliche Ergebnisse erhält man für andere Kombinationen der Bedienraten mit obiger Nebenbedingung.

Ändern wir die Bedienraten noch stärker von den optimalen ab , zum Beispiel zu

$\mu_1 = 8\ ,\mu_2 = 2\ ,\mu_3 = 1\ ,\mu_4 = 2,\ \mu_5 = 1$

so fällt der Durchsatz noch weiter ab auf $\lambda = 2$.

Das Beispiel zeigt , daß mit Hilfe der Systemgleichung relativ einfach Optimierungsprobleme behandelt werden können . Im Falle einer linearen Kostenbeschränkung lassen sich die optimalen Bedienraten μ_i^* sogar explizit angeben. Bei nichtlinearer Kostenfunktion können einfache Iterationsverfahren mit gutem Konvergenzverhalten angegeben werden.

α	1	1.25	1.5	2.0	2.5	3
μ_1^*	6.912	4.911	3.884	2.860	2.361	2.067
μ_2^*	1.689	1.242	1.017	0.806	0.716	0.674
μ_3^*	3.812	2.727	2.173	1.625	1.363	1.214
μ_4^*	1.096	0.921	0.821	0.735	0.706	0.704
μ_5^*	1.236	0.998	0.883	0.781	0.744	0.732
λ^*	6.189	4.438	3.532	2.625	2.180	1.918
λ	6.596	4.711	3.752	2.790	2.319	2.039

Tab.7: Durchsatzmaximierung (α = Exponent aus nichtlinearer Kostenfunktion)

Die obige Tabelle enthält Ergebnisse für die Durchsatzmaximierung mit nichtlinearer Kostenfunktion /SCHR 86/. Die ermittelten optimalen Bedienraten, welche den Durchsatz maximieren , wurden zum Vergleich als Eingabeparameter bei der exakten Mittelwertanalyse verwendet und den iterativ berechneten Werten gegenübergestellt.

4. Schlußbemerkung

Die Vorteile der hier beschriebenen Summationsmethode liegen im geringen Zeit- und Speicherplatzbedarf und der Transparenz des Verfahrens. Wie bei der Mittelwertanalyse erhält man unmittelbar Mittelwerte der Leistungsgrößen und es werden auch keine komplizierten Zwischengrößen wie etwa Normalisierungskonstante und Zustandswahrscheinlichkeiten benötigt. Es zeigt sich , daß das Verfahren gute Näherungswerte für die Leistungsgrößen in allgemeinen geschlossenen Warteschlangennetzen liefert. Obwohl durch die Einführung eines einzigen Nicht-BCMP-Knotens in ein Netzwerk die Markoveigenschaft des Ankunftsprozesses für alle Knoten des Modelles verloren geht, werden, wie die Beispiele zeigen, brauchbare Werte erzielt. Dies ist auf das Aussehen der Funktion $f_i(\lambda_i)$ zurückzuführen , die auch für allgemeine Ankunftsprozesse einen Verlauf wie in Abbildung 1. dargestellt aufweist.

Die Ergebnisse lassen sich vermutlich durch Veränderung der Funktion $f_i(\lambda_i)$ noch verbessern. Bei Nicht-BCMP-Netzen ist eine weitere Verbesserung möglich, wenn für alle Knoten allgemeine Ankunftsverteilungen angenommen werden. Die Zahl der Benutzer in G/G/m-FCFS-Stationen kann zum Beispiel mit folgender Approximationsformel geschätzt werden /ALLE 78/ .

$$k = m\rho + \frac{\rho}{1-\rho} * \frac{c_b^2 + c_a^2}{2} * P_m$$

Die Größe c_a bezeichnet den Variationskoeffizienten des Ankunftsprozesses , der mit dem Verfahren von Kühn /KÜHN 79/ geschätzt und iterativ verbessert werden kann .

Mit Hilfe der Systemgleichung lassen sich sehr gut Optimierungsprobleme behandeln, da dadurch der Zusammenhang zwischen der Zielgröße (z.B. Durchsatz) und den Entscheidungsgrößen (Bedienraten) in einfacher Weise festgelegt wird . In allen Fällen ergeben sich geschlossene Formeln für die optimalen Bedienraten. Bei linearer Kostenbeschränkung lassen sie sich sogar explizit angeben. Durch Einführung der Korrekturfaktoren aus Kapitel 2 ist eine weitere Verbesserung der Optimierungsmethode möglich.

5. Literaturverzeichnis

/ALLE 78/ Allen O.A., Probability, Statistics and Queueing Theory with Computer Science Applications, Academic Press, New York, 1978

/ABS 84/ Agrawal S.C.,Buzen J.F.,Shum W., Response Time Preservation, ACM Sigmetrics Conference Proceedings, Cambridge, MA, 1984

/BARD 79/ Bard Y., Some Extension to Multiclass Queueing Network Analysis, Fourth International Symposium on Modelling and Performance Evaluation of Computer Systems , Feb. 1979, vol. 1, Vienna

/BCMP 75/ Baskett F.,Chandy K.M., Muntz R.R.,Palacios F.G., Open , Closed , and Mixed Networks of Queues with Different Classes of Customers, Journal of ACM, vol. 22, no. 2, pp. 248 -260, 1975

/BOLC 83/ Bolch G., Approximation von Leistungsgrößen symetrischer Mehrprozessorsysteme, Computing 31,305-315 , Springer-Verlag 1983

/BRSE 80/ I.N. Bronstein und K.A. Semendjajew, Taschenbuch der Mathematik, Band 1 , Verlag Harri Deutsch , Thun 1980

/CHNE 82/ Chandy K.M. and Neuse D.,Linearizer: A Heuristic Algorithm for Queueing Network Models of Computing Systems, Communications of ACM, vol. 25, no. 2, pp. 126-134, February 1982

/HETO 85/ H.-U. Heiß und G. Totzauer Optimizing Utilization under Response Time Constraints, Computing 35,1-12, Springer-Verlag 1985

/KLEI 75/ L. Kleinrock, Queuing Systems, vol. I: Theory, John Wiley & Sons, Inc., New York, 1975

/KÜHN 79/ P.J. Kühn, Approximate Analysis of General Queueing Networks by Decompostion, IEEE Transactions on Communications, vol.Com-27,NO.1, Jan. 1979, pp. 113-126

/MARI 79/ Marie R., An Approximate Analytic Method for General Queueing Networks, IEEE, TSE, 1979

/NECH 81/ Neuse D.,Chandy K.M., SCAT: A Heuristic Algorithm for Queueing Network Models of Computing Systems, ACM Sigmetrics Conference Proceedings, vol. 10,no. 3. pp. 59-79,1981

/RELA 80/ M. Reiser and S. S. Lavenberg , Mean Value Analysis of Closed Multichain Queueing Networks, Journal of ACM. vol. 27. no. 2, April 1980, pp. 313-322

/SCHR 86/ Schreppel R., Ermittelung optimaler Leistungsgrößen mit Hilfe analytischer Warteschlangenmodelle, Diplomarbeit am IMMD IV, Erlangen 1986

/SCHW 79/ Schweitzer P, Approximate Analysis of Multiclass Closed Networks of Queues , Int. Conf. Stochastic Control and Optimiziation, Amsterdam, 1979

Experimente mit Ersatzdarstellungen unter Berücksichtigung der Verweilzeitverteilung

Heinz Beilner, Peter Buchholz, Bruno Müller-Clostermann
Informatik IV, Universität Dortmund
Postfach 50 05 00, 4600 Dortmund 50

Zusammenfassung:
Die Aggregierung separabler Teilnetze zu flußäquivalenten Ersatzstationen mit populationsabhängigen Bediengeschwindigkeiten und deren Einbettung sowohl in separabler als auch in nicht-separabler Umgebung wird in der Modellierung breit eingesetzt. Bekanntlich liefert die Verwendung von Aggregaten im Falle von separabler (bzw. nicht-separabler) Umgebung exakte (bzw. approximative) Mittelwerte(!) von Leistungsmaßen. Diese Methode der Ersatzdarstellungen verfälscht aber sogar in rein separablem Kontext die Verweilzeitverteilung(!) der Aufträge sowohl am betrachtetem Teilnetz als auch in der Umgebung. Diese Arbeit untersucht durch numerische Analyse geeigneter reduzibler Markovketten, wie sich die Variation diverser Netzparameter auf die Paßgüte der resultierenden Verweilzeit- verteilungen auswirkt. Zur Verbesserung der Paßgüte werden, alternativ zur üblichen exponentiellen Ersatzstation, Cox-Stationen und Ersatzdarstellungen in Form von Zwei-Stationen-Netzen erprobt.

1. Motivation und Einführung

Das Konzept der Ersatzdarstellung in separablen Netzen, die auch als Produktformnetze oder BCMP-Netze bekannt sind, wurde 1975 von Chandy, Herzog und Woo CHHW75 eingeführt und in BABR83 und KRWK82 für Mehr-Ketten-Netze weiterentwickelt. Stellvertretend für viele Arbeiten, die sich mit "flow equivalent servers", "composite queues" oder "substitute representations" beschäftigen, seien genannt STLZ77 und CHSA78, wo Konstruktion und Verwendung von Ersatzdarstellungen für nicht-separable Netze auf heuristischer Basis diskutiert werden. Die Vorzüge heterogener Simulation - auch hybride Simulation genannt - sind beeindruckend z.B. in HOYA86 illustriert. Große Bedeutung hat die Verwendung von aggregierten Ersatzdarstellungen auch bei der numerischen Lösung Markovscher Modelle; die durch die Aggregierung bewirkte Verringerung der Zahl der Modellzustände ist aus Gründen von Speicher- und Rechenzeitaufwand von erheblicher Bedeutung (BABG85,MURO87). Bekanntlich werden die zweiten und höheren Momente von Zeiten (Verweilzeiten, Zykluszeiten, Wartezeiten, etc.) sowie natürlich die zugehörigen Verteilungsfunktionen bei Verwendung von Ersatzdarstellungen anstelle des Original-Teilnetzes i.a. verfälscht im Vergleich zu den nicht-aggregierten Originalnetzen. Dies gilt auch für Einbettungen aggregierter Teilnetze in separable Umgebungen: Die Verweilzeitverteilung eines Auftrags wird bei Verwendung eines aggregierten Ersatzbedieners anders aussehen als bei Verwendung des Originals.

Ziel der Arbeit ist es, diese Thematik exemplarisch vorzustellen, Approximationsgüte und Verbesserungsmöglichkeiten für Ersatzdarstellungen zu illustrieren und das

Bewußtsein des Lesers bzgl. dieser Problematik zu schärfen. Nach einem kleinen Einblick in Analysetechniken zur Bestimmung von Momenten und approximativen Dichte- und Verteilungsfunktionen auf der Basis absorbierender Markovketten (Abschn.2), werden verschiedene Möglichkeiten zur Wahl von Ersatzdarstellungen und ihre Auswirkungen auf die Ergebnisse von Modellanalysen vorgestellt. Zunächst werden Ersatzstationen mit neg. exponentieller Bedienzeitverteilung und ihre Einbettung in separable und nicht-separable Umgebungen betrachtet (Abschn.3). Die z.T. erheblichen Approximationsfehler motivieren die Verwendung von detaillierteren Ersatzdarstellungen, nämlich Ersatzstationen mit Cox´scher Bedienzeitverteilung bzw. Ersatznetze (Abschn.4). In Abschn.5 wird ein kurzes Resumée gezogen. Die vorgestellten Resultate enstanden im Rahmen der Diplomarbeit BUCH86, in der eine Fülle von weiterem Material zusammengetragen ist.

2. Zur Analyse von Verweilzeitverteilungen

Quantitative Bewertungen von (nicht-trivialen) Warteschlangennetzen unter Berücksichtigung der Verweilzeitverteilung sind aus prinzipiellen Gründen weder einfach noch effizient durchzuführen. Simulative Untersuchungen (wie in BUCH86 ebenfalls erprobt) sind im gegebenen Kontext von beschränktem Wert und kommen für unsere Ziele kaum in Frage: Die Schätzung höherer Momente ist äußerst aufwendig bei sehr unpräzisen Ergebnissen. Analytisch-algebraische Techniken sind nur für eine beschränkte Modellklasse entwickelt und müssen sich auf Mini-Netze oder auf sehr spezielle Modellklassen beschränken. Für Zwei-Stationen-Netze gibt es eine Reihe von Ergebnissen, hier seien die Arbeiten BOXM84 und DADU85 genannt. Bzgl. einer Übersicht über die wenigen exakten Resultate in allgemeineren Netzen sei auf SCHA85 verwiesen. Aus prinzipiellen Gründen ist auch in Zukunft nicht mit wesentlichen Erweiterungen zu rechnen (SCHA85, S.126). Wir verwenden daher analytisch-numerische Techniken, deren Grundlagen im folgenden Abschnitt 2.1 kurz skizziert werden.

2.1. Analytisch-numerische Techniken

Die verwendeten numerischen Techniken zur Ermittlung von Verweilzeitverteilungen basieren auf absorbierenden Markovketten. Die grundlegende Idee, hier geschildert für den Ein-Ketten-Fall, besteht darin, einen Kunden der Kette zu "markieren" und als einzigen Auftrag einer speziellen Kette (ansonsten völlig identischen Verhaltens) zu betrachten. Dieser markierte Auftrag wird bei seinem Durchlauf durch das zu untersuchende Teilnetz bzgl. seiner Verweilzeitverteilung bzw. einiger ihrer Charakteristika (wie Momente) analysiert. Die hierzu eingesetzten Modelle sind (endliche) reduzible Markovsche Ketten, deren Zustandsraum in eine Menge von transienten sowie in (mindestens) eine Menge von rekurrenten Zuständen klassifiziert werden kann.

Wir erläutern diese Methode exemplarisch. Es bewegen sich N Aufträge gemäß eines bestimmten Verhaltensmusters (d.h. nach gleichem Routingverhalten und mit gleichen

Bedienforderungen) durch ein Warteschlangennetz. Um einen einzelnen Repräsentanten unter diesen Aufträgen identifizieren zu können, sollen (N-1) Aufträge in ihrer Kette verbleiben, der N-te Auftrag wird als einziger einer neu eingeführten, aber, bis auf nachstehend erläuterten Punkt, verhaltensgleichen Kette zugeordnet. Dieser markierte Auftrag verzweigt nämlich, im Gegensatz zu den anderen N-1 Aufträgen, nach dem Netzdurchlauf in eine künstlich und eigens zu diesem Zweck eingeführte Senke.

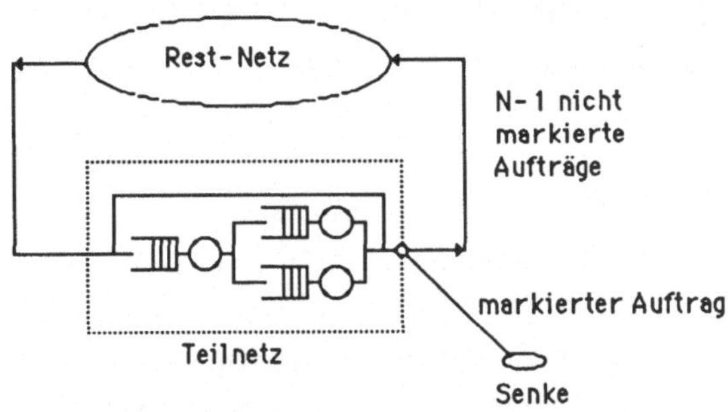

Abb. 2.1. : Prinzip des markierten Auftrags

Die aus diesem Modell resultierende Markovkette beinhaltet als rekurrente ("absorbierende") Zustandsmenge alle diejenigen Zustände, bei denen sich der markierte Auftrag in der Senke befindet, während alle anderen Aufträge beliebig auf die Stationen des Netzes verteilt sind. Die Verweilzeit des markierten Auftrags im Teilnetz, die identisch ist mit der Zeit, die der Prozeß in der Menge der transienten Zustände verbringt, ist natürlich abhängig vom Startzustand der Markovkette. Betrachten wir hierzu Abb. 2.1. Wo soll der markierte Auftrag zur Zeit t=0 starten, und wie sollen die anderen Aufträge plaziert sein, damit wir von einem repräsentativen Anfangszustand sprechen können? Denkbar wären z.B. best-case oder worst-case Situationen, wo sich alle nicht-markierten Aufträge hinter bzw. vor dem markierten Auftrag befinden. Im Hinblick auf die Konstruktion von Ersatzdarstellungen, die in nachfolgender stationärer Analyse anstelle des Original-Netzteils eingesetzt werden sollen, ist die Annahme naheliegend, daß der markierte Auftrag das zu durchlaufende Teilnetz in Stationarität (mit N-1 Aufträgen) antreffen soll. Bekanntlich trifft dies im Falle separabler Netze zu (dies ist ja gerade eine charakterisierende Eigenschaft separabler Netze, siehe z.B. SEMI81); im Falle nicht-separabler Teilnetze oder nicht-separabler "Rest-Netze" gilt diese Annahme bestenfalls in approximativem Sinne.

Wir skizzieren nun im folgenden zwei Analysetechniken, die auf Basis des Teilnetzes von Abb. 2.1 höhere Momente bzw. die approximative Verteilungsfunktion des markierten Auftrags bestimmen. Diese Techniken sind sowohl im Kurzschlußfall (d.h. unter Überbrückung des Rest-Netzes) als auch für allgemeine Rest-Netze einsetzbar. Die Transitionsmatrix wird dabei notiert wie folgt.

$$Q = \begin{bmatrix} T & X \\ \emptyset & R \end{bmatrix}$$

T bezeichnet die Transitionsmatrix der transienten Zustände, R bezeichnet die Transitionsmatrix der rekurrenten ("absorbierenden") Zustände.

Zur Berechnung des k-ten **Momentes** der Verweilzeitverteilung, hier identisch mit der Absorptionszeitverteilung für die Markovkette mit Transitionsmatrix Q, sind jeweils lineare Gleichungssysteme der Art $\underline{T} \cdot \underline{E}_k = -k \cdot \underline{E}_{k-1}$ (mit $\underline{E}_0 = \underline{1}$) zu lösen; dabei bezeichnet $\underline{E}_k$ den Vektor der bedingten k-ten Momente, wobei die Bedingung darin besteht, zur Zeit t=0 im Zustand i, i=1,2,...,ord(T), gestartet zu sein, vgl. GAED77. Die absoluten Momente E_k, k=1,2,.., werden aus den bedingten Momenten und dem Vektor der Startwahrscheinlichkeiten $\underline{p}^{(0)}$ bestimmt gemäß $E_k = \underline{E}_k \cdot \underline{p}^{(0)}$. Der Variationskoeffizient der Verteilung wird dann bestimmt gemäß $C = \sqrt{VAR}/E_1$, wobei $VAR = E_2 - E_1^2$.

Diverse Möglichkeiten zur Berechnung der **approximativen Verweilzeitverteilung** wurden versuchsweise erprobt (vgl. BUCH86). Als besonders effizient erwies sich die sog. Randomízation-Methode MEYA84. Bei diesem Verfahren wird die Transitionsmatrix Q, evtl. nach Zusammenfassung der rekurrenten Zustandsmenge zu einem einzigen absorbierenden Zustand, zunächst in eine stochastische Matrix transformiert (Q wird "randomisiert"). Bei einem nachfolgenden Iterationsverfahren wird pro Iterationsschritt (,der eine Zeiteinheit repräsentiert,) die jeweils neu in den absorbierenden Zustand geflossene Wahrscheinlichkeitsmasse als Dichtefunktionswert, bzw. per Akkumulation als Verteilungsfunktionswert berechnet und registriert. Auf diese Weise erhält man eine Folge von Stützstellen für Dichte- und Verteilungsfunktion.

Die o.g. Algorithmen wurden neben anderen Verfahren, wie z.B. Runge-Kutta-Verfahren und Diskretisierung-Techniken, implementiert und in dem auf Markovketten basierenden Analysetool NUMAS (MUEL84) integriert. Die in den Abschnitten 3 und 4 beschriebenen Experimente wurden mit NUMAS durchgeführt, wobei zur Gewinnung von Mittelwerten die durch NUMAS ebenfalls angebotenen Algorithmen zur Bestimmung stationärer Grenzverteilungen (basierend auf irreduziblen Markovketten) eingesetzt wurden. (Für "Tool-Interessierte" sei angemerkt, daß stationäre numerische Analysemethoden auch in dem Modellierungs- und Analysetool HIT (BESC85, BEST87) zugreifbar sind.)

3. Ersatzstationen mit exponentiell verteilter Bedienzeit

Die inzwischen als "klassisch" zu apostrophierende Aggregierungstechnik (etwa auf Basis von CHHW77,KRWK82,BABR83) untersucht das Kurzschlußverhalten eines zu aggregierenden Netzteils unter allen interessierenden Kundenpopulationen; die resultierenden (populationsspezifischen) Durchsätze werden zur Konstruktion eines Aggregats in Form einer Ersatzstation populationsabhängiger Bedienrate/Bediengeschwindigkeit verwendet; der Ersatzstation wird i. allg. eine exponentielle Bedienzeit zugeordnet. Die Auswirkungen dieser Technik auf die Verweilzeitverteilung an der Ersatzstation (im Vergleich zu jener am ursprünglichen Netzteil) werden in diesem Abschnitt anhand eines Beispiels aufgezeigt; die Vergleiche berücksichtigen in Abschnitt 3.1 den Kurzschlußfall selbst, in Abschnitt 3.2 die Einbettung des Netzteils/Aggregats in eine separable Netzumgebung sowie in Abschnitt 3.3 die Einbettung in eine nicht-separable Umgebung.

3.1. Ersatzstationen und Kurzschlußverhalten

Als Beispiel für einen zu aggregierenden Netzteil diene ein "central-server"-artiges Subsystem mit Einketten-Belastung:

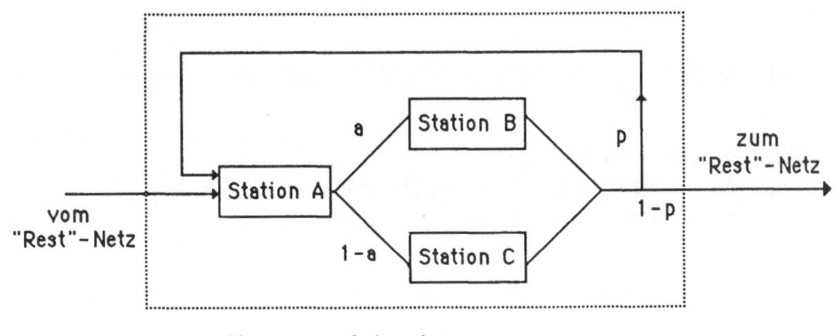

Abb. 3.1.: Struktur des Beispiel-Teilnetzes

Als weitere Festlegungen mögen gelten:

Bedienstrategien: - PS für alle Stationen

Bedienzeitverteilungen: - exponentiell für B, C mit identischen Mittelwerten $1/\mu_B = 1/\mu_C$ (variabel);
- variabel für A mit Mittelwert $1/\mu_A$ und Variationskoeffizient C_A

Die variabel gehaltenen Parameter werden in 3 Konfigurationen (Netze 1-3) untersucht:

Parameter	Netz 1	Netz 2	Netz 3
p	0	0	0.9
Verteilung A	exponentiell	hyperexp.	exponentiell
μ_A (*ZE)	2	2	20
C_A	1	10	1
$\mu_B=\mu_C$ (*ZE)	5/3	5/3	50/3

Tab. 3.2.: Parameterkonfigurationen des Subsystems; ZE : beliebige Zeiteinheit

Der klassischen Vorgehensweise zur Konstruktion von Ersatzstationen folgend, wird das Subsystem im Kurzschluß mit jeweils fester Population betrieben, wobei in unseren Beispielen die Populationen $n \varepsilon \{1,2,3,4,5\}$ Berücksichtigung finden. Wie man aus den Parameterkonfigurationen der drei Netze leicht abliest, erhält man in allen Fällen identische (populationsabhängige) Durchsätze und daraus resultierend identische Werte für die geforderten (populationsabhängigen) Bedienraten einer zu konstruierenden Ersatzstation, insgesamt also dieselben Anforderungen an eine Ersatzstation exponentieller Bedienzeit:

Population n	1	2	3	4	5
Durchsatz	0,909	1,341	1,583	1,729	1,823
Bedienrate	0,909	0,671	0,528	0,432	0,365

Tab. 3.3.: Populationsabhängigkeit
- der Subsystem-Durchsätze (Netz 1,2,3) im Kurzschluß
- der geforderten Bedienraten der Ersatzstation

Vergleichen wir die Kurzschlußverhalten der Netze 1,2,3 einerseits und der zu konstruierenden Ersatzstation andererseits, dann ergibt sich (bekannterweise!):

- für alle Populationen n sind die Durchsätze der Netze 1,2,3 und der Ersatzstation identisch und damit auch die mittleren Verweilzeiten (je Kurzschluß-"Umlauf");
- für jede Population sind ansonsten die Verweilzeit**verteilungen**der Netze 1,2,3 untereinander verschieden und auch verschieden von jener der Ersatzstation.

Die Verweilzeitverteilung an der Ersatzstation liegt fest, wenn wir die bisher noch offene Struktur dieser Station festlegen. Zur Wahl stehen die separablen Stationstypen PS und IS (diese sind hier bis auf Benennungen gleichwertig), FCFS und LCFS-PR. Offensichtlich erhalten wir für einen PS/IS-Bediener exponentieller Bedienzeit im Kurzschluß für jede Population eine exponentiell verteilte Verweilzeit; für FCFS ergäbe sich andererseits (bei Population n) eine Erlang(n)-verteilte Verweilzeit; wir konzentrieren uns im folgenden auf die klassische PS/IS-Struktur der Ersatzstation.

Die Natur der Verweilzeitverteilungen der kurzgeschlossenen Netze 1,2,3 konnte in Form ihrer approximativen Dichtefunktion sowie der zugehörigen Variationskoeffizienten (gemäß Abschn.2.2.) wie folgt bestimmt werden. Als Benennungen mögen dienen:

- $VK_{(i,n)}$ Verweilzeit im Kurzschluß; Netz i, $i \varepsilon \{1,2,3\}$; Population n, $n \varepsilon \{1,2,3,4,5\}$;
- fVK., FVK., E[VK.], C[VK.] zugeordnete Dichtefunktion, Verteilungsfunktion, Erwartungswert, Variationskoeffizient.

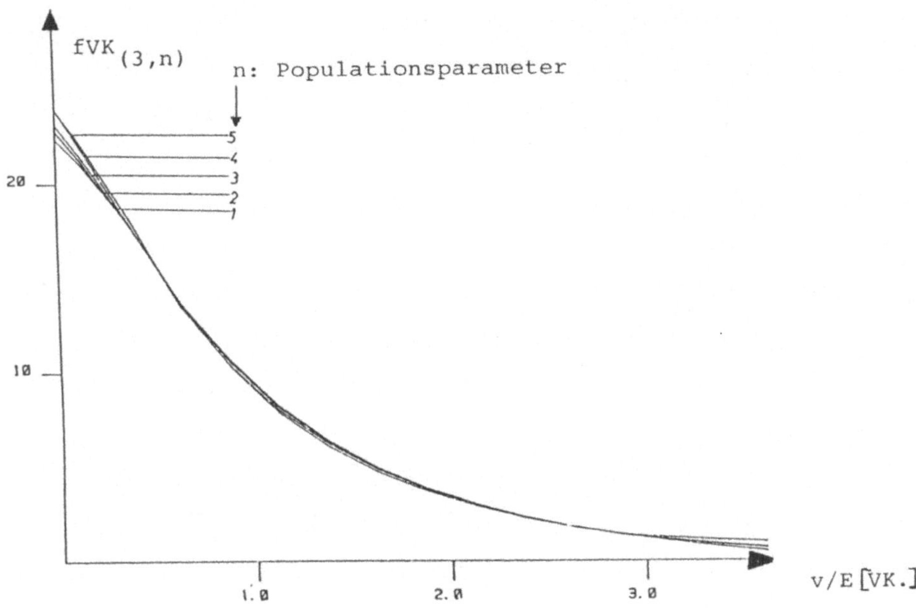

Abb. 3.4c: Verweilzeitdichte im Kurzschluß, Netz 3

Population n	1	2	3	4	5
Netz 1	0.71	0.74	0.76	0.77	0.79
Netz 2	4.58	4.97	5.35	5.71	6.05
Netz 3	0.98	0.98	0.98	0.98	0.98

Tab. 3.5.: Variationskoeffizienten $C[VK_{(i,n)}]$ der Verweilzeitverteilungen der kurzgeschlossenen Netze i bei variierender Population n

Bezeichne zusätzlich $VK_{(E,n)}$ die Verweilzeit (im Kurzschluß) einer konstruierten Ersatzstation vom PS/IS-Typ mit exponentieller Bedienzeit, dann sind zum Vergleich der Kurzschlußverhalten die Abb. 3.4 a)-c) zu vergleichen mit der bekannterweise exponentiellen Verweilzeit-Dichte $fVK_{(E,n)}(v)$, bzw. Tab. 3.5 mit deren Variationskoeffizienten $C[VK_{(E,n)}] \equiv 1$. Zur zusätzlichen Quantifizierung der Unterschiede in Dichte- und Verteilungsfunktionen zwischen Ausgangsnetz und Ersatzdarstellung definieren wir folgende Distanzmaße

$$d(A,B) = \sum_{t_j \varepsilon T} |fA(t_j)-fB(t_j)|, \qquad D(A,B) = \sum_{t_j \varepsilon T} |FA(t_j)-FB(t_j)|,$$

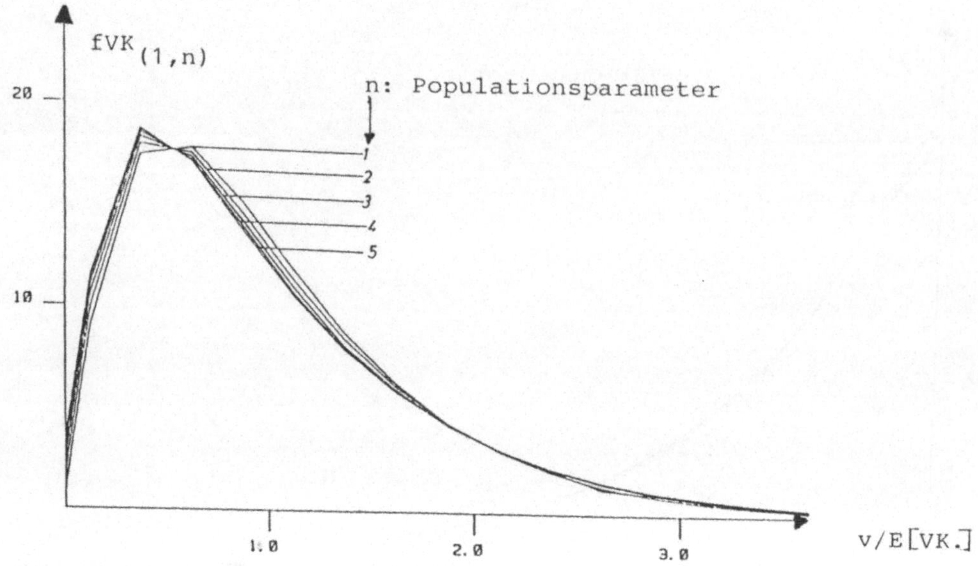

Abb. 3.4a: Verweilzeitdichte im Kurzschluß, Netz 1

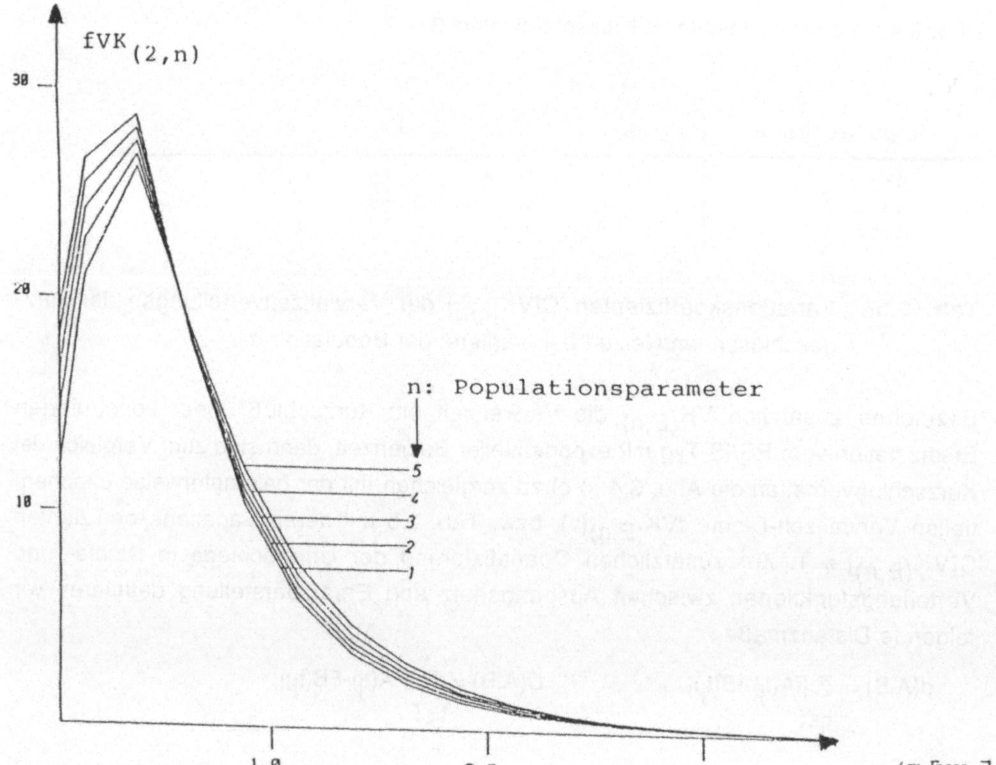

fVK(2,n)
30
20
10
n: Populationsparameter
5
4
3
2
1
1.0
2.0
3.0
v/E [VK.]

Abb. 3.4b: Verweilzeitdichte im Kurzschluß, Netz 2

wo T eine geeignete Menge von Stützstellen umfaßt; konkret verwendet werden im weiteren die 10 äquidistant liegenden Stützstellen $i \cdot 0.25 \cdot E[V.]$, i=1..10 (E[V.] ist in allen Fällen identisch).Sei nun $dK_{(i,n)} := d(VK_{(i,n)}, VK_{(E,n)})$ das Distanzmaß der Dichten und entsprechend $DK_{(i,n)}$ das Distanzmaß der Verteilungsfunktionen, dann ergeben sich die Werte

Population n	1	2	3	4	5	Zeilen-Σ
$dK_{(1,n)}$	30,9	28,4	26,5	25,3	24,4	135,5
$dK_{(2,n)}$	25,9	27,0	29,8	32,5	35,0	150,2
$dK_{(3,n)}$	3,5	3,2	3,0	3,0	2,5	15,1
$DK_{(1,n)}$	56,2	50,7	46,0	42,8	40,6	195,7
$DK_{(2,n)}$	98,5	106,4	115,3	124,4	133,0	577,7
$DK_{(3,n)}$	4,5	4,1	3,7	3,4	3,1	18,8

Tab.3.6.: Distanzmaße $dK_{(i,n)}$ und $DK_{(i,n)}$

Zusammenfassend läßt sich festhalten:

- Die Verweilzeitverteilungen der beiden nicht zyklisch durchlaufenen (p=0) Netze 1 und 2 unterscheiden sich deutlich von jener der PS/IS-Ersatzstation exponentieller Bedienzeit; dies erwartetermaßen trotz der wechselseitig identisch übereinstimmenden Durchsätze ("flow-equivalence") und Verweilzeit-Erwartungswerte;
- Die Verweilzeitverteilung des zyklisch durchlaufenen (p=0.9) Netzes 3 unterscheidet sich kaum von jener der Ersatzstation; auch dies wäre von vornherein zu erwarten gewesen, da bekanntermaßen (z.B. SALA81,COUR77) die Verweilzeit in zyklisch durchlaufenen Netzteilen sich mit der mittleren Zykluszahl einer Exponentialverteilung annähert (die ja andererseits für den Ersatzbediener über seine Struktur "gesetzt" wurde).

Es steht zu erwarten, daß sich eine mangelhafte Paßgüte der Verweilzeitverteilung der Ersatzstation vom Kurzschlußfall auf den Fall der Einbettung in ein umgebendes Netz fortsetzt.

3.2. Einbettung in separable Umgebungen

Das Subsystem der Abb. 3.1 wird jetzt in eine Umgebung so eingebettet, daß das entstehende Gesamtnetz insgesamt separabel ist. Für eine solche Umgebung ist die in Abschnitt 3.1 konstruierte klassische Ersatzstation exaktes Äquivalent des Subsystems im Sinne der stationären Populationsverteilungen: Bei Ersetzung des Subsystems durch die Ersatzstation werden die stationären Durchsätze durch Umgebung und

Subsystem exakt reproduziert, ebenso die Aufteilung der Kunden auf Umgebung/ Subsystem und, daraus folgend, die mittleren Verweilzeiten in sowohl Umgebung als auch Subsystem. **Keine** Äquivalenz dagegen ist zu erwarten zwischen den Verweilzeit-**verteilungen** am Subsystem und an der Ersatzstation. Als Beispiel einer separablen Einbettung diene die Netzstruktur in Abb. 3.7. Dieses Netz 3.7 wird untersucht

- für eine konstante Population von 5 Kunden
- alternativ für populationsunabhängige s $\varepsilon \{1,4,8\}$ sowie
  für zwei populationsabhängige Fälle s(i)=i und s(i)=5-i, i $\varepsilon \{1,2,3,4\}$
- alternativ für die Subsystemkonfigurationen Netz 1, Netz 2, Netz 3, exponentielle PS/IS-Ersatzstation.

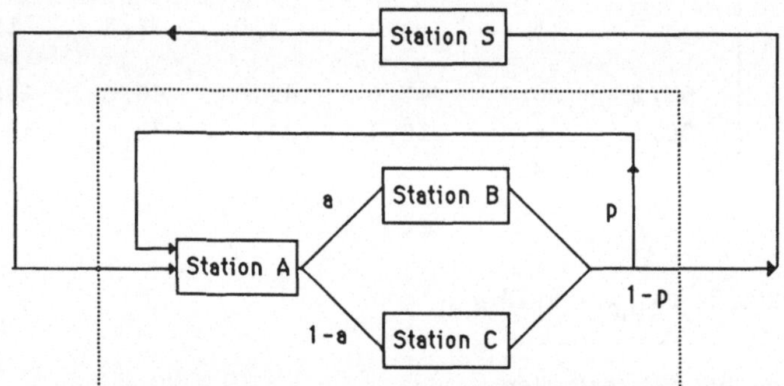

Station S: Struktur FCFS, Bediengeschwindigkeit s (variabel)
Bedienzeit exponentiell, Erwartungswert 1

Abb. 3.7.: Einbettung des Teilnetzes aus Abb. 3.1

Ermittelt werden Charakteristika der Verteilungen von

$VS_{(i,s)}$	Verweilzeit im Netz i, i $\varepsilon \{1,2,3\}$, bei Einbettung gemäß Abb. 3.7 mit 5 alternativen s-Fällen
$VS_{(E,s)}$	Verweilzeit im Ersatzbediener (PS/IS, exp. Bedienzeit), bei Einbettung gemäß Abb.3.7 mit alternativen s

im Hinblick auf die auftretenden Unterschiede zwischen $VS_{(i,s)}$, i=1,2,3 einerseits und $VS_{(E,s)}$ andererseits.

Erwartungsgemäß (vgl. Tab. 3.8) spiegeln sich für Netze 1 und 2 die in Abschn. 3.1. für den Kurzschlußfall ermittelten großen Abweichungen der Verweilzeitverteilungen zu jener der Ersatzstation (Tab.3.6) in großen Abweichungen der Verweilzeitverteilungen bei separabler Einbettung wider, wohingegen für Netz 3 die gute Paßgüte der Ersatzstation im Kurzschlußfall sich auch in einer guten Entsprechung bei separabler Einbettung niederschlägt.

s	populationsunabhängig 1.0	4.0	8.0	populationsabhängig steigend 1.0->4.0	fallend 4.0->1.0
$C[VS_{(1,s)}]$	0,82	0,79	0,79	0,81	0,79
$C[VS_{(2,s)}]$	5,52	5,94	6,00	5,94	5,23
$C[VS_{(3,s)}]$	1,03	0,99	0,98	1,01	0,99
$C[VS_{(E,s)}]$	1,05	1,01	1,00	1,02	1,01
$dS_{(1,s)}$	26,7	24,4	24,4	24,4	25,2
$dS_{(2,s)}$	48,6	81,1	86,4	77,4	69,9
$dS_{(3,s)}$	2,8	2,5	2,0	3,4	3,7
$DS_{(1,s)}$	43,5	40,4	40,7	39,1	37,4
$DS_{(2,s)}$	110,0	128,7	131,2	120,5	126,3
$DS_{(3,s)}$	3,6	3,4	3,2	3,4	2,2

Tab.3.8.: Unterschiede zwischen $VS_{(i,s)}$, i=1,2,3, und $VS_{(E,s)}$

C[ ] : Variationskoeffizienten; d( ), D( ) : Distanzmaße der Verteilungen

3.3 Einbettung in nicht-separable Umgebungen

Als Beispiel einer Einbettung in eine nicht-separable Umgebung wird erneut die Netzstruktur der Abb. 3.7 verwandt. Im Gegensatz zu Abschnitt 3.2 ist für Station S eine FCFS-Station **nicht**- exponentieller Bedienzeit gewählt. Das resultierende Gesamtnetz verletzt damit in allen drei Konfigurationen (Netze 1,2,3 als Subsystem) die Voraussetzungen separabler Netze. Damit ist auch die Grundlage für eine Flußäquivalenz des in Abschnitt 3.1 konstruierten Ersatzbedieners beseitigt und die Annahme einer Ersetzungsmöglichkeit der Netze i durch den Ersatzbediener (unter Erhaltung auch nur des Durchsatzes, der Populationsverteilungen und der mittleren Verweilzeiten) durch nichts mehr gerechtfertigt.

Bei genauerer Betrachtung ergibt sich allerdings für Netz 3 als Subsystem doch eine Rechtfertigung für die Annahme zumindest approximativer Flußäquivalenz: Die Wahl p=0.9 für den Zyklusparameter des Netzes bewirkt nämlich eine NCD-artige Ankopplung von Netz 3 an Station S und damit die Rechtfertigung der Annahme approximativer Flußäquivalenz der Ersatzstation (bzgl. **N**ear-**C**omplete **D**ecomposability siehe COUR77). Da wir nicht wie bei separablen Umgebungen eine Startverteilung auf Basis stationärer Teilnetz-Verteilungen zugrunde legen können, haben wir auf eine Bestimmung approximativer Dichte- oder Verteilungsfunktionen verzichtet und präsentieren im folgenden lediglich Mittelwerte von Leistungsmaßen, die via numerisch ermittelter stationärer Grenzverteilung gewonnen werden, vgl. Ende Abschn. 2.1.

Im Beispiel wurde für Station S angenommen:
- FCFS-Disziplin,
- zweiphasige COX-Bedienzeit mit Mittelwert 1 und Variationskoeffizient 10,
- Bediengeschwindigkeit in 5 Alternativen:
  populationsunabhängig: 1,4 und 8 sowie
  populationsabhängig: i und 6-i bei i ε {1,..,5} Kunden

Tab. 3.9 gibt die stationären Durchsätze und die mittleren Verweilzeiten bezüglich des Subsystems wieder, wobei als Subsystem alternativ die Netze 1,2,3 sowie der in Abschn. 3.1 konstruierte Ersatzbediener Verwendung finden und die Umgebungsstation S über die 5 Geschwindigkeitsalternativen variiert.

Bedien-geschwindigkeit von S	Leistungs-größe	Einbettung von Netz 1	Netz 2	Netz 3	Ersatz-station
1	Durchsatz	0.78	0.83	0.78	0.78
	Verweilzeit	2.41	2.18	2.40	2.40
4	Durchsatz	1.39	1.51	1.40	1.41
	Verweilzeit	2.67	2.53	2.65	2.65
8	Durchsatz	1.60	1.68	1.61	1.61
	Verweilzeit	2.70	2.62	2.68	2.68
1->5	Durchsatz	1.42	1.48	1.43	1.43
	Verweilzeit	2.52	2.37	2.50	2.50
5->1	Durchsatz	0.84	1.02	0.84	0.85
	Verweilzeit	2.68	2.51	2.66	2.65

Tab. 3.9.: Stationäre Durchsätze und Verweilzeitmittel am Subsystem in seinen diversen Varianten, bei verschiedenen nicht-separablen Umgebungsvarianten

Erwartungsgemäß ergibt sich nur für die Ersetzung von Netz 3 durch die exponentielle Ersatzstation weitgehende Übereinstimmung der ermittelten Leistungsgrößen, während bei Ersetzung der Netze 1 und 2 Verfälschungen diverser Art auftreten.

4. Ersatzsysteme höherer Paßgüte

Die nicht allgemein befriedigenden Ergebnisse des Abschnitts 3 mit Ersatzstationen des exponentiellen PS/IS-Typs geben Anlaß, nach "besseren" Ersatzsystemen/Aggregaten zu fahnden. In diesem Abschnitt werden Aggregate vorgeschlagen und beispielhaft untersucht, die nach folgenden Maximen konstruiert werden:

- Wegen der Annehmlichkeit, die separable Teilnetze zumindest bei Einbettung in gewisse Umgebungen mit sich bringen, werden nur Aggregate berücksichtigt, die (für sich alleine) separabel sind;

Es werden Aggregate gesucht, die im Kurzschlußbetrieb die Verweilzeitverteilung des zu ersetzenden Subsystems (im Kurzschluß, für interessierende Populationen) möglichst gut reproduzieren.

Kandidaten für solche Aggregate sind offensichtlich:

(i) separable Stationen nicht-exponentieller Bedienzeit; hier bieten sich zustandsabhängige PS/IS-Stationen mit allgemeiner COX-Bedienzeit unmittelbar an;

(ii) separable Subsysteme, die mehr als eine Station umfassen; diese sollten natürlich "kleiner/einfacher" sein als das zu ersetzende Subsystem, weshalb hier (im Hinblick auf das vorliegende 3-Stationen-Beispiel) ausschließlich 2-Stationen-Aggregate in Betracht kommen.

Dabei hat Vorgehen (ii) von vornherein bessere Chancen auf Erfolg, da ja die Aggregate des Typs (i) mit ihrer Verweilzeit im Kurzschluß die gewählte Bedienzeitverteilung (bis auf eine populationsabhängige Dehnung/Stauchung der Zeitachse) exakt reproduzieren, während sich bei den zu ersetzenden Subsystemen auch die Form der Verweilzeitverteilung populationsabhängig verändert (vgl. Abb. 3.4). Andererseits ist das Vorgehen im Falle (ii) insofern schwieriger, als zunächst Ersatzsysteme gefunden werden müssen, die im Kurzschlußfalle die Verweilzeitverteilungen populationsabhängiger Form "gut" nachvollziehen. Dieser Schritt wurde im wesentlichen heuristisch vollzogen, indem zunächst ein breites Spektrum separabler Zwei-Stationen-Subsysteme im Kurzschluß bzgl. ihrer Verweilzeitverteilungen und deren Populationsabhängigkeit untersucht wurden und die konkrete Auswahl eines Ersatzsystems auf Basis dieses "Katalogs" erfolgte (weitere Details s. BUCH86).

In jedem Falle aber ist das Aggregat gemäß (i) oder (ii) letztlich so parametrisiert, daß es im Kurzschluß (und daher auch bei **jeder** separablen Einbettung) die stationären Durchsätze, die Populationsverteilungen Subnetz/Restnetz und (folglich!) die mittleren Verweilzeiten in Sub- und Restnetz exakt wiedergibt.

4.1. Aggregierung von Netz 1

Folgende Ersatzdarstellungen für Netz 1 werden präsentiert:

E1.2: PS/IS-Ersatzstation mit Erlang(2)-Bedienzeitverteilung und populationsabhängiger Bediengeschwindigkeit.

AGG1: Zwei seriell angeordnete PS/IS-Stationen mit identischen exponentiellen Bedienzeiten und identisch populationsabhängigen Bediengeschwindigkeiten.

Zur Beurteilung der Ersetzungen wurden (wie in Abschn. 3.2) die Verweilzeitverteilungen bei separabler Einbettung überprüft; als Ergebnis hier nur kurz die Feststellung, daß erwartungsgemäß E1.2 und AGG1 sich beide "besser" verhalten, als die exponentielle Ersatzstation E (Details s. BUCH86). Ferner wurden nichtseparable Einbettungen untersucht (analog Abschn. 3.3 und mit denselben "Umgebungen"). Tab. 4.1 gibt die entsprechenden Ergebnisse wieder.

Bedien-geschwindigkeit von S	Leistungs-größe	Original-Netz	Netz 1 ersetzt durch E	E1.2	AGG1
1	Durchsatz	0.78	0.78	0.78	0.78
	Verweilzeit	2.41	2.40	2.41	2.41
4	Durchsatz	1.39	1.41	1.40	1.39
	Verweilzeit	2.67	2.65	2.66	2.67
8	Durchsatz	1.60	1.61	1.60	1.60
	Verweilzeit	2.70	2.68	2.69	2.70
1->5	Durchsatz	1.42	1.43	1.42	1.42
	Verweilzeit	2.52	2.50	2.51	2.52
5->1	Durchsatz	0.84	0.85	0.83	0.84
	Verweilzeit	2.68	2.65	2.66	2.67

Tab. 4.1.: Stationäre Durchsätze und Verweilzeitmittel an Netz 1 und seinen Ersetzungen, bei verschiedenen nicht-separablen Einbettungen ("Bediengeschwindigkeit von S")

Der erwartete Erfolg stellt sich tatsächlich ein: Bei Ersetzung des Netzes 1 durch sowohl Ersatzstation E1.2 als auch Aggregat AGG1 ist das entstehende Gesamtnetz bezüglich Durchsatz und Verweilzeit "ähnlicher" dem Originalnetz als dies bei Ersetzung durch die klassische Ersatzstation E der Fall ist; dabei ist, ebenfalls erwartungsgemäß, AGG1 (etwas) besser als E1.2.

4.2. Aggregierung von Netz 2

Für Netz 2 werden folgende Ersatzdarstellungen präsentiert:

E2.2: PS/IS-Ersatzstation mit 2-phasiger COX-Bedienzeitverteilung, Variationskoeffizient 5.0, populationsabhängige Bediengeschwindigkeit.

AGG2: Serienschaltung aus der (unveränderten) Station A des Netzes 2 (μ_A=2, C_A=10) und einer exponentiellen, populationsabhängigen PS/IS-Station.

Zur Beurteilung der Ersetzung hier wieder nur der Vergleich stationärer Durchsätze und Verweilzeitmittel bei nicht-separabler Einbettung analog Abschnitt 3.3.

Bedien-geschwindigkeit von S	Leistungs-große	Original Netz	Netz 2 ersetzt durch E	E2.2	AGG2
1	Durchsatz	0.83	0.78	0.83	0.83
	Verweilzeit	2.18	2.40	2.15	2.18
4	Durchsatz	1.51	1.41	1.53	1.51
	Verweilzeit	2.53	2.65	2.51	2.53
8	Durchsatz	1.68	1.61	1.70	1.68
	Verweilzeit	2.62	2.68	2.61	2.62
1->5	Durchsatz	1.48	1.43	1.49	1.48
	Verweilzeit	2.37	2.50	2.35	2.37
5->1	Durchsatz	1.02	0.85	1.06	1.02
	Verweilzeit	2.51	2.65	2.48	2.51

Tab. 4.2.: Stationäre Durchsätze und Verweilzeitmittel an Netz 2 und seinen Ersetzungen, bei verschiedenen nicht-separablen Einbettungen

Deutlicher noch als für das Netz 1 des Abschnitts 4.1 zeigt sich hier für Netz 2 die Überlegenheit der Ersatzstation E2.2 und des Aggregats AGG2 gegenüber der klassischen Ersatzstation E; deutlicher auch ist die Überlegenheit von AGG2 gegenüber E2.2. Das heuristische Vorgehen der Auswahl von Aggregaten auf der Basis ihrer Verweilzeitverteilung im Kurzschluß ist insgesamt, zumindest im Beispiel, bestätigt.

5. Resumée

Die skizzierte Art der Untersuchung von Teilen von Warteschlangennetzen und diesbezüglicher Aggregate läßt sich in der vorgetragenen Detaillierung sicher nur für größenmäßig deutlich beschränkte Teilnetze durchführen. Andererseits sehen wir uns durch die erzielten Ergebnisse in der Vermutung bestärkt, daß die Beachtung der Charakteristika der Verweilzeitverteilung an Teilnetzen im Kurzschlußbetrieb und der Versuch einer weitgehenden Reproduktion dieses Kurzschlußverhaltens durch zu konstruierende Aggregate zur Verringerung von Aggregationsfehlern führt. Eine heuristische Nutzung dieses Gedankens auch in komplexeren Fällen ist sicherlich erprobenswert (womit auch naheliegende Fortsetzungsarbeiten charakterisiert sind); ebenso sollte die Möglichkeit der Ersetzung eines Teilnetzes durch ein einfaches Zwei- (oder auch Mehr-) Stationen-Aggregat (anstatt der üblichen "Ersatzstation") als hoffnungsträchtig in Erinnerung bleiben. Letztlich sei darauf hingewiesen, daß angesichts des Mangels an exakten Resultaten im betrachteten Problemfeld die Verfügbarkeit numerischer Löser für Markov-Ketten eine nahezu zwingende Voraussetzung für unsere Arbeiten war (und bleibt).

6. Literatur

BABG85 Balbo,G.; Bruell, S.C.; Ghanta,S.: Combining Queueing Network and Generalized Stochastic Petri Net Models for the Analysis of a Software Blocking Phenomenon, Proc. of the Int. Workshop on Timed Petri Nets, Turin 1985 (IEEE Computer Society Press)

BABR83 Balbo,G.; Bruell, S.C.: Computational Aspects of Aggregation in multiple class queueing Networks, Performance Evaluation, vol.3, Nr.3, 1983

BESC85 Beilner,H.; Scholten,H.: Strukturierte Modellbeschreibung und strukturierte Modellanalyse: Konzepte des Modellierungswerkzeugs HIT, in: BEIL85

BEST87 Beilner ,H.; Stewing, F.J.: Concepts and Techniques of the Performance Modelling Tool HIT, European Simulation Multiconference, Wien, Juli 1987

BEIL85 Beilner,H.(Editor): 3.GI/NTG-Fachtagung "Messung, Modellierung und Bewertung von Rechensystemen", Dortmund 1985,Informatik Fachberichte 110 (Springer)

BOXM84 Boxma,O.J.: Analysis of succesive cycles in a cyclic queue, in Bux,W.; Rudin, H.(Editors), Performance of Computer-Communication Systems, 1984, North Holland

BUCH86 Buchholz,P.: Aggregierung von Warteschlangennetzen unter Berücksichtigung der Durchlaufzeitverteilung, Diplomarbeit, Fachbereich Informatik, Universität Dortmund 1986

CHSA78 Chandy,K.M.; Sauer,C.H.: Approximate Methods for analyzing queueing network models of computing systems, Computing surveys, vol. 10, No. 3, September 1978

CHHW75 Chandy,K.M.; Herzog,U., Woo,L.: Parametric analysis of queueing networks, IBM Journal Res. Dev. 19, 1, Januar 1975

COUR77 Courtois,P.J.: Decomposability: Queueing and Computer System Applications, Academic Press 1977

DADU85 Daduna,H.:The distribution of Residence Times and Cycle Times in a Closed Tandem of Processor Sharing Queues, in BEIL85

GAED77 Gaede,K.W.: Zuverlässigkeit: Mathematische Modelle, Carl Hanser 1977

GRAS76 Grassmann,W.: Transient solutions in Markovian queueing systems, Computers and Operations Research, vol. 4, 1977

HOYA86 Ho,Y.-C.; Yang,P.Q.: Equivalent Network, Load Dependent Servers and Perturbation Analysis - An Experimental Study, in Boxma, O.J.; Cohen,J.W.; Tijms,H.C. (Editors), Teletraffic Analysis and Computer Performance Analysis, 1986, North Holland,

KRWK82 Kritzinger, P.S.; Wyk, van S.; Krzesinski, A.E.: A Generalisation of Norton´s Theorem for Multiclass Queueing networks,Performance Evaluation vol 2, Nr. 2, 1982, North Holland

KUEH83 Kühn,P.: Analysis of Busy Period and Response Time Distributions in Queueing Networks, in KUSC83

KUSC83 Kühn,P.; Schulz,K.M. (Editors): 2.GI/NTG-Fachtagung "Messung, Modellierung und Bewertung von Rechensystemen", Stuttgart 1983, Informatik Fachberichte 61, Springer

MEYA84 Melamed,B; Yadin,M.: Numerical computations of soujourn time distributions in queueing networks, Journal of the ACM, vol. 31, 1984

MUEL84 Müller-Clostermann ,B.: NUMAS - A tool for the numerical analysis of computer systems, in POTI84

MURO87 Müller-Clostermann,B.; Rosentreter,G.: Synchronized Queueing Networks - Concepts, Examples and Evaluation Techniques, this volume

POTI84 Potier,D.(editor): Proc. of the Int. Conference on Modeling Techniques and Tools for Performance Analysis, Paris 1984, North Holland

SALA81 Salza,S.; Lavenberg,S.S.: Approximating response time distributions in closed queueing models of computer performance, in: Kylstra,F.J.(ed.), Proceedings of Performance·81, 1981, North Holland 1981

SCHA85 Schassberger,R.: Exact Results on Response Time Distributions in Networks of Queues, in BEIL85

SEMI81 Sevcik,K.C.; Mitrani, I.: The distribution of queueing network states at input and output instants, Journal of the ACM 28(1981), p.358-371

STLZ77 Sevcik,K.C.; Levy,A.I., Tripathi, J.L.; Zahorjan, J.L.: Improving approximations of aggregated queueing network subsystems, in: Chandy,K.M.; Reiser,M.(eds.), Computer Performance, 1977, North Holland

Kopplung der Kettendurchsätze in geschlossenen

Warteschlangennetzwerken

Gunter Totzauer
Nixdorf Computer AG
Pontanusstr. 55
D - 4790 Paderborn

Zusammenfassung: Im industriellen Alltag treten oft Aufgabenstellungen auf, die eine Analyse des Verhaltens eines Rechensystems bei vorgegebenen Durchsätzen der verschiedenen Anwendungen oder deren Verhaltnissen zueinander erfordern. Die Modellierung als geschlossenes Netz erlaubt normalerweise diese Vorgaben der Durchsätze nicht. Die Erganzung eines geschlossenen Warteschlangennetzwerkes um eine Verzögerungsstation ermoglicht jedoch die Einhaltung dieser Vorgaben in vielen Fällen. Ein aus der Bard/Schweitzer-Approximation abgeleiteter Algorithmus zur Berechnung der erforderlichen Verzogerungszeiten der verschiedenen Ketten wird angegeben.

1. Einleitung

Für die Klasse separabler WS-Netze /BCMP75/ existieren effiziente Losungsverfahren /ReKo75/ und fur die Behandlung großerer Modelle geeignete Approximationsverfahren /Schw79,NeCh81/.

Die Auftragsketten eines separablen Netzes heißen offen oder geschlossen je nachdem, ob Auftragsubergänge zwischen Außenwelt und System möglich sind, oder ob die Auftragszahl in einer Kette konstant ist. Fur offene Ketten sind die Durchsätze als Parameter vorzugeben und die mittlere Auftragsanzahl dieser Kette ergibt sich als Ergebnis. Bei geschlossenen Ketten wird die Auftragsanzahl vorgegeben und der Durchsatz ist eine Ergebnisgroße.

Die begrenzte Terminalanzahl eines Rechensystems oder der festgelegte Multiprogrammiergrad legen die Modellierung des Systems als geschlossenes WS-Netz häufig nahe.

Folgende Fragestellungen, die mit Hilfe von Modellen beantwortet werden sollen, treten in der Industrie auf.

- Vor dem Kauf einer Anlage möchte der Kunde wissen, wie sich das System bei einer fest vorgegebenen Last verhalt. Die Durchsatze fur die verschiedenen Anwendungstypen sind also fest vorgegeben.

- Vor der Erweiterung eines Systems mochte der Kunde wissen, welchen Durchsatzfaktor die Systemerweiterung mit sich bringt. Dabei sollen die Durchsatze der verschiedenen Anwendungen im gleichen Verhältnis bleiben wie bisher.

Diese Probleme sind wegen der Durchsatzvorgaben nicht direkt auf geschlossene WS-Netze abbildbar. Nimmt man eine zusatzliche Verzogerungsstation hinzu, deren Bedienzeiten nicht vorgegeben sind, so gewinnt man neue Freiheitsgrade. Die Wahl der Bedienzeiten soll nun so erfolgen, daß die Durchsatzvorgaben erfullt werden. Das Verfahren gestattet die Behandlung geschlossener separabler WS-Netze.

Die Verzögerung von Aufträgen wurde schon von Gelenbe und Kurinckx /GeKu76/ zur Verhinderung von Überlastsituationen im virtuellen Speicher verwendet. Ein bereits lange im Einsatz befindliches Simulationsmodell bei der Nixdorf Computer AG erlaubt auch die Vorgabe von Durchsatzen fur verschiedene Auftragstypen.

Burke /Burk85/ fordert, daß bei Analyseverfahren jede Variable als Ergebnisgröße möglich sein soll. Die Verwendung herkommlicher Methoden erfordert die Iteration über vollständige Analysen, fur die die Eingabeparameter zielgerichtet verandert werden, bis der gewünschte Effekt eingetreten ist. Hier liefert die Durchsatzsteuerung und -kopplung einen Beitrag zur direkten Behandlung solcher Fragestellungen.

Dieser Aufsatz stellt zunächst noch einmal das Schema des Bard/Schweitzer-Verfahrens kurz vor, um dann die erforderlichen Anderungen zur Behandlung der Durchsatzsteuerung und -kopplung dort einzubringen. Die Darstellung verschiedener Einsatzmöglichkeiten und Sichtweisen des vorgestellten Verfahrens beschließt die Arbeit.

2. Durchsatzbeeinflussung in geschlossenen WS-Netzen

2.1 Terminologie

Der Formalismus des Bard/Schweitzer-Verfahrens braucht hier nicht in Details erwahnt zu werden. Deshalb werden im folgenden auch nur die notwendigen Bezeichner aufgelistet.

S Stationsanzahl

K Kettenanzahl

h(k,s) mittlere Anzahl der Teilaufträge der Kette k an Station s pro Auftrag der Kette k

b(k,s) mittlere Bedienzeit der Teilaufträge der Kette k an Station s

v(k,s) mittlere Verweilzeit der Teilaufträge der Kette k an Station s

V(k) mittlere Gesamtverweilzeit eines Auftrages der Kette k

N(k) Anzahl der Aufträge der Kette k

D(k) Durchsatz der Aufträge der Kette k

n(k,s) mittlere Anzahl der Aufträge der Kette k an Station s

u(s) Auslastung der Station s

2.2 Bard/Schweitzer-Approximation

Die Bard/Schweitzer-Approximation lauft nach folgendem Schema ab:

```
Bard/Schweitzer-Algorithmus
{
    Initialisiere;
    while ( Ergebnisse nicht genau genug ) do
    {
        setze_Verweilzeiten;
        setze_Durchsätze;        /* nach Little */
        setze_Anzahlen;
    }
}
```

Die einzelnen Prozeduren ubernehmen dabei folgende Aufgaben:

setze_Verweilzeiten	berechnet und setzt die Verweilzeiten v(k,s) für alle Ketten und Stationen sowie die V(k) gemaß der Bard/Schweitzer-Formel
setze_Durchsätze	berechnet und setzt die Kettendurchsatze D(k) fur alle Ketten nach Little
setze_Anzahlen	berechnet und setzt die mittleren Anzahlen n(k,s) für alle Ketten und Stationen nach Little

2.3 Durchsatzsteuerung

Hier sollen nun fest vorgegebene Durchsätze D(k) in einem Netz eingestellt werden, indem die Bedienzeiten an einer zusatzlichen Verzögerungsstation, die den Stationsindex 0 erhält, geeignet gewahlt werden. Die Verweilzeiten der normalen Stationen lassen sich durch die Lösung von linearen Gleichungssystemen /ScTo85/ oder auch durch Formeln /ToSc87/ angeben, falls keine zustandsabhängigen Bedienraten auftreten.

Eine weitere Möglichkeit, die Verzögerungszeiten zu bestimmen, besteht darin, gemäß dem Iterationsschema vorzugehen, die Durchsatze jedoch am Anfang einmal auf die vorgegebenen Werte zu setzen und dann unverandert zu lassen, was dazu führt, daß im Gegensatz zum ursprünglichen Schema das Setzen der Durchsatze aus der Schleife herausgenommen und an den Anfang gesetzt wird.

Dies liefert dann für die ursprünglichen Stationen des Netzes die Verweilzeiten. Es fehlen noch die der Verzogerungsstation 0. Nach Little gilt fur jede Kette k:

$$\sum_{s=0}^{S} h(k,s)\ v(k,s) = N(k)\ /\ D(k) \qquad (1)$$

O.B.d.A. gelte an der Verzögerungsstation h(k,0)=1. Damit erhält man aus (1) für die Verzögerungszeiten v(k,0):

$$v(k,0) = N(k)\ /\ D(k) - \sum_{s=1}^{S} h(k,s)\ v(k,s) \qquad (2)$$

Die Berechnung dieser Verzögerungszeiten erfolge durch die Prozedur setze_Verzögerungen.

Das Schema zu Berechnung der Verzogerungszeiten fur ein Netz mit gegebenen Durchsatzen hat dann die Form:

```
Durchsatzsteuerungs-Algorithmus
{
    Initialisiere;
    setze_Durchsatze;          /* auf vorgegebene Werte */
    while ( Ergebnisse nicht genau genug ) do
    {
        setze_Verweilzeiten;
        setze_Anzahlen;
    }
    setze_Verzögerungen;
}
```

Nicht immer ist ein Netz mit den geforderten Durchsatzen realisierbar. Zwei Bedingungen mussen erfullt sein. Fur eine Station s mit nur einem einzigen Bediener ist die Auslastung auf 1 begrenzt. Wegen der vorgegebenen Durchsätze D(k) kann die Auslastung u(s) bestimmt werden und es muß gelten:

$$u(s) = \sum_{k=1}^{K} D(k)\ h(k,s)\ b(k,s) \quad <= \quad 1 \qquad (3)$$

Eine weitere Bedingung stellt die Forderung dar, daß die Verweilzeiten nicht negativ sein durfen:

$$v(k,s) \quad >= \quad 0 \qquad (4)$$

Insbesondere bei der Bestimmung der Verzogerungszeiten laut Formel (2) konnen rein rechnerisch jedoch negative Terme auftreten. Ist eine der Forderungen (3),(4) verletzt, so ist das Netz mit den geforderten Durchsatzen nicht realisierbar.

2.4 Durchsatzkopplung

Bei der Durchsatzkopplung ist es moglich, die Verhaltnisse der Kettendurchsätze zueinander gemäß vorgegebener Werte einzustellen.

Zunächst wird beschrieben, wie die Abhängigkeit ausgedrückt werden kann. Soll der Durchsatz der Kette kk zu dem der Kette k in Beziehung stehen, so sei dies durch

$$bezug(kk) \quad = \quad k \qquad (5)$$

ausgedrückt.

Soll der Durchsatz einer Kette absolut vorgegeben werden, so gilt:

$$\text{bezug}(kk) = 0 \qquad (6)$$

Die absolut vorgegebenen Teile der Last definieren einen fixen Teil der Last. Dieser wird im folgenden als Fixanteil bezeichnet. Ist fur alle Ketten der Durchsatz absolut vorgegeben, so liegt ein Problem der Durchsatzsteuerung als Spezialfall der Durchsatzkopplung vor.
Soll fur die durch (5) ausgedrückte Beziehung der Ketten k und kk

$$D(kk) / D(k) = c \qquad (7)$$

gelten, so sei

$$\text{koeff}(kk) = c \qquad (8)$$

Wird eine Kette kk zur Kette k in Beziehung gesetzt, so darf die Kette k sich nicht auf eine weitere Kette beziehen. Es muß gelten:

$$\text{bezug}(kk) = k \Rightarrow \text{bezug}(k) = k \qquad (9)$$

Es können also verschiedene Gruppen von Ketten gebildet werden, die sich jeweils auf ein Gruppenmitglied beziehen, untereinander aber nicht weiter in Beziehung stehen.

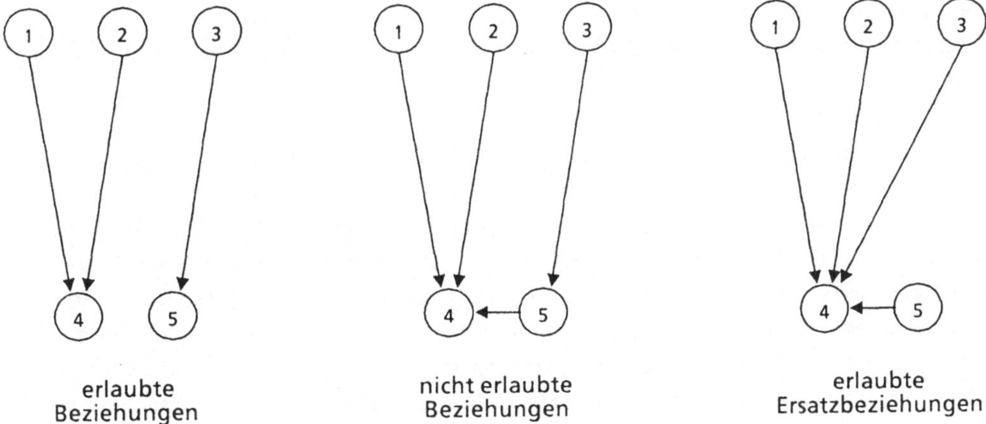

Abb.1: Beziehungen zwischen den Ketten

Abbildung 1 zeigt eine erlaubte, eine nicht erlaubte und die äquivalente Ersatzdarstellung der nicht erlaubten Beziehung. Gruppen die untereinander in Beziehung stehen, konnen zu einer Gruppe zusammengefaßt werden.

Fur die Berechnung der zugehorigen charakteristischen Größen bildet die Prozedur setze_Durchsätze wieder den Ansatzpunkt. In jedem Iterationsschritt werden zunächst alle Durchsatze D(k) wie beim Bard/Schweitzer-Algorithmus bestimmt. Anschließend werden die Durchsatze so verandert, daß die geforderten Verhältnisse eingehalten werden. Durchsätze dürfen dabei nur verringert werden, denn dies ist in der Realität durch eine Verzögerung zu erreichen. Die Durchsätze in einer Gruppe von Ketten werden also durch die "schwächste" Kette bestimmt. Seien D(1),...,D(k),...,D(K) die Durchsätze, wie sie laut der Formel der normalen Bard/Schweitzer-Iteration entstehen. Folgende Operationen werden dann jeweils anschließend durchgeführt.
Sei k eine Kette auf die sich andere beziehen.

Setze

$$D(k) = \min_{kk:bezug(kk)=k} \{ D(kk) / koeff(kk) \} \quad (10)$$

und anschließend für alle kk mit bezug[kk] = k

$$D(kk) = D(k)\ koeff(kk) \quad (11)$$

Die Gleichungen (2) und (3) liefern wieder die Aussage, ob das WS-Netz realisierbar ist. Ist es realisierbar, so sind die absolut vorgegebenen Durchsätze eingestellt und die vorgegebenen Verhältnisse sind ebenfalls realisiert.

Hat diese erste Berechnung nicht zu einem realisierbaren Netz geführt, so sind noch mehrere Fälle zu unterscheiden. Die Durchsätze wurden so gesetzt, daß in jeder Gruppe von Ketten der Durchsatz der jeweils "schwächsten" nicht künstlich verringert wurde. Das führt dazu, daß nach der obigen Berechnung in jeder Gruppe eine Kette existiert, die an der Verzögerungsstation nicht verzögert wird. Das kann insgesamt zu zu hohen Durchsätzen und damit Auslastungen führen, so daß das Netz so nicht realisierbar ist. Eine Verzögerung auch der "schwächsten" Kette in einer Gruppe kann da Abhilfe schaffen. Im folgenden werden die verschiedenen Möglichkeiten aufgezählt und deren weitere Behandlung dargestellt.

- Fixanteil sowie alle Kettengruppen mit jeweils unverzögerter "schwächster" Kette sind realisierbar
  => einfachster Fall, gemaß vorgestelltem Algorithmus zu berechnen

- Fixanteil nicht realisierbar
  => Es gibt kein zugehoriges realisierbares Netz.

- Fixanteil realisierbar und es gibt außerdem nur eine Gruppe von Ketten
  => Alle Ketten sind so zu verzögern, daß das Netz gerade noch realisierbar ist. Die Beschreibung der erforderlichen Berechnungen erfolgt unten.

- Fixanteil realisierbar und es gibt außerdem noch mehr als eine Gruppe von Ketten
  => Es lassen sich viele verschiedene realisierbare Netze angeben. Es ist jedoch nicht vorgegeben, welche Gruppe in welcher Weise zu bevorzugen ist. Die Aufgabenstellung ist in diesem Fall nicht eindeutig. Der Algorithmus endet hier mit der Aufforderung, die Aufgabenstellung eindeutig zu machen. Dies kann dadurch geschehen, daß die verschiedenen Kettengruppen zu einer Gruppe zusammengefaßt werden.

Der letzte noch nicht abschließend behandelte Fall ist der mit realisierbarem Fixanteil und einer einzigen Gruppe von Ketten. Die Mindestverzögerungszeit an der Verzögerungsstation geht in die Durchsatzberechnung in setze-Durchsätze mit ein, da sie zur Gesamtverweilzeit eines Auftrages beiträgt. Die Mindestverzögerungszeit stellt die kleinst mogliche Zeit dar, bei der ein realisierbares Netz auftritt. Zunachst wird eine genügend große Mindestverzögerungszeit angenommen, für die das Netz realisierbar ist. Durch binäre Schachtelung wird diese soweit wie möglich verringert.

Der Basisalgorithmus zur Durchsatzkopplung hat das gleiche Schema wie die Bard/Schweitzer-Approximation. Lediglich in setze_Durchsätze sind die Verhältnisse gemäß (10) und (11) sowie eventuell eine Mindestverzögerungszeit zu berücksichtigen.

Damit wird das Verfahren zur Behandlung von Netzen mit Durchsatzkopplung folgendermaßen darstellbar:

```
Durchsatzkopplungs-Algorithmus
{
    Mindestverzogerung = 0.0;
    Basisalgorithmus;
    if ( realisierbar )
    {
        Ergebnisausgabe;
    }
    else
    {
        Durchsatzsteuerungsalgorithmus fur Fixanteil
        if ( Fixanteil nicht realisierbar )
        {
            Fehlermeldung: Fixanteil nicht realisierbar
        }
        else
        {
            if ( Kettengruppenanzahl > 1 )
            {
                Fehlermeldung: Aufgabenstellung nicht eindeutig
            }
            else
            {
                do
                {
                    vergroßere Mindestverzogerung;
                    Basisalgorithmus;
                } while ( Netz nicht realisierbar ) ;
                while ( Intervall fur Mindestverzogerung zu groß )
                {
                    setze Mindestverzogerung testweise;
                    Basisalgorithmus;
                    verkleinere Intervall;
                }
                Ergebnisausgabe;
            }
        }
    }
}
```

Dieser Algorithmus liefert konsistente Ergebnisse in dem Sinne, daß Netze in denen die Verzogerungsstation explizit mit den berechneten Verzogerungszeiten modelliert werden, exakt dieselben Ergebnisse liefern wie die, in denen die Verzogerungszeiten bestimmt wurden.

2.5 Sichtweisen und Einsatzmoglichkeiten

Die Prozessor- oder Betriebsmittelvergabestrategien werden immer ausgefeilter, um vorgegebene Ziele erreichen zu konnen. Ausgeklügelte Strategien mit dynamischen Prioritaten sind ein Beispiel dafur. Es ist hoffnungslos, diese mit mathematischen Methoden nachbilden zu wollen. Traut man ihnen jedoch zu, ihr Ziel zu erreichen, so reicht es in einem großeren Zusammenhang moglicherweise aus, lediglich den erzielten Effekt im Modell wiederzufinden. Soweit die Ziele die Kapazitatsaufteilung unter verschiedenen Ketten sind, bieten Durchsatzsteuerung und -kopplung erfolgversprechende Anwendungsmoglichkeiten. Diese Sicht wird auch von Lazowska et al. /LZGS84/ unter dem Thema zielorientiertes Scheduling erwahnt.

In Rechnernetzen mit einer oder wenigen ausgezeichneten Stationen gibt es viele Symmetrien. Jede Anwendung nutze die ausgezeichneten Stationen in der gleichen Weise, erledige den Rest des Auftrages jedoch an einem privaten System. Dann entstehen bei der Modellierung sehr viele gleichartige Ketten. Es reicht jedoch aus, eine einzige Kette detailliert zu modellieren. Der Rest kann zu einer weiteren Kette zusammengefaßt werden, deren Durchsatz wegen der Symmetrien zu dem der detailliert modellierten Kette in Beziehung gesetzt wird. Man spart hier die muhselige mehrfache Modellierung der gleichen Details. Aus den Formeln in /ToSc87/ ist ersichtlich, daß die charakteristischen Großen der detailliert modellierten Ketten dadurch nicht beeinflußt werden.

Aus einer Gruppe von Ketten bzw. Anwendungstypen kann die "schwachste" bestimmt werden als die, die am wenigsten verzogert wird. Handelt es sich z.B. um Dienstprozesse, so kann es schon genugen, ihre Anzahl zu erhohen, um den Gesamtdurchsatz zu steigern.

Die Interpretation der erforderlichen Verzogerungszeiten kann durchaus problematisch sein. Handelt es sich dabei um zusatzliche Denkzeit am Terminal oder um Wartezeiten innerhalb des Systems? Bei der reinen Durchsatzsteuerung können die Verzögerungszeiten der einzelnen Ketten als Denkzeiten an den Terminals interpretiert werden. Bei der Durchsatzkopplung hat es eher den Anschein als müßten Auftrage warten, bis andere den ihnen zugedachten Durchsatz erreicht haben. Die Interpretation sollte also sehr umsichtig erfolgen.

2.6 Beispiel

In einem Point-of-Sale-Projekt sollen sehr viele intelligente Terminals über Datex-L bzw. Datex-P an ein Nixdorfsystem angeschlossen werden, wo zentral alle Umsatze verbucht werden. Zwei Auftragstypen bewegen sich im Netz. Die PoS-Anwendungen beschreiben im Mittel n-mal eine Spooldatei, die durch einen speziellen Spooloutprozeß, der durch Kette 2 modelliert wird, asynchron abgearbeitet wird. Das Originalmodell umfaßt eine Vielzahl Stationen und eine recht umfangreiche Lastbeschreibung. Zur Verdeutlichung der dort beobachteten Softwareengpässe und der Demonstration der Durchsatzkopplung reicht ein stark vereinfachtes Netz aus.

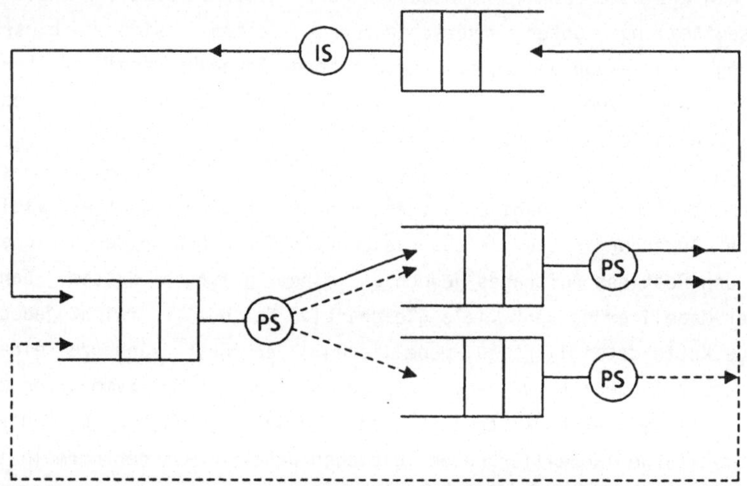

Abb. 2: Struktur des Beispielnetzes (—— Kette 1,- - - Kette 2)

Es besteht aus 4 Stationen (Abb.2), den Terminals, die durch eine Infinite-Servers-Station dargestellt werden, der CPU und zwei Platten, die als Processor-Sharing-Stationen modelliert werden. Die Bedienzeiten und Besuchshäufigkeiten sind gegeben durch:

$$\begin{bmatrix} b(1,1) & b(1,2) & b(1,3) & b(1,4) \\ b(2,1) & b(2,2) & b(2,3) & b(2,4) \end{bmatrix} = \begin{bmatrix} 2.00 & 0.06 & 0.04 & 0.00 \\ 0.00 & 0.02 & 0.04 & 0.04 \end{bmatrix}$$

$$\begin{bmatrix} h(1,1) & h(1,2) & h(1,3) & h(1,4) \\ h(2,1) & h(2,2) & h(2,3) & h(2,4) \end{bmatrix} = \begin{bmatrix} 1 & 3 & 3 & 0 \\ 0 & 4 & 1 & 3 \end{bmatrix}$$

Im eingeschwungenen Zustand müssen die Spoolprozesse genau den n-fachen Durchsatz wie die PoS-Anwendungen haben. Hier sei n=3 angenommen. Dies kann im Modell durch die Kopplung der Durchsatze ausgedruckt werden. Im Detail:

bezug(1) = 1 bezug(2) = 1

koeff(1) = 1 koeff(2) = 3

Dieses Beispiel wurde für verschiedene Auftragsanzahlen modelliert. Die Anzahl der PoS-Anwendungen betrug in allen Fallen 15, die der Spoolprozesse wurde von 1 bis 15 variiert. Abbildung 3 zeigt charakteristische Größen in Abhängigkeit von der Anzahl der zur Verfügung stehenden Spoolprozesse. Es wurde die Auslastung der Station mit der höchsten Auslastung ausgewählt, hier die CPU. Wegen der Durchsatzkopplung ist diese proportional zu allen anderen Auslastungen und den Durchsätzen der Ketten. Von der erforderlichen Verzögerungszeit der PoS-Anwendungen wurde ein Zehntel eingezeichnet, um den Maßstab gut zu nutzen.

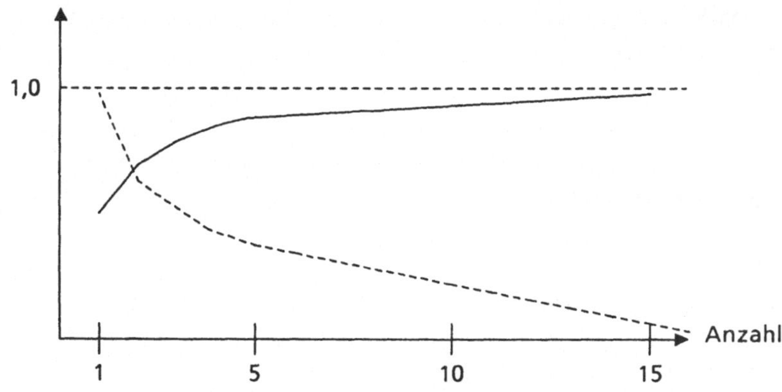

Abb.3: CPU-Auslastung (——) und PoS-Verzögerungszeiten (- - -) über der Auftragsanzahl der Spoolprozesse

Die Auslastung der CPU erreicht bei nur einem Spoolprozeß lediglich ca. 50%. Es wird ersichtlich, daß die Hardware hier keinen Engpaß darstellt. Die Struktur der Last selbst beinhaltet die Begrenzung. Ein Spoolprozeß muß allein den dreifachen Durchsatz erreichen, wie alle PoS-Anwendungen zusammen. Da bei der

Prozessorzuteilung alle Prozesse gleichberechtigt sind, mussen die PoS-Anwendungen an anderer Stelle zusatzlich je Auftrag um ca. 10 Sekunden verzogert werden, damit sie nicht ofter aktiviert werden, als es durch die Durchsatzkopplung vorgegeben ist. Mit wachsender Spoolprozeßanzahl steigt die CPU-Auslastung, die proportional zum Durchsatz ist und die erforderliche Verzogerung fur die PoS-Anwendung sinkt.

Nach der Analyse des Originalmodells konnten Vorschlage erarbeitet werden, die eine Umstrukturierung der Software zum Gegenstand hatten und laut weiteren Modellanalysen ein besseres Durchsatzverhalten zeigten.

3. Zusammenfassung und Ausblick

Es wurde ein Verfahren vorgestellt, das es gestattet in geschlossenen separablen Netzen Durchsatze beziehungsweise ihre Verhaltnisse zueinander einzustellen. Dieses Verfahren liefert konsistente Ergebnisse in dem Sinne, daß Netze, in denen die Verzogerungsstation explizit mit den berechneten Verzogerungszeiten modelliert werden, exakt dieselben Ergebnisse liefern wie die, in denen die Verzogerungszeiten bestimmt wurden.

Neben der Vorgabe von einfachen Koeffizienten sind komplexere funktionale Zusammenhange denkbar. Die Implementierung ware auf die Anderung der Prozedur setze-Durchsatze beschrankt.

Dieser Ansatz ist auf Netze mit zustandsabhangigen Bedienraten leicht ubertragbar, da die zur Durchsatzkopplung erforderlichen Eingriffe sich auf die Prozedur setze_Durchsatze beschranken.

Problemstellungen, die bisher nur durch Iteration mehrerer Modellierungslaufe lösbar waren, lassen sich nun mit Hilfe der Durchsatzkoplung innerhalb eines einzigen Modellierungslaufes behandeln. In zahlreichen Verfahren werden separable Netze innerhalb mehrerer Iterationsschritte verwendet /LZGS84/, wobei Ergebnisgroßen verwendet werden, um Parameter im Folgemodell zu verandern. Sofern dies Durchsatze oder Auslastungen sind, erubrigt die Verwendung der Durchsatzsteuerung die Iteration.

4. Literatur

BCMP75 F. Baskett, K. Chandy, R. Muntz, G. Palacios
Open, Closed and Mixed Networks of Queues with
Different Classes of Customers
J. ACM 22 (1975)

Burk85 W. Burke
Q-Net-Application: Performanceprognose im Bereich
Vertrieb mit Hilfe angewandter Techniken aus der
Warteschlangentheorie
Proceedings Messung, Modellierung und Bewertung von
Rechensystemen 1985
Inf. Fachbericht 110, Springer Verlag

GeKu76 E. Gelenbe, A. Kurinckx
Random Injection Control of Multiprogramming in
Virtual Memory
in Modelling and Performance Evaluation of Computer Systems
Gelenbe (ed.)
North Holland Publishing Company (1976)

LZGS84 E. Lazowska, J. Zahorjan, G. Graham, K. Sevcik
Quantitative System Performance
Computer System Analysis Using Queueing Network Models
Prentice-Hall Inc., Englewood Cliffs, New Yersey (1984)

NeCh81 D. Neuse, K. Chandy
SCAT: A Heuristic Algorithm for Queueing Network
Models of Computing Systems
ACM Sigmetrics Vol. 10 No. 3 Fall 81 (1981)

ReKo75 M. Reiser, H. Kobayashi
Queueing Networks with Multiple Closed Chains:
Theory and Computational Algorithms
IBM Journal of Research and Development (May 1975)

Schw79 P. Schweitzer
Approximate Analysis of Multichain Closed
Queueing Networks
Int. Conf. Stochastic Control and Optimization
Amsterdam 1979

ScTo85 A. Schatter, G. Totzauer
Aufteilung von Rechenkapazitat durch Steuerung von
Durchsatzen
Proceedings Messung, Modellierung und Bewertung von
Rechensystemen 1985
Inf. Fachbericht 110, Springer Verlag

ToSc87 G. Totzauer, A. Schätter
Controlling Throughput in Closed Multiclass
Queueing Networks
zur Veröffentlichung eingereicht

ARTIFICIAL INTELLIGENCE IN SIMULATION

Tuncer I. Ören

Computer Science Department
University of Ottawa
Ottawa, Ontario, Canada K1N 9B4

ABSTRACT

Advances of the applications of artificial intelligence in modelling and simulation are viewed within the continuum of tool making. To have a proper perspective, the spectrums of the developments of physical tools as well as abstract tools such as software tools and modelling and simulation tools are presented. Then some promising directions for advanced research are highlighted under the following categories:

1. AI in modelling and simulation environments,
2. AI in simulation models,
   - 2.1 Simulation with models having time-varying structures,
   - 2.2 Simulation with goal-directed models,
   - 2.3 Simulation with models having perception abilities,
   - 2.4 Behaviorally anticipatory simulation,
3. AI in simulation query systems,
4. Machine learning in simulation,
5. Simulation in computer-embedded machines, and
6. Quality assurance in modelling and simulation in AI era.

KEY WORDS: artificial intelligence in simulation, AI in simulation environments, AI in simulation models, time-varying model, goal-directed model, models with perception ability, bahaviorally anticipatory simulation, AI in simulation query, computer-embedded machines, quality assurance

1. INTRODUCTION

Simulation can be conceived from several points of view: i) from the point of view of the *execution* of the simulation programs and ii) from the point of view of *describing* a knowledge generation problem based on experimentation with a model. In continuous simulation (i.e., simulation with continuous-change models), for example, the program execution point of view necessitates conceiving simulation programs in three sections, i.e.. initial, dynamic, and terminal sections. This view was necessitated in early days of computers due to their limited abilities. A model-based problem specification point of view necessites the description of the following parts: a parametric model, values of the parameters (a parameter set), experimental conditions, and simulation run control conditions.

Th shift of paradigm in simulation from execution oriented view to a model-based activity has several implications. An important one is easing the synergy of simulation with other model-based disciplines such as artificial intelligence, cybernetics, and general system theories (Ören 1984).

The intersection of artificial intelligence and simulation has two aspects: i) implications of simulation in artificial intelligence and ii) implications of artificial intelligence in simulation. The first aspect has been the use of simulation in artificial intelligence studies, i.e., simulation of cognitive processess and is known as cognitive simulation. The second aspect which is the application of artificial intelligence in simulation constitutes the subject matter of this article.

The state of the art of artificial intelligence in simulation is developing with an increasing acceleration. Several books already exist on the topic: Birtwistle (1985), Elzas, Ören, and Zeigler (1986), Holmes (1985), Kerkhoffs, Vansteenkiste, and Zeigler (1986), Luker and Adelsberger (1986), Luker and Birtwistle (1987), Widman, Loparo, and Helman (1987). Several other books are being written. For example, the author is involved in the preparation of two such books: Zeigler, Ören, and Reddy (1988) and Zeigler, Ören, and Elzas (1988). In this article, I would like to highlight some desirable directions of research and development in the application of artificial intelligence in modelling and simulation.

2. PROGRESS IN TOOL MAKING

Realization of advanced applications of artificial intelligence in modelling and simulation can be perceived as part of the progress in tool making. Hence, exploring the analogy between different types of tools can be useful in discovering new dimensions. Tools can be conceived as physical tools and abstract tools. The latter consist of the software tools and modelling and simulation tools.

2.1 Physical Tools

As seen in Figure 1, three basic levels of physical tools are: manual tools, power tools, and cybernetic tools. To go from one level to the next one, one has to have an additional fundamental feature. *Energy* is the additional feature to advance first level tools to the second level. Simillarly, *knowledge processing ability* is the additional feature which distinguishes third level tools from the second level ones.

Manual tools started at the beginning of civilization, with different types of stone tools and continued with tools made from other materials and were upgraded to metallic tools.

Power tools consists of simple power tools, machine tools (which increased tremendeously the effectiveness of manual tools), machines in general, and integrated machines (or transfer machines).

Cybernetic tools are tools (or machines, or systems) with knowledge processing ability. There are two fundamental types: stand-alone computers and computer-embedded machines.

Stand-alone *computers*, as knowledge processing machines, are instruments to execute programs. A computer *program* is a knowledge transducer. Both input and output of a computer program can be numerical, textual (one- or two-dimensional, hand- or machine-written), graphical (two- or three-dimensional, still or moving, monochrome or in color), audio (sound or speech, programming or a natural language), tactile, or other types of signals (coming from sensing devices or going to control devices).

Computer-embedded machines or systems are machines or systems whose primary goal is not knowledge processing. However, the knowledge processing ability provided by the embedded computer(s), permits the machine or the system to satisfy its primary goal. Computer-embedded machines can perform their knowledge processing either by hardware connections (as in fixed-wited machines or automata) or by stored programs (as in programmable machines including robots). The latter group includes preprogrammable machines, reprogrammable machines, and autoprogrammable machines.

Based on the type of knowledge processing ability several possibilities exist such as: knowledge-based machines, optimizing machines, reasoning machines (or rule-based

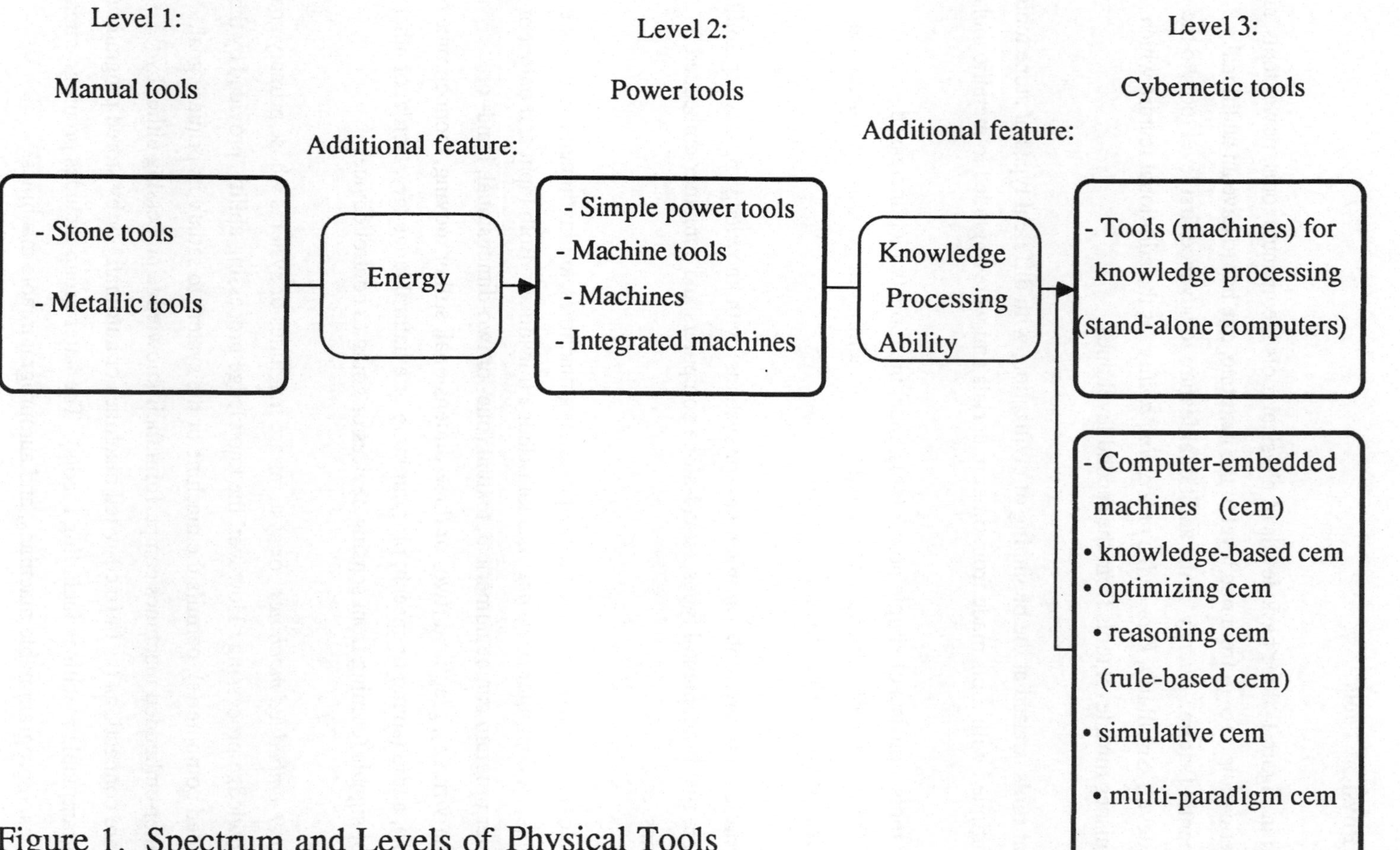

Figure 1. Spectrum and Levels of Physical Tools

machines including reprogrammable and autoprogrammable robots), simulative machines (machines or systems which can perform simulation studies using models of other systems), autosimulative machines (machines or systems which can perform simulation studies using their own models), and multi-paradigm knowledge processing machines (which can have a combination of the above mentionad characteristics). The importance of computer-embedded machines is increasing and their widespread use as household items, or industrial machines, or as advanced weapons may be the beginning of the era of advanced machines which can be also labelled as intelligent machines or autonomous machines.

2.2 Software Tools

As seen in Figure 2, the three basic levels of software tools are manual tools, power tools, and cybernetic tools.

Manual tools are basically hand-coded software. The jump from the first to the second level of software tools requires the additional feature which is computer-aided programming.

Power tools benefit from the computer-aided programming abilities of computers and consists of software ttols, software toll kits, and software environments (or integrated software tools). The analogy between software power tools and physical power tools is apparent.

Cybernetic tools require an additional feature over the second level of software tools. This additional feature is *advanced knowledge processing ability*. A fashionable term would be "intelligent software tools." However, to raise the standards, the term *intelligent* can be saved to indeed advanced knowledge processing ability such as *goal-processing* and *goal-directed knowledge processing* and the term *cognizant* can be used to denote knowledgeable, or knowledge-based, or expert software tools (Ören 1987a).
Basically, two types of third level software tools can be identified: i) artificial intelligence in software environments (or in software life cycle) and ii) artificial intelligence in software. Additional information can be found in Barstow (1979), Johnson (1986), Manna and Waldinger (1977), Ören (1981, 1988), Rich and Waters (1986).

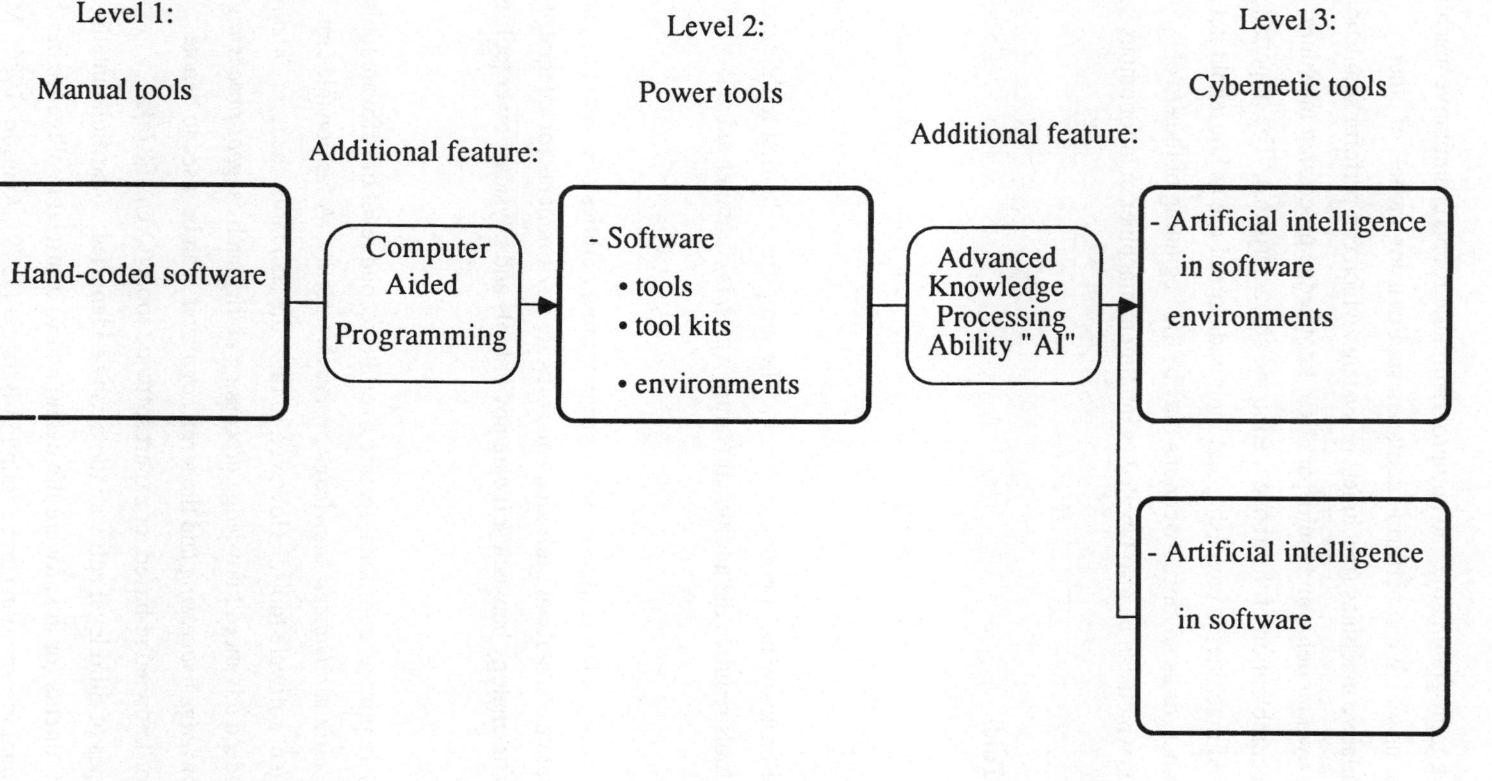

Figure 2. Spectrum and Levels of Software Tools

The importance of the third level of software tools in modelling and simulation is obvious since any relevant advance in software tools can be used to boost modelling and simulation environments as well as modelling and simulation software.

2.3 Modelling and Simulation Tools

As seen in Figure 3, three levels of modelling and simulation tools can be identified. They are: manual tools, power tools, and cybernetic tools.

Manual tools are hand-coded modelling and simulation software regardless of the type of the simulation language used (Ören 1987b). As it is the case in software tools, in modelling and simulation tools, the jump from first level to second level tools requires the additional feature which is *compuer-aided* modelling and simulation (Ören 1979, 1982).

Power tools in modelling and simulation are basically modelling and simulation tools (including program generators and computer-aided modelling and simulation tools) and modelling and simulation environments which are integrated tools for modelling and simulation. The latter also includes symbolic model processing functions (Ören 1987c).

Cybernetic tools in modelling and simulation, similar to the respective case in software, requirean additional feature over the second level of modelling and simulation tools. This additional feature is *advanced knowledge processing ability* which is currently also called artificial intelligence. The following possibilities can be identified: i) artificial intelligence in modelling and simulation environments, ii) artificial intelligence in simulation models, and iii) artificial intelligence in simulation query systems.

3. ARTIFICIAL INTELLIGENCE IN SIMULATION: SOME PROMISING DIRECTIONS FOR RESEARCH

My comments are centered in the following categories:

- AI in modelling and simulation environments,
- AI in simulation models,
- AI in simulation query systems,
- Machine learning in simulation,

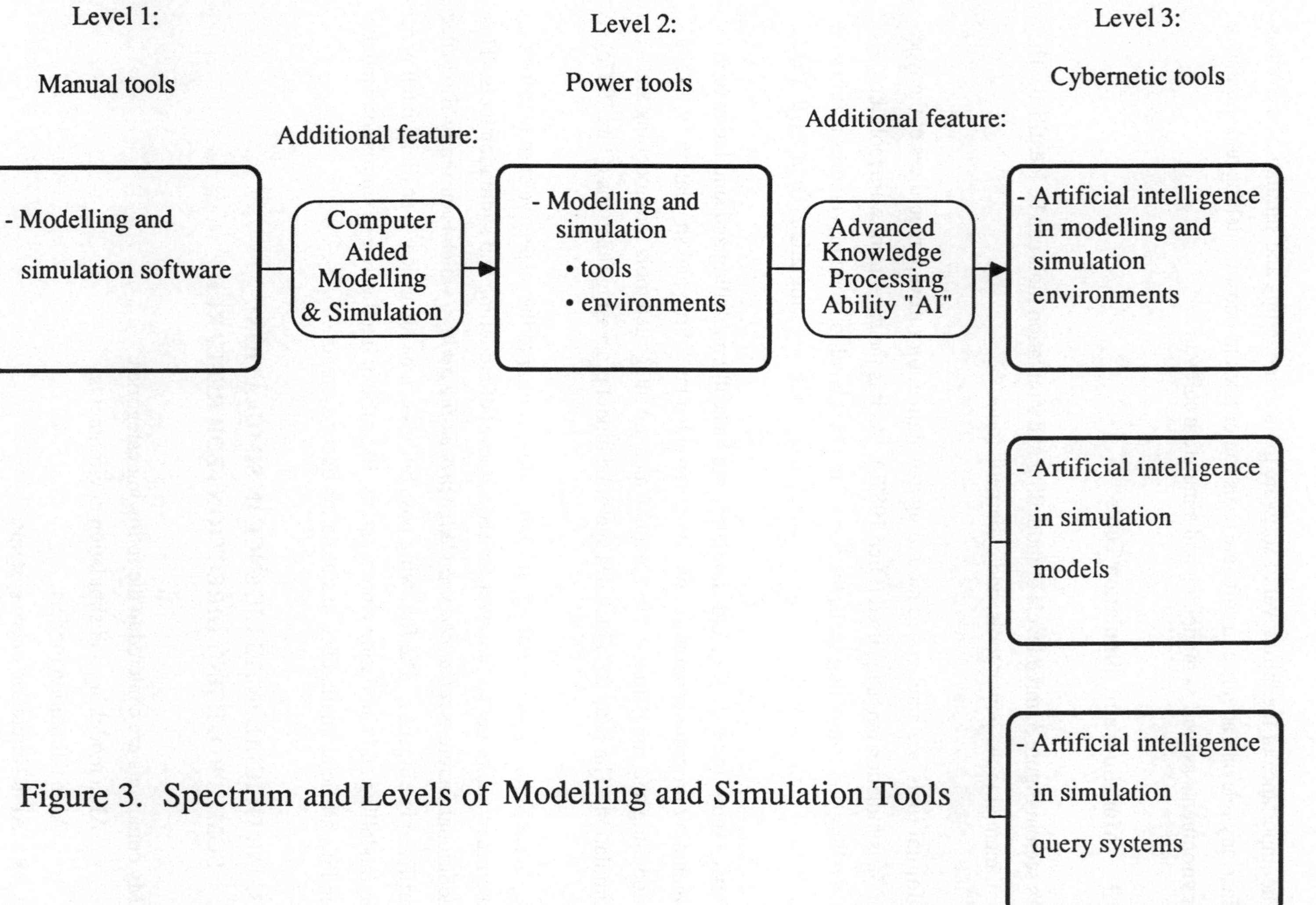

Figure 3. Spectrum and Levels of Modelling and Simulation Tools

- Simulation in computer-embedded machines, and
- Quality assurance in modelling and simulation in AI era.

3.1 AI in Modelling and Simulation Environments

Developers of modelling and simulation environments have a definite advantage over developers of software environments since there are two categories of knowledge which can be formulated to became part of the knowledge of a modelling and simulation environment. These are domain-specific knowledge of an application area of interest and domain-independent but methodology-based knowledge. An example to the latter is knowledge provided in the Magest system (Aytaç and Ören 1986).

Therefore, a basic level of realization of artificial intelligence in modelling and simulation environment is to tap domain-specific and/or domain-independent (but methodology-based knowledge) in a modelling and simulation environment which may then become a knowledge-based environment or an expert system. Still there are tremendeous useful realization possibilities in this are. Some advanced features for such modelling and simulation environments may include machine learning abilities as well as goal-directed knowledge processing abilities.

3.2 AI in Simulation Models

One of the contributions of general system theories to advanced simulation is to provide advanced systemic concepts to build models (Ören 1984). Some possibilities are highlighted in the sequel.

3.2.1 Simulation with Models Having Time-varying Structures

Simulation has traditionally been used for the generation of trajectories of descriptive variables of models and not to study structural changes of system. However, structural simulation can be beneficial for several types of systems such as growth systems (cell, plant, crystal, etc.), evolutionary systems, social systems, etc. Some early studies are done by Hogeweg (1979). Graphic abilities (both for input and output) are assets for structural simulation. Some references are Angell and Goldstein (1984) and Zeleny (1980).

3.2.2 Simulation with Goal-Directed Models

"Intelligence is a goal-directed and often adaptive knowledge processing ability. Intelligence of a system allows its knowledge processing elements to perform: (1) goal processing, including setting, selecting, sequencing, and modification of its goals; and (2) goal realization by knowledge processing including identification, creation, compilation, acquisition, and assimilation of task oriented new knowledge required for goal processing or for goal realization" (Ören 1987a, p. 130).

Viewed from this point of view, goal-directed models become prime candidates for advanced simulation. A taxonomy of goal-directed models is givenin Ören (1987d). Some books on goal-directed models and goal processing are Laird, Rosenbloom, and Newell (1986), vo Foerster et al. (1968), and Weir (1984).

3.2.3 Simulation with Models Having Perception Abilities

Object-oriented simulation, which has been a reality since Simula, has been revitalized with Smalltalk. Message passing capability is an important paradigm for communication for object-oriented simulation. Message sending is active, i.e., an object can send message(s) to itself, to another object, to some other objects, to a class of objects, to some members of some classes of objects, to all objects, to a common area accessable by all the objects. Therefore, on can have a broad spectrum from message sending to message broadcasting. Message reception is usually passive, i.e., an object which has a message or a message notice will receive it when certain conditions are met.

An important type of objects has not yet been elaborated on in advanced modelling and simulation. These are objects with perception capability. Some input channels of an object with perception capability may be connected to an internal pre-processor which may act like a sensory organ of a living organism. The pre-processor may not be limited to the usual senses of living organisms. The inputs of a pre-processor provide observed knowledge. The output of a pre-processor may be the perceived knowledge of the object. The pre-processor may transform observed knowledge as a memory function or a memoryless function and may depend on straightforward knowledge transformation or on advanced knowledge processing such as pattern recognition as it is the case in perceptrons.

Loosely-coupled autonomous systems can be represented by objects with perception abilities. Interesting behavior may be expected from the simulation studies with objects with perception abilities.

3.2.4 Behaviorally Anticipatory Simulation

In causal models, next state depends on current state and current input. Normally, future value(s) of input variable(s) do not affect next state. Since future value(s) of input variable(s) is(are) not known, estimated value(s) (or the current images of the future values of input variables) can be taken into consideration in computing next value of a state variable. Most human decision making is anticipatory by nature and similarly, some simulation studies can be made behaviorally anticipatory to put more realism in them.

3.3 AI IN SIMULATION QUERY SYSTEMS

A simulation query system may benefit from artificial intelligence applications at several levels: i) query/response can be in a (written or spoken) natural language, ii) computer-assissted query formulation, processing, or management can be made available.

3.4 MACHINE LEARNING IN SIMULATION

Machine learning is an ability of knowledge processing systems which enhances system's knowledge processing ability. A detailed taxonomy of about thirthy types of machine learning is given by Ören (1986). Some important possibilities are: i) simulation of objects with learning abilities such as robots, ii) simulation used in discovering fatal operating conditions (and learning these conditions) in monitoring complex systems such as space station and advanced weapons, iii) machine learning through simulated scenarios for simulative machines and autosimulative machines.

3.5 SIMULATION IN COMPUTER-EMBEDDED MACHINES

Some advanced machines or systems such as a submarine, can be enhanced by using simulation to have on-line predictive displays.

3.6 QUALITY ASSURANCE IN MODELLING AND SIMULATION IN AI ERA

Several issues of quality assurance in modelling and simulation are of primordial nature. In addition to these issues, one has new quality assurance issues which became important from the artificial intelligence perspective. In a recent article, Ören provided a classification of quality assurance problems and three out of four categories are relevant from artificial intelligence point of view (Ören 1987e). The following four categories are identified: i) qa in modelling and simulation, ii) qa in cognizant modelling and simulation, iii) cognizant qa in modelling and simulation, and iv) cognizant qa in cognizant modelling and simulation. Contrary to intuitive expectation, application of artificial intelligence techniques may not guaranty quality assurance requirements of a modelling and simulation study.

4. CONCLUSION

Advances in computers, software engineering, artificial intelligence, cybernetics, general system theories, and modelling and simulation have produced useful synergies to realize advanced modelling and simulation systems. The term "AI in modelling and simulation" captures some of these possibilities. Within these possibilities, some promising directions for research and development are highlighted after a a view of tool development paradigms for physical as well as abstract tools such as software tools and modelling and simulation tools.

REFERENCES

Angell, C.A., Goldstein, M. (Eds.) (1986), Dynamic aspects of structural change in liquids and gasses, The New York Academy of Sciences, New York, NY

Aytaç, K.Z., Ören, T.I. (1986), Magest: A model-based advisor and certifier for Gest programs, In: Modelling and simulation methodology in the artificial intelligence era, M.S. Elzas, T.I. Ören, B.P. Zeigler (Eds.), North-Holland, Amsterdam, pp. 299-307.

Barstow, D.R. 91979), Knowledge-based program construction, North-Holland, New York, NY.

Birtwistle, G. (1985), Artificial intelligence, graphics, and simulation, SCS, San Diego, CA.

Elzas, M.S., Ören, T.I., Zeigler, B.P. (Eds.) (1986), Modelling and simulation methodology in the artificial intelligence era, North-Holland, Amsterdam.

Hogeweg, P., Hesper, B. (1979), Heterarchical, selfstructuring simulation systems: Concepts and applications in biology, In: Methodology in systems modelling and simulation, B.P. Zeigler et al. (Eds.), North-Holland, Amsterdam, pp. 221-232.

Holmes, W.M. (Ed.) (1985), Artificial intelligence and simulation, SCS, San Diego, CA.

Johnson, W.L. (1986), Intention-based diagnosis of novice programming errors, Pitman, London.

Kerckhoffs, E.J.H., Vansteenkiste, G., Zeigler, B.P. (Eds.) (1986), Artificial intelligence applied to simulation, SCS, San Diego, CA.

Lair, J., Rosenbloom, P., Newell, A. (1986), Universal subgoaling and chunking - The automatic generation and learning of goal hierarchies, Kluwer Academic Publ., Boston.

Luker, R.A., Adelsberger, H.H. (Eds.), Intelligent simulation environments, SCS, San Diego, CA.

Luker, R.A., Birtwistle, G. (Eds.), Simulation and AI, SCS, San Diego, CA.

Manna, Z., Waldinger, R. et al. (1977), Studies in automatic programming logic, North-Holland, New York, NY

Ören, T.I. (1979), Concepts for advanced computer-assisted modelling, In: Methodology in systems modelling and simulation, B.P. Zeigler et al. (Eds.), North-Holland, Amsterdam, pp. 29-55.

Ören, T.I. (1981), Proceedings of Cybersoft 80 - International symposium on cybernetics and software, Sept. 9, 1980, Namur, Belgium, International Association for Cybernetics, Namur, Belgium

Ören, T.I. (1982), Computer-aided modelling systems, In: Progress in modelling and simulation, F.E. Cellier (Ed.), Academic press, London, England, pp. 189-203.

Ören, T.I. (1984), Model-based activities: a paradigm shift, In: Simulation and model-based methodologies: an integrative view, T.I. Ören, B.P. Zeigler, M.S. Elzas (Eds.), Springer-Verlag, Heidelberg, 3-40.

Ören, T.I. (1986), Implications of machine learning in simulation, In: Modelling and simulation methodology in the artificial intelligence era, M.S. Elzas, T.I. Ören, B.P. Zeigler (Eds.), North-Holland, Amsterdam, pp. 41-57.

Ören, T.I. (1987a), Artificial intelligence and simulation: From cognitive simulation toward cognizant simulation, Simulation, 48:4 (April), 129-130.

Ören, T.I. (1987b - in press), Simulation and model-oriented languages: Taxonomy, In: Systems and control encyclopedia, M.G. Shing (Ed.), Pergamon Press, Oxford, England.

Ören, T.I. (1987c - in press), Simulation models - symbolic processing: Taxonomy, In: Systems and control encyclopedia, M.G. Shing (Ed.), Pergamon Press, Oxford, England.

Ören, T.I. (1987d - in press), Simulation models: Taxonomy, In: Systems and control encyclopedia, M.G. Shing (Ed.), Pergamon Press, Oxford, England.

Ören, T.I. (1987e), Quality assurance paradigms for artificial intelligence in modelling and simulation, Simulation, 48:4 (April), 149-151.

Ören, T.I. (1988), Advances in artificial intelligence in software engineering, vol. 1, JAI Press, Greenwich, Connecticut. (in press)

Ören, T.I., Zeigler, B.P. (1987), From stone tools to cognizant tools: The quest continues, In: Proceedings of the 2nd European simulation congressm, Antwewrp, Belgium, Sept. 9-12, 1986, G.C. Vansteenkiste, E.J.H. Kerkhoffs, L. Dekker, J.C. Zuidervaart (Eds.), SCS, San Diego, CA, pp. 801-807.

Rich, C., Waters, R.C. (Eds.) (1986), Readings in artificial intelligence and software engineering, Morgan Kaufmann Publ. co, Los Altos, CA.

Von Foerster, H. et al. (Eds.) (1968), Purposive systems, Spartan Books, New York, NY.

Weir, M. (1984), Goal-directed behavior, Gordon and Breach Science Publ., New York, NY.

Wertz, H. (1985), Intelligence artificielle - Application à l'analyse de programmes, Masson, Paris.

Widman, L.A., Loparo, K.A., Helman, D.H. (Eds.) (1987), Artificial intelligence, simulation, and modelling, D. Reidel publ. co.

Zeigler, B.P., Ören, T.I., Elzas, M.S. (Eds.) (1988 - In preparation), Modelling and simulation methodology: Knowledge systems paradigms, North-Holland, Amsterdam.

Zeigler, B.P., Ören, T.I., Reddy, R.V. (Eds.) (1988 - In preparation),Knowledge representation in simulation environments: Expert systems and simulation models, Academic Press, USA.

Zeleny, M. (Ed.) (1980), Autopoiesis, dissipative structures, and spontaneous social order, American Association for the Advancement of Science, Washington, DC.